Recombinant Gene Expression

METHODS IN MOLECULAR BIOLOGY™

John M. Walker, SERIES EDITOR

METHODS IN MOLECULAR BIOLOGY™

Recombinant Gene Expression

Reviews and Protocols

SECOND EDITION

Edited by

Paulina Balbás

Centro de Investigación en Biotecnología,
Cuernavaca, México

and

Argelia Lorence

Virginia Polytechnic Institute and State University,
Blacksburg, VA

HUMANA PRESS ✳ TOTOWA, NEW JERSEY

Cover illustration:Background: HSA accumulation as inclusion bodies in transgenic chloroplasts. *See* Fig. 6 on p. 381. Foreground: WT and *35S::AtTPS1-NOS* seedlings. *See* Fig. 1 on p. 390.

Production Editor: Mark J. Breaugh.

Cover design by Patricia F. Cleary.

For additional copies, pricing for bulk purchases, and/or information about other Humana titles, contact Humana at the above address or at any of the following numbers: Tel.: 973-256-1699; Fax: 973-256-8341; E-mail: humana@humanapr.com; or visit our Website: www.humanapress.com

Printed in the United States of America. 10 9 8 7 6 5 4 3 2 1

Library of Congress Cataloging in Publication Data

E-ISBN: 1-59259-774-2

Recombinant gene expression : reviews and protocols / edited by Paulina Balbás and Argelia Lorence.-- 2nd ed.
    p. ; cm. -- (Methods in molecular biology ; 267)
  Rev. ed. of: Recombinant gene expression protocols / edited by Rocky S. Tuan. c1997.
  Includes bibliographical references and index.
  ISBN 1-58829-262-2 (alk. paper)
  1. Genetic recombination--Laboratory manuals.
  [DNLM: 1. Gene Expression. 2. Recombination, Genetic. QH 443 R3109 2004] I. Balbás, Paulina. II. Lorence, Argelia. III. Recombinant gene expression protocols. IV. Series.
  QH443.R36 2004
  572.8'77--dc22
                              2004001171

Preface

Since newly created beings are often perceived as either wholly good or bad, the genetic alteration of living cells impacts directly on a symbolic meaning deeply imbedded in every culture. During the earlier years of gene expression research, technological applications were confined mainly to academic and industrial laboratories, and were perceived as highly beneficial since molecules that were previously unable to be separated or synthesized became accessible as therapeutic agents. Such were the success stories of hormones, antibodies, and vaccines produced in the bacterium *Escherichia coli*.

Originally this bacterium gained fame among humans for being an unwanted host in the intestine, or worse yet, for being occasionally dangerous and pathogenic. However, it was easily identified in contaminated waters during the 19th century, thus becoming a clear indicator of water pollution by human feces. Tamed, cultivated, and easily maintained in laboratories, its fast growth rate and metabolic capacity to adjust to changing environments fascinated the minds of scientists who studied and modeled such complex phenomena as growth, evolution, genetic exchange, infection, survival, adaptation, and further on—gene expression.

Although at the lower end of the complexity scale, this microbe became a very successful model system and a key player in the fantastic revolution kindled by the birth of recombinant DNA technology.

Without the information provided by years of basic research on *E. coli*, the successful application of gene expression would not have taken place with such extraordinary speed. *E. coli* is still unmatched for gene expression work; foreign genes can be introduced into its genome, or that of its plasmids and viruses, with relative ease and predictability; several of its primary metabolic regulatory networks have been unveiled; and its complete genomic DNA sequence is available since 1996. Although it suffers from deficiencies and limitations in the biosynthesis of complex proteins, after 30 years it remains the preferred model system in which to try out new strategies.

Other more complex biological systems for gene expression were developed in parallel to surpass the limitations of *E. coli*. Several bacteria, fungi, plant, and animal cells, as well as complete eukaryotic organisms are today in full use in laboratories and industries throughout the word as protein factories, and gene expression is now understood as a more integrated process. A protein expression system contains at least four general components: (i) the genetic elements necessary for transcription/translation and selection; (ii) in the case of vector-based systems, a suitable replicon: plasmid, virus, bacteriophage, etc.; (iii) a host strain containing the appropriate genetic traits needed to function with the specific expression signals and selection scheme; and (iv) the culturing conditions of the transformed cells or organisms. Bioengineering-related downstream operations are also considered in the original expression system design, so that purification of the product is made easy.

Nowadays, expression systems have evolved well beyond the era of the early high-level "one gene makes one protein" expression schemes. Well-characterized, efficient, and flexible expression vectors for these applications are available from companies, whereas novel approaches by the scientific community point toward a finer and more precise regulation of gene expression in the production of nonprotein molecules via metabolic engineering: chromosomal editing, promoter replacement vectors, and chloroplast transformation/ expression systems are just some of the promising examples of emerging applications of older technologies.

Since understanding and engineering the metabolism of a cell offers an unparalleled flexibility for the use of biological systems as environment-friendly factories for biomolecules, enormous efforts toward this goal are underway. Moreover, gene expression is also developing beyond living cells, with in vitro systems for gene expression and protein folding now evolving in parallel as novel strategies to overcome certain very specific problems.

Public concern about the biosafety of genetically engineered cells and organisms is also impacting the design of molecules and expression strategies. Therefore, chromosomal editing for the removal of undesirable DNA, the use of safer selection schemes and replicons, are some of the actively developing fields of application and research.

But the increasing number and varieties of both applications and transgenic organisms, especially plants, needed to be tested out of containment, an issue that continues to raise serious concerns about their possible environmental impact. To date, these concerns remain only partially answered. There are as yet no clear indicators that transgenic plants impose health risks on people and animals, or even threaten the environment farther than what regular agricultural practices do. Tons of herbicides, insecticides, and fertilizers poured yearly on behalf of productivity have indeed poisoned the environment and originated degenerative diseases in millions of people. Modern biotechnology, through gene expression, offers just such an alternative to help overcome some of the negative consequences that are the inevitable byproducts of industrialization. Transgenic plants may not be the ultimate answer, and long-term monitoring of secondary effects is absolutely necessary. However, the field of research and development is one of the most active and promising in terms of applications.

Transgenic animals are also generally confined to closed research environments, although concerns remain about the potential risk from escapes into the wilderness and also their probable impact on the biological chain and environment.

Scientific proof and experimentation is the best tool humans have developed to help reduce the fear and anxiety raised by the unknown consequences of using recombinant organisms sparingly and without custody in open spaces. Thus, the relevance of *Recombinant Gene Expression* becomes enormous, because it offers the many different views of scientific experts about how best to enhance the biosafety of recombinant organisms, without compromising their efficiency and productivity.

History has demonstrated that significant advances in scientific endeavors are always accomplished faster if, and only if, the work is supported by efficient tools and methodologies. Tools and methods are usually subject to evolution themselves, so their refinement also becomes a very important aspect of scientific progress. Consequently, the relevance of publishing a collection of protocols dealing with the different systems and applications of gene expression becomes enormous, because these are the key creation pathways to which biological scientific growth is anchored.

We are indebted to our authors, all of them experts on a particular expression system. Thanks to this truly interdisciplinary group of international scientists, the present book contains an original collection of protocols for gene expression as well as some overviews and troubleshooting guides for the biological systems addressed. Thus, the main objective of *Recombinant Gene Expression* has been enlarged, since it contains much more than a first quality collection of hands-on protocols. It will capture the attention of a wider variety of experts and experts-to-be, impelled by the curiosity of looking through the experts' eyes at the evolution and advancement of different fields of application.

While organizing a book on such an extensive topic as gene expression, it proved indispensable to pick and choose from the endless variety of strategies, vectors, promoters, and so on, so our coverage is far from complete. Some expression systems were omitted because of size limitations, and even within the areas presented, unavoidably, some research approaches were unevenly treated.

The information we provide in *Recombinant Gene Expression* is organized in sections by biological host: bacteria, fungi, plants and plant cells, and animals and animal cells, presenting a review and several protocol chapters for each. In the accompanying table, the reader will find a comprehensive glimpse of the contents and organization of our book, beyond the subject index. Although every chapter refers to the basic components of an expression system, as mentioned above, only those chapters containing detailed protocols for transformation or selection schemes are quoted in the table. Finally, every single chapter offers the valuable expertise of scientists and their personal views of strategy planning, as well as a variety of approaches that will surely be useful and inspiring to you, the reader.

Paulina Balbás
Argelia Lorence

Table 1
***Recombinant Gene Expression* Contents by Chapter**

		Vectors					Transformation		
Group	General Info	Chromosomal editing/ delivery	Chromosomal promoter replacement	Translational fusions	Extra-chromosomal expression	Chloroplast delivery	Transfection Encapsulation	Selection schemes	Growth
Prokaryotes	1–4, 12,13	7–10	7	5–7	5,6,11,12, 14,15	—	13–15	6,7,12, 13,15	2,3
Fungi	1–3, 16,21	20,22	17	18,19	18,19	—	18–20,22	20,22	2,3
Plants and plant cells	3,23, 24,27	24,26	—	24	—	25	24–26	25,26	3,24
Animal cells and animals	1–3, 28,33	29–32	—	29–31	—	—	29–32	29–32	2,3

Contents

Contributors

GERARDO ACOSTA-GARCÍA • *Department of Genetic Engineering, CINVESTAV, Irapuato Guanojuato, México*

AMALIA S. AFENDRA • *Department of Chemistry, University of Ioannina, Ioannina, Greece*

MIKHAIL F. ALEXEYEV • *Department of Pharmacology, University of South Alabama, Mobile, AL*

DAPHNÉ AUTRAN • *Department of Genetic Engineering, CINVESTAV, Irapuato, Guanojuato, México*

NELSON AVONCE • *Centro de Investigación en Biotecnología, UAEM, Cuernavaca, Morelos, México*

PAULINA BALBÁS • *Centro de Investigación en Biotecnología, UAEM, Cuernavaca, Morelos, México*

FRANCISCO BOLÍVAR • *Instituto de Biotecnología, UNAM, Cuernavaca, Morelos, México*

MEWES BÖTTNER • *Berlin Technical University, Institute for Biotechnology, Berlin, Germany*

JEAN-CLAUDE BOULAIN • *Département d'Ingénierie et d'Études des Protéines, CEA Saclay, Gif-sur-Yvette Cedex, France*

SERGIO CASAS-FLORES • *Department of Plant Genetic Engineering, CINVESTAV, Irapuato, Guanojuato, México.*

CARLOS CORTÉS-PENAGOS • *Department of Plant Genetic Engineering, CINVESTAV, Irapuato, Guanojuato, México*

CAROLE CRAMER • *Ralin Biotechnology Center, Virginia Polytechnic Institute and State University, Blacksburg, VA*

HENRY DANIELL • *Department Molecular Biology and Microbiology, University of Central Florida, Orlando, FL*

CONSTANTIN DRAINAS • *Department of Chemistry, University of Ioannina, Ioannina, Greece*

FRÉDÉRIC DUCANCEL • *Département d'Ingénierie et d'Études des Protéines, CEA Saclay, Gif-sur-Yvette Cedex, France*

ANGELA DUILIO • *Dipartimento di Chimica Organica e Biochimica, Università di Napoli "Federico II" - Complesso Universitario MS, Napoli, Italia*

SANDINO ESTRADA-MONDACA • *Instituto de Biotecnología, UNAM, Cuernavaca, Morelos, México*

BEATRIX FAHNERT • *Department of Process and Environmental Engineering, University of Oulu, Finland*

NOEMÍ FLORES • *Instituto de Biotecnología, UNAM, Cuernavaca, Morelos, México*

MARTIN FUSSENEGGER • *Institute of Biotechnology, Swiss Federal Institute of Technology, Zurich, Switzerland*

JEAN-LOUIS GOERGEN • *Laboratoire des Sciences du Génie Chimique - CNRS, Institut National Polytechnique de Lorraine, Vandoeuvre-lès-Nancy, France*

GUILLERMO GOSSET • *Instituto de Biotecnología, UNAM, Cuernavaca, Morelos, México*

JAMES J. GREENE • *Catholic University of America, Washington DC*

BÄRBEL HAHN-HÄGERDAL • *Department of Applied Microbiology, Lund University, Lund, Sweden*

HANSJÖRG HAUSER • *Department of Gene Regulation and Differentiation, German Research Centre for Biotechnology, Braunschweig, Germany*

ALFREDO HERRERA-ESTRELLA • *Department of Plant Genetic Engineering, CINVESTAV, Irapuato, Guanojuato, México*

CATERINA HOLZ • *Berlin Technical University, Institute for Biotechnology, Berlin, Germany*

LOUIS-MARIE HOUDEBINE • *Biologie du Développement et Reproduction, Institut National de la Recherche Agronomique, France*

GABRIEL ITURRIAGA • *Centro de Investigación en Biotecnología, UAEM, Cuernavaca, Morelos, México*

MICHAEL C. JEWETT • *Department of Chemical Engineering, Stanford University, Stanford, CA*

BJÖRN JOHANSSON • *Departamento de Biologia, Universidade do Minho, Campus de Gualtar, Braga, Portugal*

SHASHI KUMAR • *Department of Molecular Biology and Microbiology, University of Central Florida, Orlando, FL*

CHRISTINE LANG • *Berlin Technical University, Institute for Biotechnology, Berlin, Germany*

SYLVIE LE BORGNE • *Departamento de Biotecnología, Instituto Mexicano del Petróleo, México DF, México*

BARBARA LEYMAN • *Institute of Botany and Microbiology, Katholieke Universiteit Leuven; and Department of Molecular Microbiology, Flemish Interuniversity Institute of Biotechnology (VIB), Leuven-Heverlee, Belgium*

ARGELIA LORENCE • *Virginia Polytechnic Institute and State University, Blacksburg, VA*

GENNARO MARINO • *Università di Napoli "Federico II," Facoltà di Scienze Biotecnologiche, Dipartimento di Chimica Organica e Biochimica, Complesso Universitario, M.S. Angelo, Napoli, Italia*

DIETHARD MATTANOVICH • *Institute of Applied Microbiology, University of Natural Resources and Applied Life Sciences, Vienna, Austria*

FABRICIO MEDINA-BOLÍVAR • *Fralin Biotechnology Center, Virginia Polytechnic Institute and State University, Blacksburg, VA*

LUCÍA MONACO • *Keryos SpA, San Donato Milanese, Italia*

JOAQUÍN J. NIETO • *Department of Microbiology and Parasitology, Faculty of Pharmacy, University of Seville, Sevilla, España*

VIANEY OLMEDO-MONFIL • *Department of Plant Genetic Engineering, CINVESTAV, Irapuato, Guanojuato, México*

LAURA A. PALOMARES • *Instituto de Biotecnología, UNAM, Cuernavaca, Morelos, México*

VIKTORIYA V. PASTUKH • *Department of Pharmacology, University of South Alabama, Mobile, AL*

DANILO PORRO • *Department of Biotechnology and Bioscience, University of Milano-Bicocca, Milano, Italia*

OCTAVIO T. RAMÍREZ • *Instituto de Biotecnología, UNAM, Cuernavaca, Morelos, México*

MATTHEW RAMON • *Institute of Botany and Microbiology, Katholieke Universiteit Leuven; and Department of Molecular Microbiology, Flemish Interuniversity Institute of Biotechnology (VIB), Leuven-Heverlee, Belgium*

FÉLIX RECILLAS-TARGA • *Instituto de Fisiología Celular, UNAM, México DF, México*

HÉCTOR RINCÓN-ARANO • *Instituto de Fisiología Celular, UNAM, México DF, México*

TERESA ROSALES-SAAVEDRA • *Department of Plant Genetic Engineering, CINVESTAV, Irapuato, Guanojuato, México*

INNA N. SHOKOLENKO • *Department of Cell Biology and Neuroscience, University of South Alabama, Mobile, AL*

JAMES R. SWARTZ • *Department of Chemical Engineering, Stanford University, Stanford, CA*

WACLAW SZYBALSKI • *McArdle Laboratory for Cancer Research, University of Wisconsin Medical School, Madison, WI*

JOHAN M. THEVELEIN • *Institute of Botany and Microbiology, Katholieke Universiteit Leuven; and Department of Molecular Microbiology, Flemish Interuniversity Institute of Biotechnology (VIB), Leuven-Heverlee, Belgium*

MARIA LUISA TUTINO • *Università di Napoli "Federico II," Facoltà di Scienze Biotecnologiche, Dipartimento di Chimica Organica e Biochimica, Complesso Universitario, M.S. Angelo, Napoli, Italia*

FERNANDO VALLE • *Department of Metabolic Pathway Engineering, Genencor International Inc., Palo Alto, CA*

PATRICK VAN DIJCK • *Institute of Botany and Microbiology, Katholieke Universiteit Leuven; and Department of Molecular Microbiology, Flemish Interuniversity Institute of Biotechnology (VIB), Leuven-Heverlee, Belgium*

CARMEN VARGAS • *Department of Microbiology and Parasitology, Faculty of Pharmacy, University of Seville, Sevilla, España*

ROBERT VERPOORTE • *Leiden/Amsterdam Center for Drug Research, Leiden University, Leiden, The Netherlands*

KEVIN J. VERSTREPEN • *Centre for Malting and Brewing Science, Katholieke Universiteit Leuven, Leuven (Heverlee), Belgium*

JEAN-PHILIPPE VIELLE-CALZADA • *Department of Genetic Engineering, CINVESTAV, Irapuato, GTO, México*

BARRY L. WANNER • *Department of Biological Sciences, Purdue University, West Lafayette, IN*

WILFRIED WEBER • *Institute of Biotechnology, Swiss Federal Institute of Technology, Zurich, Switzerland*

JADWIGA WILD • *McArdle Laboratory for Cancer Research, University of Wisconsin Medical School, Madison, WI*

GLENN L. WILSON • *Department of Cell Biology and Neuroscience, University of South Alabama, Mobile, AL*

DAGMAR WIRTH • *Department of Gene Regulation and Differentiation, German Research Centre for Biotechnologie, Braunschweig, Germany*

KIM A. WOODROW • *Department of Chemical Engineering, Stanford University, Stanford, CA*

KE ZHANG • *Department of Biological Sciences, Purdue University, West Lafayette, IN*

LU ZHOU • *Department of Biological Sciences, Purdue University, West Lafayette, IN*

I

GENERAL ISSUES ABOUT RECOMBINANT GENE EXPRESSION

1

Host Cell Compatibility in Protein Expression

James J. Greene

Summary

The expression of cloned genes in prokaryotic or eukaryotic host cells provides the means not only for the study of gene function but also for the production of substantial amounts of protein and nonprotein molecules for commercial and investigational use. In the case of proteins, strategies for determining the most appropriate vector–host combination for the expression of an exogenous gene depend on a diverse range of factors that relate ultimately to the properties of the gene and its product. The approach used in the downstream purification of the product is another factor that impinges on this selection. However, among the most important considerations in the choice of vector and host in ensuring the maximal amount of expression is the compatibility of the host cells to translate the RNA transcript, to ensure the proper folding of the product, and to sustain the protein in the intact and functional state.

Key Words: Translation; protein folding; codon usage; redox state.

1. Introduction

By its ability to place a gene within a new genetic environment, recombinant DNA technology frees genes from the normal regulatory constraints of their natural environment and allows for their expression in a controlled manner. This can be exploited to transcribe and translate prokaryotic or eukaryotic genes within prokaryotic cells to achieve high-level expression of gene products. Moreover, eukaryotic genes can be expressed in homologous eukaryotic hosts for functional studies that examine relationships between gene expression and cellular response. They can also be expressed in other eukaryotic cells optimized for expression of gene products that require post-translational and post-transcriptional processing.

From: *Methods in Molecular Biology, vol. 267:*
Recombinant Gene Expression: Reviews and Protocols, Second Edition
Edited by: P. Balbás and A. Lorence © Humana Press Inc., Totowa, NJ

The selection of the most appropriate host cell and corresponding vector for expression of the desired gene ultimately depends on the study's objective. However, myriad other factors influence this selection; these have been discussed in several comprehensive reviews (*1–5*). Initially, vector-host selection was predicated primarily on the premise that if transcription could be achieved, then the gene would be expressed. Now it has become increasingly clear that the compatibility of the host's biochemical environment with the ability to process and translate the RNA transcript, along with its ability to modify and sustain the translated protein are equally important. Indeed, the considerations that relate to these posttranscriptional host properties are intimately related to the selection of the vector and its transcription-promoter system (**Fig. 1**). This chapter addresses some of the issues that determine the compatibility of the host cells in ensuring the successful production of a functional product.

2. Translational Compatibility

The ability of the host's translational machinery to cope with the foreign gene's transcription product underlies much of the difficulty in achieving optimal expression once the gene has been transcribed. This is particularly the case for heterogous expression of eukaryotic genes in prokaryotic cells. In addition for the requirement that the RNA transcript have a properly positioned Shine-Delgarno sequence (*1,3*), there needs to be a significant overlap in the codon usage of host and donor cells. Depending on the gene, this is not always the case. As shown in **Table 1**, there are considerable differences in the frequencies of various codons for *E. coli* K12 and humans. Codon usage can vary significantly between species as well as between genome types such as nuclear DNA and mitochondrial DNA (*6*, and Internet sites such as http://www.kazusa.or.jp/codon/; http://www3.ncbi.nlm.nih.gov/Taxonomy/). Codons that occur in high frequency in humans but in low frequencies in the host cells will normally result in a condition whereby the pool of tRNA for that codon will be so low as to become depleted (*6,7*). Depletion of rare tRNAs during translation of foreign mRNA is the principal reason for the reduction in growth of the host cells that can be manifested as an overall toxic response (**Table 2**).

This can have significant consequences in protein yield if the exogenous gene is expressed using a constitutive promoter. Transcription of the gene before the culture has reached optimal density may cause depletion of the hosts' rare tRNAs, resulting in growth reduction and even complete cessation of cell growth. Under these conditions, culture densities are always low, with a commensurate reduction in protein yield (**Fig. 2**). This problem can be largely circumvented through the use of an inducible expression vector, whereby the promoter is normally inactive unless induced. When using an inducible expression system, the culture of transformed host cells is allowed to reach maximal density and then the cells are induced. The rationale here is that once the culture is saturated, the cells' translational resources presumably do not have to be used for cell proliferation, so they can be dedicated to expression of the foreign gene product. As a general rule, inducible expression vectors offer the greatest flexibility and best chance of achieving maximal expression. Large selections of inducible promoters whose char-

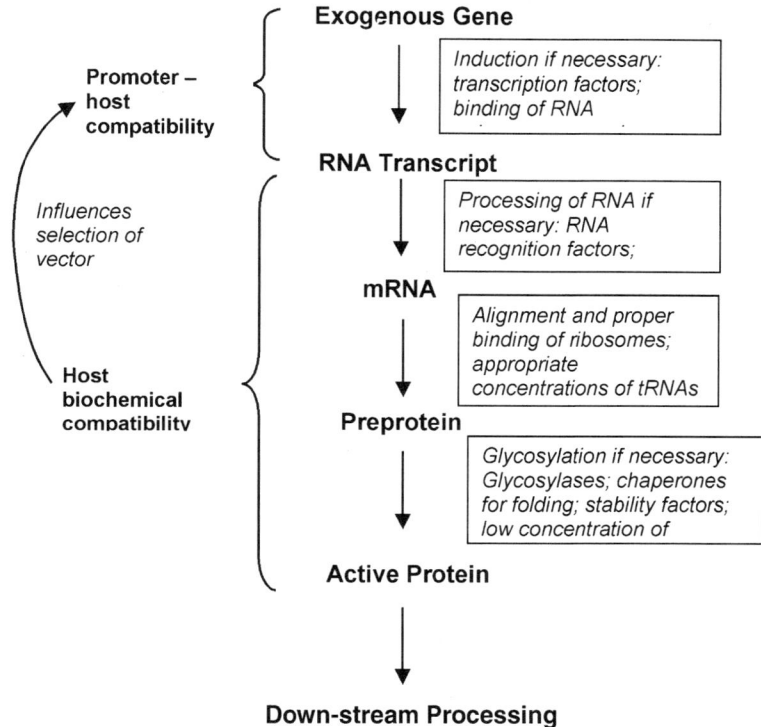

Fig. 1. Diagram showing the relationship among the host compatibility factors and principal steps in gene expression.

acteristics are matched to the host cells are available for this purpose (*2*) (for examples *see* Chapters 8, 17, 24, 25, 29, 30); vectors with inducible copy numbers are also available (*see* Chapters 10 and 11).

While inducible expression vectors can often be a satisfactory solution to the codon bias problem, there may be situations in which optimal expression is still not achieved. This occurs in cases in which the foreign gene being expressed contains stretches or a large number of rare codons (*8*). When this occurs, there are insufficient pools of cognate rare tRNAs to allow for translation even when the culture is at maximal density. Substitution of the rare codons in the foreign gene being expressed with prevalent codons can increase expression but is not always reliable (*8*). Another approach is the use of host cells that co-express rare tRNAs. Cells transformed with rare tRNA genes make ideal hosts for expression of gene products for which codon bias may be problematic. Several such cell lines are commercially available, such as the Rosetta™ family of cells (Novagen, Darmstadt, Germany). Alternatively, custom host cells can be created with plasmids such as the pRARE expression plasmids (also from Novagen) that contain a number of rare tRNA genes.

Table 1
Codon Usage Table Showing Several Codons
and Frequency of Occurrence in Genes of *E. coli* Strain K-12 and *H. sapiens*

Codon	Encoded amino acid	Frequency in *E. coli*	Frequency in *H. sapiens*
AGG	Arg	1.2	11.4
AGA	Arg	2.1	11.5
CGA	Arg	3.6	6.3
CUA	Leu	3.9	7.8
CUG	Leu	52.7	39.8
AUA	Ile	4.3	7.7
CCC	Pro	5.5	20.1

Frequency is given in terms of occurrence per 1000 codons.

Table 2
Common Problems, Probable Causes,
and Possible Solutions for Protein Expression in Heterologous Hosts

Problem	Possible cause	Solution
Poor cell growth	Codon usage problem	Use rare tRNA-supplied host cells
Cell death	Toxicity problem	Use stringent control plasmid
No protein production	Condon usage problem	Use rare tRNA-supplied host cells
Truncated protein	Condon usage problem	Use rare tRNA-supplied host cells
Inactive protein	Improper folding	Use chaperone-supplied host cells; cells with more favorable redox environment
Insoluble protein	Overexpression; denatured protein	Used controlled expression vector; cells with more favorable redox environment

3. Redox State and Protein Folding Compatibility

Expression of the full polypeptide is not an assurance that the final protein product is functional. For the protein to be functional, it has to be folded properly (*see* Chapters 2 and 3). Two factors work in concert to ensure that a protein folds properly: chaperones and foldases that collectively are in the family of oxidoreductases, and the cells' oxidation-reduction state, also termed the redox state. Each oxidoreductase has an optimal redox state to execute proper folding. An environment that is either too reducing or not reducing enough will cause the protein to either misfold on its own or have it incorrectly folded by a chaperone. The redox environment of cells can vary sufficiently as to cause a protein that normally folds properly in its native environment to misfold in a different cellular environment (*3,9*). Indeed, a difference as little as ± 20 mV is sufficient to cause misfolding.

While eukaryotic cells are more likely to present the most compatible environment for the folding of eukaryotic proteins, prokaryotic hosts such as *E. coli* are still the first

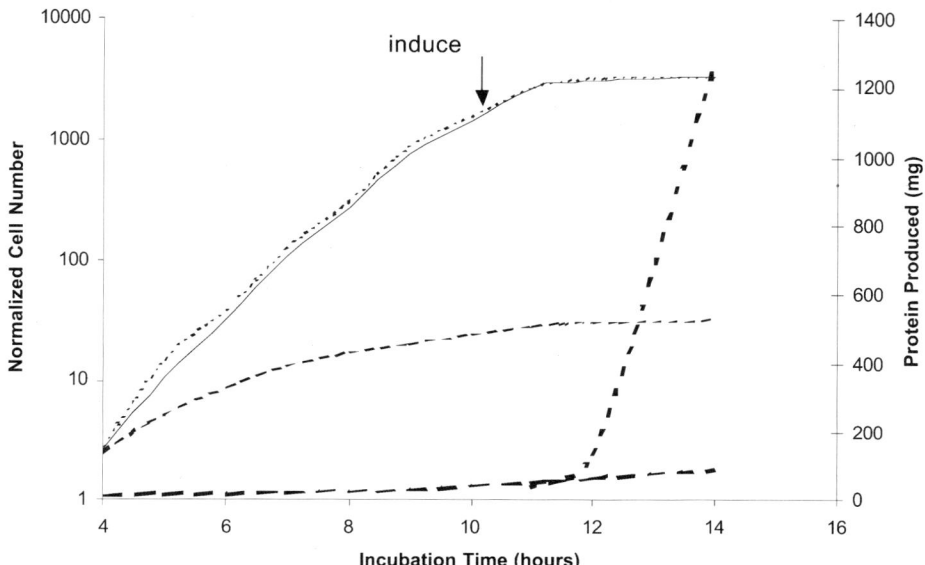

Fig. 2. Cell density and production of a human transcription factor protein. Light lines show normalized cell density in culture while dark lines show amount of accumulated protein (light solid line = density of nontransformed *E coli*; light dotted line = density of cell transformed with IPTG-inducible expression vector; light dashed line = density of cells transformed with constitutive expression vector; dark dotted line = accumulated protein from cultures of induced cells transformed with IPTG-inducible expression vector; dark dashed line = accumulated protein from cultures of cell transformed with constitutive expression vector).

preference for high-level production. There are many reasons for this, including the considerably lower cost of culture, the higher overall efficiency of product expression, and the versatility of *E. coli* to be modified (**9**). One example of modification is expression of ancillary genes such as eukaryotic chaperones that modify their cellular environment to accommodate the foreign gene protein. For some heterologous gene products that require complex folding, the host cells' natural chaperones can aid in proper folding, but more often they cannot (**9,10**). If the hosts' internal environment and chaperones cannot induce proper folding, this problem can be overcome by producing soluble fusion proteins in which an oxidoreductase, such as thioredoxin, is fused to the protein product of interest (**11**). A compatible redox environment is created by introducing mutations in the thioredoxin reductase and glutathione reductase genes that improve disulfide bond formation and encourage correct protein folding. This approach has now been used to express several products that require complex folding (**12,13**).

4. Protein Solubility Compatibility

Another consideration in the production of a functional product is that the protein from the foreign gene, normally referred to as the "target" protein, has to be soluble. Either misfolding or overexpression of the target protein can result in an insoluble product. A statistical model for predicting the solubility of proteins in *E. coli* has been developed by Wilkinson and Harrison (*14*). Their model considers several parameters, primarily average charge, turn-forming residue fraction, hydrophilicity (from the fraction of certain polar and nonpolar amino acids), and total number of residues. These elements are incorporated into calculating a canonical variable (CV) or composite parameter for the protein for which the solubility is being predicted:

$$CV = \alpha \left[\frac{N + G + P + S}{n} \right] + \beta \left[\left(\frac{(R + K) - (D + E)}{n} \right) - 0.3 \right]$$

where:

  n = number of amino acids in protein
  N, G, P, S = number of Asn, Gly, Pro, or Ser residues, respectively
  R, K, D, E = number of Arg, Lys, Asp, or Glu residues, respectively
  α, β = empirical coefficients (15.43 and −29.56, respectively)

This resulting CV can be used to calculate the probability that the protein is insoluble or soluble (*see 14*); probabilities for several proteins are shown in **Table 3**. It is apparent that many higher-order eukaryotic target proteins have a low probability of being soluble in *E. coli*, while the native *E. coli* GrpE and NusA proteins are highly soluble. This led to the concept that expression of the target protein as a fusion with GrpE, NusA, or another highly soluble protein bacterioferritin (BFR) would yield a more soluble product (*15*). Indeed, many mammalian proteins originally expressed in *E. coli* as insoluble inclusion bodies (aggregates of insoluble protein) have now been expressed as soluble product through coupling with a soluble native protein, even though the size of the fusion product can be substantially larger than the target protein alone.

5. Posttranslational Modifications and Host Compatibility

By virtue of their extreme versatility, prokaryotic hosts can accommodate the expression of many, if not most, eukaryotic proteins. However, if from a higher-order eukaryote, the protein of interest often requires substantial posttranslational modification to yield the final product. These modifications include proteolytic cleavage, glycosylation, and amino acid modifications such as phosphorylation, acetylation, amidation, sulfation, isoprenylation, and fatty acid addition (summarized in **Table 4**). Consequently, it becomes a necessary to use hosts other than *E. coli*. Another consideration pertinent to the pharmaceutical industry is the potential for contamination of the product by bacterial endotoxins.

Eukaryotic systems can fulfill these posttranslational prerequisites to produce a biologically active product that is free of bacterial toxins. A number of eukaryotic expression systems are available that vary in their degree of "eukaryotic complexity," including mammalian, amphibian, insect, plant, and yeast cells (Chapters 16–33). The choice of the most appropriate eukaryotic host is then normally determined by the efficiency and cost-effectiveness of the system. In general, the ease, speed and costs in growing the cells along with overall gene expression efficiency are the principal determinants. Consistent with this notion is that the host with the simplest eukaryotic complexity is selected first. The one exception to this is for gene functional studies. Such studies usually entail the use of homologous hosts or at least hosts very similar to the cells from which the exoge-

Table 3
Predicted Probability of Solubilities Based on Wilkinson and Harrison Model

Protein	Amino acid length	MW (kDal)	Probability
Interleukin-3	133	15.1	−73%
Bovine growth hormone	189	21.6	−60%
Human interferon γ	146	17.1	−58%
Thioredoxin	109	11.7	+73%
GrpE	197	21.7	+92%
NusA	495	54.8	+95%

Negative values represent the probability that the protein will be insoluble, while positive values represent the probability that the protein will be soluble in *E. coli*.

Table 4
Capabilities of Various Host Cells as Hosts for Gene Expression

Host cells	Growth efficiency	Expression efficiency	Proteolytic processing[c]	Phosphor-ylation[c]	Glycos-ylation[c]	Amino acid modifi-cation[c]
Bacteria	Very high	Very high	No	Possible[a]	No	No
Yeast/ Fungi	High	Moderate-high	Some	Some	Yes[b,c]	Some
Insect	Moderate	Moderate	Some	Some	Moderate	Some
Plants	High	High	Moderate	Some	Moderate	Some
Plant cells	Moderate	High	Moderate	Some	Moderate	Some
Animals	Moderate	Moderate	High	High	High	Moderate
Animal cells	High	Low	High	High	High	Full
Mam-malian	Low	Low	Yes	Full	Full	Full

These are very general qualitative considerations, which may not be accurate for every single gene/host combination.

[a] For *E. coli*, "possible" refers to the potential for modification of host properties through the co-expression of exogenous kinase and other genes.

[b] In yeast, insect, and *Drosophila*, the glycosylation pattern is different from that in mammalian cells, exemplified by the lack of scialic acid and mannose.

[c] Type and degree of modification is dependent upon species.

nous gene of interest was obtained (**Table 4**). This chapter provides an example of two eukaryotic hosts—yeasts and insect cells—and their particular properties.

Yeasts are the simplest of the eukaryotic hosts. Yeasts that that have been used to express eukaryotic genes include *Saccharomyces cerevisiae, Schizosaccharomyces pombe, Pichia pastoris, Hansela polymorpha, Kluyveromyces lactis,* and *Yarrowia lipolytica*. Of these, the first three are the most widely used. The choice of yeast as a host for high-level expression of heterologous protein can be attributed to the large body of existing knowledge on yeasts' biochemistry and genetics as well as the fact that they exhibit some of the posttranslational properties of more complex eukaryotes (*16–18*). Moreover, they can be rapidly grown inexpensively, making them suited for large-scale fermentation; recombinants can be easily selected by complementation; and expressed proteins can be specifically engineered for cytoplasmic localization or for extracellular export (*17,18*) (*see* Chapters 2, 3, 16–20).

The budding yeast *S. cerevisiae* was the first of the yeasts to be used as hosts for eukaryotic protein production, as it was the best characterized. However, its limitations became apparent with the discovery that it could not process gene transcripts containing introns and that it had primitive glycosylation and amino acid modification systems. This led to the interest in the development of non-*Saccharomyces* yeast species as expression systems. The methylotrophic budding yeasts *H. polymorpha* and *P. pastoris* and the fission yeast (multiplies by fission instead of budding) *Schizosaccharomyces pombe* have become effective alternatives (*19*). These yeasts are unique in that they will grow using methanol as the sole carbon source, a property that can be exploited to achieve high expression yields of a target protein. Growth in methanol is mediated in part by an alcohol oxidase, an enzyme whose synthesis is tightly regulated by the alcohol oxidase (AOX) promoter. The high efficiency of target gene expression in these yeasts is related to the use of AOX promoter in corresponding expression vectors. The protein expression in these yeast strains growing in methanol to induce the AOX1 promoter is 10–100 times that in *S. cerevisiae*. In the case of *S. pombe* (*pombe* means "beer" in Swahili, a most logical name for this yeast isolated from beer and used to brew it), it has many features reminiscent of mammalian cells. For example, *S. pombe* has glycosylation and amino acid modification systems more similar to those of mammalian cells than those of other yeast species. It has an endogenous galactosyltransferase activity allowing for a more complex glycosylation than other yeast, N-terminal acetylation and polyisoprenylation capabilities. Additionally, it can support the function of some mammalian promoters.

The need for hosts capable of more mammalian-like posttranslational processing led to the use of insect cells as hosts. In this case, the development of insect host cells was driven by the realization that the baculovirus *Autographa california* multicapsid nuclear polyhedrosis virus (AcMNPV) replicates very efficiently and has a extremely strong promoter to drive the expression of the polyhedrin protein in insect cells (*20–22*). Expression of target proteins by baculovirus is based on the introduction of the foreign gene into the polyhedron region of the viral genome via homologous recombination with a transfer vector containing the cloned gene, an event that occurs in the co-transfected insect cells. The transfected cells generate infectious recombinant virus that is used to express the target protein using the polyhedrin or other promoters upon infection of a second culture of cells or insect larvae. Lepidopteran cells—particularly those of the fall army worm (*Spodoptera frugiperda*) and the cabbage looper (*Trichoplusia ni*)—proved to be the most effective hosts. Both these cells are capable of extensive but still incomplete glycosylation as shown in **Fig. 3**, and have been used to produce a variety of active mammalian glycoproteins (*22*).

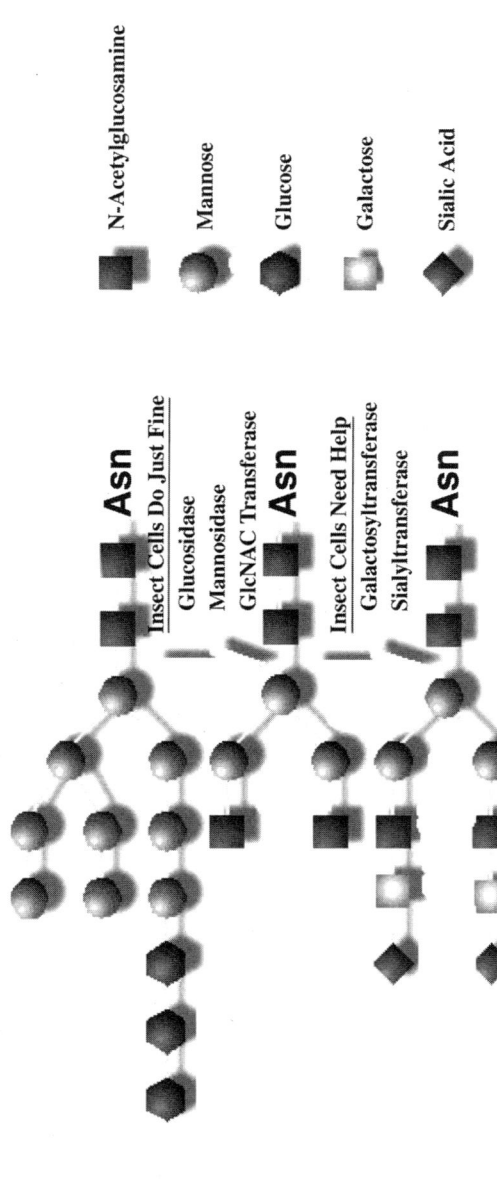

Fig. 3. N-glycosylation during posttranslational processing and the capabilities of lepidopteran cells. This scheme shows trimming of intermediates, particularly the GlcNAc residues by glycosidases do some trimming of these structures, particularly from the intermediate structure. Lepidopteran cells have minimal or no capacity for galactose or sialic acid addition. Figure courtesy of Dr. Jim Litts.

11

While the insect cells that serve as hosts for baculoviral-based vectors can provide some eukaryotic protein modifications, most recombinant glycoproteins produced by the baculovirus-insect cell system are not sialylated (*23*). This is not always required for bioactivity of some products. Nevertheless, the absence of terminal sialic acids on baculovirus-expressed recombinant glycoproteins can be a significant problem, as these terminal sugars often contribute critically to the functional activity of the glycoprotein. Sialylation of newly synthesized *N*-glycoproteins is the last step in an elaborate biosynthetic pathway that begins with the co-translational transfer of a glycan precursor to a nascent polypeptide. The glycan precursor is subsequently trimmed and elongated by various enzymes in the endoplasmic reticulum and the Golgi. In mammalian cells, elongation of *N*-glycan precursors yields products known as complex *N*-glycans.

Galactosyltransferases and sialyltransferases are the enzymes involved in the final elongation steps in mammalian cells. Insect cells have an analogous *N*-glycan processing pathway but fail to produce terminally sialylated *N*-glycans due to extremely low or nonexistent levels of galactosyltransferase and sialyltransferase activities (**Fig. 3** (*23*). Most recently, baculoviral systems have been modified to allow galactosyl and sialyl transfer (*24*). Despite these advances in insect-based systems, authentic mammalian-like glycosylation remains a problem. This deficiency makes it necessary to express proteins that require these modifications in even high-order cells such as *Drosophila melanogaster* (*25*) and mammalian cells. A wide variety of vector-expression systems, some viral and others plasmid-based, is available using these hosts (*see* Chapters 2, 3, 28–33).

6. Conclusions

Bacterial cells are generally the first choice as hosts for the expression of heterologous target proteins. Their extremely efficient growth rate, amenability to be grown in large fermentation cultures, ease of genetic manipulation, and the availability of highly effective expression vectors with strong promoters ensure that the target protein can be expressed in the most efficient and cost-effective manner. Although bacteria have significant differences from eukaryotic cells, some of these that have proven to be a detriment to producing functional eukaryotic protein can be overcome in part. The dilemma represented by an entirely different biochemical background that may impede the proper folding of protein and the codon bias differences in donor and host cells are two examples. Prokaryotic host cells are now available that co-express chaperones and foldases to allow for proper folding and rare tRNAs to suppress the codon bias problem. However, the requirement of some eukaryotic proteins to require posttranslational modifications such as glycosylation, phosphorylation, and amino acid modification presents a barrier than can be bridged only by the use of eukaryotic hosts. As a general rule, the simpler the host, the more efficient are its protein expression and cell growth. The basic tenet in choosing the most appropriate host is that the simplest cell that can provide a functional product in a time- and cost-effective manner is the best host.

The gene expression field of applications has widened its range of possibilities recently with the advent of metabolic engineering. Efficient tools and schemes for chromosomal editing by insertion and deletion of specific genes and control regions (promoters) is a very active field of research. The foreseen advantages of finely tuning the expression of homologous or heterologous enzymes to modify the fluxes of metabolites in the cells include a better management of the cell's biosynthetic, breathing, transporting, and growing capabilities; therefore, gene expression together with metabolism enhancement are the most promising trends for the next successful chapter of gene expression.

References

1. Baneyx, F. (1999) Recombinant protein expression in *Escherichia coli*. *Curr. Opin. Biotechnol.* **10**, 411–421.
2. Balbás, P. (2001) Understanding the art of producing protein and nonprotein molecules in *Escherichia coli*. *Mol. Biotechnol.* **19**, 251–267.
3. Swartz, J. R. (2001) Advances in *Escherichia coli* production of therapeutic proteins. *Curr. Opin. Biotechnol.* **12**, 195–201.
4. Jonasson, P., Lijeqvist, S., Nygren, P., and Ståhl, S. (2002) Genetic design for facilitated production and recovery of recombinant proteins in *Escherichia coli*. *Biotechnol. Appl. Biochem.* **35**, 91–105.
5. Kaufman, R. J. (2000) Overview of vector design for mammalian gene expression. *Mol. Biotechnol.* **16**, 151–160.
6. de Boer, H. A. and Kastelein, R. A. (1986) Biased codon usage: an exploration of its role in optimization of translation, in *Maximizing Gene Expression* (Reznikoff, W. and Gold, L., eds.), Buttersworth, Boston.
7. Akashi, H. (2001) Gene expression and molecular evolution. *Curr. Opin. Genet. Dev.* **11**, 660–666.
8. Kane J. F. (1995) Effects of rare codon clusters on high-level expression of heterologous proteins in *Escherichia coli*. *Curr. Opin. Biotechnol.* **6**, 494–500.
9. Hannig, G. and Makrides, S. C. (1998) Strategies for optimizing heterologous protein expression in *Escherichia coli*. *Trends in Biotechn.* **16**, 54–60.
10. Prinz, W. A., Aslund, F., Holmgren, A., and Beckwith, J. (1997) The role of the thioredoxin and glutaredoxin pathways in reducing protein disulfide bonds in the *Escherichia coli* cytoplasm. *J. Biol. Chem.* **272**, 15661–15667.
11. LaVallie, E. R., DiBlasio-Smith, E. A., Collins-Racie, L. A., Lu, Z. and McCoy, J. M. (2003). Thioredoxin and related proteins as multifunctional fusion tags for soluble expression in *E. coli*. *Methods Mol. Biol.* **205**, 119–140.
12. Lobel, L., Pollak, S., Klein, J., and Lustbader, J. W. (2001) High-level bacterial expression of a natively folded, soluble extracellular domain fusion protein of the human luteinizing hormone/chorionic gonadotropin receptor in the cytoplasm of *Escherichia coli*. *Endocrine* **14**, 205–212.
13. Lobel, L., Pollak, S., Lustbader, B., Klein, J., and Lustbader, J. W. (2002) Bacterial expression of a natively folded extracellular domain fusion protein of the hFSH receptor in the cytoplasm of *Escherichia coli*. *Protein Express. Purif.* **25**, 124–133.
14. Wilkinson, D. L. and Harrison, R. G. (1991) Predicting the solubility of recombinant proteins in *Escherichia coli*. *Biotechnology* **9**, 443–448.
15. Davis, G. D., Elisee, C., Newham, D. M. and Harrison, R. G. (1999) New fusion protein systems designed to give soluble expression in *Escherichia coli*. *Biotechnol Bioeng.* **65**, 382–388.
16. Romanos, M. A., Scorer, C. A. and Clare, J. J. (1992) Foreign gene expression in yeast: a review. *Yeast* **8**, 423–488.

17. Sudbery, P. E. (1996) The expression of recombinant proteins in yeasts. *Curr. Opin. Biotechnol.* **7**, 517–524.
18. Ostergaard, S., Olsson, L., and Nielsen, J. (2000) Metabolic engineering of *Saccharomyces cerevisiae. Microbiol. Mol. Biol. Rev.* **64**, 34–50.
19. Giga-Hama, Y. and Kumagai, H. (1999) Expression system for foreign genes using the fission yeast *Schizosaccharomyces pombe. Biotechnol. Appl. Biochem.* **30**, 235–244.
20. Luckow, V. A. (1993) Baculovirus systems for the expression of human gene products. *Curr. Opin. Biotechnol.* **4**, 564–572.
21. Jarvis, D. L. and Guarino, L. A. (1995) Continuous foreign gene expression in transformed lepidopteran insect cells. *Methods Mol. Biol.* **39**, 187–202.
22. McCarroll, L. and King, L. A. (1997) Stable insect cell cultures for recombinant protein production. *Curr. Opin. Biotechnol.* **8**, 590–594.
23. Jarvis, D. L. and Finn, E. E. (1995) Biochemical analysis of the N-glycosylation pathway in baculovirus-infected lepidopteran insect cells. *Virology* **212**, 500–511.
24. Jarvis, D. L., Howe, D. and Aumiller, J. J. (2001) Novel baculovirus expression vectors that provide sialylation of recombinant glycoproteins in lepidopteran insect cells. *J. Virol.* **75**, 6223–6227.
25. Condreay, J. P., Witherspoon, S. M., Clay, W. C. and Kost, T. A. (1997) Transient and stable gene expression in mammalian cells transduced with a recombinant baculovirus vector. *Proc. Natl. Acad. Sci. USA* **96**, 127–132.

2

Production of Recombinant Proteins

Challenges and Solutions

Laura A. Palomares, Sandino Estrada-Mondaca, and Octavio T. Ramírez

Summary

Efficient strategies for the production of recombinant proteins are gaining increasing importance, as more applications that require high amounts of high-quality proteins reach the market. Higher production efficiencies and, consequently, lower costs of the final product are needed for obtaining a commercially viable process. In this chapter, common problems in recombinant protein production are reviewed and strategies for their solution are discussed. Such strategies include molecular biology techniques, as well as manipulation of the culture environment. Finally, specific problems relevant to different hosts are discussed (*see* Chapters 1 and 3).

Key Words: Fermentation; prokaryotes; yeasts; fungi; animal cells.

1. Common Problems Encountered During Production of Recombinant Proteins

The demand of recombinant proteins has increased as more applications in several fields become a commercial reality. Recombinant proteins have been utilized as tools for cellular and molecular biology. Various application areas have experienced substantial advances thanks to the possibility of producing large amounts of recombinant proteins by an increasing availability of genetically manipulated organisms. For instance, uncountable lives have been saved because of the almost unlimited accessibility of therapeutic and prophylactic proteins that before the era of modern biotechnology could be obtained only in very small amounts from unsafe sources. Today, more than 75 recombinant proteins are utilized as pharmaceuticals, and more

From: *Methods in Molecular Biology, vol. 267:*
Recombinant Gene Expression: Reviews and Protocols, Second Edition
Edited by: P. Balbás and A. Lorence © Humana Press Inc., Totowa, NJ

than 360 new medicines based on recombinant proteins are under development (www.phrma.org). The impact of the production of recombinant proteins has also extended to the development of bioinsecticides, diagnostic kits, enzymes with numerous applications, and bioremediation processes, among many others. In particular, areas such as detergent production and food processing have been among the most notable success.

Even when hundreds of proteins are produced at commercial scale, the production of recombinant proteins still constitutes a challenge in many cases. Moreover, many applications would benefit with higher production efficiencies and consequent lower costs of the final product. In this chapter, typical problems encountered during recombinant protein production are reviewed and strategies to solve them and increase productivity are discussed.

1.1. Loss of Expression

A necessary condition for adequate recombinant protein production is the efficient expression of the gene of interest. However, expression can be lost due to structural changes in the recombinant gene or disappearance of the gene from host cells. Loss of expression will be discussed here, with emphasis on the three alternative locations of the gene of interest: in plasmids, integrated to the host's chromosome, or delivered by a virus.

1.1.1. Plasmid-Based Systems

Plasmids are extrachromosomal self-replicating cytoplasmic DNA elements that are found in prokaryotes and eukaryotes. They have been used as molecular vehicles for recombinant genes since the dawn of genetic engineering. Plasmid-based expression is the most popular choice when using prokaryotes as hosts, as genetic manipulation of plasmids is easy. Furthermore, gene dose, which depends on plasmid copy number, is higher than when the recombinant gene is integrated into the host's chromosome. Plasmid copy number is an inherent property of each expression system and depends on the plasmid, the host, and the culture conditions (*1*). In particular, plasmid copy-number is regulated by copy-number control genes (*2*). Plasmid copy number can range from a few up to 200. Plasmids impose a metabolic load on the host, as cellular resources must be utilized for their replication as well as for the expression of plasmid-encoded genes and production of recombinant protein. The metabolic load increases with an increase in the size of the insert, temperature, expression level, recombinant protein yield, and toxicity of the expressed protein toward the host (*3,4*). Such a metabolic load often results in a decrease in the growth rate of plasmid-bearing cells. As copy number increases, the metabolic load increases. Consequently, growth rate decreases (*2*) and faster-growing plasmid-free cells eventually overtake the culture.

Plasmid loss is the main cause of reduced recombinant protein productivity in plasmid-based systems. An unequal plasmid distribution upon cell division will eventually lead to plasmid-free cells. This is called plasmid segregational instability. Plasmid copy number depends on the number of plasmid copies at the time of cell division and their random distribution between daughter cells (*2*). If plasmid number is high (>10), the probability that a plasmid-free daughter cell will emerge is extremely low (*4*). Another factor that increases plasmid instability is plasmid multimerization. As plasmid copies have the same sequence, they can recombine and form a single dimeric circle with two origins of replication. This results in fewer independent units to be segregated between daughter cells, and consequently plasmid loss can increase (*4*). In addition, cells bearing multimers grow more slowly than those bearing monomers, even at the same copy numbers (*4*). Other parameters that influence plasmid stability are plasmid size (larger plasmids are less stable) (*5*), the presence of foreign DNA (*3*), cell growth rate, nutrient availability, temperature, and mode of culture, which will be further discussed in **Subheading 2**.

Several natural mechanisms exist to ensure plasmid survival in cell populations (*6*). For example, low-copy-number plasmids guarantee their persistence by multimer resolution through site-specific recombination systems (*cer* sequence) or active partition mechanisms, such as the *par* sequences (*2*). Genes responsible for both mechanisms have been incorporated in man-made plasmids to increase their stability (*7*). Plasmid instability is prevented if plasmid-bearing cells have a competitive advantage over plasmid-free cells. Thus, selective pressure can be utilized to select for plasmid-containing cells. The strategy most commonly used is to introduce into the plasmid a gene or genes that provide resistance to particular antibiotics. Selective pressure is then applied by supplementing the antibiotic to the culture medium. This approach can be ineffective if antibiotics are degraded or inactivated, or if periplasmic detoxifying enzymes leak from plasmid-containing cells (*5*). Moreover, antibiotics are expensive, and their presence is undesirable in food and therapeutic products as well as in the exhausted culture broth that is discharged to wastewater treatment facilities of large-scale fermentation operations. Accordingly, other forms of selective pressure have been explored, such as deletion of an essential gene from the bacterial chromosome and its inclusion in the plasmid, or the introduction of a growth repressor in the bacterial genome and its antidote in the plasmid (*8*).

Plasmid structural instability is another form in which foreign gene expression can be lost (*1*). In this case a genetic reorganization of the plasmid structure occurs, yielding a nonproductive vector (*3*). Structural instability is less common than segregational instability and cannot be prevented through selective pressure. On the contrary, strong positive selection at the time of foreign gene expression can induce structural instability (*3*). Structural instability can result either in a complete elimination of recombinant protein production or in the accumulation of aberrant recombinant proteins with minor changes in the original amino acid sequence (deletions, additions, or substitutions). The latter situation can be even more insidious than the former because its presence is usually not evident, as selection markers can remain unchanged. Thus, complete amino acid sequencing of the recombinant protein or DNA sequencing of the gene of interest must be performed to detect such a problem.

Another important issue in plasmid-based systems is plasmid copy number. Although high plasmid copy numbers are generally desired for improving recombinant protein yield, this might not always be true. For instance, high copy numbers may drive high protein production rates, which can result in protein aggregation and deficient posttranslational modification (*8*). Low recombinant protein yields can also occur in cells with a high plasmid copy number, possibly because of a reduction in translation efficiency (*9*). Accordingly, different production strategies should be chosen for different plasmid copy numbers in order to obtain a productive process. For applications such as DNA production for gene therapy, high plasmid copy number is an important objective function (*10*).

1.1.2. Chromosomal Integration

Chromosomal integration of the gene of interest is a powerful alternative for overcoming problems of expression stability in plasmid-based systems. In addition, the host does not bear the burden of plasmid maintenance and replication. Chromosome integration is especially suitable for metabolic engineering of the host (*11*) (*see* Chapters 7–10, 20, 22, 24, 26, 29–33). However, several disadvantages over plasmid-based systems exist for recombinant protein production. Adequate integration of a foreign gene in the chromosome is labor-intensive and time-consuming. Moreover, chromosome integration typically results in lower production rates than with plasmid-based systems due to a low copy number of the recombinant gene (*12*). Nonetheless, Olson et al. (*13*) have described methods for obtaining multiple gene integration into the chromosome that yield similar expression levels to those achieved by plasmid systems. The recombinant cells obtained are able to grow in the absence of antibiotics without any reduction of recombinant protein yields. This approach also had the advantage of not infringing patents. Other strategies for achieving chromosome integration in *E. coli* have been discussed by Balbás and Gosset (*11*).

Chromosome integration is the strategy of choice for the commercial expression of recombinant proteins by animal cells. In this case, the long and intricate procedure invested in host development is easily compensated with a stable host. Several strategies to obtain chromosomally integrated genes in animal cells have been developed and are summarized by Twyman and Whitelaw (*14*). Still, a major problem encountered with chromosomal integration is the possibility that the gene of interest will become integrated into an inactive region of chromatin. Among the various strategies used to overcome such a problem (*14,15*) is the use of locus control regions (LCRs), which ensures transcriptional regulation of the transgene (*see* **Subheading 3.3**).

1.1.3. Viral Vectors

An easy and very effective way of delivering the gene of interest is through viral vectors. Viruses have evolved to deliver their genetic material to the host in an efficient and nondestructive way. Some viral vectors, such as retroviruses, promote integration of the viral genome into the cell's chromosome. Many others are used for transient expression. In these cases, recombinant protein production occurs only during certain stages of the life cycle of the virus. Common viral vectors are summarized in **Table 1**, and are described with more detail in Twyman and Whitelaw (*14*). The simplicity of virus-driven protein expression makes it useful for production in higher eukaryotes, as obtaining stable recombinant animal cells may be a tedious and long procedure. Transient

Table 1
Common Viral Expression Vectors for Recombinant Protein Production

Viral vectors	Genetic material	Observations
Adenovirus	Double-stranded linear DNA	Reach high titers (10^{12}–10^{13} pfu/mL); some subgroups are oncogenic; has a wide host range; gene transfer is very efficient; are easy to manipulate in vitro.
Adeno-associated virus	Single-stranded DNA that stably integrates into the host's genome	Naturally defective viruses; can enter in a latent infection that results in long-term transgene expression.
Alphavirus (Semiliki forest virus and Sindbis virus)	Single-stranded positive sense RNA	Host range includes insects and mammals; high recombinant protein concentration; RNA will not integrate into the host's chromosome. They are not pathogenic.
Baculovirus	Double-stranded circular DNA	Hosts are arthropods; may deliver genetic material to mammalian cells (*see* **ref. 16**). Safe, easy to manipulate, and highly productive.
Herpes virus	Double-stranded linear DNA	Broad host range; can infect neurons; can carry up to 50 kbp of foreign DNA.
Poxvirus (vaccinia)	Double-stranded linear DNA	Wide host range; strong expression levels; cytoplamic transcription.
Retrovirus	Single-stranded RNA	Integrates DNA into host's genome; easy to manipulate; some are oncogenic; infection efficiencies close to 100%.

Data from ref. *14.*

expression is often utilized for rapidly generating sufficient amounts of protein for laboratory scale applications or for preliminary testing of drug candidates. Once a promising molecule is identified, a stable cell line can be generated. Viral expression systems may also find a niche for industrial protein production. For example, the insect cell-baculovirus expression vector system (BEVS) is utilized to commercially produce several recombinant proteins (*16*). Moreover, BEVS is especially suitable for the production of vaccines. A relatively new field of application for viral vectors is gene therapy, but this will not be discussed here because requirements and characteristics are different from those for recombinant protein production.

Recombinant gene expression from viral vectors comprises specific issues that are different from those of plasmid- or chromosome-based systems. The use of viral vectors involves a process with two different phases: first, cells are grown to a desired cell density, and then they are infected with the virus of interest. In addition, a virus-free product must be guaranteed for most applications; thus, special considerations are required during purification operations. Virus infection can be comparable to induction in other systems. One of the most important limitations of expression systems based on viral vectors is the quality of the viral stock. Serial in vitro passaging of stocks can result in the appearance of mutant viruses known as defective interfering particles (DIP). The genome of DIP has several deletions that make their replication faster than that of intact viruses. Therefore, DIP compete for the cellular machinery and can drastically reduce recombinant protein yields (*17*). As DIP replication requires a helper virus, in this case the complete virus, their accumulation can be avoided by using multiplicities of infection (MOI) lower than 0.1 plaque-forming unit (pfu) per cell. At such low MOI, the probability that both an intact virus and a DIP will infect the same cell is very low (*16*).

Two parameters of particular relevance during expression with viral vectors are the MOI and the time of infection (TOI). Time of infection refers to the cell concentration at which virus is added to the culture. The TOI should be late enough to allow for sufficient accumulation of cells, but should be early enough for nutrients to remain in an abundant concentration to sustain recombinant protein production. The MOI utilized defines the fraction of the population that is infected at the TOI. At MOI higher than 5 pfu/cell, a synchronous infection can be expected. In contrast, only a fraction of the population will be initially infected when employing MOI lower than 5 pfu/cell, whereas the remaining uninfected cells will be infected during a later stage by the viral progeny generated from the primary infection (*16*). If infection is analogous to induction in other systems, then MOI is equivalent to the strength of induction and gene copy number, and TOI corresponds to the time of induction. The MOI and TOI are closely related and should be selected carefully depending on the particular characteristics of the system of interest. When a high TOI is utilized, then a high MOI should also be employed for maximizing protein yield. On the other hand, low MOI and TOI increase the time of exposure of the recombinant protein to the culture environment, which can be deleterious to labile proteins. Further discussion on this topic can be found in **Subheading 2.1**.

In addition to the mode of infection (MOI and TOI), culture conditions can also affect the infection process (*16,18*). A direct relation between the amount of virus attached to cells and recombinant protein concentration has been observed (*19*). Thus, infection strategies should be aimed at increasing virus attachment, which in turn depends on cell concentration, medium composition, temperature, viscosity, and amount of cell surface available for infection (*18,19*).

1.2. Posttranslational Processing

1.2.1. Folding, Aggregation, and Solubility

Protein folding is a complex process in which two kinds of molecules play an important role: foldases, which accelerate protein folding; and chaperones, which prevent the formation of non-native insoluble folding intermediates (*20*). On occasions, folding does not proceed adequately. This results in misfolded proteins that accumulate in intracellular aggregates known as inclusion bodies. One of the main causes of incorrect protein folding is cell stress, which may be caused by heat shock, nutrient depletion, or other stimuli (*21,22*). Cells respond to stress by increasing the expression of various chaperones, some of them of the *hsp70* and *hsp100* families (*22*). Of particular importance to eukaryotic cells is the "unfolded protein response" that activates transcription of genes encoding chaperones and foldases when unfolded proteins accumulate in the endoplasmic reticulum (*23*). Production of inactive proteins represents an energetic drain and metabolic load, while accumulation of inclusion bodies can cause structural strains to the cell. Accordingly, incorrect protein folding has adverse consequences. For instance, several human pathologies, such as Alzheimer's disease, Parkinson's disease, and Huntington's disease, are characterized by intracellular protein aggregation and accumulation (*21*).

The overexpression of heterologous proteins often results in the formation of inclusion bodies. This phenomenon is still not fully understood, but several explanations have been proposed. For instance, as reviewed by Carrió and Villaverde (*24*), heterologous proteins often reach nonphysiological concentrations, which may promote aggregation. Aggregation can also result from the lack of disulfide bond formation due to the reducing environment of the bacterial cytosol (*5*). Additionally, overexpression of heterologous genes is stressful *per se* and may cause the saturation of the cellular folding machinery (*22*). During heterologous protein production, high rates of expression are required. Proteins may also be larger than those typical of the host, as is the case of mammalian proteins expressed in bacteria. Rapid intracellular protein accumulation (*8*) and expression of large proteins (*22*) increase the probability of aggregation. Accordingly, inclusion body formation is likely to occur during production of recombinant proteins.

Protein aggregation has been observed in bacteria, yeast, insect, and mammalian cells (*20,21,24*) (*see* Chapter 3). Aggregation protects proteins from proteolysis and can facilitate protein recovery by simply breaking the cells and centrifuging the inclusion bodies (*8*). In addition, when the expressed protein is toxic to the host, its deleterious effect can be prevented by producing the heterologous product as inclusion bodies (*25*). In many cases, as with those of the first recombinant proteins that reached the market (insulin and growth hormone), recovery and renaturation operations can be performed in an economically feasible manner (*8*). Accordingly, inclusion body formation not only is desirable, but also can be promoted through molecular biology and/or operation strategies, such as the use of protease-deficient strains, culturing at high temperatures, or designing suitable fusion peptides and amino acid sequences through protein engineering approaches (*24*). If production in inclusion bodies is preferred, solubilization and renaturation can be performed in different ways (*25*). However, the refolding step is an empirical process that on occasions is very inefficient, with yields usually lower than 10% (*24,26*). Thus, in many cases it may be difficult and expensive to obtain a soluble functional protein after downstream operations. For instance, Datar et al. (*26*) have shown that the overall costs for producing tissue plasminogen activator by an *E. coli* process are higher than those for a mammalian cell-based bioprocess. This is because of the higher expenses incurred during the solubilization and renaturation steps required in the *E. coli* process.

It is impossible to predict whether a protein will aggregate or not in a particular expression system, or how easily it will be solubilized and renaturated (*8,25*). Thus, a soluble protein is generally preferred. Several strategies have been proposed for reducing protein aggregation. Various chaperones and foldases have been stably cloned into hosts to facilitate protein folding (**Table 2;** *8,20,22*). However, this strategy is not always successful. It is not possible to predict which chaperone will facilitate folding of a particular protein, or whether more than one chaperone or cofactor will be required. Overexpression of more than one chaperone has been explored with satisfactory results (*22,24*). Protein engineering can also reduce aggregation (*20,24*); changing the extent of hydrophobic regions or using fusion proteins are two successful strategies. Fused proteins often contain a peptide native to the host used. For example, fusing single chain antibodies to an *E. coli* maltose-binding protein allows the production of soluble functional protein in *E. coli* cytoplasm (*27*). Interestingly, it has been observed that proteins accumulated as inclusion bodies can naturally solubilize when heterologous protein production ceases (*24*). Finally, certain additives may facilitate protein folding both in vivo and in vitro. These have been summarized by Fahnert in Chapter 3 of this book.

1.2.2. Proteolytic Processing

Signal peptides, needed to direct proteins to the various cellular compartments, must be cleaved to obtain a functional protein. Upon membrane translocation, the signal peptide is removed by a signal peptidase complex that is membrane-bound to the endoplasmic reticulum in eukaryotes or to the cellular membrane in prokaryotes (*28*). Inefficient removal of the signal peptide may result in protein aggregation and retention within incorrect compartments, such as the endoplasmic reticulum (*29*). Consequently, the yields of secreted proteins can be drastically reduced. To solve this problem, the *E. coli* signal peptidase I and the *Bacillus subtilis* signal peptidase have been overexpressed in *E. coli* and insect cells, respectively (*29,30*). Signal peptidase overexpression increased the release of mature beta-lactamase (*30*) and the processing of antibody single-chain fragments (*29*). Such results demonstrate that low signal peptidase activity can limit the production of recombinant proteins. Despite these promising results, signal peptidase overexpression has rarely been used.

Other proteins, such as proteases, insulin, or penicillin acylase, must be expressed as proproteins because prodomains act as folding catalysts (*31*). In these cases, cells utilize endoproteases to produce the mature active protein (*32*). Accordingly, low endoprotease activity may limit the concentration of a correctly folded mature protein.

Table 2
Some Chaperones or Foldases Utilized to Facilitate Protein Folding

Chaperone/foldase	Host	References
Human hsp70	Insect cells	*20*
BiP	Insect cells	*20*
Calnexin and calreticulin	Insect cells	*20*
Bacterial protein disulfide isomerase (PDI)	Insect cells, *E. coli*	*8,20*
Peptidylprolyl *cis-trans* isomerase	Insect cells	*20*
Trigger factor (TF)	*E. coli*	*22*
DnaK	*E. coli*	*22*
GroEL/ES	*E. coli*	*22,24*
ClpB	*E. coli*	*22*
Skp	*E. coli*	*5*
DegP	*E. coli*	*101*
ClpG	*E. coli*	*101*
HtbG	*E. coli*	*101*
Human PDI	CHO cells	*157*
Polyubiquiton	*Kluyveromyces lactis*	*158*
Kluyveromyces lactis PDI	*Kluyveromyces lactis*	*158*

Overexpression of the mammalian endoprotease furin in mouse mammary gland and insect cells increased the concentration of correctly folded product up to eightfold (*33,34*). Similarly, overexpressing yeast's Kex2p increased processing of proopiomelanocortin by baby hamster kidney (BHK) cells (*32*).

Another type of proteolytic processing is the removal of the N-terminal methionine. This processing is performed by a methionine aminopeptidase (MAP) and occurs only in proteins in which the second amino acid is alanine, glycine, proline, serine, threonine, or valine (*35*). Removal of N-terminal methionine is a common problem during expression by *E. coli*. Overexpression of recombinant proteins may saturate MAP or deplete required metal cofactors (*35*). Similarly to other enzymes, MAP has been overexpressed in *E. coli* to solve such a problem. Using this strategy, Hwang et al. (*36*) were able to increase *N*-methionine removal by 40%, but recombinant gluthatione S-transferase concentration was reduced 10%. Since Vassileva-Atanassova et al. (*37*) did not find a correlation between the extent of *N*-methionine removal and recombinant protein concentration in two strains with different intrinsic *N*-methionine removal ability, the reduced yield observed by Hwang et al. (*36*) could be attributed to the higher metabolic load that results from overexpression of two recombinant genes. Another alternative for *N*-methionine removal is the construction of fusion proteins, where the -*N*-methionine is removed along with the fusion peptide either intracellularly or during a later in vitro enzymatic removal stage (*35*).

1.2.3. Glycosylation

Glycosylation is a very complex posttranslational modification that requires several consecutive steps and involves tens of enzymes and substrates (**Figs. 1, 2**). It usually occurs in the endoplasmic reticulum and Golgi apparatus of eukaryotic cells, although - *N*-glycosylation has been detected in proteins produced by bacteria (*38*). Three types of glycosylation exist: *N*-(glycans linked to an Asn of an AsnXaaSer/Thr consensus sequence, where Xaa is any amino acid), *O*-(glycans linked to a Ser or Thr), and C (attached to a tryptophan) linked. Of these, C-linked glycosylation has hardly been studied and little is known about its biological significance (*39*). *N*-linked glycosylation is the most studied and is considered as the most relevant for recombinant protein production. In many cases, glycosylation determines protein stability, solubility, antigenicity, folding, localization, biological activity, and circulation half-life. Glycosylation profiles are protein-, tissue-, and animal-specific (*40*). Nonauthentic glycosylation may trigger immune responses when present in proteins for human or animal use (*40*). Therefore, authentic glycosylation is especially relevant for recombinant proteins to be utilized as drugs.

The *N*-glycosylation pathway is depicted in **Figs. 1** and **2**. Several bottlenecks can be expected from the complexity of the process. Moreover, different glycosylation sites are often glycosylated in different ways (*41*). Recombinant proteins may present macroheterogeneous (differences in site occupancy) or microheterogeneous (differences in the structures of oligosaccharides between glycosylation sites) glycosylation (*42*). First, the synthesis of the dolicholphosphate oligosaccharide can limit the extent of glycosylation. This can occur from a reduction of the lipid pool. In addition, the concentration of lipid-linked oligosaccharides has been reported to be cell-cycle-dependent (*43,44*). As an attempt to solve this, dolicholphosphate has been fed to Chinese hamster ovary (CHO) cells, producing recombinant proteins (*45,46*). Although dolicholphosphate was internalized (*46*), no increase in site occupancy was observed upon its addition (*45,46*). On the other hand, a reduced pool of sugar nucleotides, the activated sugar donors required for oligosaccharide synthesis, limits the buildup of the G3M9N2Dol PP precursor (where G is glucose, M is mannose, and N is *N*-acetylglucosamine) and reduces the glycosylation site occupancy (*47,48*). Limitation of sugar nucleotide donors occurs upon prolonged glucose or glutamine starvation (*47,48*). The availability of sugar nucleotide donors also affects microheterogeneity, as each step of the building of oligosaccharide chains in the Golgi apparatus requires nucleotide sugars (**Fig. 2**). To alleviate such a problem, sugar nucleotide precursors have been added to the culture medium. With this approach, sialylation by Chinese hamster ovary (CHO) and genetically engineered insect cells has been increased through feeding of *N*-acetylmannosamine (*49,50*). Another factor that can affect glycosylation is the transport of sugar nucleotides to the endoplasmic reticulum or Golgi apparatus. Gu and Wang (*49*) and Hills et al. (*51*) proposed this when an increase in nucleotide sugar pool did not result in a proportional increase of the extent of protein glycosylation.

Another possible factor affecting glycosylation is the presence of glycosidases, either intracellularly or in the culture medium. This can be a major problem when

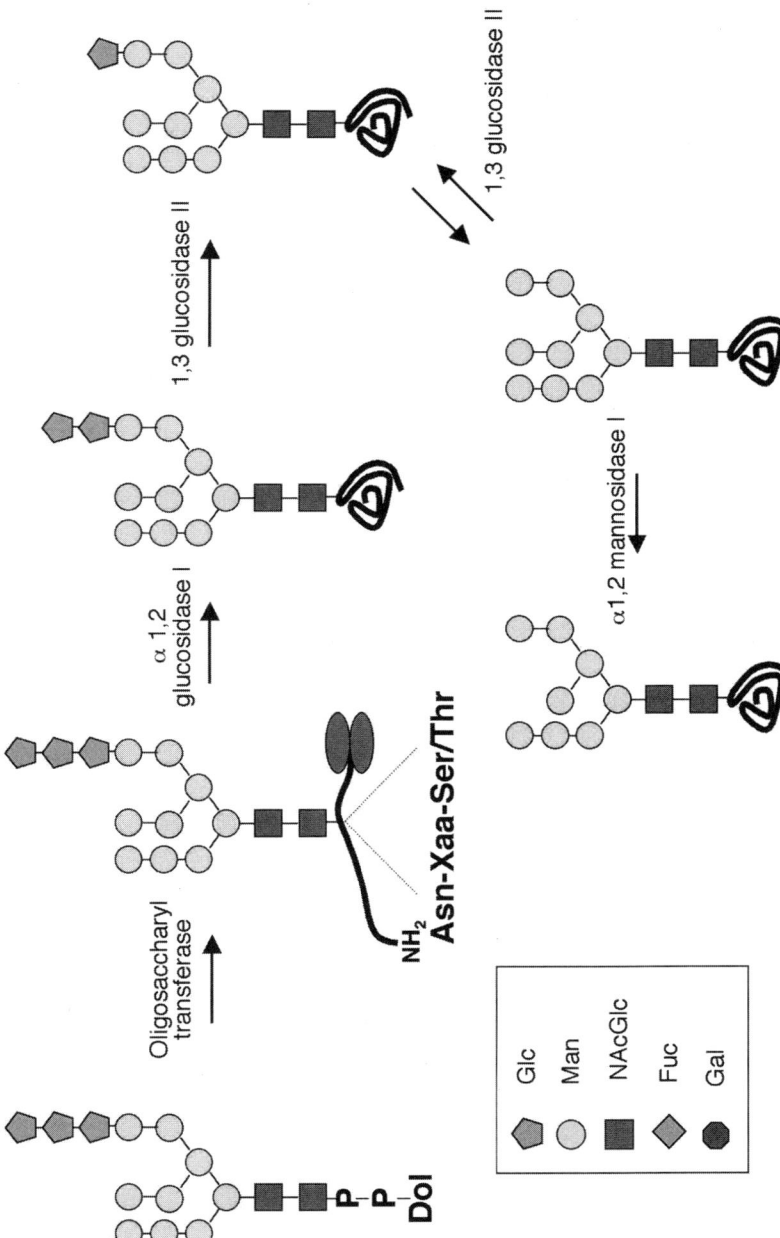

Fig. 1. *N*-glycosylation in the endoplasmic reticulum. First, a lipid-linked oligosaccharide is synthesized in the endoplasmic reticulum (depicted as the Dol-P-P-oligosaccharide). Glycans are then transferred to the nascent peptides. Thereafter, glycan processing proceeds as depicted. Dol = dolichol; Glc = glucose; Man = mannose; NAcGlc = *N*-acetylglucosamine; Fuc = fucose; Gal = galactose.

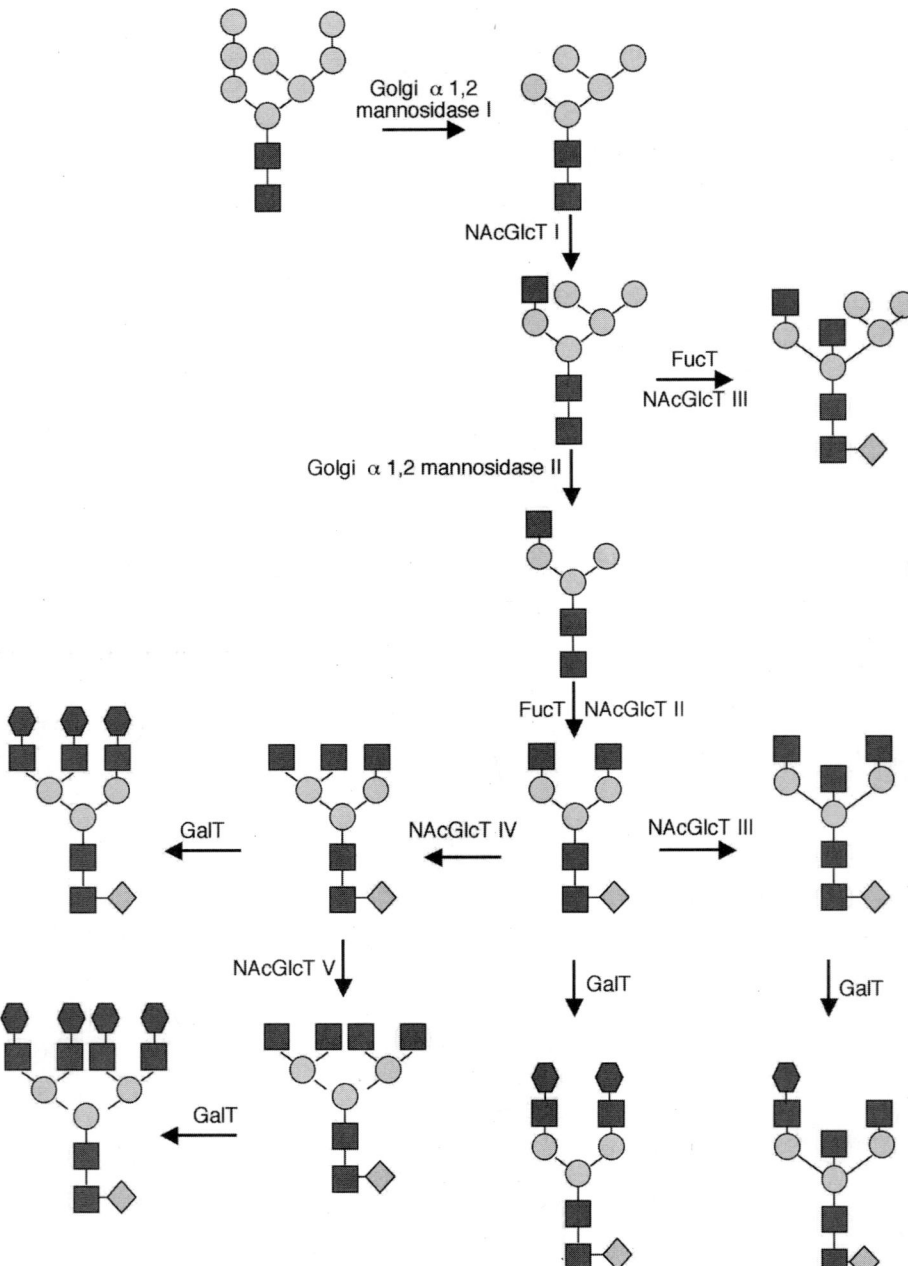

Fig. 2. *N*-glycan processing in the Golgi. The protein is not depicted for clarity. Pathways shown are only typical processing routes, others may occur. T refers to transferase. Symbols are the same as those in **Fig. 1**. FucT may act in the different sites shown. Galactosylated glycans are substrates for sialyltransferase, and thus can be sialylated (pathway not shown).

expressing proteins in insect cells, as an intracellular hexosaminidase activity results in the accumulation of paucimannosidic glycans (containing three or fewer mannose residues and the NacGlc core) (*52*). Moreover, Gramer and Goochee (*53*) have detected and characterized the activities of sialidase, β-galactosidase, α-hexosaminidase, and fucosidase in supernatants of CHO cell cultures. Sialidase activity increased upon cell lysis due to the release of cytoplasmic enzymes (*54*). High extracellular sialidase activity resulted in decreased sialylation of glycans attached to recombinant human antithrombin III (*54*). This problem can be solved through the addition of glycosidase inhibitors or by harvesting the product before extensive cell lysis occurs (*53,54*).

Culture conditions can also affect glycosylation. For example, pH can affect the activity of extracellular glycosidases. The concentration of toxic byproducts, such as ammonia, CO_2, and hyperosmotic conditions, can reduce sialylation and the extent of *N*- and *O*-glycosylation (*55,56*). Cell growth rate and protein production rate also influence glycosylation. For instance, Andersen et al. (*45*) observed a direct relation between site occupancy and the fraction of cells in the G0/G1 phases of the cell cycle. Moreover, decreasing temperature of CHO cell cultures significantly increased the degree of sialylation of secreted alkaline phosphatase (*57*). Reduced growth rate may result in a reduced protein production rate, which in turn increases the extent of glycosylation (*42*). Protein glycosylation is a dynamic phenomenon that changes as culture time progresses. Andersen et al. (*45*) and Yuk and Wang (*46*) have found that glycosylation levels increase with increasing culture time until the onset of cell death. As with *N*-glycosylation, nonauthentic *O*-glycosylation profiles can also elicit an immune response toward the recombinant product (*58*). Strategies proposed for *N*-glycosylation can also improve the amount of sialylated and galactosylated *O*-glycans.

1.2.4. Other Posttranslational Modifications

Other posttranslational modifications, such as myristoylation, palmitoylation, isoprenylation, phosphorylation, sulfation, C-terminal amidation, β-hydroxylation, and methylation, are less common than glycosylation, but may be important for certain recombinant proteins. In general, the extent of modification depends on the host utilized, being the modifications performed by higher eukaryotic cells closer to those found in human proteins (*see* **Subheading 3**).

1.3. Transport and Localization

As already discussed, recombinant proteins may be directed to different cellular compartments by signal peptides or through fusion proteins. Different sites of protein localization have different advantages and disadvantages, which are summarized in **Table 3**. Intracellular accumulation often results in high protein amounts and allows an easy recovery of concentrated protein along with cells (*35*). Nonetheless, purification of the product from the protein-rich cell extract may be difficult. In contrast, the product of interest usually constitutes the major component when it is secreted to a low-protein or protein-free medium. This can greatly facilitate its purification. Nonetheless, secreted proteins will be highly diluted and bottlenecks in the secretion pathway can further

Table 3
Possible Locations and Conditions of Recombinant Protein Accumulation

Protein design	Location	Soluble	Advantages	Disadvantages
Native sequence	Cytoplasm	yes	Direct purification with high yield recovery. High level of expression.	Susceptible to proteolysis. High cellular native protein content.
	Cytoplasm	no	High-level expression. May prevent proteolysis. Toxicity effects of protein to cell may be avoided. Easy partial purification.	Protein folding must be carried out. Recovery purified native protein can be low or even zero.
Fusion protein	Cytoplasm	yes	High-level expression. Purification may be aided with affinity-tagged protein. Solubility and stability may be enhanced by fusion partners.	Site-specific cleavage of fusion peptide required. Overall yield may be low.
Fusion protein directed to secretion	Cytoplasm	no		Signal peptide unprocessed, purification usually not attempted.
	Periplasmic space[a]/medium	yes	Ease of purification	Expression level and recovery may be low. Diluted product.

Adapted from Wingfield (*35*).
[a]In Gram-negative bacteria.

reduce their accumulation in the culture medium (*20*). Accordingly, concentration operations, such as ultrafiltration, are always used prior to other purification stages when dealing with secreted proteins.

Protein localization is especially relevant when expressing recombinant proteins in *E. coli*. Accumulation in the periplasm often results in soluble and correctly folded proteins, whereas cytoplasmic localization yields an inactive and insoluble product (*35*).

The characteristics of the protein should be considered when deciding the site of accumulation. Small proteins susceptible to proteolysis should be produced in *E. coli* as inclusion bodies. Apart from intra- or extracellular accumulation, certain applications may require recombinant proteins to be targeted to the cell membrane, usually through fusion proteins. This is the case of virus or phage display, where protein localization in the virus surface allows for rapid screening and isolation of the desired phenotype, which is coupled to the corresponding genotype (*59*). The transport efficiency of the protein of interest depends on the signal peptide utilized, which should be chosen according to the host. Nonoptimal selection of the signal peptide results in intracellular protein accumulation and aggregation (*20*).

2. Bioengineering Approaches to Solve Common Problems Associated With Heterologous Gene Expression

Bioprocess engineering plays a crucial role when the goal of recombinant protein production is to obtain as great amounts as possible of a high-quality product. As already discussed, bioprocess conditions affect not only the amount of protein obtained, but also its solubility and its posttranslational modifications. The biology of the host, and the molecular biology tools utilized for its modification, should be taken into account when defining bioprocess conditions. For example, different approaches are required when employing either high- or low-copy-number plasmids. Similarly, animal cells have very different requirements from bacteria or fungi. Experience has shown that the best results are obtained when both molecular biology and bioengineering approaches are used. Some common strategies for improving recombinant protein production through manipulation of the culture conditions will be discussed in the following sections.

2.1. Induction Strategies

Recombinant genes can be placed under a variety of promoters. The promoter selected will determine whether gene expression is constitutive or inducible (*60*). Constitutive gene expression may increase plasmid instability because the metabolic load of recombinant protein production is constantly present (*see* **Subheading 1.1.1**). Thus, constitutive promoters are normally chosen when recombinant gene expression does not significantly affect the growth rate of the host. In many situations the best conditions for cell growth are different from those for recombinant protein production (*15*). In such cases, inducible systems are preferred—i.e., systems in which induction is performed after a particular cell density has been obtained. Different types of stimuli can be utilized for induction (*12*). Induction may depend on starvation of a nutrient and/or the addition of an alternative nutrient that turns on specific molecular machinery, such as the *lac* operon. Other inducers include osmolarity, pH, or temperature shifts, anaerobiosis, antibiotic addition, and the like. Several considerations should be made when choosing an inductive system. Induction should be simple, economical, and efficient. In addition, the inducer should not have negative effects on cell viability and recombinant product quality, and should not complicate downstream operations. Finally, the chosen system should be efficiently repressed in the absence of the inducer (*12*). An advantage of physical or physicochemical induction, such as temperature, pH, dissolved oxygen tension, and osmolarity shifts, is that chemicals, which may be undesirable in the final product, are not added. Moreover, these induction methods are easy to implement and are inexpensive at laboratory and pilot-plant scales.

Industrial recombinant protein production requires additional considerations of the type of inducer employed. Among these are the deficient mass, heat, and momentum transfers often observed in large-scale bioreactors (*61*). For instance, mixing times (time required to achieve homogeneity) in large-scale animal or plant cell culture vessels (10,000 L) can be in the order of 10^3s. Thus, up to 16 min would be required for the inducer to be homogeneously distributed in the reactor under this extreme situation. This can be solved by using several feeding ports, if the inducer is a chemical added to the vessel (*62*). However, such an approach cannot be utilized for other type of induction, such as temperature changes. Reducing or increasing the temperature of a large-scale vessel may be very expensive and ineffective. Moreover, the rate of temperature change can affect recombinant protein yield (*63*). In conclusion, the dynamics of the production process should be considered.

Once a system of induction has been chosen, induction strategies must be planned. The first consideration should be the effect of recombinant gene expression on cell growth and physiology. In some cases, usually when low plasmid copy numbers are present, recombinant protein production does not affect the specific growth rate (e.g., *64,65*). Thus, higher recombinant protein yields are obtained by inducing foreign gene expression as early as possible, even at the time of inoculation (*64,65*). When cell growth is significantly inhibited by the expression of a recombinant gene, sufficient buildup of biomass should be allowed before induction (*66*). However, extreme cell concentrations may reduce production of the recombinant protein, as nutrient limitation may occur. The importance of adequate nutrient feeding after induction was investigated by Yazdani and Mukherjee (*66*). They observed a 10-fold increase in recombinant streptokinase concentration when concentrated medium was fed after induction (*see* **Subheadings 2.2** and **2.3**). A similar effect has been observed when expressing recombinant proteins through the BEVS. In this case, very little recombinant protein is produced if infection with the recombinant baculovirus is performed above an optimum cell concentration. This phenomenon has been called the "cell-density effect" and can be overcome through adequate nutrient feeding strategies (*16*). It should be noted, however, that infection at extremely high cell concentrations (in insect cells, above 14×10^6 cells/mL) may drastically reduce recombinant protein yields even when nutrients are available (*67*). The reasons for this are still unknown, but it is possible that a trace element that is not fed limits yields.

The mode of induction can also affect the solubility of the recombinant product (*68*). Cells must be actively growing at the time of induction to reduce protein aggregation. Accordingly, Eriksen et al. (*65*) observed an increase in solubility of a recombinant protein when expression was induced in the very early exponential phase, in comparison with induction at later times or in the lag phase. As discussed previously, a common strategy to increase protein solubility is by reducing the culture temperature after induction. This gives an advantage to cold-shock over heat-shock promoters when aggregation must be avoided. However, maintaining the culture at low temperatures drastically decreases growth rate. Therefore, cell concentration should be as high as possible at the time of induction when cold-shock promoters are used (*68*).

The magnitude and length of induction also affect recombinant protein yields. Low inducer concentration may result in an inefficient induction (low recombinant protein yields), whereas expensive inducers added in excess can result in an important economic loss or in toxic effects, including reduced cell growth and/or recombinant protein concentration. A saturation-type relationship between inducer concentration and maximum recombinant protein volumetric or specific yields has been reported for the *lacZ* promoter induced with IPTG (*64*). Thus, inducer concentration should be maintained at or slightly higher than the critical concentration (the concentration below which recombinant protein yield becomes a function of inducer concentration). As observed by Ramírez et al. (*64*), IPTG concentration between 0 and 1 mM did not affect *E. coli* specific growth rate or maximum cell concentration. However, such a behavior must be characterized for the particular host/vector/protein employed. In the case of temperature-induced promoters, the temperature of induction and the duration of the temperature shift have an important effect. For example, Gupta et al. (*69*) expressed *lacZ* under the T7 system using the λP_L> heat-shock promoter. Several choices exist for inducing such a system. Namely, temperature can be increased to 42°C for a given period of time, or maintained at 42°C until the end of the culture (*69*). As cell growth ceases at 42°C, a high cell concentration must be present before induction for the culture to be productive when the temperature is maintained at 42°C. In contrast, Gupta et al. (*69*) found that a heat shock of 2 min did not arrest cell growth and was optimal for recombinant protein production. Therefore, maximum recombinant protein yields were obtained when induction was performed in the early growth phase.

2.2. Growth Control

Growth rate affects several parameters that determine recombinant protein accumulation rate. Among them are the percentage of substrate utilized for cellular maintenance, RNA polymerase activity, ribosome number, plasmid stability, plasmid copy number, plasmid multimerization, and the distribution of cells in the cell-cycle phases (*70–73*). Thus, it is possible to control recombinant protein production through growth rate. Growth rate can be manipulated through nutrient availability. Namely, the main carbon or nitrogen source can be maintained at a predetermined concentration to obtain the desired growth rate. Such a manipulation can be achieved through fed-batch or continuous cultures (*74,75*). Dissolved oxygen, an essential nutrient for aerobic cells, can also be utilized to control growth rate. Temperature also affects growth rate by changing the rate of the reactions occurring in the culture vessel. Temperature is an especially effective tool for arresting growth in animal cells, as at low temperatures cells remain viable, mostly in the G1 phase of the cell cycle (*57*). It should be noted that all these factors can have additional particular effects besides modifying growth rate. For instance, reducing the growth rate by limiting nutrient concentration may reduce the production of undesirable metabolites by increasing the metabolic efficiency (*8,76,77*). Molecular biology approaches can also be utilized to manipulate growth rate. For example,

Kaufmann et al. (*57*) introduced the cell-cycle-arresting gene p27 under the control of a tetracycline-repressible promoter in CHO cells. In this manner, they divided the process in two: first, a stage of active cell growth, and second, after tetracycline decomposition, a stage in which recombinant protein is produced and cell growth is arrested in the G1/S restriction point. As a result, the concentration of recombinant-secreted alkaline phosphatase (SeAP) increased 17 times.

Contradictory information can be found in the literature, where protein production rate has been reported to increase (*72*) or decrease with specific growth rate (*73*), or show no relation at all (*7*). Therefore, the effect of growth rate on protein productivity cannot be generalized. The first consideration to be made is whether protein accumulation is associated with cell growth. This is usually the case, except when the recombinant protein is toxic to the cell, when it severely reduces growth, or when growth drastically decreases plasmid stability. When recombinant protein production is growth-associated, sustained growth should result in higher protein concentrations and induction can be performed in the beginning of the culture (*see* **Subheading 2.1.**). On the contrary, when protein production is not growth-associated, an optimized process should be divided into a growth phase and a production phase. In the latter phase, cell survival and plasmid maintenance should be promoted instead of cell growth. The host and the plasmid construction determine the effect of growth rate, as has been shown by Saraswat et al. (*73*). In many cases, the relation between growth rate and the amount of recombinant protein cannot be explained. In many others, changes in recombinant protein yields have been correlated with plasmid copy number, stability, or multimerization. Generally, growth rate is inversely related to plasmid stability (*73,78*), although the contrary has also been reported (*7*), suggesting that each particular case should be evaluated individually. It has been proposed that high growth rates and protein production rates represent a stressful condition that may affect plasmid replication and multimerization (*73*). On the other hand, reduced growth rates increase the plasmid copy number of continuously replicating plasmids (*78*).

In addition, cells in different stages of the cell cycle produce different amounts of recombinant protein or are less susceptible to infection. Thus, suitable growth rate control strategies must be imposed during the protein production phase. For example, Leelavatcharamas et al. (*79*) used control of the cell cycle to improve production of interferon γ, a growth-associated product.

2.3. Bioreactor and Operation Strategies

The main objective of a bioreactor, besides containment, is the control of environmental parameters in predetermined values. The number of parameters that can be manipulated depends on the complexity of the bioreactor. It can range from only temperature, when static culture flasks are introduced in an incubator, to several parameters in a fully instrumented vessel. Among the conditions that can be controlled are dissolved oxygen, pH, temperature, agitation rate, redox potential, dissolved carbon dioxide, cell concentration, cell growth, substrate concentration, inlet gas flow and composition, volume, pressure, fluid dynamics, and power input. Lidén (*80*) proposed to call the set of environmental conditions present in a bioreactor the "envirome." The envirome results not only from the action of process parameters manipulated by the operator, but also from the direct interaction of cells with their environment. The envirome interacts with several steps of the recombinant protein production process, namely cell growth, cellular metabolic state, transcription, translation, and posttranslational modification. From the importance of the envirome, it can be seen that bioreactors have an immense potential for increasing recombinant protein productivity.

Of the parameters listed above, dissolved oxygen tension (DOT) has received special attention, because oxygen has a low solubility in water and is difficult to deliver to the culture broth (*81*). The problem aggravates at very high cell concentrations, as higher amounts of oxygen must be transferred to the culture medium to satisfy demand. Cultures need to be fully aerated and homogeneous to avoid alcoholic or acid fermentation in bacteria, yeast, and animal cell cultures (*82,83*). Consequently, bioreactors are designed to increase the oxygen transfer rate (OTR) as much as possible. In the case of bioreactors employing suspended cells, homogeneity is achieved by both the action of the impellers and the liquid motion induced by gas sparging. Nonetheless, bioreactors employing cells immobilized to a variety of supports are needed in some circumstances. This is the case of anchorage-dependent animal cells. In other cases, immobilizing cells that would otherwise grow freely in suspension is needed to attain high cell concentrations and high productivities. In these situations, homogeneity can be achieved by agitation if cells are immobilized in supports that become suspended during operation, such as microcarriers. When a fixed matrix configuration is employed, homogeneity can be achieved by increasing medium flow rate and by suitable bioreactor design. Several strategies for operating bioreactors with immobilized cells are described elsewhere (*61*). Palomares and Ramírez (*61*) have discussed the characteristics and problems of the different types of bioreactors upon process scale-up.

In general, a DOT higher than 20% (with respect to air saturation) does not limit growth, unless transfer from the liquid to the cells is restrained by diffusion through additional resistances, such as when cells form aggregates or pellets or are immobilized. In these cases, a 50% DOT in the bulk liquid may be required to sustain growth of agglomerated cells (*84*). In addition to its effects on cell growth, oxygen privation can drastically increase plasmid instability (*10,85*). For instance, Li et al. (*86*) observed an increase in plasmid content at higher DOT, but no significant effect on recombinant protein yields was detected. Thus, they hypothesized that plasmid replication is suppressed and gene expression increased in anaerobic conditions. It should be noted that the effect of DOT on recombinant protein yield was strain-dependent. Oxygen is also required for maturation of proproteins, as penicillin acylase (*87*). Among the strategies utilized to cope with the problem of poor oxygenation of cultures is the expression of *Vitreoscilla* hemoglobin in the host. This allows efficient growth at limiting dissolved oxygen concentrations and improves recombinant protein yields (*88*). In contrast to oxygen limitation, an oversupply of oxygen can cause oxidative stress to cells or oxidative damage to proteins (*85*). Some proteins, such as cylohexanone monooxygenase, are very susceptible to oxidation and should be produced at DOT of 0% (*89*). Special considerations should be made when utilizing temperature-inducible promoters, as oxygen is less soluble in water as temperature increases. If a reactor is near its maximum OTR capacity and temperature is increased, the resulting OTR may not be sufficient to sustain recombinant protein production.

Oxygenation of cultures employing fragile cells, such as animal cells and filamentous fungi, is often problematic, as sparging and agitation are limited to shear stresses that are not harmful. Typical energy dissipation rates in bioreactors are usually below those deleterious to animal cells; thus, damage from agitation should not be expected (*90*). However, bubble rupturing in sparged cultures liberates very high amounts of energy that kill almost every cell in the surrounding area (*91*). As the energy liberated from bubble bursting is inversely related to bubble size, large bubbles should be utilized in fragile cultures (*92*). The area for oxygen transfer decreases as bubble size increases. Hydrodynamic stress can be lethal to cells, or may only infringe sublethal damages that may trigger apoptosis, arrest the cell cycle, increase nutrient consumption rates, change intracellular pH, and reduce recombinant protein yields (*16*). A strategy for reducing shear damage to cells is the use of shear-protective additives, such as Pluronic F68® (BASF), which yields stronger cells by decreasing their membrane fluidity and reduces their attachment to bubbles (*90,93,95*). The situation in cultures of filamentous fungi is different. The morphology of fungi, either dispersed or in pellets, depends on culture conditions (*90*). As agitation speed increases, fungi acquire the form of pellets. Moreover, pellet size decreases as agitation increases. Such changes in morphology are often accompanied by changes in product production, which often decreases as the power applied to the bioreactor increases (*90,96*).

The bioreactor operation mode is another approach to control the environment. Fed-batch cultures are utilized for increasing cell concentration and obtaining high product titers (*see* **Subheading 2.2.**). The control of nutrient concentration can increase metabolic efficiency. For example, maintenance of low glucose concentration can be used to avoid the Crabtree effect (alcoholic or acid fermentation in aerobic conditions due to high concentrations of glucose). The Crabtree effect results in a waste of glucose and the generation of toxic byproducts that often limit recombinant protein yields (*77,97*). On the other hand, nutrient-deprived cultures are more drastically affected by the metabolic burden of foreign gene expression. Glucose, magnesium, phosphate, or oxygen limitation decrease plasmid stability (*9,85*). Meanwhile, the carbon-to-nitrogen ratio also affects plasmid loss and the burden that plasmids impose on cells (*98*). In animal cell cultures, nutrient privation may trigger apoptosis (*99*).

Bioreactor operation mode also influences plasmid stability. High-density cultures and continuous operation are prone to plasmid segregation due to the high number of generations in the culture (*5*). Similarly, large-scale operation increases plasmid instability for the same reason (*81*). Cell immobilization has been observed to reduce plasmid instability (*9*). As mentioned before, the design of two-stage processes, in which cells grow on one stage and are induced and produce recombinant protein in the other, is an interesting alternative to reduce plasmid instability. This can be performed in two-stage systems consisting of chemostats in series. In this arrangement, cell growth is optimized in the first chemostat, and recombinant protein concentration in the second. For example, Sayadi et al. (*9*) utilized such a system to produce cathecol 2,3-dioxygenase in *E. coli*, where plasmid stability was guaranteed by immobilizing the cells. Two-stage arrangements may be especially useful in systems such as the BEVS and expression in *Bacillus subtilis*, where protein production starts after an infection phase or close to the sporulation phase, respectively.

3. Specific Problems and Their Solutions in Different Expression Systems

Recombinant protein production requires integrated bioprocesses that include considerations spanning from molecular biology to downstream processing. Under this notion, the host undoubtedly has a prominent role. Many characteristics of the product are endowed by the host and are influenced by protein concentration and site of accumulation. In general, protein concentration is inversely related to the extent of protein posttranslational processing, and a compromise between quality and productivity must be made. Moreover, the host dictates the molecular biology techniques to be used, production mode, and product recovery strategies. As a rule of thumb, the most simple host expression system that delivers the required quality should be chosen for recombinant protein production. Animal cells, fungi, yeast, and bacteria are commonly used nowadays for the expression of recombinant products.

3.1. Prokaryotes

The Gram-negative bacterium *E. coli* was the first organism utilized for the production of recombinant human proteins. It is still extensively used for industrial applications, as evidenced by a market, only for recombinant pharmaceutical proteins, of $2.9 billion in 1999 (*100*). A large amount of knowledge has been generated about its molecular biology, biochemistry, and physiology (*5,8,101*). *E. coli* is easy to grow to high cell densities (over 100 g/L), and has simple nutritional requirements that can be satisfied with fully defined simple media (*60*). Despite its proven success, recombinant protein production in *E. coli* has several drawbacks that have been addressed through different approaches (*5,8,102,103*). *E. coli* is usually not capable of efficiently producing very long or short proteins, although the successful expression of a 210 kDa protein has been achieved (*104*). Proteolytic cleavage and disulfide bond formation seldom occur, and posttranslational modifications, including glycosylation, acylation, and isoprenylation, are not performed. In many cases, neither of these modifications is required for obtaining an adequate product, and bacteria are the host of choice. In addition, bacteria possess pyrogens and endotoxins that must be totally eliminated from proteins to be injected in animals or humans. Other concerns about the expression of recombinant proteins in *E. coli* include variability in the level of expression, protein solubility, and protein purification. Most of these inconveniences have been approached through genetic manipulations (*103,105*). **Figure 3** summarizes some strategies used for enhancing recombinant protein expression in *E. coli*. (*see* Chapters 5–12 for new applications).

The problem of expression levels traditionally has been solved by using strong promoters and/or intervening on the pathways that include possible rate-limiting steps—namely, novel "metabolic optimization" strategies can be used to finely tune the expression of genes in particular pathways that modulate the final product yield, as well as the expression of the gene of interest (*106*). Fine-tuning can be accomplished by utilizing artificial promoters with different strengths. Such artificial promoters have been constructed by

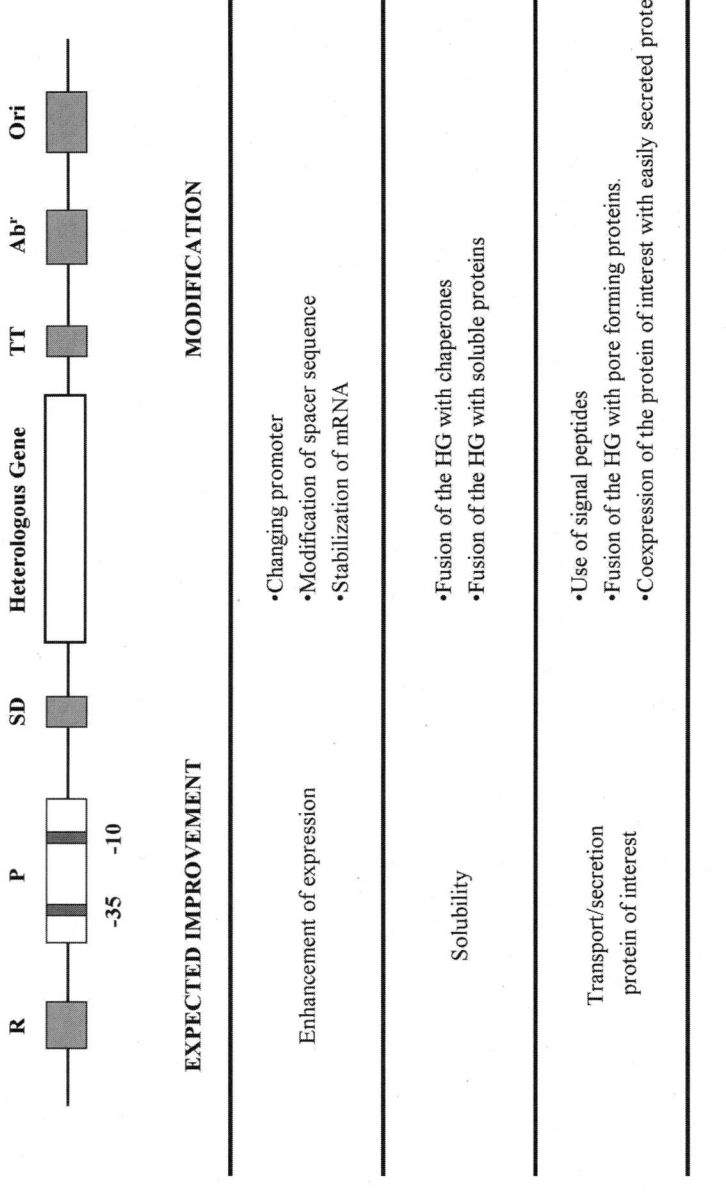

Fig. 3. On top is shown the classic organization of a prokaryotic expression vector where (R) is the regulator that exerts its effect on the (P) promoter, whose −35 and −10 sequences are separated by spacer sequences. The Shine Dalgarno (SD) sequence precedes the heterologous gene (HG). Transcription termination (TT), antibiotic resistance marker (Abʳ) and origin (Ori) of replication of the plasmid are necessary for mRNA stabilization, adequate selection and vector copy number, respectively. The table below the diagram indicates some of the modifications of the plasmid that can lead to specific improvements.

modifying the spacing sequences between the –10 and –35 regions of constitutive promoters from *Lactococcus lactis*, which are also useful in *E. coli* (*107*). In this way, libraries of artificial promoters with different strengths can be generated for each host. Such a technology has been patented (*108*). Often, metabolic optimization requires the simultaneous regulation of expression of various genes. Different promoters can be utilized for each gene, or various genes can be placed under the same promoter in an operon. In the latter case, expression can be regulated by increasing the stability of each coding region through the introduction of stabilizing sequences, such as those forming hairpins (*109*).

Plasmid copy number is directly related to recombinant protein productivity (*see* **Subheading 1.1.1.**), and is regulated by plasmid replication. ColE1-type plasmids are found in Gram-negative bacteria and are part of most cloning vehicles used today. Their replication requires an RNA preprimer called RNA II (*110*). RNA II must be cleaved by the host's RNAase H to release the 3' OH that is used by the DNA polymerase I to initiate replication. Control of the initiation of ColE1 replication is mediated by the interaction of RNA II with an antisense RNA, RNA I, that impedes cleavage of RNA II (*111*). However, control of plasmid replication can be lost when a recombinant protein is overproduced. This is due to an increased pool of uncharged tRNAs provoked by high amino acid consumption rates. Such uncharged tRNAs bind to RNA I, disturbing the natural plasmid replication control mechanism. To avoid this, Grabherr et al. (*112*) modified the nucleotide sequence of RNA I, preventing the binding of tRNA. Such a strategy allowed a better control of the recombinant protein production process, and reduced the metabolic burden that occurs upon uncontrolled plasmid replication. Plasmid copy number can be modified by mutating RNA I or RNA II, or by altering their expression rates. Further control of plasmid replication can be obtained by altering the structure of RNA I and RNA II (*112*) (*see* Chapter 4).

Apart from the strategies discussed in **Subheading 1.2.1.**, many molecular biology approaches have been employed to deal with the problem of recombinant protein accumulation in inclusion bodies when using *E. coli* (*24*). Non-membrane-bound proteins that are correctly folded should be reasonably soluble in aqueous solution, and it is believed that the amino acid sequence at the amino and carboxy termini play a role in their solubility (*113*). Likewise, recognition by proteases is dependent on the polarity of the residues at these termini. Accordingly, Sati et al. (*113*) analyzed the overexpression of a cytoplasmic protein from *Plasmodium falciparium* in *E. coli*. Various constructs bearing extra amino acids at the N- and C-termini were designed. Results indicated that the presence of polar amino acids in the C-terminus and the length of the additional sequence enhanced solubility and stability of the recombinant protein. Similar results on the stability of other recombinant proteins by addition of C-terminal tails have been summarized by Sati et al. (*113*; **Table 3**). As already discussed, fusion proteins are a strategy commonly used to increase recombinant protein solubility. Davis et al. (*114*) proposed a rational strategy for the identification of possible fusion partners that could confer solubility to proteins expressed in *E. coli*. Possible fusion partners were identified from a statistical solubility model. Proteins predicted to be highly soluble were useful to increase the solubility of recombinant proteins when fused to them.

The disulfide bond (Dsb) protein A and DsbB, from the oxidizing pathway, and DsbC and DsbG, from the isomerizing pathway, are found in the periplasmic space of *E. coli* (*115*). These enzymes catalyze the formation of disulfide bonds in nascent proteins. Moreover, both DsbC and DsbG have been shown to have chaperone activity, promoting reactivation and folding and suppressing aggregation (*116,117*). Maskos et al. (*118*) have recently shown that coexpression of DsbC with a complexly folded protein can improve disulfide bond formation in the periplasm.

Gram-positive bacteria, such as *B. subtilis*, have also been utilized for recombinant protein production, with the advantage that they can secrete large amounts of properly folded product and contain low concentrations of pyrogens (*119*). However, recombinant plasmids are not stable in *B. subtilis* and chromosomal integration is the only way to obtain a stable recombinant cell. Yields are lower than those of Gram-negative bacteria, due mostly to the high activity of endogenous proteases. In fact, one of these proteases, subtilisin (produced by *B. subtilis*), is produced in very large amounts to satisfy the detergent industry. Protein engineering has been utilized to produce subtilisin with new properties. *B. subtilis*, in contrast to *E. coli*, is generally recognized as safe (GRAS) and can be used for the production of proteins for the food industry (*120*). However, *B. subtilis* responds to stress by producing proteases and sporulating, consequently reducing recombinant protein concentration (*121*). Medium composition, specifically the concentration of some salts and peptone, can prevent sporulation and increase the concentration of recombinant protein (*122*). Strains of *B. subtilis* that produce lower concentrations of proteases have been utilized for the production of recombinant proteins. Moreover, asporogenous mutants have been isolated (*123*). The utilization of Gram-positive bacteria may find a niche for the production of recombinant proteins, most probably for the production of proteins for nonpharmaceutical industrial applications. Additionally, their use in the synthesis of correctly posttranslationally modified nonribosomal peptide synthetases proves to be appealing, as Doekel et al. (*104*) stably expressed these complex enzymes in *B. subtilis*.

3.2. Yeast and Fungi

Yeasts have been utilized by humans since the Neolithic age (*124*). Their various applications in the food industry and for single-cell protein production has taken yeast fermentations to the largest volumes ever performed (*81*) (*see* Chapters 16–22). The yeast *Saccharomyces cerevisiae* was the first yeast species to be manipulated for recombinant protein expression (*125*), and many proteins have been produced in it. Due to its many applications, excellent knowledge of *S. cerevisiae* molecular biology and physiology has accumulated (*125*). *S. cerevisiae* is GRAS and, like other yeasts, can secrete recombinant proteins to the culture medium. Moreover, intracellular proteins are usually properly folded. As other eukaryotes, yeasts are also capable of performing most posttranslational processing typical of mammalian cells. However, extracellular proteases and differences in glycosylation in proteins expressed in yeast, compared to those of mammalian cells, limit their use. *N*-glycosylation of proteins produced by yeasts are high-mannose (with more than 3 mannose residues) or hypermannose (more than 6 mannose residues) types, with terminal α-1,3 linkages (*126*). Such forms are very immunogenic to mammals (*127,128*). Moreover, *O*-glycosylation by yeasts contains only mannose residues (*126*). Cell engineering has been utilized for obtaining nonimmunogenic glycoproteins from yeasts. Namely, Chiba et al. (*129*) introduced the gene of an α1,2-mannosidase with an ER retention signal in a *S. cerevisiae* mutant that had disrupted the genes of several mannosyltransferases. Such a manipulation resulted in recombinant and native glycoproteins with the structure M5N2. This structure is not found in glycoproteins produced by *S. cerevisiae* and is the substrate for further processing to yield complex glycans as in mammalian cells.

Unmodified yeasts are suitable for the production of proteins that do not require mammalian-type glycosylation and are resistant to proteases. One of these proteins is insulin, which has been commercially produced in *S. cerevisiae* after enhancing its folding and secretion capacities through genetic engineering (*125*). A promising strategy for enhancing secretion has been published by Tan et al. (*130*), who succeeded in the universal application of a 15-residue secretion signal from bacterial endotoxin. Using such a secretion signal on constructs destined for expression of recombinant proteins both in prokaryotes and eukaryotes, *S. cerevisiae* among them, the model protein was secreted in all cases.

Facultative methylotrophic yeasts, such as *Pichia pastoris*, *P. methanolica*, *Candida boidinii*, and *Pichia angusta* (formerly known as *Hansenula polymorpha*), are hosts with great potential and with various recombinant proteins within reach or already in the market (*126*). Some of these proteins are hepatitis B vaccine, human serum albumin, phytase, and insulin-like growth factor (*126*). Industrial application of methylotrophic yeasts started when they were utilized for single-cell protein production. Very large fermentations of methylotrophic yeasts were performed in the 1970s. As single-cell protein production was not economically attractive, *Pichia pastoris* was proposed as a host for recombinant protein production in the 1980s (*131*). Very high cell densities have been obtained, up to 100 g of dry weight per liter, and also high protein concentrations, up to 1 g/L of secreted recombinant protein (*127*). Additionally, *N*-glycosylation proceeds differently than in *S. cerevisiae*, with hypermannosylation being less elaborate (*132*) and occurring less frequently (*126*). Importantly, *P. pastoris* does not produce the immunogenic terminal α-1,3-linked mannoses (*132*). Similarly to what has been performed in *S. cerevisiae*, Callewaert et al. (*132*) constructed a recombinant *P. pastoris* expressing the α1,2-mannosidase gene with a retention signal that targets the enzymatic activity to the ER-Golgi transit region. As a result, M5N2 glycans were the most common structures attached to the recombinant protein.

A potential disadvantage of *P. pastoris* and *P. methanolica* is that transgenes are placed under the promoter of the alcohol oxidase I (AOX1) gene, which requires methanol to induce gene expression. This has three implications for the process. First, large tanks of flammable methanol are needed in the production facilities, and second, methanol, which is toxic to humans, must be thoroughly removed from the final product. Methanol is also toxic to the cells; thus, a third consideration is that specifically designed methanol feeding strategies must be implemented to guarantee its continuous supply during the induction stage but avoiding its accumulation to inhibitory levels. An alternative is the use of the promoter of the MOX1 gene, which is induced either by methanol or derepressed by glycerol in *P. angusta* (*133*).

Filamentous fungi have been utilized for a long time for the production of a wide variety of substances with various applications. Fungi fermentations at large scales have been performed since the first half of the 20th century, mostly for the production of antibiotics or ascorbic acid (*81*). Fungi can secrete large amounts of homologous proteins (up to 30 g/L), and up to 3 g/L of heterologous proteins, although typically only tens of milligrams per liter are obtained (*134*). Such a difference is a consequence of RNA instability or incorrect processing and of high protease activity (*135*). Recombinant protein concentration has been increased by fusing the gene of interest with genes of fungal origin (*134,135*). Fungi produce proteases; this limits their utility for recombinant protein production. Promoting growth in pellets and controlling pH can reduce protease activity more than fourfold (*136*). Such strategies have been utilized for the commercial production of chymosin. Additionally, many homologous fungal proteins (mostly enzymes) have been engineered to obtain some desirable characteristics not present in the original counterpart.

As in yeasts, filamentous fungi produce high-mannose-type glycans, easily recognized by mammalian lectins; therefore, recombinant proteins intended for therapeutic use and expressed in fungi can be rapidly and inconveniently cleared from blood. Trying to palliate for this inconvenience, Maras et al. (*137*) first demonstrated that glycoproteins from *Trichoderma reesei* could be converted in vitro to mammalian-like hybrid oligosaccharides. Later on, Maras et al. (*138*) expressed in *Trichoderma reesei* the human *N*-acetylglucosaminiltransferase I that transfers an N-acetylglucosamine residue to an α-1,3-linked mannose of the M5N2 oligosaccharide. Efforts like this indicate that mammalian *N*-glycans expressed in filamentous fungi are not far away.

3.3. Animal Cells

Animal cells have been cultured in vitro for more than a hundred years. For a long time they have been used for the production of viruses as vaccines, or for synthesizing endogenous proteins, such as interferon. Their complexity delayed their genetic manipulation to the time when manipulation of bacterial genomes was performed almost routinely. The first recombinant proteins approved for human use were produced in bacteria, but of 33 products approved by the FDA between 1996 and 2000, 21 are produced by animal cells. It is expected that this situation will continue as more proteins with pharmaceutical applications have complex glycosylation that cannot be practically produced in prokaryotes or lower eukaryotes (*see* Chapters 28–33). In the mean time, animal cell culture has become routine, with several reactors operating worldwide at the 10,000-L scale. However, successful recombinant protein production in animal cells had to overcome many hurdles, such as the cellular fragility and the complex nutritional requirements of cells (*93,94,139*). Animal cells require hormones and growth factors that were initially supplied by bovine serum. Possible contamination of the final product with virus or prions, and the difficulty of recovering extracellular proteins from serum-containing media, have resulted in the development of serum-free media that are used for large-scale production.

Gene transfer is a particularly relevant issue in cell culture. The development of mammalian cell culture methodologies included designing a variety of vector systems (*see* **Table 4** for a synthesis of their components) and of gene transfer methods (*140*). A

Table 4
Summary of the Required Elements in a Mammalian Expression Vector

Viral-based	Plasmid-based	Promoter/enhancer	Locus control Regions[a]	Transient Expression Reporter gene	Stable expression Chromosomal or episomal	Selection markers Gene-deficient cells	Selection markers Cytotoxic drugs
• Adenovirus • Epstein–Barr • Herpes simplex • Papilloma • Polyoma • Retrovirus • SV40 • Vaccinia	• Prokaryotic, eukaryotic, and viral sequences	• Adenovirus inverted terminal repeats (ITR). • Citomegalo virus • Mouse mammary tumor virus • Murine leukemia virus (MuLV) long terminal repeat (LTR) • Rous sarcoma virus (RSV) LTR • SV40 • β-actin • α-fetoprotein • γ-globin/- • β-globin • β-interferon • Metallo-tionein II	• Human β-globin locus • Human adenosine deaminase gene • Human apolipoprotein E/C-1 gene locus • Human T cell receptor α/δ locus • Human CD2 gene • Human S100 β gene • Human growth hormone gene • Human apolipoprotein B gene	• Chloramphenicol Acetyltransferase (CAT) • β-Gal • Firefly luciferase (Luc) • Human placental alkaline phosphatase (AP) • β-glucoronidase (GUS) • Green Fluorescent Protein (GFP)	• BKV-based vectors[b] • Bovine papilloma virus-based vectors[b] • Epstein-Barr virus-based vectors[b]	• Herpes simplex virus thymidine kinase • Dihydrofolate reductase (dhfr) • Hypoxantine guanine phosphoribosyl transferase (hprt) • Adenyl phosphoribosyl transferase (aprt)	• Hygromycin B phosphotransferase (HygB) • Xanthineguanine phospho ribosyl transferase (XGPRT) • Zeocin (Zeo) • Blasticidin (Bsd) • Aminoglycoside phosphotransferase (aph)

[a]Shows only a limited selection of the available loci listed in Li et al. (*159*). [b]Some of the most frequently used viral-based episomal vectors (*160*). Data from **ref. *140***.

recent development in gene transfer is the use of baculovirus vectors in cultured mammalian cells. Baculoviruses used with this objective carry promoters that are efficiently transcribed in mammalian cells, such as those from Rous sarcoma virus or cytomegalovirus. This methodology has been tested (*141*) of rat hepatic stellate cells, showing a 100% efficiency of heterologous gene expression (*lacZ*) using elevated multiplicities of infection (500 plaque-forming units per cell) in an *Autographa californica* multiple-nucleocapsid polyhedrovirus. This report as well as the increasing number of references relating to the use of baculovirus for gene transfer in mammalian cells, both in vitro and in vivo, show the promises of this approach.

Other issues that arise when expressing proteins in mammalian cells can be solved through cell engineering. For example, when large scale production is engaged, the cells suffer metabolic pressures, such as oxygen depletion and toxic metabolite accumulation, which affect final yields. An interesting approach with CHO cells (*142*) consisted of engineering their mRNA translation initiation machinery with the aim of leaving it on, despite the prevalence of stressful conditions derived from large-scale production schemes. Traditional strategies for productivity optimization involve manipulation of cell division as well as cell longevity, supported by the increasing knowledge of cell cycle control (*143*). Such is the case of the manipulation of a myeloma cell line that constitutively expresses a chimeric antibody. The cell line was modified to express, upon induction, an inhibitor of cyclin E-dependent kinase that causes cell cycle arrest. With this manipulation, Watanabe et al. (*144*) arrested cell proliferation, thus preventing accumulation of deleterious metabolites. Additionally, with this operation the yield of a recombinant hybrid antibody was enhanced 4-fold. A somewhat similar strategy was used by Meents et al. (*145*), who arrested *dhfr*-deficient CHO cells in G1 by inducibly expressing the cyclin-dependent kinase inhibitor p27[Kip1], being able to enhance specific productivities by fivefold (*see* **Subheading 2.2**). Apoptosis represents a major inconvenience in cultures intended for production (*146*), but overexpression of the antiapoptotic gene *bcl2* (*147,148*) leads to sustained growth and therefore sustained protein production (*149,150*).

An important aspect to be considered when expressing recombinant proteins in mammalian cells, and part of the reason that these cells are used as an expression vehicle, is glycosylation. A major drawback that emerges from altering the glycosylation machinery in vivo is the resulting heterogeneity of products (*151*), given the variety of pathways that can be followed. In spite of this, and given the subtle differences that exist between glycans obtained in commonly used mammalian cell lines and those associated with glycoproteins synthesized in human cells, cloning glycosyl-transferases into common mammalian cell lines has proved useful for the expression of humanized *N*-glycoproteins (*148*) and *O*-glycoproteins (*152*).

A rather new and exciting application of mammalian cell culture is gene-function analysis (*153*). The key tool for this application is RNA interference (RNAi), which occurs by sequence-specific gene silencing initiated by double-stranded RNA (dsRNA) homologous to the gene to be silenced. The mediators for mRNA degradation are small 21 to 22 nucleotide-interfering RNAs (siRNAs) that result from the enzymatic activity of dicer, a cellular ribonuclease III. This specific process seems to have emerged as a defense against aberrant or unwanted gene expression (*154*). Although somehow differently, this phenomenon also silences genes in mammalian cells, and has been reported in neurogenesis and neuronal differentiation studies *(155)*. Gene silencing was achieved in cultured mouse P19 cells by means of synthetic duplex RNAs as well as with hairpin siRNAs. Hamada et al. *(156)* proved the system useful in mammalian cells by targeting mRNA of Jun dimerization protein expressed in mouse RAW264.7 and NIH3T3 cells with both duplex RNAs and the sense strand of the synthetic siRNA. RNA silencing can become a powerful technique for improving recombinant protein production.

Acknowledgments

Excellent technical support by K. Levy, V. Hernández, A. Martínez Valle, and R. Ciria is gratefully acknowledged. Financial support was provided by CONACyT NC230 and 33348 and DGAPA UNAM IN218202.

References

1. Margaritis, A. and Bassi, A. S. (1991) Plasmid stability of recombinant DNA microorganisms, in *Recombinant DNA Technology and Applications.* (Prokop, A., Bajpai, R. K., and Ho, C., eds.) McGraw-Hill, New York, NY, pp. 316–332.
2. Paulsson, J. and Ehrenberg, M. (2001) Noise in a minimal regulatory network: plasmid copy number control. *Q. Rev. Biophys.* **34,** 1–59.
3. Corchero, J. L. and Villaverde, A. (1998) Plasmid maintenance in *Escherichia coli* recombinant cultures is dramatically, steadily, and specifically influenced by features of the encoded proteins. *Biotechnol. Bioeng.* **58,** 625–632.
4. Summers, D. (1998) Timing, self-control and a sense of direction are the secrets of multicopy plasmid stability. *Mol. Microbiol.* **29,** 1137–1145.
5. Baneyx, F. (1999) Recombinant protein expression in *Escherichia coli. Curr. Op. Biotechnol.* **10,** 411–421.
6. Lewin, B. (1997) *Genes VI.* Oxford University Press, New York, NY, pp. 460–463.
7. Ramírez, O. T., Flores, E., and Galindo, E. (1995) Products and bioprocesses based on genetically modified organisms: review of engineering issues and trends in the literature. *Asia-Pacific J. Mol. Biol. Biotechnol.* **3,** 165–197.
8. Swartz, J. R. (2001) Advances in *Escherichia coli* production of therapeutic proteins. *Curr. Op. Biotechnol.* **12,** 195–201.
9. Sayadi, S., Nasri, M., Berry, F., Barbotin, J. N., and Thomas, D. (1987) Effect of temperature on the stability of plasmid pTG201 and productivity of *xylE* gene product in recombinant *Escherichia coli*: development of a two-stage chemostat with free and immobilized cells. *J. Gen. Microbiol.* **133,** 1901–1908.
10. Prazeres, D. M. F., Ferreira, G. N. M., Monteiro, G. A., Cooney, C. L., and Cabral, J. M. S. (1999) Large-scale production of pharmaceutical-grade plasmid DNA gene therapy: problems and bottlenecks. *Trends Biotechnol.* **17,** 169–174.
11. Balbás, P. and Gosset, G. (2001) Chromosomal editing in *Escherichia coli.* Vectors for DNA integration and excision. *Mol. Biotechnol.* **19,** 1–12.
12. Balbás, P. (2001) Understanding the art of producing protein and nonprotein molecules in *Escherichia coli. Mol. Biotechnol.* **19,** 251–267.
13. Olson, P., Zhang, Y., Olsen, D., Owens, A., Cohen, P., Nguyen, K., et al. (1998) High-level expression of eukaryotic polypeptides from bacterial chromosomes. *Protein Express. Purif.* **14,** 160–166.

14. Twyman, R. M. and Whitelaw, B. (2000) Genetic engineering: Animal cell technology, in *The Encyclopedia of Cell Technology* (Spier, R. E., ed.), John Wiley and Sons, New York, NY, pp. 737–819.

15. Sanders, P. G. (1990) Protein production by genetically engineered mammalian cell lines, in *Animal Cell Biotechnology Vol. 4* (Spier, R. E. and Griffiths, J. B., eds.) Academic Press, London, pp. 15–70.

16. Palomares, L. A., Estrada-Mondaca, S., and Ramírez, O. T. (2004) Principles and applications of the insect-cell-baculovirus expression vector system, in *Cell Culture Technology for Pharmaceutical and Cellular Applications* (Ozturk, S. and Hu, W. S., eds.), Marcel Dekker, New York, NY, in press.

17. Cohen, D. M. and Ramig, R. F. (1999) Viral genetics, in *Fields Virology* (Fields, B. N., Knipe, D. M., Howley, P. M., et al., eds.), Lippincott-Raven, Philadelphia, PA, pp. 113–151.

18. Dee, K. U., and Shuler, M. L. (1997) A mathematical model of the trafficking of acid-dependent enveloped viruses: application to binding, nuclear accumulation and uptake of baculovirus. *Biotechnol. Bioeng.* **54**, 468–490.

19. Petricevich, V. L., Palomares, L. A., González, M., and Ramírez, O. T. (2001) Parameters that determine virus adsorption kinetics: toward the design of better infection strategies for the insect-cell baculovirus expression system. *Enzyme Microb. Technol.* **28**, 52–61.

20. Ailor, E. and Betenbaugh, M. J. (1999) Modifying secretion and posttranslational processing in insect cells. *Curr. Op. Biotechnol.* **10**, 142–145.

21. Kopito, R. R. (2000) Aggresomes, inclusion bodies and protein aggregation. *Trends Cell Biol.* **10**, 524–530.

22. Schlieker, C., Bukau, B., and Mogk, A. (2002) Prevention and reversion of protein aggregation by molecular chaperones in the *E. coli* cytosol: implications for their applicability in biotechnology. *J. Biotechnol.* **9**, 13–21.

23. Chapman, R., Sidrauski, C., and Walter, P. (1998) Intracellular signaling from the endoplasmic reticulum to the nucleus. *Annu. Rev. Cell. Dev. Biol.* **14**, 459–485.

24. Carrió, M. M. and Villaverde, A. (2002) Construction and deconstruction of bacterial inclusion bodies. *J. Biotechnol.* **96**, 3–12.

25. De Bernadez Clark, E. (2001) Protein refolding for industrial processes. *Curr. Op. Biotechnol.* **12**, 202–207.

26. Datar, R. V., Cartwright, T., and Rosen, C. G. (1993) Process economics of animal cell and bacterial fermentations: a case study analysis of tissue plasminogen activator. *Bio/Technol.* **11**, 349–357.

27. Bach, H., Mazor, Y., Shaky, S., Berdichevsky, A. S. Y., Gutnick, D. L., and Benhar, I. (2001) *Escherichia coli* maltose binding protein as a molecular chaperone for recombinant intracellular cytoplasmic single chain antibodies. *J. Mol. Biol.* **312**, 79–93.

28. Paetzel, M., Karla, A., Strynadka, N. C. J., and Dalbey, R. E. (2002) Signal peptidases. *Chem. Rev.* **102**, 4549–4579.

29. Ailor, E., Pathmanathan, J., Jongbloed, J. D. H., and Betenbaugh, M. J. (1999) A bacterial signal peptidase enhances processing of a recombinant single chain antibody fragment in insect cells. *Biochem. Biophys. Res. Comm.* **255**, 444–450.

30. van Dijl, J. M., de Jong, A., Smith, H., Bron, S., and Venema, G. (1991) Signal peptidase I overproduction results in increased efficiencies of export and maturation of hybrid secretory proteins in *Escherichia coli. Mol. Gen. Genet.* **227**, 40–48.

31. Bryan, P. N. (2002) Prodomains and protein folding catalysis. *Chem. Rev.* **102**, 4085–4815.

32. Nakayama, K. (1997) Furin: a mammalian subtilisin/Kex2p-like endoprotease involved in processing of a wide variety of precursor proteins. *Biochem. J.* **327**, 25–35.

33. Laprise, M. H., Grondin, F., and Dubois, C. M. (1998) Enhanced TGFβ1 maturation in High Five cells coinfected with recombinant baculovirus encoding the convertase furin/pace: improved technology for the production of recombinant proproteins in insect cells. *Biotechnol. Bioeng.* **58**, 85–91.

34. Drews, R., Paleyanda, R. K., Lee, T. K., Chang, R. R., Rehemtulla, A., Kaufman, R. J., et al. (1995) Proteolytic maturation of protein C upon engineering the mouse mammary gland to express furin. *Proc. Natl. Acad. Sci. USA* **92**, 10462–10466.

35. Wingfield, P. T. (1997) Purification of recombinant proteins, in *Current Protocols in Protein Science* (Coligan, E., Dunn, B. M., Ploegh, H. L., Speicher, D. W., and Wigfield, P. T., eds.), John Wiley and Sons, New York, NY, pp. 6.0.1–6.1.22.

36. Hwang, D. D. H., Liu, L. F., Kuan, I. C., Lin, L. Y., Tam, T. C. S., and Tam, M. F. (1999) Coexpression of glutathione S-transferase with methionine aminopeptidase: a system of producing enriched N-terminal processed proteins in *Escherichia coli*. *Biochem. J.* **338**, 335–342.

37. Vassileva-Atanassova, A., Mironova, R., Nacheva, G., and Ivanov, I. (1999) N-terminal methionine of recombinant proteins expressed in two different *Escherichia coli* strains. *J. Biotechnol.* **69**, 63–67.

38. Benz, I. and Schmidt, M. A. (2002) Never say never again: protein glycosylation in pathogenic bacteria. *Mol. Microbiol.* **45**, 267–276.

39. Furmanek, A. and Hofsteenge, J. (2000) Protein C-mannosylation: facts and questions. *Acta Biochimica Polonica* **47**, 781–789.

40. Lisowska, E. (2002) The role of glycosylation in protein antigenic properties. *Cell. Mol. Life Sci.* **59**, 445–455.

41. Gu, X., Harmon, B. J., and Wang, D. I. C. (1998) Monitoring and characterization of glycoprotein quality in animal cell cultures, in *Advances in Bioprocess Engineering II* (Galindo, E. and Ramírez, O. T., eds.), Kluwer Academic Publishers. Dordrecht, The Netherlands, pp. 1–24.

42. Shelikoff, M., Sinskey, A. J., and Stephanopoulos, G. (1994) The effect of protein synthesis inhibitors on the glycosylation site occupancy of recombinant human prolactin. *Cytotechnol.* **15**, 195–208.

43. Ohkura, T., Fukushima, K., Kurisaki, A., Sagami, H., Ogura, K., Ohno, K., et al. (1997) A partial deficiency of dehydrodolichol reduction is a cause of carbohydrate-deficient glycoprotein syndrome type I. *J. Biol. Chem.* **272**, 6868–6875.

44. Fukushima, K., Ohkura, T., and Yamashita, K. (1997) Synthesis of lipid-linked oligosaccharides is dependent on the cell cycle in rat 3Y1 cells. *J. Biochem. (Tokyo)* **121**, 415–418.

45. Andersen, D. C., Bridges, T., Gawlitzek, M., and Hoy, C. (2000) Multiple cell culture factors can affect the glycosylation of Asn184 in CHO-produced tissue-type plasminogen activator. *Biotechnol. Bioeng.* **70**, 25–31.

46. Yuk, I. H. Y. and Wang, D. I. C. (2002) Glycosylation of Chinese hamster ovary cells in dolichol phosphate-supplemented cultures. *Biotechnol. Appl. Biochem.* **36**, 141–147.

47. Rearick, J. I., Chapman, A., and Kornfeld, S. (1981) Glucose starvation alters lipid-linked oligosaccharide biosynthesis in Chinese hamster ovary cells. *J. Biol. Chem.* **256**, 6255–6261.

48. Nyberg, G. B., Balcarcel, R., Follstad, B. D., Stephanopoulos, G., and Wang, D. I. C. (1999) Metabolic effects on recombinant interferon-γ glycosylation in continuous culture of Chinese hamster ovary cells. *Biotechnol. Bioeng.* **62**, 336–347.

49. Gu, X. and Wang, D. I. C. (1998) Improvement of interferon-γ sialylation in Chinese hamster ovary cell culture by feeding of N-acetyl-mannosamine. *Biotechnol. Bioeng.* **58**, 642–648.

50. Lawrence, S. M., Huddleston, K. A., Pitts, L. R., Nguyen, N., Lee, Y. C., Vann, W. F., et al. (2000) Cloning and expression of the human N-acetyl-neuraminic acid phosphate synthase

gene with 2-keto3-deoxy-D-glycero-D-galactonononic acid biosynthetic ability. *J. Biol. Chem.* **275**, 17869–17877.

51. Hills, A. E., Patel, A., Boyd, P., and James, D. C. (2001) Metabolic control of recombinant monoclonal antibody N-glycosylation in GSNS0 cells. *Biotechnol. Bioeng.* **75**, 239–251.

52. Palomares, L. A. and Ramírez, O. T. (2002) Complex N-glycosylation of recombinant proteins by insect cells. *Bioprocessing.* **1**, 70–73.

53. Gramer, M. J. and Goochee, C. F. (1993) Glycosidase activities in Chinese hamster ovary cell lysate and cell culture supernatant. *Biotechnol. Prog.* **9**, 366–373.

54. Munzert, E., Heidermann, R., Büntemeyer, H., Lehmann, J., and Múthing, J. (1997) Production of recombinant human antithrombin III on 20L bioreactor scale: correlation of supernatant neuraminidase activity, desialylation, and decrease of biological activity of recombinant glycoprotein. *Biotechnol. Bioeng.* **56**, 441–448.

55. Yang, M. and Butler, M. (2000) Effects of ammonia on the glycosylation of human recombinant erythropoietin in culture. *Biotechnol. Prog.* **16**, 751–759.

56. Schmelzer, A. E. and Miller, W. M. (2002) Effects of osmoprotectant compounds on NCAM-polysialylation under hyperosmotic stress and elevated pCO_2. *Biotechnol. Bioeng.* **77**, 359–368.

57. Kaufmann, H., Mazur, X., Marone, R., Bailey, J. E., and Fussenegger, M. (2001) Comparative analysis of two controlled proliferation strategies regarding product quality, influence on tetracyclineregulated gene expression, and productivity. *Biotechnol. Bioeng.* **72**, 592–602.

58. Lopez, M., Tetaert, D., Juliant, S., Gazon, M., Cerruti, M., Verbert, A., et al. (1999) O-glycosylation potential of lepidopteran insect cell lines. *Biochim. Biophys. Acta* **1427**, 49–61.

59. Grabherr, R., Ernst, W., Oker-Blom, C., and Jones, I. (2001) Developments in the use of baculovirus display of complex eukaryotic proteins. *Trends Biotechnol.* **19**, 231–236.

60. Palomares, L. A., Kuri-Breña, F., and Ramírez, O. T. (2002) Industrial recombinant protein production, in *The Encyclopedia of Life Support Systems.* EOLSS Publishers, Oxford. 6.58.3.8., www.eolss.net.

61. Palomares, L. A. and Ramírez, O. T. (2000) Bioreactor scaleup, in *The Encyclopedia of Cell Technology* (Spier, R. E., ed.), John Wiley and Sons, New York, NY, pp. 174–183.

62. Ozturk, S. S. (1996) Engineering challenges in high density cell culture systems. *Cytotechnol.* **22**, 3–16.

63. Vasina, J. A., Peterson, M. S., and Baneyx, F. (1998) Scale-up and optimization of the low-temperature inducible *cspA* promoter system. *Biotechnol. Prog.* **14**, 714–721.

64. Ramírez, O. T., Zamora, R., Espinosa, G., Merino, E., Bolívar, F., and Quintero, R. (1994) Kinetic study of penicillin acylase production by recombinant *E. coli* in batch cultures. *Process Biochem.* **29**, 197–206.

65. Eriksen, N. T., Kratchmarova, I., Neve, S., Kristiansen, K., and Iversen, J. J. L. (2001) Automatic inducer addition and harvesting of recombinant *Escherichia coli* cultures based on indirect on-line estimation of biomass concentration and specific growth rate. *Biotechnol. Bioeng.* **75**, 355–361.

66. Yazdani, S. S. and Mukherjee, K. J. (1998) Overexpression of streptokinase using a fed-batch strategy. *Biotechnol. Lett.* **20**, 923–927.

67. Elias, C. B., Zeisea, A., Bédard, C., and Kamen, A. A. (2000) Enhanced growth of Sf9 cells to a maximum cell density of 5.2×10^7 cells per mL and production of b-galactosidase at high cell density by fed-batch culture. *Biotechnol. Bioeng.* **68**, 381–388.

68. Schein, C. H. (1999) Protein expression, soluble, in *Encyclopedia of Bioprocess Technology. Fermentation, Biocatalysis and Bioseparation* (Flickinger, M. C. and Drew, S. W., eds.), John Wiley and Sons, New York, NY, pp. 2156–2169.

69. Gupta, J. C., Jisani, M., Pandey, G., and Mukherjee, K. J. (1999) Enhancing recombinant protein yields in *Escherichia coli* using the T7 system under the control of heat inducible λPL promoter. *J. Biotechnol.* **68**, 125–134.
70. Farewell, A. and Neidhardt, F. C. (1998) Effect of temperature on in vivo protein synthetic capacity in *Escherichia coli. J. Bacteriol.* **180**, 4704–4710.
71. Martínez, A., Ramírez, O. T., and Valle, F. (1998) Effect of growth rate on the production of β-galactosidase from *Escherichia coli* in *Bacillus subtilis* using glucose-limited exponentially fed-batch cultures. *Enzyme Microb. Technol.* **22**, 520–526.
72. Sandén, A. M., Pytz, I., Tubelakas, I., Föberg, C., Le, H., Hektor, A., et al. (2003) Limiting factors in *Escherichia coli* fed-batch production of recombinant proteins. *Biotechnol. Bioeng.* **81**, 158–166.
73. Saraswat, V., Kim, D. Y., Lee, J., and Park, Y. H. (1999) Effect of specific production rate on multimerization of plasmid vector and gene expression level. *FEMS Microbiol. Lett.* **179**, 367–373.
74. Ramírez, O. T., Zamora, R., Quintero, R., and López-Munguía, A. (1994) Exponentially fed-batch cultures as an alternative to chemostats: the case of penicillin acylase production by recombinant *E. coli. Enzyme Microb. Technol.* **16**, 895–903.
75. Yazdani, S. S. and Mukherjee, K. J. (2002) Continuous-culture studies on the stability and expression of recombinant streptokinase in *Escherichia coli. Bioprocess Biosyst. Eng.* **24**, 341–346.
76. Åkesson, M., Karlsson, E. N., Hagander, P., Axelsson, J. P., and Tocaj, A. (1999) Online detection of acetate formation in *Escherichia coli* cultures using dissolved oxygen responses to feed transients. *Biotechnol. Bioeng.* **69**, 590–598.
77. Suzuki, H., Kishimoto, M., Kamoshita, Y., Omasa, T., Katakura, Y., and Suga, K. (2000) Online control of feeding of medium components to attain high cell density. *Bioproc. Eng.* **22**, 433–440.
78. Bailey, J. E. (1993) Host-vector interactions in *Escherichia coli*, in *Advances in Biochemical Engineering and Biotechnology* (Fietcher, A., ed.), Springer Verlag. Berlin, **48**, pp. 29–52.
79. Leelavatcharamas, V., Emery, A. N., and Al-Rubeai, M. (1999) Use of cell cycle analysis to characterize growth and interferon production in perfusion culture of CHO cells. *Cytotechnol.* **30**, 59–69.
80. Lidén, G. (2002) Understanding the bioreactor. *Bioproc. Biosyst. Eng.* **24**, 273–279.
81. Shuler, M. L. and Kargi, F. (2002) *Bioprocess engineering. Basic concepts.* 2nd ed. Prentice Hall, Upper Saddle River, NJ, USA.
82. O'Beirne, D. and Hamer, G. (2000) Oxygen availability and the growth of *Escherichia coli* W-3110: a problem exacerbated by scaleup. *Biprocess Eng.* **23**, 375–380.
83. Palomares, L. A. and Ramírez O. T. (1996). The effect of dissolved oxygen tension and the utility of oxygen uptake rate in insect cell culture. *Cytotechnol.* **22**, 225–237.
84. Yegneswaran, P. K., Thompson, B. G., and Gray, M. R. (1991) Effect of dissolved oxygen control on growth and antibiotic production by *Strepmomyces clavuligerus* fermentations. *Biotechnol. Prog.* **7**, 246–250.
85. Konz, J. O., King, J., and Cooney, C. L. (1998) Effects of oxygen on recombinant protein expression. *Biotechnol. Prog.* **14**, 393–409.
86. Li, X., Robbins, J. W., and Taylor, K. B. (1992) Effect of the levels of dissolved oxygen on the expression of recombinant proteins in four recombinant *Escherichia coli* strains. *J. Ind. Microbiol.* **9**, 1–10.
87. De León, A., Galindo, E., and Ramírez, O. T. A post–fermentative stage improves penicillin acylase production by a recombinant *E. coli. Biotechnol. Lett.* **18**, 927–932.

88. Bollinger, C. J. T., Bailey, J. E., and Kallio, P. T. (2001) Novel hemoglobins to enhance micro-aerobic growth and substrate utilization in *Escherichia coli*. *Biotechnol. Prog.* **17**, 798–808.

89. Doig, S. D., O'Sullivan, L. M., Patel, S., Ward, J. M., and Woodley, J. M. (2001) Large scale production of cyclohexanone monooxygenase from *Escherichia coli* TOP10 pQR239. *Enzyme Microb. Technol.* **28**, 265–274.

90. Thomas, C. R. and Zhang, Z. (1998) The effect of hydrodynamics on biological materials. In *Advances in Bioprocess Engineering II* (Galindo, E. and Ramírez, O. T., eds.) Kluwer Academic Publishers, Dordrecht, pp. 137–170.

91. Garcia-Briones, M. A., Brodkey, R. S., and Chalmers, J. J. (1994) Computer simulation of the rupture of a gas bubble at a gas-liquid interface and its implications in animal cell damage. *Chem. Eng. Sci.* **49**, 2301–2320.

92. Chisti, Y. (2000) Animal cell damage in sparged reactors. *Trends Biotechnol.* **18**, 420–423.

93. Ramírez, O. T. and Mutharasan, R. (1990) The role of plasma membrane fluidity on the shear sensitivity of hybridomas grown under hydrodynamic stress. *Biotechnol. Bioeng.* **36**, 911–920.

94. Palomares, L. A., González, M., and Ramírez, O. T. (2000) Evidence of Pluronic F68 direct interaction with insect cells: impact on shear protection, recombinant protein and baculovirus production. *Enzyme Microb. Technol.* **26**, 324–331.

95. Wu, J., Ruan, Q., and Lam, H. Y. P. (1997) Effects of surface-active medium additives on insect cell surface hydrophobicity relating to cell protection against bubble damage. *Enzyme Microb. Technol.* **21**, 341–348.

96. Galindo, E., Flores, C., Larralde-Corona, P., Corkidi-Blanco, G., Rocha-Valadez, J. A., and Serrano-Carreón, L. (2004) Production of 6-pentyl-a-pyrone by *Trichoderma harzianum* cultured in unbaffled and baffled shake flasks. *Biochem. Eng. J.* In press.

97. Ferreira, B. S., Calado, C. R. C., van Keulen, F., Fonseca. L. P., Cabral, J. M. S., and da Fonseca, M. M. R. (2003) Towards a cost effective strategy for cutinase production by a recombinant *Saccharomyces cerevisiae*: strain physiological aspects. *Appl. Microbiol. Biotechnol.* **61**, 69–76.

98. Bryers, J. D. and Huang, C. T. (1995) Recombinant plasmid retention and expression in bacterial biofilm cultures. *Water Sci. Technol.* **31**, 105–115.

99. Laken, H. A. and Leonard, M. W. (2001) Understanding and modulating apoptosis in industrial cell culture. *Curr. Op. Biotechnol.* **12**, 175–179.

100. Demain, A. (2000) Small bugs, big business: The economic power of a microbe. *Biotechnol. Adv.* **18**, 499–514.

101. Andersen, D. C. and Krummen, L. (2002) Recombinant protein expression for therapeutic applications. *Curr. Op. Biotechnol.* **13**, 117–123.

102. Hannig, G. and Makrides, S. (1998) Strategies for optimizing heterologous protein expression in *Escherichia coli*. *Trends Biotechnol.* **16**, 54–60.

103. Jonasson, P., Liljeqvist, S., Nygren, P. Å., and Ståhl, S. (2002) Genetic design for facilitated production and recovery of recombinant proteins in *Escherichia coli*. *Biotechnol. Appl. Biochem.* **35**, 91–105.

104. Doekel, S., Eppelmann, K., and Marahiel, M. A. (2002) Heterologous expression of non-ribosomal peptide synthetases in *B. subtilis*: construction of a bifunctional *B. subtilis/E. coli* shuttle vector system. *FEMS Microbiol. Letters.* **216**, 185–191.

105. Makrides, S. C. (1996) Strategies for achieving high-level expression of genes in *Escherichia coli*. *Microbiol. Rev.* **60**, 512–538.

106. Jensen, P. R. and Hammer, K. (1998) Artificial promoters for metabolic optimization. *Biotechnol. Bioeng.* **58**, 191–195.
107. Solem, C. and Jensen, P. R. (2002) Modulation of gene expression made easy. *Appl. Environ. Microbiol.* **68**, 2397–2403.
108. Jensen, P. R. and Hammer, K. (1998) The sequence of spacers between the consensus sequences modulates the strength of prokaryotic promoters. *Appl. Environ. Microbiol.* **64**, 82–87.
109. Smolke, C. D. and Keasling, J. D. (2002) Effect of gene location, mRNA secondary structures, and RNase sites on expression of two genes in an engineered operon. *Biotechnol. Bioeng.* **80**, 762–776.
110. Tolmasky, M. E., Actis, L. A., and Crosa, J. H. (1999) Plasmid DNA replication, in *Encyclopedia of Bioprocess Techology* (Flickinger, M. C. and Drew, S. W., eds), John Wiley and Sons Inc., New York, NY, pp. 2004–2019.
111. Grabherr, R. and Bayer, K. (2002) Impact of targeted vector design on ColE1 plasmid replication. *Trends Biotechnol.* **20**, 257–260.
112. Grabherr, R., Nilsson, E., Striedner, G., and Bayer, K. (2002) Stabilizing plasmid copy number to improve recombinant protein production. *Biotechnol. Bioeng.* **77**, 142–147.
113. Sati, S. P., Singh, S. K., Kumar, N., and Sharma, A. (2002) Extra terminal residues have a profound effect on the folding and solubility of a *Plasmodium falciparium* sexual stage-specific protein overexpressed in *Escherichia coli*. *Eur. J. Biochem.* **269**, 5259–5263.
114. Davis, G. D., Elisee, C., Newman, D. M., and Harrison, R. G. (1999) New fusion protein systems designed to give soluble expression in *Escherichia coli*. *Biotechnol. Bioeng.* **65**, 382–388.
115. Collet, J. F. and Bardwell, J. C. A. (2002) Oxidative protein folding in bacteria. *Mol. Microbiol.* **44**, 1–8.
116. Chen, J., Song, Jl., Zhang, S., Wang, Y., Cui, D. F., and Wang, C. C. (1999) Chaperone activity of DsbC. *J. Biol. Chem.* **274**,19601–19605.
117. Shao, F., Bader, M. W., Jakob, U., and Bardwell, J. C. A. (2000) DsbG, a protein disulfide isomerase with chaperone activity. *J. Biol. Chem.* **275**, 13349–13352.
118. Maskos, K., Huber-Wunderlich, M., and Glockshuber, R. (2003) DsbA-catalyzed oxidative folding of proteins with complex disulfide bridge patterns in vitro and in vivo. *J. Mol. Biol.* **325**, 495–513.
119. Harwood, C. R. (1992) *Bacillus subtilis* and its relatives: Molecular biological and industrial workhorses. *Trends Biotechnol.* **10**, 247–256.
120. Sánchez, M., Prim, N., Rández-Gil, F., Pastor, F. I. J., and Diaz, P. (2002) Engineering of baker's yeast, *E. coli* and *Bacillus* hosts for the production of *Bacillus subtilis* lipase A. *Biotechnol. Bioeng.* **78**, 339–345.
121. Huang, H., Ridgway, D., Gu, T., and Moo-Young, M. (2003) A segregated model for heterologous amylase production by *Bacillus subtilis*. *Enzyme Microb. Technol.* **32**, 407–413.
122. El-Helow, E. R., Abdel-Fattah, Y. R., Ghanem, K. M., and Mohamad, E. A. (2000) Application of the response surface methodology for optimizing the activity of an *aprE*-driven gene expression system in *Bacillus subtilis*. *Appl. Microbiol. Biotechnol.* **54**, 515–520.
123. Oh, M. K., Kim, B. G., and Park, S. H. (1995) Importance of spore mutants for fed-batch and continuous fermentation of *Bacillus subtilis*. *Biotechnol. Bioeng.* **47**, 696–702.
124. Dequin, S. (2001) The potential of genetic engineering for improving brewing, wine making and baking yeasts. *Appl. Microbiol. Biotechnol.* **5**, 577–588.
125. Kjeldsen, T. (2000) Yeast secretory expression of insulin precursors. *Appl. Microbiol. Biotechnol.* **54**, 277–286.

126. Gellisen, G. (2000) Heterologous protein production in methylotrophic yeasts. *Appl. Microbiol. Biotechnol.* **54**, 741–750.

127. Lin Cereghino, G. P., Lin Cereghino, J., Ilgen, C., and Cregg, J. M. (2002) Production of recombinant proteins in fermenter cultures of the yeast *Pichia pastoris*. *Curr. Op. Biotechnol.* **13**, 329–332.

128. Ko, J. H., Hahm, M. S., Kang, H. A., Nam, S. W., and Chung, B. H. (2002) Secretory expression and purification of *Aspergillus niger* glucose oxidase in *Saccharomyces cerevisiae* mutant deficient in *PMR1* gene. *Prot. Expr. Purif.* **25**, 488–493.

129. Chiba, Y., Suzuki, M., Yoshida, S., Yoshida, A., Ikenaga, H., Takeuchi, M., Jigami, Y., and Ichishima, E. (1998) Production of human compatible high mannose-type (Man$_5$GlcNAc$_2$) sugar chains in *Saccharomyces cerevisiae*. *J. Biol. Chem.* **273**, 26298–26304.

130. Tan, N. S., Ho, B., and Ding, J. L. (2002) Engineering a novel secretion signal for cross-host recombinant protein expression. *Prot. Eng.* **15**, 337–345.

131. Lin Cereghino, J. and Cregg, J. M. (2000) Heterologous protein expression in the methylotrophic yeast *Pichia pastoris*. *FEMS Microbiol. Rev.* **24**, 45–66.

132. Callewaert, N., Laroy, W., Cadirgi, H., Geysens, S., Saelens, X., Jou, W. M., et al. (2001) Use of HDEL-tagged *Trichoderma reesei* mannosyl oligosaccharide 1,2-a-D-mannosidase for *N*-glycan engineering in *Pichia pastoris*. *FEBS Letters*. **503**, 173–178.

133. Houard, S., Heinderyckx, M., and Bollen, A. (2002) Engineering of non-conventional yeasts for efficient synthesis of macromolecules: the methylotrophic genera. *Biochimie* **84**, 1089–1093.

134. Punt, P. J., Van Biezen, N., Conesa, A., Albers, A., Mangnus, J., and van den Hondel, C. (2002) Filamentous fungi as cell factories for heterologous protein production. *Trends Biotechnol.* **20**, 200–206.

135. Gouka, R. J., Ount, P. J., and van den Hondel, C. A. M. J. J. (1997) Efficient production of secreted proteins by *Aspergillus*: progress, limitations and prospects. *Appl. Microbiol. Biotechnol.* **47**, 1–11.

136. Gyamerah, M., Merichetti, G., Adedato, O., Scharer, J. M., and Moo-Young, M. (2002) Bioprocessing strategies for improving hen egg-white lysozyme (HEWL) production by recombinant *Aspergillus niger* HEWL WT1316. *Appl. Microbiol. Biotechnol.* **60**, 403–407.

137. Maras, M., Saelens, X., Laroy, W., Piens, K., Claetssens, M., Fiers, W., et al. (1997) *In vitro* conversion of the carbohydrate moiety of fungal glycoproteins to mammaliantype oligosaccharides. Evidence for *N*-acetylglucosamyniltransferase-I-accepting glycans from *Trichoderma reesei*. *Eur. J. Biochem.* **249**, 701–707.

138. Maras, M., De Bruyn, A., Vervecken, W., Uusitalo, J., Penttilä, M., Busson, R., et al. (1999) In vivo synthesis of complex *N*-glycans by expression of human N-acetylglucosamyniltransferase I in the filamentous fungus *Trichoderma reesei*. *FEBS Letters*. **452**, 365–370.

139. Ramírez, O. T., Sureshkumar, G. K., and Mutharasan, R. (1990) Bovine colostrum or milk as a serum substitute for the cultivation of a mouse hybridoma. *Biotechnol. Bioeng.* **35**, 882–889.

140. Colosimo, A., Goncz, K. K., Holmes, A. R., Kunzelmann, K., Novelli, G., Malone, R. W., et al. (2000) Transfer and expression of foreign genes in mammalian cells. *BioTechniques*. **29**, 314–331.

141. Gao, R., McCormick, C. J., Arthur, M. J. P., Rudell, R., Oakley, F., Smart, D. E., et al. (2002) High efficiency gene transfer into cultured primary rat and human hepatic stellate cells using baculovirus vectors. *Liver*. **22**, 15–22.

142. Underhill, M. F., Coley, C., Birch, J. R., Findlay, A., Kallmeier, R., Proud, C. G., et al. (2003) Engineering mRNA translation initiation to enhance transient gene expression in Chinese hamster ovary cells. *Biotechnol. Prog.* **19**, 121–129.

143. Fussenegger, M. and Bailey, J. E. (1998) Molecular regulation of cell cycle progression and apoptosis in mammalian cells: implications for biotechnology. *Biotechnol. Prog.* **14**, 807–833.

144. Watanabe, S., Shuttleworth, J., and Al-Rubeai, M. (2002) Regulation of cell cycle and productivity in NS0 cells by the overexpression of p21^{CIP1}. *Biotechnol. Bioeng.* **77**, 1–7.

145. Meents, H., Enenkel, B., Werner, R. G., and Fussenegger, M. (2002) p27^{Kip1} mediated controlled proliferation technology increases constitutive sICAM production in CHODUKX adapted for growth in suspension and serumfree media. *Biotechnol. Bioeng.* **79**, 619–627.

146. Meneses-Acosta, A., Mendonça, R. Z., Merchant, H., Covarrubias, L., and Ramírez, O. T. (2001) Comparative characterization of cell death between Sf9 insect cells and hybridoma cultures. *Biotechnol. Bioeng.* **72**, 441–457.

147. Fussenegger, M., Bailey, J. E., Hauser, H., and Mueller, P. P. (1999) Genetic optimization of recombinant glycoprotein production by mammalian cells. *Trends Biotechnol.* **17**, 35–42.

148. Fussenegger, M., and Betenbaugh, M. J. (2002) Metabolic engineering II. Eukaryotic systems. *Biotechnol. Bioeng.* **79**, 509–531.

149. Tey, B. T., Singh, R. P., Piredda, L., Piacentini, M., and Al-Rubeai, M. (2000) Influence of Ccl2 on cell death during the cultivation of a Chinese hamster ovary cell line expressing a chimeric antibody. *Biotechnol. Bioeng.* **68**, 31–43.

150. Isahque, A. and Al-Rubeai, M. (2002) Role of vitamins in determining apoptosis and extent of suppression by bcl2 during hybridoma cell culture. *Apoptosis* **7**, 231–239.

151. Davis, B. G. (2002) Synthesis of glycoproteins. *Chem. Rev.* **102**, 579–601.

152. Prati, E. G. P., Matasci, M., Suter, T. B., Dinter, A., Sburlati, A. R., and Bailey, J. E. (2000) Engineering of coordinated up and down-regulation of two glycosyltransferases of the O-glycosylation pathway in Chinese hamster ovary (CHO) cells. *Biotechnol. Bioeng.* **68**, 239–244.

153. Elbashir, S. M., Harborth, J., Lendeckel, W., Yalcin, A., Weber, K., and Tuschi, T. (2001) Duplexes of 21-nucleotide RNAs mediate RNA interference in cultured mammalian cells. *Nature.* **411**, 494–498.

154. Fire, A. (1999) RNA-triggered gene silencing. *Trends in Genetics* **15**, 358–363.

155. Yu, J. Y., DeRuiter, S. L., and Turner, D. (2002) RNA interference by expression of short-interfering RNAs and hairpin RNAs in mammalian cells. *Proc. Natl. Acad. Sci. USA* **99**, 6047–6052.

156. Hamada, M., Ohtsuka, T., Kawaida, R., Koizumi, M., Morita, K., Furukawa, et al. (2002) Effects on RNA interference in gene expression (RNAi) in cultured mammalian cells of mismatches and the introduction of chemical modifications at the 3' ends of siRNAs. *Antisense Nucleic Acid Drug Dev.* **12**, 301–309.

157. Davis, R., Schooley, K., Rasmussen, B., Thomas, J., and Reddy, P. (2000) Effect of PDI overexpression on recombinant protein secretion in CHO cells. *Biotechnol. Prog.* **16**, 736–743.

158. Bao, W. G. and Fukuhara, H. (2001) Secretion of human proteins from yeast: simulation by duplication of polyubiquitin and protein disulfide isomerase genes in *Kluyveromyces lactis*. *Gene* **272**, 103–110.

159. Li, Q., Peterson, K.R., Fang, X., and Stamatoyannopoulos, G. (2002) Locus control regions. *Blood* **100**, 3077–3086.

160. Van Craenenbroeck, K., Vanhoenacker, P., and Haegeman, G. (2000) Episomal vectors for gene expression in mammalian cells. *Eur. J. Biochem.* **267**, 5665–5678.

3

Folding-Promoting Agents in Recombinant Protein Production

Beatrix Fahnert

Summary

Recombinant protein production has become an essential tool for providing the necessary amounts of a protein of interest to either research or therapy. The target proteins are not in every case soluble and/or correctly folded. That is why different production parameters, such as host, cultivation conditions, and co-expression of chaperones and foldases, are applied in order to gain functional recombinant proteins. Furthermore, the addition of folding-promoting agents during the cultivation is increasingly performed. The impact of all these strategies cannot be predicted and must be analyzed and optimized for the corresponding target protein. In this chapter recent cases of using folding-promoting agents in recombinant protein production are reviewed and discussed with respect to their in vivo applicability. Their effects in the cells are mostly not known in detail but at least partially comparable with the in vitro mode of action. The corresponding in vitro effects are also included in the chapter in order to facilitate a decision about their potential in vivo use.

Key Words: Recombinant protein; expression; folding-promoting agents; chemical chaperones; osmolytes, solutes

1. How Can the Folding of Recombinant Proteins Be Promoted?

This part of the chapter introduces the role and action of folding-promoting agents in cells in general and in recombinant protein production in particular. The second part gives a detailed overview of the use of the agents and approaches for screening their effects so readers new to the field can easily find some necessary information on how to get started. Experienced readers might appreciate the amount of data to choose from while assessing and planning new experiments.

From: *Methods in Molecular Biology, vol. 267:*
Recombinant Gene Expression: Reviews and Protocols, Second Edition
Edited by: P. Balbás and A. Lorence © Humana Press Inc., Totowa, NJ

1.1. General Considerations

In the last two decades recombinant protein production became a prerequisite in both basic and applied research by providing the adequate amounts of the proteins for studying protein structure and function. Moreover, scaled-up facilities produce recombinant vaccines, hormones, antibodies, growth factors, blood components, and enzymes. Despite these successful cornerstones, many of the target proteins are not soluble and/or correctly folded in the first attempt. Various strategies—including host, induction conditions, temperature, compartment, proteinaceous fusion partners, and co-expression of chaperones and foldases—are embarked on in order to gain functional recombinant protein (*1–3*). The addition of folding-promoting agents during cultivation is also performed more and more often for it not only enhances solubility but also has a beneficial effect on the folding. That approach is the subject of this chapter. The corresponding impact of all these strategies is not predictable and has to be analyzed and optimized for every target protein. Therefore sufficient information is needed about the protocols, the folding of the protein in question, the host cell, and all interactions.

Years of investigation have led to vast knowledge about the effects of different chemicals on protein folding in vitro, providing one basis for their consideration for an in vivo performance. Substances and protocols instrumental in in vitro refolding have been found worth testing in vivo with prokaryotes (e.g., *Escherichia coli* as recombinant host [*4,5*]) and eukaryotes (e.g., animal cells in the context of diseases [*6,7*]; yeast as recombinant host [*8*]).

Agents added to the medium affect the prokaryotic periplasm more efficiently than they affect the cytoplasm because of the high permeability of the outer membrane of *E. coli* for molecules smaller than 600 Da (*9*). The pH of the periplasmic space varies in response to the extracellular fluid as well (*10*). The gel-like consistency leads to slower diffusion (*11*) and concentration gradients can be used. So the periplasm is to be considered as a test tube of 50-nm width (about 30% of the cell volume) (*12*).

Folding-promoting agents have better access to recombinant proteins accumulated in the prokaryotic periplasm than in the cytoplasm. Substances entering the cell can accumulate to concentrations causing effects comparable to their in vitro ones. For example, reduced L-glutathione (GSH) (*13*) can enter the periplasm, whereas sucrose cannot (*3*). Glycerol can accumulate in the cell up to molar concentrations (*14*) and glycine betaine (GB) to 50 mM to 1 M (*15,16*). This also results in changes of the cellular stress gene response to recombinant protein overexpression (*17*). The stated outer membrane's permeability has caused the evolution of different mechanisms to protect the content against the environment (*16*). Folding-promoting agents are one such mechanism. Others are physiological changes in the cell (i.e., protein-protective stress responses, such as activation of molecular chaperones (*14*)). Such responses can be induced, (e.g., by sucrose (*18*) or ethanol (*19,20*)).

The more direct effect of the additives on the recombinant protein—and, of course, also on the host proteins—has to be seen in the physicochemical context of protein folding. Every change in the thermodynamic parameters of the environment of a protein potentially affects the protein's configuration since, in general, protein folding is dependent on both entropy and enthalpy. Inter- and intramolecular interactions between residues up to structural elements follow pathways to either the correct or the wrong folding state. Aggregation is a competing side reaction in this regard. This has to be prevented. In vivo cellular proteins support adopting the native state. Although there are differences between folding in the crowded cytosol of a host cell and the studied optimized in vitro folding, the mechanisms seem to be comparable. (*21*) Thus, the effects of low-molecular-weight additives can be specified, which is done in the following sections.

1.2. Media

Influencing the recombinant protein production starts with the culture medium (*see* Chapters 1, 2, and 24). Either complex media or defined mineral salt media can be used. Defined media enable high cell-density cultivations, but the recombinant product yield is often low. Growth on glycerol (*2*) or complex medium (*22–24*) can be advantageous for solubility and folding of the recombinant product. This benefit of complex media may be caused by an increased expression of foldases (*25*) or different osmotic pressure (*26*). T4-phage deoxycytidylate deaminase could be accumulated in a soluble active form to at least 20% of cellular protein using a rich growth medium, whereas inclusion bodies were found in less-rich media (*22*). Another reason for this can be the misincorporation of certain amino acids during growth in minimal media. By this, not only the product quality, but also the folding can be affected, possibly leading to aggregation. Muramatsu et al. reported incorporation of β-methylnorleucine instead of isoleucine in recombinant hirudin produced by *E. coli* in a medium not enriched with amino acids (*27*). On the other hand, the higher growth rates in complex media can lead to aggregation of the product.

1.3. Alcohols

Alcohols are alkyl groups substituted with a hydroxyl group. Of these mainly the short-chained ones are used and discussed here. Polyols are the subject of **Subheading 1.4.4**.

The addition of ethanol to the medium can be beneficial for the yield of functional product. It was reported to increase the amount of soluble, native DsbA-proinsulin in *E. coli* about threefold (*24*).

Ethanol in the medium leads to changes in the lipid and fatty acids composition of the cell membrane (*28*). Moreover, it inhibits the peptidoglycan assembly due to weakened hydrophobic interactions (*29*) being the main effect on protein structure at low alcohol concentrations in addition to hydrogen-bonding, thus causing a twist in the peptide backbone (*30*). By this mode of action the whole tertiary structure can be completely modified as reported for β-lactoglobulin A, because α-helix structure is increased followed by a transformation rich in beta-sheets (*31*). Organic solutes can both stabilize and destabilize proteins due to their favoring the role of the peptide backbone over that of side chains (*32*). Using predominantly organic solutions, the folding-promoting effect can be dramatically increased by adding salts such as lithium chloride or bromide and sodium acetate or perchlorate, because the solubility of a charged species should be higher in the presence of another charged or polar species in solution (*33*).

In vivo, the addition of alcohols leads, after incubation for 1 h to an induction of heat-shock protein (e.g., 70 and 60 kDa homologous to DnaK and GroEL of *E. coli*, respectively) synthesis in C6 rat glioma cells (*19*) and *Leuconostoc mesenteroides* (*20*). The critical step for this is the interaction of the alcohols with the cell's lipophilic compounds (e.g., membranes, lipophilic core regions of proteins) (*19*). Isopropanol increases the solubility of human lymphoma-derived antibody Fab fragments (*34*).

1.4. Compatible Solutes

The effects of compatible (not affecting the biological activity of proteins) solutes must be discussed with respect to their being osmoprotective toward the cells in the natural habitat. These substances (amino acids and their derivatives, sugars, polyols, quaternary amines and their sulfur analogs, sulfate esters, *N*-acetylated diamino acids, peptides) are highly water-soluble, uncharged at a neutral pH value, and can be accumulated in high amounts in the cell by either uptake or synthesis. For instance, betaine is imported via ProU and ProP. The expression of these transport proteins is increased by osmotic stress (*16*). The solutes are compatible with macromolecular structure and function, do not interact with substrates and cofactors, have favorable effects on macromolecular-solvent interactions, and stabilize proteins by perturbing unfolded states (*35, 36*). Thus, they can be applied as chemical chaperones.

1.4.1. Osmoprotectants and Osmotic Stress

Osmotic stress is a cultivation parameter exploited in order to optimize the import of folding-promoting agents in the cells (*16*). The combination has been successfully applied frequently. α1-Antitrypsin (*37*) was reported being protected against thermally induced polymerization and inactivation in vitro by sarcosine, GB, and trimethylamine *N*-oxide (TMAO), also stabilizing RNase S proteins (*36*).

A hyperosmotic culture medium, for instance, supports the import of exogenously supplied GB. GB is then amassed up to 1 *M* in stressed cells of *E. coli* (*4,15*). The cellular concentration of osmolytes depends on the organism, osmolyte availability in the medium, and the type, severity, and duration of the osmotic stress (*14*). Imported GB is also capable of activating a mutant diaminopimelate decarboxylase in vitro and in vivo. As this activation was correlated with a conformational change in the mutant, GB may have actively assisted in vivo protein folding in a chaperone-like manner (*4*). GB and choline protect citrate synthase comparable to DnaK (*15*). Intracellular levels of DnaK itself are increased after addition of NaCl (*18*). For creating an appropriate periplasmic microenvironment for the generation of high concentrations of correctly folded recombinant proteins (immunotoxins in this case), even inhibitory osmotic stress conditions (NaCl and sorbitol) can be used, since adding GB also allows the bacteria to grow (*38*). The sorbitol/GB combination is very effective. In rich medium not only inclusion bodies of a transferase disappeared, but the active yield was also increased by up to 427-

fold. In minimal medium it was only 44-fold, being at least partially due to the lower osmotic pressure compared to rich medium (*26*).

1.4.2. Amino Acids

The effect of adding amino acids has been studied both in vitro and in vivo. Different modes of action were found.

Less than 0.5 *M* of glycine and 1 *M* of proline improved the in vitro refolding yields of creatine kinase, but higher concentrations decreased the recovery. In contrast to glycine, proline was able to inhibit aggregation of creatine kinase (*39,40*). This is in accord with the findings by Kumar and coworkers reporting proline to inhibit effectively the aggregation of bovine carbonic anhydrase during in vitro refolding, contrarily to glycine (*41*). The same authors also showed proline preventing precipitation in case of hen egg-white lysozyme in vitro at even high concentrations (>4.0 *M*) due to an ordered supramolecular assembly (*42*). These higher-order aggregates might be crucial for proline to function as a protein-folding aid. Again this additive promotes folding (*43*) in addition to its role in osmoregulation (compatible osmolyte (*44*)). Nevertheless, glycine also decreases aggregation of certain proteins (e.g., ribonuclease-A) (*45*). Twenty different amino acids were added to *E. coli* cultures and analyzed with respect to their in vivo effects on a periplasmic recombinant cytochrome b5. Glycine, and to a lesser extent histidine, doubled the synthesis of the protein and its discharge into the medium (*46*).

The labilizing agent L-arginine (*21,47*) increases the yield of native product (proinsulin fused to the C-terminus of DsbA (*24*), native tissue-type plasminogen activator variant, single-chain antibody fragment (*5*)) because of its solubilizing effect (*21,47*). Furthermore, it prevents the death of animal cell cultures producing recombinant protein or compensates the arginine consumption needed for recombinant protein production (*48*). The addition of single amino acids can both cope with limits and directly affect the proteins.

1.4.3. Sugars

Sugars are not only supplemented in order to establish osmotic stress conditions having an indirect effect on the recombinant protein. Some affect proteins directly as well.

β-Lactamase aggregation is inhibited by growing *E. coli* in the presence of nonmetabolizable sugars (e.g., sucrose) due not to osmotic but to direct effects on the in vivo pathway of folding, analogous to the well-analyzed effect of sugars in vitro (*49,50*). Depending on the concentration, sucrose inhibits an increase in protein surface area exposure by affecting the hydrogen exchange rates of the amide protons with intermediate rates (*51*). In vivo, some stress-related sigma32- and sigmaE-dependent promoters were rapidly but transiently induced in midexponential-phase *E. coli* cells treated with sucrose. For instance, *dnaK* and *ibp* promoters were transiently induced 15 minutes after adding sucrose, or somewhat later using PEG. This actual emergency response required to repair protein misfolding and to facilitate the proper folding of proteins synthesized following loss of turgor can be exploited during the production of recombinant proteins (*18*).

The disaccharide trehalose stabilizes the native state of proteins in vivo in yeast cells during heat shock comparable to sucrose and maltose and suppresses the aggregation of denatured proteins, maintaining them in a partially folded state from which they can be activated by molecular chaperones. In vitro it protects membranes in addition to proteins both in solution and in a dry environment (*8,52*).

Xylose has been proven to significantly increase the rate of folding and unfolding of staphylococcal nuclease. It stabilizes the folded state of proteins through surface-tension effects (*53*).

1.4.4. Polyols

Comparable to sugars, polyols act on different sites during the production of the target protein. Glycerol and sorbitol are the most frequently used polyols.

The presence of glycerol in the culture medium improved the yield of wild-type and mutant forms of human recombinant phenylalanine hydroxylase and increased the specific activity of the purified enzymes (*54*). Nevertheless, glycerol is not only an additive but also a substrate. Thus, a possible decrease in the concentration must be considered. It is also applicable in eukaryotes since glycerol enhances the level and posttranslational stability of human P-glycoprotein expressed in yeast (*55*). Glycerol was reported to restore the mutant TP53-activated factor 1 to the wild-type conformation in human cell lines as well (*56*). In addition to the application in recombinant production, there is also a pharmaceutical interest. Antigen presentation is enhanced by glycerol (*57*). Even a mouse model for determining maximum serum glycerol concentrations has been established in order to test the in vivo efficacy of chemical chaperones for therapeutic use (*58*). Glycerol is also used as a cryoprotectant and for stabilizing proteins in solution. In vitro the glycerol effect was studied in detail. Glycerol increases the α-helix structure of horse heart cytochrome C and induces slight changes at the active-site environment without significantly perturbing the tertiary structure or changing the redox potential (*59*). It promotes renaturation of denatured citrate synthase (*60*).

Polyols stabilizing yeast iso-1-ferricytochrome C were proven to be primarily caused by nonspecific steric repulsions (excluded volume effects) and not by binding interactions (*61*). Polyols also accelerate protein folding in general and disrupt hydrogen-bridged water bridges due to their ability to compete with water (*62*). Furthermore, polyhydric compounds oppose the effect of denaturants on water structure (*63*). Due to the presence of D-sorbitol, soluble expression of the N gene of Chandipura virus could be achieved in *E. coli*, otherwise leading to large insoluble aggregates both in COS cells and in bacteria. The aggregation was also significantly reduced in vitro (*64*).

1.5. Counteracting Solutes

Contrary to compatible osmolytes (protective only against extremes of temperature, dehydration, high salt), counteracting osmolytes alter the biological activity of proteins. The stabilizing or destabilizing ability of organic solutes depends on their effect on the peptide backbone compared to that of the side chains.

TMAO is one example. Its advantage is the stabilization of both hydrophobic and hydrophilic proteins. Thus, its interaction with the backbone is of concern here, and not the interaction with the side chains (*32*). By this mode of action, TMAO has the general ability to counteract the denaturing effects of urea on protein structure and function (*65*), because urea is unfolding proteins because of its favorable interaction with the backbone. This is then opposed by TMAO's unfavorable interaction with the backbone (*32*).

Tests with eukaryotic cell cultures have been performed, because there is also therapeutic interest in TMAO. Exposure of scrapie-infected mouse neuroblastoma cells to TMAO (also to glycerol and DMSO) reduced the rate and extent of the formation of the pathogenic isoform of the cellular prion protein. The chemical chaperone interfered with the conversion of the alpha helices into β-sheets in the newly synthesized proteins (*7*). Deficient mitochondrial branched-chain α-ketoacid dehydrogenase complex causes maple syrup urine disease. Adding TMAO restored up to 50% of the wild-type activity (*66*).

TMAO should also be applicable for *E. coli* since this organism can cope with TMAO being a substrate for anaerobic respiration. Thus, a certain tolerance can be expected.

1.6. Reducing and Oxidizing Agents

Disulfide-bridge isomerization is insufficient in the periplasm because DsbA is highly oxidizing and the isomerases cannot correct bridging sufficiently. Using a glutathione redox buffer in the medium allows a modified formation and breakage of disulfide bridges in the periplasm of the bacterial host (*67*). Varying the redox potential of the medium establishes different redox states within the periplasm. Thus, comparable to already known in vitro effects (*47*), the addition of redox components allows reshuffling of wrongly formed disulfide bridges in vivo in the periplasm. This does not affect the prokaryotic cytoplasm. GSH and oxidised L-glutathione (GSSG) are contained there in a ratio of 50 to 200:1 (*68*). In redox mutants only this ratio is altered, influencing protein folding, including recombinant ones, by disulfide bridge formation.

Wunderlich and Glockshuber (*13*) reported a fivefold increase in correctly folded α-amylase/trypsin inhibitor from Ragi after adding GSH/GSSG to the medium. This effect was even enhanced by co-expressing DsbA, which catalyzes disulfide bridge formation. This approach was confirmed successfully by others (*69*). GSH and other agents (acetamide, ethylurea) added to the culture medium increased the yield of a native tissue-type plasminogen activator variant and a single-chain antibody fragment up to 10- and 37-fold, respectively (*5*). In the same work it was also shown that the concentration of those supplements remained constant during cultivation. The ratio of GSH/GSSG changed only after 20 h. Effects of the tested substances are comparable to those in vitro and are due to folding-enhancing activities rather than secondary osmolyte effects. Nevertheless, GSH/GSSG can also negatively affect the production of correctly folded target proteins (*24*).

Beneficial effects have also been reported for eukaryotic systems. Oculopharyngeal muscular dystrophy is caused by the expansion of an alanine stretch in the intranuclear poly(A)-binding protein 2, leading to aggregations. The added chemical chaperone DMSO reduced aggregation of this protein and cell death (*6*).

1.7. Metal Ions

The addition of metal ions is beneficial because of their importance as cofactors for both the recombinant and the host proteins. The cofactors can interact specifically with the unfolded polypeptide. This is sometimes essential for the folding and thus dramatically accelerating formation of the functional protein. Cofactors stabilize the native protein as well (*70*).

Cu^{2+} was reported to induce and enhance the formation of an α-helix conformation (*71*). The amount of an active, shortened form of the alkaline phosphatase was increased by addition of magnesium to the medium. Zinc was not supportive in this case (*72*) but can inhibit proteases in the periplasm (*73*). Calcium was also shown to be beneficial (*23*).

1.8. Artificial Chaperones

Positive effects of artificial chaperones on protein folding have been found both in vitro and in vivo.

Cycloamylose supported the complete recovering of the enzymatic activity of chemically denatured citrate synthase, carbonic anhydrase B, and a reduced form of lysozyme. This is due to its accommodation of detergents, preventing aggregation (*74*). The artificial chaperone system cetyltrimethylammonium bromide and beta-cyclodextrin enhanced lysozyme renaturation in vitro as well (*75*). There is also a synthetic dithiol, vectrase P [(+/−)-*trans*-1,2-*bis*(2-mercaptoacetamido)cyclohexane]. It was successfully used as a folding promoting agent both in vivo and in vitro, showing a partial isomerase-like function (*76,77*). The agent can enter the eukaryotic endoplasmic reticulum and the prokaryotic periplasm. In vitro vectrase P catalyzed the activation of denatured bovine ribonuclease A (*76*) and doubled the yield of native proinsulin during refolding (*77*). When 0.4 to 1.5 m*M* of this dithiol was added to the growth medium, the heterologous secretion of *Schizosaccharomyces pombe* acid phosphatase from *Saccharomyces cerevisiae* increased threefold (*76*), whereas 2 to 50 μ*M* added at induction time increased the formation of native proinsulin in the *E. coli* periplasm by about 60% (*77*).

2. Experimental Approaches to Promoting the Folding of Recombinant Proteins

2.1. General Considerations

Many different protocols and variations thereof have been published during recent years. By reviewing them it becomes clear quite soon that as every target protein is different, one cannot predict the effect of a certain approach. Most of the time one cannot even speculate. Thus, in the majority of cases it is necessary to adapt the whole protocol to the target protein. The procedure has to be tested either randomly or according to a certain rationale (if the protein's parameters are known or there are in vitro findings).

Methods for faster screening of the additive's benefit have also been developed and established (*see* **Subheading 2.2.**).

2.1.1. General Cultivation Strategies

The cultivation regime should be planned depending on the most likely conducive medium, temperature, times of adaptation to the supplement, and production time (*see* Chapter 2). These parameters depend not only on the target protein but also on the context and purpose of the production. Solubility, origin and monitoring of additives, the medium compounds, and even economy have to be considered with respect to the production scale.

There are four main approaches:

1. Both precultures and main culture with additives at inoculation (e.g., *26*).
2. Precultures without any additives, main culture with additives at inoculation (e.g., *78*).
3. Precultures without any additives, main culture with additives after inoculation, induction of recombinant protein production after a corresponding adaptation time (e.g., *79*).
4. Precultures without any additives, main culture with additives after inoculation, induction of recombinant protein production without any adaptation time (e.g., *80*).

In **Fig. 1** a scheme of successful cultivation regimes cited in this review is depicted. Overviews of used folding-promoting agents (**Table 1**) and combinations thereof (**Table 2**) with the corresponding concentrations might support the choice of the strategy. The supplements can be used in rich media (e.g., Terrific Broth [TB] [*38*]) or minimal media (e.g., M9 [*26*]). In some cases both were tested (e.g., LB and M63 [*4*]).

The cultivation temperature must be chosen according to the protein's aggregation tendency and economical growth and production rates. Lower temperatures often promote solubility (*2*); thus, the temperature should be as low as possible. In order to find a compromise in some cases the temperature is lowered at certain points, such as at the beginning of the second preculture (*78*) or at induction (*26*). Nevertheless it might be more favorable not to change the temperature during the cultivation because the host cells are not additionally stressed then. Chosen temperatures in the references cited in this review are 22°C (*15,77*), 24°C (*5,78*), 25°C (*26*), 26°C (*80*), and 30°C (*18,81*). Thomas and Baneyx (*81*) found 30°C to be more beneficial than 25, 20, 15, or 37°C. However, this is not a common standard, because, in general, lower cultivation temperatures favor solubility.

2.1.2. Aspects to Be Considered

There are some limits in the use of certain substances of which the experimenter should be aware because the growth and viability of the host cells are concerned.

More than 0.4 M of L-arginine leads to growth inhibition (*24*). Organic solvents can cause cell lysis depending on their chain length and substitution (*28,82*). A glycine concentration of 1% reduced bacterial growth to 50% (*46*). Trehalose might be difficult to handle as a supplement, because it is mostly degraded in the periplasm as a carbon source. Only endogenously synthesized trehalose as a response to a stressful environment can have a protective function on the recombinant protein (*16*). That is why

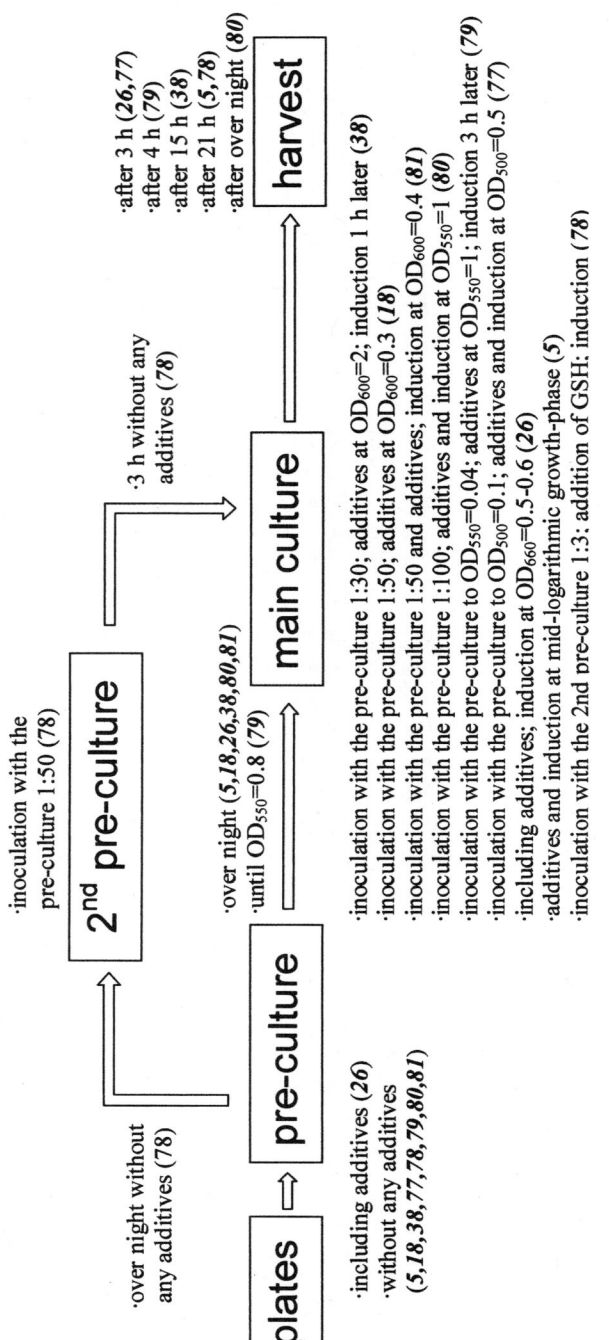

Fig. 1. Overview of successful cultivation regimes cited in this review.

plates ⇨ **pre-culture**

·including additives (*26*)
·without any additives
(*5,18,38,77,78,79,80,81*)

·over night without
any additives (*78*)

2ⁿᵈ pre-culture

·inoculation with the
pre-culture 1:50 (*78*)

·over night (*5,18,26,38,80,81*)
·until OD$_{550}$=0.8 (*79*)

·3 h without any
additives (*78*)

main culture

·inoculation with the pre-culture 1:30; additives at OD$_{600}$=2; induction 1 h later (*38*)
·inoculation with the pre-culture 1:50; additives at OD$_{600}$=0.3 (*18*)
·inoculation with the pre-culture 1:50 and additives; induction at OD$_{600}$=0.4 (*81*)
·inoculation with the pre-culture 1:100; additives and induction at OD$_{550}$=1 (*80*)
·inoculation with the pre-culture to OD$_{550}$=0.04; additives at OD$_{550}$=1; induction 3 h later (*79*)
·inoculation with the pre-culture to OD$_{500}$=0.1; additives and induction at OD$_{500}$=0.5 (*77*)
·including additives; induction at OD$_{660}$=0.5-0.6 (*26*)
·additives and induction at mid-logarithmic growth-phase (*5*)
·inoculation with the 2nd pre-culture 1:3; addition of GSH: induction (*78*)

harvest

·after 3 h (*26,77*)
·after 4 h (*79*)
·after 15 h (*38*)
·after 21 h (*5,78*)
·after over night (*80*)

Table 1
Overview of Folding-Promoting Agents With Their Corresponding Concentrations for In Vivo Approaches as Stated in the References Cited in This Review

Folding-Promoting agent	Concentration	References
Acetamide	1 M	*5*
Dimethylsulfonioacetate	1 mM	*4*
Dimethylsulfoniopropionate	1 mM	*4*
Ectoine	1 mM	*4*
Ethanol	3%	*81*
Ethylurea	0.6 M	*5*
Formamide	1 M	*5*
GB	1 mM	*4*
	2.5 mM	*26*
	10 mM	*38*
Glycerol	10%	*55*
	0.6–1.2 M	*56*
Glycine	1%	*46*
GSH	5mM	*5,78,80*
Hydroxyectoine	10 mM	*38*
Isopropanol	5%	*34*
L-arginine	0.4 M	*5,24*
	1.6–4.2 g/L	*48*
Methylformamide	0.6M	*5*
Methylurea	0.6 M	5
NaCl	4%	*38*
	300 mM	*15*
	0.5 M	*4*
	0.6 M	*18*
PEG	0.464 M	*18*
Potassium glutamate	0.5 M	*87*
Proline	1 mM	*4*
Raffinose	0.3–0.7 M	*80*
Sorbitol	0.5 M	*38*
	1 M	*26*
Sorbose	0.3–0.7 M	*80*
Sucrose	0.9 M	*4*
	2%	*26*
	0.464 M	*18*
	0.3–0.7 M	*80*
TMAO	1 M	*66*
Vectrase P	0.4–1.5 mM	*76*
	2–50 μM	*77*
Zinc	0.5 mM	*38*

Table 2
Overview of Combinations
of Folding Promoting Agents With Their Corresponding Concentrations
for In Vivo Approaches as Stated in the References Cited in This Review

Combination of folding-promoting agents	References
Combination 1	
0.5 *M* NaCl	
(or 0.5 *M* KCl, 0.9 *M* sucrose)	
1 m*M* GB	
(or ectoine, proline, dimethylsulfonioacetate,	
dimethylsulfoniopropionate)	*4*
Combination 2	
4% NaCl	
0.5 *M* sorbitol	
10 m*M* GB	
(or hydroxyectoine)	*38*
Combination 3 (redox system of reducing	
and oxidizing thiols 5:1 to 10:1)	
GSH	
cystein	
N-acetylcystein	
cysteamin	
β-mercaptoethanol	
0.3 *M*–0.7 *M* saccharides	
(sorbose, sucrose, raffinose)	*80*

trehalose is degraded in the cells after stress to not inhibit reactivation of denatured proteins (*8*).

Thus, the applied concentration should be found in the range between starting effectiveness and inhibiting the cell. Adaptation times are needed either for accumulation in the host or inducing a physiological answer of the host. Sometimes adaptation is not performed. This might not be worthwhile. In case of ethanol supplementation, less than 3% had no impact at all and more caused a growth inhibition. Ethanol added at the induction time or 30 min before induction was less effective than when added at inoculation of the main culture in this experiment (*81*).

Another important ratio to be considered is the appropriate one of reducing and oxidizing thiols. Most efficient are 3 to 12 m*M* of one or various thiol reagents. Less than 0.1 m*M* of it is useless and more than 20 m*M* decrease the growth rate and the yield of the target protein. Both the application of a single reducing thiol reagent and combinations of reducing and oxidizing ones (ratio 5:1 to 10:1) can improve the yield of correctly disulfide-bridged proteins (*80*).

Concentrations of more than 1 m*M* vectrase P resulted in an adverse effect on the target protein (*77*).

2.1.3. Strategies Including Foldases

Foldases assist in folding of newly synthesized proteins and prevent misfolding and aggregation of proteins in the cells (reviewed in *83*). They are frequently applied in recombinant protein production (e.g., co-expression of foldases (reviewed in *84*)). The aspect discussed here is regarding the induction of foldases in the host cells belonging to the physiological changes that can be caused by the addition of folding-promoting agents. The use of folding-promoting agents in combination with co-expressed foldases will be considered shortly.

Diamant et al. (*14*) reported low physiological concentrations of proline, glycerol, and especially GB to activate GroEL, DnaK, and ClpB. The increased yield of native target protein might then be likely due to promotion of local refolding within the chaperone and stabilization of the product. Osmolytes also have an impact on the chaperone-substrate interaction. Release and reactivation of proteins from the GroEL–GS complex in the cell is due to ATP binding, but the addition of glycerol (and other polyols such as sucrose, 1,2-propanediol, or 1,3-propanediol, but not GB, sarcosine, or high salt) had the same result even in the absence of the nucleotide. As dextran or Ficoll failed to reactivate a GroEL–GS-bound model protein, the effect cannot be attributed to viscosity or molecular crowding. Thus glycerol may alter the chaperonin structure similar to ATP (*62*).

Osmolyte concentrations that are too high, on the other hand can have negative effects, especially in the case of trehalose-inhibited DnaK-dependent chaperone networks, as high viscosity affects dynamic chaperone-substrate interactions and even stabilizes protein aggregates. Different osmolytes also have different effects on chaperones but the protection mechanism against the aggregation was similar for all analyzed osmolytes. The nature of the osmolyte determines the partitioning and commitment of the unfolded species to the proper refolding or improper misfolding pathways. Furthermore the osmolyte's viscosity is not always correlated with the protection (e.g., not against urea, but against heat) (*14*).

In order to increase the positive impact, some authors additionally coexpressed certain foldases (e.g., PDI [*80*], DnaK and J [*81*]). Coexpression and cosecretion of PDI combined with simultaneous addition of GSH significantly increased the yield compared with either GSH or protein disulfide isomerase (PDI) alone (*80*). Also, DnaJ plus the additives caused a dramatic increase, which was not found with DnaJ alone (*5*). This beneficial effect again might depend on the nature of the target protein, because it was also found that osmoprotectants accumulated in the cell lead to a low production of chaperones. Thus, the protection of destabilized proteins is mainly ensured by osmoprotectants (*4,85*).

2.1.4. Recommended Basic Approach

As stated in **Subheading 2.1.1**, the cultivation regime depends on the requirements of the protein. Every parameter has to be optimized. So far this is mostly a time-consuming linear approach involving shaking flasks. Moreover, already optimized parameters can require reoptimization later due to interrelations. This is depicted as a recommended basic approach in **Fig. 2** with respect to medium, temperature, inducer, additives, and timing. Additionally, the host strain, occurring proteolysis, foldases, and codon usage might have to be optimized in the same manner. It is impossible to realize a weighting in the process. Thus, all the approaches are suboptimal successive approximations. An alternative is the use of the powerful statistic technique "design of experiment" (DOE). This methodology is implemented in modeling toolboxes. Even though this shows progress, the next demanding task is the efficient analysis of the additive's effect. This is a problem in most cases. Recent strategies are discussed in the following section.

2.2. Screening the Impact of Additives on the Folding of Recombinant Proteins

The benefit of folding-promoting agents added to the culture medium is as unpredictable as the general success of other approaches in producing the recombinant protein and depends on the interactions of the target protein and the host as well.

Schaeffner et al. (*5*) achieved different results for three model proteins (native tissue-type plasminogen activator variant, a single-chain antibody fragment, proinsulin) produced in *E. coli*. They applied both culture additives and cosecreted molecular chaperones. Not only was the success unpredictable, but some strategies even led to negative effects. For example, although ethanol increased the yield of S2-S'-β-galactosidase, in the same publication Thomas and Baneyx reported that ethanol promotes the aggregation of human secreted protein acidic and rich in cysteine (SPARC). Both proteins are similarly dependent on chaperones (*81*). Thus the beneficial effect of ethanol depends on more than the folding pathway.

Finding an effective culture supplement and the corresponding concentration and adaptation time is, in most cases, a time-consuming process. That is why fast approaches are of outstanding importance. Both in vivo and in vitro methods for screening the influences of osmolytes on the folding of the target protein in helping to find the optimal conditions have been developed.

One strategy was published by Voziyan and coworkers. There the effects of a broad array of folding promoting agents (e.g., osmolytes, detergents, gradients of ionic strength, and pH) can be conveniently analyzed in vitro by using a stable complex of the chaperone GroEL and the misfolded target protein (protein folding intermediate, substrate). Aggregations are avoided by means of this. As protein conformations are stabilized in general, proper folding reactions are enhanced in the presence of osmolytes (in some cases, nevertheless, the initial aggregation reaction can also be accelerated, depending on the protein and the osmolyte). Using this method, the superior folding conditions necessary for release of a truncation mutant of bacterial glutamine synthetase were rapidly identified. Only glycerol and sucrose helped with folding in this case (*62,86,87*).

Another in vitro screening method is also based on a cell-free system. A polypeptide having a defective three-dimensional conformation is contacted with a test agent in order to identify the one correcting the conformation. Determining the efficiency of the test agent is conducted by detecting a change in fluorescence intensity of a combination

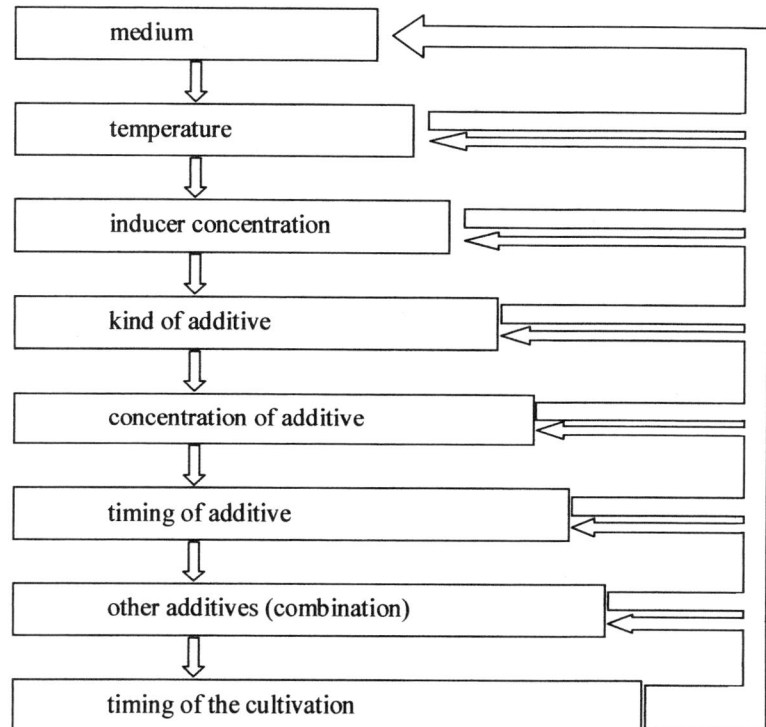

Fig. 2. Recommended basic approaches for cultivation with additives. The phases of optimization are interrelated and can require reoptimization.

of the peptide and a fluorescent compound to more closely approximate that of a corresponding wild-type peptide. Analyzing changes in the nuclear magnetic resonance spectrum or the circular dichroism, as well as tracing the specific binding of an antibody to the peptide, are other means of assessing the test agent (*88*).

The impact of additives depends not only on the folding pathway (*81*) in case of certain parameters of the target protein, so in vitro screening is not sufficient. The wrong chaperones might be induced or not enough or additional cofactors are needed, for instance. Thus the specific needs of the protein in question should be investigated. A convenient and time-saving in vivo approach can provide some useful information on this. In case of a soluble but incorrectly folded fusion of maltose-binding protein and human bone morphogenetic protein 2 produced in *E. coli*, such a strategy was undertaken (*85*). Thirteen different systems of either a single folding-promoting agent or combinations were tested in vivo (cultures) with respect to their beneficial effect. In order to screen for correct folding of the target protein, a fast and efficient test was established. After cleaving the fusion partners, native bone morphogenetic protein 2 bound to a nitrocellulose membrane can be detected colorimetrically, exploiting its binding to the biotinylated ecto-domain of its receptor (Alk3) (*85*). The signal is specific since two of the receptor molecules can bind only to native bone morphogenetic protein 2 dimer by contacting both monomers (*89,90*). Useful information concerning the host cells was gained by means of gene expression data (*91*). Producing the fusion protein without any additives led to an increased expression of *groEL*, ES, *dnaJ*, *dnaK*, *rpoE*, and *degP*, proving the stress. As the 13 applied strategies were not beneficial to bone morphogenetic protein 2 folding, the impact on the host was also analyzed for some of them. It was found that due to the addition of GB under osmotic stress, less proteases and *rpoE* were expressed, clearly indicating the supportive effect in the periplasm. Increased expression of proteases and *dnaK, J* was seen to be caused by the addition of GSH/GSSG (redox system) in combination with salt stress. In the case of bone morphogenetic protein 2, a detailed analysis implied that its very hydrophobic surface was the main problem. Thus the additives could not succeed. The general applicability of the entire culture-additives/protein/host analysis approach was proven using a model system (incorrectly folded recombinant proteins in the periplasm of *E. coli* during a fermentation process). There the aggregation preventing and folding promoting influence of sorbitol and GB were analyzed concerning the transcription of cellular chaperones, proteases, and stress response elements. The expression of periplasmic factors and cytoplasmic *groEL,ES* was not changed after the addition. (*85*). This is in accordance with the finding of Bourot et al. (*4*) that accumulated osmoprotectants in the cell cause a low production of chaperones. Hence the addition of sorbitol and GB may stabilize the newly synthesized recombinant protein during its folding in such a way that the natural capacity of GroEL, ES within the cytoplasm and of the foldases within the periplasm is sufficient. On the other hand the expression of some unfolding proteins of the cytoplasm (DnaK, ClpA, ClpX) was increased. This might be due to the fact that the recombinant protein aggregated preliminarily to the addition of sorbitol and GB is thereafter degraded by DnaK and the Clps (*85*).

3. Concluding Remarks

As apparent in all these examples, the benefit of adding folding-promoting agents to insoluble or non-native target proteins during cultivation cannot be predicted. Despite many successful cases, some approaches are not working or may even be disadvantageous. Due to the importance of recombinant protein production, it is worth it to optimize the conditions for the corresponding target protein. For this purpose substances and protocols already proven instrumental in both in vitro refolding and in vivo approaches ought to be considered. However the parameters have different priorities depending on the production scale (laboratory, pilot, industry). This chapter was intended as guidance through the plethora of recent applications of folding-promoting agents in recombinant protein production. The provided and discussed overview of strategies should help to find a suitable one. Traditional shaking-flask approaches so far allow only a linear optimization without considering interrelations directly. A multidimensional "design of experiment" procedure would be more efficient. Furthermore, high-throughput technology, including automation and miniaturization, might contribute to more economic strategies in the future. Recombinant protein expression and purification have already been subjected to high-throughput approaches (*92*). Promising folding-promoting agents can be an input afterward for either an in vitro or in vivo screening and the subsequent in vivo optimizing. As an intermediate stage one might also use in vitro translation methods.

Acknowledgments

The author thanks Prof. P. Neubauer for critical reading of the manuscript. This research has been supported by a Marie Curie Fellowship of the European Community program "Quality of Life Individual Fellowships of the Fifth Framework Programme" under contract number QLK3-CT-2001-51066.

References

1. Baneyx, F. (1999) Recombinant protein expression in *E. coli. Curr. Opin. Biotechnol.* **10**, 411–421.
2. Kopetzki, E., Schumacher, G., and Buckel, P. (1989) Control of formation of active soluble or inactive insoluble baker's yeast alpha-glucosidase PI in *Escherichia coli* by induction and growth conditions. *Mol. Gen. Genet.* **216**, 149–155.
3. Georgiou, G. and Valax, P. (1996) Expression of correctly folded proteins in *E. coli. Curr. Opin. Biotechnol.* **7**, 190–197.
4. Bourot, S., Sire, O., Trautwetter, A., Touze, T., Wu, L. F., Blanco, C., et al. (2000) Glycine betaine-assisted protein folding in a *lysA* mutant of *Escherichia coli. J. Biol. Chem.* **275**, 1050–1056.
5. Schaeffner, J., Winter, J., Rudolph, R., and Schwarz, E. (2001) Cosecretion of chaperones and low-molecular-size medium additives increases the yield of recombinant disulfide-bridged proteins. *Appl. Environ. Microbiol.* **67**, 3994–4000.
6. Bao, Y. P., Cook, L. J., O'Donovan, D., Uyama, E., and Rubinsztein, D. C. (2002) Mammalian, yeast, bacterial, and chemical chaperones reduce aggregate formation and death in a cell model of oculopharyngeal muscular dystrophy. *Biol. Chem.* **277**, 12263–12269.
7. Tatzelt, J., Prusiner, S. B., and Welch, W. J. (1996) Chemical chaperones interfere with the formation of scrapie prion protein. *EMBO J.* **15**, 6363–6373.
8. Singer, M. A. and Lindquist, S. (1998) Multiple effects of trehalose on protein folding *in vitro* and *in vivo. Mol. Cell.* **1**, 639–648.
9. Rosenbusch, J. P. (1990) Structural and functional properties of porin channels in *E. coli* outer membranes. *Experientia.* **46**, 167–173.
10. Ostermeier, M. and Georgiou, G. (1994) The folding of bovine pancreatic trypsin inhibitor in the *Escherichia coli* periplasm. *J. Biol. Chem.* **269**, 21072–21077.
11. Brass, J. M., Higgins, C. F., Foley M, Rugman, P. A., Birmingham, J., and Garland, P. B. (1986) Lateral diffusion of proteins in the periplasm of *Escherichia coli. J. Bacteriol.* **165**, 787–795.
12. Van Wielink, J. E. and Duine, J. A. (1990) How big is the periplasmic space? *Trends Biochem. Sci.* **15**, 136–137.
13. Wunderlich, M. and Glockshuber, R. (1993) *In vivo* control of redox potential during protein folding catalyzed by bacterial protein disulfide-isomerase (DsbA). *J. Biol. Chem.* **268**, 24547–24550.

14. Diamant, S., Eliahu, N., Rosenthal, D., and Goloubinoff, P. (2001) Chemical chaperones regulate molecular chaperones *in vitro* and in cells under combined salt and heat stresses. *J. Biol. Chem.* **276**, 39586–39591.
15. Caldas, T., Demont-Caulet, N., Ghazi, A., and Richarme, G. (1999) Thermoprotection by glycine betaine and choline. *Microbiology.* **145**, 2543–2548.
16. Kempf, B. and Bremer, E. (1998) Uptake and synthesis of compatible solutes as microbial stress responses to high-osmolality environments. *Arch. Microbiol.* **170**, 319–330.
17. Gill, R. T., DeLisa, M. P., Valdes, J. J., and Bentley, W. E. (2001) Genomic analysis of high-cell-density recombinant *Escherichia coli* fermentation and "cell conditioning" for improved recombinant protein yield. *Biotechnol. Bioeng.* **72**, 85–95.
18. Bianchi, A. A. and Baneyx, F. (1999) Hyperosmotic shock induces the sigma32 and sigmaE stress regulons of *Escherichia coli*. *Mol. Microbiol.* **34**, 1029–1038.
19. Neuhaus-Steinmetz, U. and Rensing, L. (1997) Heat shock protein induction by certain chemical stressors is correlated with their cytotoxicity, lipophilicity and protein-denaturing capacity. *Toxicology.* **123**, 185–195.
20. Salotra, P., Singh, D. K., Seal, K. P., Krishna, N., Jaffe, H., and Bhatnagar, R. (1995) Expression of DnaK and GroEL homologs in *Leuconostoc mesenteroides* in response to heat shock, cold shock or chemical stress. *FEMS Microbiol. Lett.* **131**, 57–62.
21. Jaenicke, R. (1998) Protein self-organization *in vitro* and *in vivo*: partitioning between physical biochemistry and cell biology. *Biol. Chem.* **379**, 237–243.
22. Moore, J. T., Uppal, A., Maley, F., and Maley, G. F. (1993) Overcoming inclusion body formation in a high-level expression system. *Protein Expr. Purif.* **4**, 160–163.
23. Kurokawa, Y., Yanagi, H., and Yura, T. (2000) Overexpression of protein disulfide isomerase DsbC stabilizes multiple-disulfide-bonded recombinant protein produced and transported to the periplasm in *Escherichia coli*. *Appl. Environ. Microbiol.* **66**, 3960–3965.
24. Winter, J., Neubauer, P., Glockshuber, R., and Rudolph, R. (2000) Increased production of human proinsulin in the periplasmic space of *Escherichia coli* by fusion to DsbA. *J. Biotechnol.* **84**, 175–185.
25. Wei, Y., Lee, J. M., Richmond, C., Blattner, F. R., Rafalski, J. A., and LaRossa, R. A. (2001) High-density microarray-mediated gene expression profiling of *Escherichia coli*. *J. Bacteriol.* **183**, 545–556.
26. Blackwell, J. R. and Horgan, R. (1991) A novel strategy for production of a highly expressed recombinant protein in an active form. *FEBS Lett.* **295**, 10–12.
27. Muramatsu, R., Negishi, T., Mimoto, T., Miura, A., Misawa, S., and Hayashi, H. (2002) Existence of beta-methylnorleucine in recombinant hirudin produced by *Escherichia coli*. *J. Biotechnol.* **93**, 131–142.
28. Ingram, L. O., Dickens, B. F., and Buttke, T. M. (1980) Reversible effects of ethanol on *E. coli*. *Adv. Exp. Med. Biol.* **126**, 299–337.
29. Ingram, L. O. (1981) Mechanism of lysis of *Escherichia coli* by ethanol and other chaotropic agents. *J. Bacteriol.* **146**, 331–336.
30. Dwyer, D. S. (1999) Molecular simulation of the effects of alcohols on peptide structure. *Biopolymers.* **49**, 635–645.
31. Barteri, M., Gaudiano, M. C., Mei, G., and Rosato, N. (1998) New stable folding of beta-lactoglobulin induced by 2-propanol. *Biochim. Biophys. Acta.* **1383**, 317–326.
32. Wang, A. and Bolen, D. W. (1997) A naturally occurring protective system in urea-rich cells: mechanism of osmolyte protection of proteins against urea denaturation. *Biochemistry* **36**, 9101–9108.

33. Rariy, R. V. and Klibanov, A. M. (1999) Protein refolding in predominantly organic media markedly enhanced by common salts. *Biotechnol. Bioeng.* **62**, 704–710.
34. Kuivila, R. (2002) University of Oulu (Finland), personal communication.
35. Yancey, P. H., Clark, M. E., Hand, S. C., Bowlus, R. D., and Somero, G. N. (1982) Living with water stress: evolution of osmolyte systems. *Science.* **217**, 1214–1222.
36. Ratnaparkhi, G. S. and Varadarajan, R. (2001) Osmolytes stabilize ribonuclease S by stabilizing its fragments S protein and S peptide to compact folding-competent states. *J. Biol. Chem.* **276**, 28789–28798.
37. Chow, M. K., Devlin, G. L., and Bottomley, S. P. (2001) Osmolytes as modulators of conformational changes in serpins. *Biol. Chem.* **382**, 1593–1599.
38. Barth, S., Huhn, M., Matthey, B., Klimka, A., Galinski, E. A., and Engert, A. (2000) Compatible-solute-supported periplasmic expression of functional recombinant proteins under stress conditions. *Appl. Environ. Microbiol.* **66**, 1572–1579.
39. Ou, W. B., Park, Y. D., and Zhou, H. M. (2002) Effect of osmolytes as folding aids on creatine kinase refolding pathway. *Int. J. Biochem. Cell. Biol.* **34**, 136–147.
40. Meng, F., Park, Y., and Zhou, H. (2001) Role of proline, glycerol, and heparin as protein folding aids during refolding of rabbit muscle creatine kinase. *Int. J. Biochem. Cell. Biol.* **33**, 701–709.
41. Kumar, T. K., Samuel, D., Jayaraman, G., Srimathi, T., and Yu, C. (1998) The role of proline in the prevention of aggregation during protein folding *in vitro*. *Biochem. Mol. Biol. Int.* **46**, 509–517.
42. Samuel, D., Kumar, T. K., Jayaraman, G., Yang, P. W., and Yu, C. (1997) Proline is a protein solubilizing solute. *Biochem. Mol. Biol. Int.* **41**, 235–242.
43. Samuel, D., Kumar, T. K., Ganesh, G., Jayaraman, G., Yang, P. W., Chang, M. M., et al. Proline inhibits aggregation during protein refolding. *Protein. Sci.* **9**, 344–352.
44. Wang, A. and Bolen, D. W. (1996) Effect of proline on lactate dehydrogenase activity: testing the generality and scope of the compatibility paradigm. *Biophys. J.* **71**, 2117–2222.
45. Kita, Y. and Arakawa, T. (2002) Salts and glycine increase reversibility and decrease aggregation during thermal unfolding of ribonuclease-A. *Biosci. Biotechnol. Biochem.* **66**, 880–882.
46. Kaderbhai, N., Karim, A., Hankey, W., Jenkins, G., Venning, J., and Kaderbhai, M. A. (1997) Glycine-induced extracellular secretion of a recombinant cytochrome expressed in *Escherichia coli*. *Biotechnol. Appl. Biochem.* **25**, 53–61.
47. Buchner, J., Pastan, I., and Brinkmann, U. (1992) A method for increasing the yield of properly folded recombinant fusion proteins: single-chain immunotoxins from renaturation of bacterial inclusion bodies. *Anal. Biochem.* **205**, 263–270.
48. Kim., T. K., Chung, J. Y., Lee, G. M., and Park, S. K. (2001) Arginine-enriched medium composition used for mass-producing recombinant protein in animal cell culture. Patent WO0144442.
49. Georgiou, G., Valax, P., Ostermeier, M., and Horowitz, P. M. (1994) Folding and aggregation of TEM beta-lactamase: analogies with the formation of inclusion bodies in *Escherichia coli*. *Protein. Sci.* **3**, 1953–1960.
50. Bowden, G. A. and Georgiou, G. (1990) Folding and aggregation of beta-lactamase in the periplasmic space of *Escherichia coli*. *J. Biol. Chem.* **265**, 16760–16766.
51. Wang, A., Robertson, A. D., and Bolen, D. W. (1995) Effects of a naturally occurring compatible osmolyte on the internal dynamics of ribonuclease A. *Biochemistry* **34**, 15096–15104.

52. Sola-Penna, M., Ferreira-Pereira, A., Lemos, A. P., and Meyer-Fernandes, J. R. (1997) Carbohydrate protection of enzyme structure and function against guanidinium chloride treatment depends on the nature of carbohydrate and enzyme. *Eur. J. Biochem.* **248**, 24–29.

53. Frye, K. J. and Royer, C. A. (1997) The kinetic basis for the stabilization of staphylococcal nuclease by xylose. *Protein. Sci.* **6**, 789–793.

54. Leandro, P., Lechner, M. C., Tavares de Almeida, I., and Konecki, D. (2001) Glycerol increases the yield and activity of human phenylalanine hydroxylase mutant enzymes produced in a prokaryotic expression system. *Mol. Genet. Metab.* **73**, 173–178.

55. Figler, R. A., Omote, H., Nakamoto, R. K., and Al-Shawi, M. K. (2000) Use of chemical chaperones in the yeast *Saccharomyces cerevisiae* to enhance heterologous membrane protein expression: high-yield expression and purification of human P-glycoprotein. *Arch. Biochem. Biophys.* **376**, 34–46.

56. Ohnishi, T., Ohnishi, K., Wang, X., Takahashi, A., and Okaichi, K. (1999) Restoration of mutant TP53 to normal TP53 function by glycerol as a chemical chaperone. *Radiat. Res.* **151**, 498–500.

57. Ghumman, B., Bertram, E. M., and Watts, T. H. (1998) Chemical chaperones enhance superantigen and conventional antigen presentation by HLA-DM-deficient as well as HLA-DM-sufficient antigen-presenting cells and enhance IgG2a production *in vivo*. *J. Immunol.* **161**, 3262–3270.

58. Bai, C., Biwersi, J., Verkman, A. S., and Matthay, M. A. (1998) A mouse model to test the *in vivo* efficacy of chemical chaperones. *J. Pharmacol. Toxicol. Methods* **40**, 39–45.

59. De Sanctis, G., Maranesi, A., Ferri, T., Poscia, A., Ascoli, F., and Santucci, R. (1996) Influence of glycerol on the structure and redox properties of horse heart cytochrome C. A circular dichroism and electrochemical study. *J. Protein. Chem.* **15**, 599–606.

60. Zhi, W., Landry, S. J., Gierasch, L. M., and Srere, P. A. (1992) Renaturation of citrate synthase: influence of denaturant and folding assistants. *Protein. Sci.* **1**, 522–529.

61. Saunders, A. J., Davis-Searles, P. R., Allen, D. L., Pielak, G. J., and Erie, D. A. (2000) Osmolyte-induced changes in protein conformational equilibria. *Biopolymers.* **53**, 293–307.

62. Voziyan, P. A. and Fisher, M. T. (2002) Polyols induce ATP-independent folding of GroEL-bound bacterial glutamine synthetase. *Arch. Biochem. Biophys.* **397**, 293–297.

63. Taylor, L. S., York, P., Williams, A. C., Edwards, H. G., Mehta, V., Jackson, G. S., et al. (1995) Sucrose reduces the efficiency of protein denaturation by a chaotropic agent. *Biochim. Biophys. Acta.* **1253**, 39–46.

64. Majumder, A., Basak, S., Raha, T., Chowdhury, S. P., Chattopadhyay, D., and Roy, S. (2001) Effect of osmolytes and chaperone-like action of P-protein on folding of nucleocapsid protein of Chandipura virus. *J. Biol. Chem.* **276**, 30948–30955.

65. Baskakov, I., Wang, A., and Bolen, D. W. (1998) Trimethylamine-N-oxide counteracts urea effects on rabbit muscle lactate dehydrogenase function: a test of the counteraction hypothesis. *Biophys. J.* **74**, 2666–2673.

66. Song, J. L. and Chuang, D. T. (2001) Natural osmolyte trimethylamine N-oxide corrects assembly defects of mutant branched-chain alpha-ketoacid decarboxylase in maple syrup urine disease. *J. Biol. Chem.* **276**, 40241–40246.

67. Samuelsson, E., Jonasson, P., Viklund, F., Nilsson, B., and Uhlen, M. (1996) Affinity-assisted *in vivo* folding of a secreted human peptide hormone in *Escherichia coli*. *Nat. Biotechnol.* **14**, 751–755.

68. Hwang, C., Sinskey, A. J., and Lodish, H. F. (1992) Oxidized redox state of glutathione in the endoplasmic reticulum. *Science* **257**, 1496–1502.

69. Bardwell, J. C. (1994) Building bridges: disulphide bond formation in the cell. *Mol. Microbiol.* **14**, 199–205.
70. Wittung-Stafshede, P. (2002) Role of cofactors in protein folding. *Acc. Chem. Res.* **35**, 201–208.
71. Zou, J. and Sugimoto, N. (2000) Complexation of peptide with Cu^{2+} responsible to inducing and enhancing the formation of alpha-helix conformation. *Biometals* **13**, 349–359.
72. Beck, R. and Burtscher, H. (1994) Expression of human placental alkaline phosphatase in *Escherichia coli*. *Protein. Expr. Purif.* **5**, 192–197.
73. Baneyx, F., Ayling, A., Palumbo, T., Thomas, D., and Georgiou, G. (1991) Optimization of growth conditions for the production of proteolytically-sensitive proteins in the periplasmic space of *Escherichia coli*. *Appl. Microbiol. Biotechnol.* **36**, 14–20.
74. Machida, S., Ogawa, S., Xiaohua, S., Takaha, T., Fujii, K., and Hayashi, K. (2000) Cycloamylose as an efficient artificial chaperone for protein refolding. *FEBS Lett.* **486**, 131–135.
75. Dong, X. Y., Shi, J. H., and Sun, Y. (2002) Cooperative effect of artificial chaperones and guanidinium chloride on lysozyme renaturation at high concentrations. *Biotechnol. Prog.* **18**, 663–665.
76. Woycechowsky, K. J., Wittrup, K. D., and Raines, R. T. (1999) A small-molecule catalyst of protein folding *in vitro* and *in vivo*. *Chem. Biol.* **6**, 871–879.
77. Winter, J., Lilie, H., and Rudolph, R. (2002) Recombinant expression and *in vitro* folding of proinsulin are stimulated by the synthetic dithiol Vectrase-P. *FEMS Microbiol. Lett.* **213**, 225–230.
78. Schwarz, E., Rudolph, R., Ambrosius, D., and Schaeffner, J. (2001) Process for the production of naturally folded and secreted proteins by co-secretion of molecular chaperones. Patent EP1077262
79. Schroeckh, V., Hortschansky, P., Fricke, S., Luckenbach, G. A., and Riesenberg, D. (2000) Expression of soluble, recombinant alphav-beta3 integrin fragments in *Escherichia coli*. *Microbiol. Res.* **155**, 165–177.
80. Glockshuber, R., Skerra, A., Rudolph, R., and Wunderlich, M. (1992) Improvement of the secretion yield of proteins with a disulfide bridge. Patent EP0510658.
81. Thomas, J. G. and Baneyx, F. (1997) Divergent effects of chaperone overexpression and ethanol supplementation on inclusion body formation in recombinant *Escherichia coli*. *Protein Expr. Purif.* **11**, 289–296.
82. Gustafson, C. and Tagesson, C. (1985) Influence of organic solvent mixtures on biological membranes. *Br. J. Ind. Med.* **42**, 591–595.
83. Hartl, F. U. and Hayer-Hartl, M. (2002) Molecular chaperones in the cytosol: from nascent chain to folded protein. *Science* **295**, 1852–1858.
84. Schlieker, C., Bukau, B., and Mogk, A. (2002) Prevention and reversion of protein aggregation by molecular chaperones in the *E. coli* cytosol: implications for their applicability in biotechnology. *J. Biotechnol.* **96**, 13–21.
85. Fahnert, B. (2001) Rekombinantes humanes BMP-2 aus *Escherichia coli*—Strategien zur Expression und Funktionalisierung., Doctorate Thesis, Friedrich-Schiller-University Jena (Germany).
86. Voziyan, P. A., Jadhav, L., and Fisher, M. T. (2000) Refolding a glutamine synthetase truncation mutant *in vitro*: identifying superior conditions using a combination of chaperonins and osmolytes. *J. Pharm. Sci.* **89**, 1036–1045.
87. Fisher, M. T. and Voziyan, P. (2002) Chaperonin and osmolyte protein folding and related screening methods. Patent US2002006636.

88. Ko, Y. H. and Pedersen, P. (2001) Methods for identifying an agent that corrects defective protein folding. Patent WO0121652.

89. Kirsch, T., Sebald, W., and Dreyer, M. K. (2000) Crystal structure of the BMP-2-BRIA ectodomain complex. *Nat. Struct. Biol.* **7**, 492–496.

90. Kirsch, T., Nickel, J., and Sebald, W. (2000) Isolation of recombinant BMP receptor IA ectodomain and its 2:1 complex with BMP-2. *FEBS Lett.* **468**, 215–219.

91. Fahnert, B., Hahn, D., and Guthke, R. (2002) Knowledge-based assessment of gene expression data from chemiluminescence detection. *J. Biotechnol.* **94**, 23–35.

92. Braun, P., Hu, Y., Shen, B., Halleck, A., Koundinya, M., Harlow, E., et al. (2002) Proteome-scale purification of human proteins from bacteria. *Proc. Natl. Acad. Sci. USA* **99**, 2654–2659.

II

PROKARYOTES

4

Back to Basics

pBR322 and Protein Expression Systems in E. coli

Paulina Balbás and Francisco Bolívar

Summary

The extensive variety of plasmid-based expression systems in *E. coli* resulted from the fact that there is no single strategy for achieving maximal expression of every cloned gene. Although a number of strategies have been implemented to deal with problems associated to gene transcription and translation, protein folding, secretion, location, posttranslational modifications, particularities of different strains, and the like and more integrated processes have been developed (*1,2*), the basic plasmid-borne elements and their interaction with the particular host strain will influence the overall expression system and final productivity (*3*) (*see* Chapters 1–3).

Plasmid vector pBR322 (*4*) is a well-established multipurpose cloning vector in laboratories worldwide, and a large number of derivatives have been created for specific applications and research purposes, including gene expression in its natural host, *E. coli,* and few other bacteria. The early characterization of the molecule, including its nucleotide sequence, replication and maintenance mechanisms, and determination of its coding regions, accounted for its success, not only as a universal cloning vector, but also as a provider of genes and an origin of replication for other intraspecies vectors (*5,6*). Since the publication of the aforementioned reviews, novel discoveries pertaining to these issues have appeared in the literature that deepen the understanding of the plasmid's features, behavior, and impact in gene expression systems, as well as some important strain characteristics that affect plasmid replication and stability.

The objectives of this review include updating and discussing the new information about (1) the replication and maintenance of pBR322; (2) the host-related modulation mechanisms of plasmid replication; (3) the effects of growth rate on replication control, stability, and recombinant gene expression; (4) ways for plasmid amplification and elimination. Finally, (5) a summary of novel ancillary studies about pBR322 is presented.

Key Words: Integration; excision; replication, Rop/Rom.

From: *Methods in Molecular Biology, vol. 267:*
Recombinant Gene Expression: Reviews and Protocols, Second Edition
Edited by: P. Balbás and A. Lorence © Humana Press Inc., Totowa, NJ

1. Replication and Maintenance of pBR322

Plasmid pBR322 contains a ColE1-type replicon derived from the natural isolate pMB1 (*see* **Note 1**). Under normal culturing conditions, the ColE1-like replicon present in pBR322 exhibits a copy number of 15–20 molecules per cell, replicates unidirectionally, and is mobilizable by conjugative plasmids under special conditions (*5–7*). The replication process can be separated into three different stages: initiation, elongation, and termination. By far, the process of initiation of plasmid replication is the best known; new findings will be thoroughly discussed in the following sections. Elongation of replication is carried out by the cell's DNA replisome and it is known that the termination process is different from the *E. coli*'s chromosome (*8,9*) (*see* Chapter 2).

The mechanisms controlling transcription initiation of the RNA II primer, the interaction of RNA II with its repressing RNA I antisense RNA, the action of the Rop/Rom protein with the hybrid duplex RNA, the decay rates of RNA I mediated by a number of endo- and exonucleolytic activities, as well as polynucleotide phosphorilases and culture growth rate, seem to be key steps in the process of copy number control and replication of ColE1-type replicons.

1.1. Replication Initiation

This process is dependent on the assembly of the primosome and replisome structures at the proper assembly sites (DnaA binding site, *pas* and *rri*), downstream of the replication origin (*10,11*). Initiation of the leading strand requires, at least, RNA polymerase holoenzyme, DNA polymerase I, topoisomerase I, RNAse H, and DNA gyrase. Initiation of the lagging strand requires other primosomal proteins, including the products of DnaA, DnaB, DnaC, DnaE, DnaG, Dna Z, ssb, the replication factor Y (protein n'), and proteins i, n, and n" (*6,12–16*).

After assembly of the primosome at the origin of replication, transcription of a 550-bases-long single-stranded RNA beginning at the region known as P2 (promoter 2) is the key step for replication initiation. The host's RNA polymerase synthesizes this molecule, termed RNA II, which contains extensive alternative hairpin secondary structures. One of such structures facilitates the RNA II base-pair bonding to the DNA template, where it undergoes specific processing by the enzyme RNAseH, therefore serving as a primer for the synthesis of the leading strand by DNA polymerase I (*5,6*).

1.2. RNA-Mediated Replication Inhibition

RNA II can interact with a smaller antisense RNA molecule known as RNA I (108 nucleotides), transcribed in 100-fold excess over RNA II from P1, a strong promoter located on the opposite overlapping DNA strand (*17–19*) (*see* **Note 2**). When RNA I hybridizes with RNA II, an interaction facilitated by the Rop/Rom protein (*see* **Subheading 1.4.**), RNA II fails to achieve its priming conformation for DNA polymerase I, and replication is aborted. RNA I is therefore, the primary agent of DNA and copy number control. Since RNA I is a trans-acting molecule that can hybridize to any RNA II molecule synthesized by plasmids in the cytoplasm, it also serves as the agent for plasmid incompatibility (*20*) (*see* **Note 3**).

Experiments with mutant plasmids lacking RNA I showed that plasmid replication becomes uncoupled from the host cell growth (*21*). Also, a number of mutations in the zone adjacent or within RNA I have been isolated where the usual phenotype is an increase of the net plasmid copy number (*22,23*), thus supporting the view that the whole region encoding RNA I is key for copy number control and coupled to growth rate.

1.3. Control of Replication by mRNA Decay Rates

RNA I and RNA II are constitutively synthesized. The turnover rate of RNA I is essential to promote the rapid changes in replication frequency and copy number control. RNA I is thus unstable, and decays with a half-life of about 2 min during exponential growth (*18,24,25*). The rate of decay is determined by the combined action of endo- and exonucleases that modify transcripts in different ways. Some of the known enzymes affecting RNA I decay, and thus plasmid copy number and replication, are RNAse E, poly(A) polymerase; polynucleotide phosphorilase; RNAse II exonuclease, and RNAse III endonuclease (*26–28*).

RNAse E is an endoribonuclease encoded by the *E. coli ams/rne* locus that is required for normal RNA turnover, and also processes rRNA transcripts to yield 5S ribosomal RNA precursor (*26*). Cleavage of RNA I by RNAse E occurs at a single site, five nucleotides from the 5'-end, producing a molecule called RNA I-$_5$, which is further degraded by the poly(A)-dependent activity of polynucleotide phosphorilase and other 3'–5' exonucleases (*29*). Several studies about the degradation of this intermediate have suggested that cleavage by RNAse E is the rate-limiting step in the degradation of RNA I (*18,24,25*). A mutation in the *rne* locus prolongs RNA I half-life at elevated temperatures, suggesting that RNAse E is virtually a nonspecific single-stranded endonuclease with preference for cutting 5' to an AU dinucleotide (*30*).

The *pcnB* gene encodes an ATP-dependent poly(A) polymerase in *E. coli,* which catalyzes the template-independent sequential addition of AMP to the 3';-terminal hydroxyl groups of RNA molecules. Mutations in this locus affect plasmid copy number by decreasing the rate of decay of products generated by RNAse E-mediated cleavage of RNA I, leading to the accumulation of decaying intermediates capable of repressing plasmid DNA replication, therefore lowering the copy number of pBR322 derivatives (*31–34*). Also, mutations of *pcnB* prevent normal polyadenylation of RNA I (*34*).

Polynucleotide phosphorilase of *E. coli* can both synthesize RNA by using nucleotide diphosphates as precursors, and act as an exonuclease to degrade RNA in the presence of inorganic phosphate. It also accounts for the residual polyadenilation in poly(A) polymerase I-deficient strains. As mentioned before, this enzyme plays a role in the degradation of RNA I-$_5$ (*28,35*).

The roles of RNAse II exonuclease and RNAse III endonuclease have also been studied. RNAse II is a 5' to 3' exonuclease, sensitive to secondary structures present in the RNA, while RNAse III cleaves double-stranded RNA structures. Both enzymes are involved in RNA decay, and studies are on their way to establish direct correlations with ColE1 replication in vivo (*28*).

1.4. Modulation of RNAII–RNAI Interaction by the Rop/Rom Protein

The gene encoding the trans-acting protein Rop (repressor of primer), also known as Rom (RNA one modulator), present in pBR322 and some of its derivatives, is dispensable for the replication process. Deletion of the coding gene, situated between the origin of replication and the Tc^R gene, increases copy number several-fold, but does not alter any other plasmid functions. The Rop/Rom protein is an acidic, trans-acting, 63-amino-acids-long protein that is active as a dimer. This protein accelerates the association of RNA I with the growing RNA II molecules as they are being synthesized, therefore acting as a modulator by favoring the RNA I–RNA II interaction and producing premature termination of replication (**5,6,20**).

Rop/Rom deletion derivatives, such as pBR327, pBR327par (**36,37**), and the pUC family of plasmid vectors (**38**), exhibit higher copy numbers because the interaction of RNA I–RNA II is weaker. It has been suggested that Rop/Rom may confer stability by protecting RNA I against RNAse E activity in *rho* mutant strains (**39**). The presence of the *rop/rom* gene is known to decrease the copy number of pBR322 in slowly growing cultured cells (**40**), and some point mutations in RNA I and RNA II have been reported to produce higher copy number derivatives where the Rop/Rom protein is involved (**21–23,41**).

2. Host-Related Modulation Mechanisms of Plasmid Replication

A replication-modulation mechanism operating on the bacterial *oriC* chromosomal, bidirectional origin of replication also affects ColE1-type replicons (**42**). The *ori* region contains three GATC sites for recognition by the *dam* methylase. Hemimethylation of the adenine residues within this tetramer inhibits replication, and membrane sequestration of the hemimethylated origin DNA has been suggested as the restricting mechanism in *dam-* mutant cells (**42–44**) (*see* **Note 4**).

3. Effects of Growth Rate on Replication Control, Stability, and Recombinant Gene Expression

Plasmid stability in cultures is one of the major determining factors of the productivity of recombinant fermentations, especially when high-level production of heterologous proteins is achieved. Therefore, understanding the various elements affecting vector maintenance has highlighted the importance of parameters that are not related to the plasmid or the host but influence replication and partition of plasmids between daughter cells during cultures (**3**). This phenomenon is known as segregation, and it has been observed in ColE1-type plasmids grown in different culture media without selection pressure. Alternatively, plasmid copy number can also be positively affected by the nutrients and culturing conditions as well.

Multicopy plasmids exert a metabolic burden on fast-growing cultured cells (**45,46**), probably due to the sequestration of substrates (nucleotides and amino acids), catalytic components (RNA polymerase, DNA polymerase, tRNAs, cofactors, and other proteins), and energy requirements (**3,47**). Also, the effect of the bacterial growth rate on replication control of plasmid pBR322 has been known for years, thus indicating a close interplay between culturing conditions and plasmid maintenance (*3,45,46,48–51*).

The concentrations of RNA I and RNA II decrease with increasing growth rates, whereas the RNA I/RNA II ratio increases, while maintaining the same half-life of both RNA species (*48,52*). Stringent response in cultures harboring ColE1-like plasmids causes inhibition of replication in amino acid-starved *E. coli* cells in the absence of Rom/Rop (*40,53–55*). Also, plasmid loss decreases with increasing dilution rates in cultures growing in complex media (*49*), and a precisely defined medium has been reported to significantly increase plasmid copy number (*50*). These reports support the view of a complex interplay among variables in cultured cells, and although some mathematical models have been designed to predict plasmid stability, other authors consider that segregation rates are quite unpredictable and chaotic (*46,56,57*). Kinetic descriptions of recombinant systems based on observations of molecular mechanisms have been devised (*58*). As useful as this may be for the rational design of an expression system, experimental verification of predicted results will have the last word about the usefulness of every incorporated parameter (*see* **Note 5**).

It has been reported that structural instability of inserts carried in pBR322 may be due to high levels of mRNA synthesis from strong promoters converging with the replication machinery (*59–61*), or due to the unlikely situation of two convergent origins of replication (*62*). In both cases, knots within the newly replicated DNA regions have been observed, which supports the view that imposed superhelical stresses may destabilize the duplex (*63*). Therefore, it is recommended to isolate transcription units situated clockwise by transcription terminators, especially in high-level expression systems (*3,61*).

4. Amplification and Elimination of Plasmids

pBR322 and its derivatives can replicate in the presence of the antibiotic chloramphenicol. This antibiotic acts as a bacteriostatic by inhibiting protein synthesis, via the interaction with the 50S ribosomal subunit and inhibition of the peptidyltransferase reaction. Addition of chloramphenicol to the culture medium in early log phase is used for amplification of the plasmid copy number (*5,6,64*). Also, amino-acid-starved cells of mutant *relA–* strains can accumulate large amounts of pBR322 plasmid DNA (*65–67*) but this is undoubtedly a poor application for high-level production of proteins.

Cells can be "cured" from pBR322 by sequential cultures without selective pressure, or by addition of specific substances that inhibit DNA gyrase. Examples include coumermycin A1 (*68,69*), complexes of platinum (II) (*70,71*) and α-santonin (*72*).

5. Ancillary Studies About pBR322

Analysis of several aspects of the pBR322 molecule *per se,* its behavior in mutant strains, topology, genes, and regulatory proteins has produced a large amount of information about the mechanics of plasmid replication, maintenance, partition, and topological aspects of the molecule. Although these are usually isolated pieces of information, with relatively little use in experimental procedures or in protein expression schemes, the information summarized in **Table 1** may be useful for both strain selection during strategy planning and troubleshooting when specific problems arise.

Table 1
Special Studies About pBR322

Contribution	References
Nucleotide sequence revisions	
Possible errors upstream of the *rop/rom* gene	*73*
High copy number and antibiotic resistant mutants	
Point mutations adjacent to RNA I	*22*
Nucleotide transversion in loop II of RNA I	*23*
Overexpression of Tc^R	*74*
Rifampin-resistant mutants	*75*
Replication in mutant strains	
DNA-A-mediated replication	*76*
Replication in RecBCD SbcB mutants	*77*
Replication in RnaseH and DNA PolI mutants	*13*
Replication in RNA polymerase mutant	*78*
Studies on the tetracycline resistance gene and protein	
Genetic analysis of *tetA(C)*	*79*
tet gene and DNA knotting	*80*
Membrane topology of the Tc^R protein	*81–84*
Tc^R gene affects stability of pBR322	*85*
Tc^R gene affects plasmid supercoiling	*86,87*
Structure and function of the Tc^R antiporter	*88*
Osmotic sensitivity due to Tc^R	*89*
Studies on the ampicillin resistance gene	
Overexpression of Ap^R and cell growth	*90*
Expression of Ap^R at 42°C	*91*
Multimer formation	
Dimerization produced by large inserts	*92*
Hotspots for DNA recombination and/or insertion	
New hot spots for *Tn5* insertion	*93*
Local supercoiling in pBR322 and *Tn5* insertion	*94*
Hotspots and host integration factors	*95*
Studies on the P4, cAMP-CAP-dependent promoter	
Kinetics of activation	*96*
Complex binding and DNA bending	*97*
Inducers of structural modifications and/or protective agents	
α-Tocopherol	*98*
α-Tocopherol monoglucoside	*99*

Table 1 (continued)

Contribution	References
Inducers of structural modifications and/or protective agents	
Aromatic amine DNA adducts	*100*
Ascorbic acid	*98*
β-Carotene	*98*
Caffeine	*101*
Chlorophyllin	*102*
Electromagnetic fields	*103*
Fructose	*104*
L-propionil-carnitine	*105*
N-hydroxylthiazole-2(3H)-thione	*106*
Pd(II) cisplatin	*107*
Pt(II) mepirizole	*107*
Transplatin	*107*

It is well known that strain choice plays a key role when expressing heterologous genes at high levels. Pinpointing the specific origin of such phenomena is too complex to be done, so trial and error with different hosts is the typical procedure when troubles with final protein yields arise. Some point mutations within key replication regions have been reported to play a role, as well as the presence of antibiotic resistance genes, dimerization and superhelicity.

Studies on the antibiotic resistance genes showed that the Tc^R determinant has more impact on the plasmid's stability and maintenance than the Ap^R determinant, not only because of its convergent direction toward the replication origin, but also because of the intrinsic nature of the protein it encodes. Also, the effects of the culturing conditions on the host-vector interaction have been systematically studied in order to improve the productivity of recombinant fermentations.

6. Concluding Remarks

Moderate- to high-copy-number plasmid vectors have been the choice systems for gene expression in *E. coli*, especially for extrachromosomal expression of proteins, protein transduction experiements, shuttle vectors, and the like, and most of these vectors contain the origin of replication of pBR322. Fragments of pBR322 have also been very popular for the construction of intraspecies shuttle or binary vectors and, recently, for the construction of vectors for the targeted integration/excision of DNA from the chromosome (*108*). These two growing areas of application of pBR322 argue in favor of regularly reviewing the basics of plasmid functioning and its interrelationships with its host strains, so better strategies for gene expression may be pursued.

7. Notes

1. The origin of replication present in pBR322 shows a strong homology to several other origins found in natural isolates; these have been termed ColE1-like origins. At least five

popular plasmids belong to this family: pMB1 (ancestor of pBR322), ColE1, CloDF13, RSF1030, and p15A.

2. The regions coding for the interacting RNA I and RNA II are very structured and several palindromes may be formed in vivo (pBR322 regions situated between nucleotides 2540 and 3100) (*5,6*). Mutations in these structures usually produce altered replicating phenotypes.

3. Incompatibility is "the inability of two closely related plasmids sharing some aspects of their replication machinery, to stably coexist in a population of growing bacteria without selective pressure for both plasmids" (*3,109*). For gene expression schemes, co-expression of genes must avoid the use of incompatible replicons. Stable chromosomal integration of regulatory genes into the chromosome is a strategy to overcome problems associated with two plasmids cohabiting the same cell (*108*).

4. A number of observations throughout the literature suggest that other strain-related factors influence the plasmid's behavior in cultures, especially when a high expression level of an encoded protein is achieved. If problems arise during protein production, trial of different strains for the same DNA construct is advisable.

5. For practical effects, culture conditions are usually determined by other factors involving the overall strategy of system design and downstream processing operations (*1*).

Acknowledgments

P. Balbás thanks the PROMEP program supported by the Secretaría de Educación Pública, México.

References

1. Balbás, P. (2001) Understanding the art of producing protein and non protein molecules in *E. coli*. *Mol. Biotechnol.* **19**, 251–267.

2. Swartz, J. R. (2001) Advances in *Escherichia coli* production of therapeutic proteins. *Curr. Opin. Biotechnol.* **12**, 195–201.

3. Balbás, P. and Bolívar, F. (1990) Design and construction of expression plasmid vectors in *Escherichia coli*. *Methods Enzymol.* **185,** 3–40.

4. Bolívar, F., Rodríguez, R. L., Greene, P. J., Betlach, M. C., Heyneker, H. L., Boyer, H. W., et al. (1976) Construction and characterization of new cloning vehicles. II. A multipurpose cloning system. *Gene* **2**, 95–113.

5. Balbás, P., Soberón, X., Merino, E., Zurita, M., Lomelí, H., Valle, F., et al. (1986) Plasmid vector pBR322 and its special-purpose derivatives: a review. *Gene* **50**, 3–40.

6. Balbás, P., Soberón, X., Bolívar, F., and Rodríguez, R.L. (1988) The plasmid, pBR322, in *Vectors. A Survey of Molecular Cloning Vectors and Their Uses* (Rodríguez, R. L. and Denhardt, D. T., eds.) Butterworth, pp. 5–41.

7. Covarrubias, L., and Bolívar, F. (1982) Construction and characterization of new cloning vehicles. VI. Plasmid pBR329, a new derivative of pBR328 lacking the 482-base-pair inverted duplication. *Gene* **17**, 79–89.

8. Mohanty, B. K., Sahoo, T., and Bastia, D. (1998) Mechanistic studies on the impact of transcription on sequence-specific termination of DNA replication and vice versa. *J. Biol. Chem.* **30**, 3051–3059.

9. Bussiere, D. E. and Bastia, D. (1999) Termination of DNA replication of bacterial and plasmid chromosomes. *Mol. Microbiol.* **31**, 1611–1618.

10. Ohmori, H., Murakami, Y., and Nagata, T. (1987) Nucleotide sequences required for a ColE1-type plasmid to replicate *in Escherichia coli* cells with or without RNase H. *J. Mol. Biol.* **20**, 223–234.

11. Chiang, C. S., Xu, Y. C. and Bremer, H. (1991) Role of DnaA protein during replication of plasmid pBR322 in *Escherichia coli. Mol. Gen. Genet.* **225**, 435–442.

12. Minden, J. S. and Marians, K. J. (1985) Replication of pBR322 DNA in vitro with purified proteins. Requirement for topoisomerase I in the maintenance of template specificity. *J. Biol. Chem.* **260**, 9316–9325.

13. Parada, C. A. and Marians, K. J. (1989) Transcriptional activation of pBR322 DNA can lead to duplex DNA unwinding catalyzed by the *Escherichia coli* preprimosome. *J. Biol. Chem.* **264**, 15120–15129.

14. Masai, H. and Arai, K. (1989) *Escherichia coli dnaT* gene function is required for pBR322 plasmid replication but not for R1 plasmd replication. *J. Bacteriol.* **171**, 2975–2980.

15. Del Solar, G., Giraldo, R., Ruiz-Echevarría, M. J., Espinosa, M. and Díaz-Orejas, R. (1998) Replication and control of circular bacterial plasmids. *Microbiol. Mol. Biol. Rev.* **62**, 434–464.

16. Lee, E. H. and Kornberg, A. (1991) Replication deficiencies in *priA* mutants of *Escherichia coli* lacking the primosomal replication n'protein. *Proc. Natl. Acad. Sci. USA* **15**, 3029–3032.

17. Lin-Chao, S. and Bremer, H. (1987) Activities of the RNA I and RNA II promoters of plasmid pBR322. *J. Bacteriol.* **169**, 1217–1222.

18. Brenner, M. and Tomizawa, J. (1991) Quantitation of ColE1-encoded replication elements. *Proc. Natl. Acad. Sci. USA* **88**, 405–409.

19. Liang, S., Bipatnath, M., Xu, Y., Chen, S., Dennis, P., Ehrenberg, M., et al. (1999) Activities of constitutive promoters in *Escherichia coli. J. Mol. Biol.* **292**, 19–37.

20. Polisky, B. (1988) ColE1 replication control circuitry: sense from antisense. *Cell* **55**, 929–932.

21. Chiang, C. S. and Bremer, H. (1991b) Maintenance of pBR322-derived plasmids without functional RNA I. *Plasmid* **26**, 186–200.

22. Ivanov, I., Yavashev, L., Gigova, I., Alexciev, K., and Christo, C. (1988) A conditional high-copy-number plasmid derivative of pBR322. *Microbiologica* **11**, 95–99.

23. Nugent, M. E., Smith, T. J. and Tacon, W. C. (1986) Characterization and incompatibility properties of ROM-derivatives of pBR322-based plasmids. *J. Gen. Microbiol.* **132**, 1021–1026.

24. Lin-Chao, S. and Cohen, S. N. (1991) The rate of processing and degradation of antisense RNA I regulates replication of ColE1-type plasmids *in vivo. Cell* **65**, 1233–1242.

25. Bouvet, P. and Belasco, J. G. (1991) Control of RNAseE-mediated RNA degradation by 5'-terminal base-pairing in *E. coli. Nature* **360**, 488–491.

26. Kushner, S. R. (1996) mRNA decay. In *Escherichia coli* and *Salmonella.* Cellular and Molecular Biology. 2nd ed. (Niedhardt *et al.*, eds.), ASM Press, Washington, D.C.

27. Jung, Y. H. and Lee, Y. (1995) RNAses in ColE1 DNA metabolism. *Mol. Biol. Rep.* **22**, 195–200.

28. Binnie, U., Wong, K., McAteer, S. and Masters, M. (1999) Absence of RNAse III alters the pathway by which RNA I, the antisense inhibitor of ColE1 replication, decays. *Microbiology* **145**, 3089–3100.

29. Kaberdin, V. R., Chao, Y. H. and Lin-Chao, S. (1996) RNAse E cleaves at multiple sites in bubble regions of RNA I stem-loops yielding products that dissociate differentially from the enzyme. *J. Biol. Chem.* **271**, 13103–13109.

30. Lin-Chao, S., Wong, T. T., McDorwall, K. J., and Cohen, S. N. (1994) Effects of nucleotide sequence on the specificity of *rne*-dependent and RNAseE-mediated cleavages of RNA I encoded by the pBR322 plasmid. *J. Biol. Chem.* **269**, 10797–10803.

31. Lopilato, J., Bortner, S., and Beckwith, J. (1986) Mutations in a new chromosomal gene of *Escherichia coli* K-12, reduce plasmid copy number of pRB322 and its derivatives. *Mol. Gen. Genet.* **205**, 285–290.

32. Liu, J. and Parkinson, J. S. (1989) Genetics and sequence analysis of the *pcnB* locus, an *Escherichia coli* gene involved in plasmid copy number control. *J. Bacteriol.* **171**, 1254–1261.

33. He, L., Söderbom, F., Wagner, G. H., Binnie, U., Binns, N. and Masters, M. (1993) PcnB is required for the rapid degradation of RNA I, the antisense RNA that controls the copy number of ColE1-related plasmids. *Molec. Microbiol.* **9**, 1131–1142.

34. Xu, F., Lin-Chao, S. and Cohen, S. N. (1993) The *Escherichia coli pcnB* gene promotes adenylation of antisense RNAI of ColE1-type plasmids *in vivo* and degradation of RNAI decay intermediates. *Proc. Natl. Acad. Sci. USA* **90**, 6756–6760.

35. Mohanty, B. K. and Kushner, S. R. (2000) Polynucleotide phosphorilase functions both as a 3' right-arrow 5' exonuclease and poly(A) polymerase in *Escherichia coli. Proc. Natl. Acad. Sci. USA* **97**, 11966–11971.

36. Soberón, X., Covarrubias, L. and Bolívar, F. (1980) Construction and characterization of new cloning vehicles. IV. Deletion derivatives of pBR322 and pBR325. *Gene* **9**, 287–305.

37. Zurita, M., Bolívar, F. and Soberón, X. (1984) Construction and characterization of new cloning vehicles. VII. Construction of plasmid pBR327par, a completely sequenced, stable derivative of pBR327 containing the par locus of pSC101. *Gene* **28**, 119–122.

38. Vieira, J. and Messing, J. (1982) The pUC plasmids, an M13mp7-derived system for insertion mutagenesis and sequencing with synthetic universal primers. *Gene* **19**, 259–268.

39. Sozhamannan, S., Morris J. G. Jr., and Stitt, B. L. (1999) Instability of pUC19 in *Escherichia coli* transcription termination factor mutant, *rho026. Plasmid* **41**, 63–69.

40. Atlung, T., Christensen, B. B., and Hansen, F. G. (1999) Role of the Rom protein in copy number control of plasmid pBR322 at diffrerent growth rates in *Escherichia coli* K-12. *Plasmid* **41**, 110–119.

41. Lin-Chao, S., Chen, W. T., and Wong, T. T. (1992) High copy number of the pUC plasmids results from a Rom/Rop-supressible point mutation in RNA II. *Mol. Microbiol.* **6**, 3385–3393.

42. Malki, A., Kern, R., Kohiyama, M., and Huges, P. (1992) Inhibition of DNA synthesis at the hemimethylated pBR322 origin of replication by a cell membrane function. *Nucleic Acids Res.* **20**, 105–109.

43. Russell, D. W. and Zinder, N. D. (1987) Hemimethylation prevents DNA replication in *E. coli. Cell* **50**, 1071–1079.

44. Patnaik, P. K., Merlin, S. and Polisky, B. (1990) Effect of altering GATC sequences in the plasmid ColE1 primer promoter. *J. Bacteriol.* **172**, 1762–1764.

45. McDermott, P. J., Gowland, P., and Gowland, P. C. (1993) Adaptation of *Escherichia coli* growth rates to the presence of pBR322. *Lett. Appl. Microbiol.* **17**, 139–143.

46. Paulson, J. and Ehrenberg, M. (1998) Trade-off between segregational stability and metabolic burden: a mathematical model of plasmid ColE1 replication control. *J. Mol. Biol.* **279**, 73–88.

47. Eisenbraun, M. D. and Griffith, J. K. (1993) Effects of plasmid pBR322 on respiratory and ATPase activities in *Escherichia coli. Plasmid* **30**, 159–162.

48. Bremer, H. and Lin-Chao, S. (1986) Analysis of the physiological control of replication of ColE1-type plasmids. *J. Theor. Biol.* **123**, 453–470.

49. Weber, A. E. and San, K. Y. (1987) Persistence and expression of the plasmid pBR322 in *Eschericha coli* K-12 cultured in complex medium. *Biotechnol. Lett.* **9**, 757–760.
50. Duttweiler, H. M. and Gross, D. S. (1998) Bacterial growth medium that significantly increases the yield of recombinant plasmid. *Biotechniques* **24**, 438–444.
51. Lin-Chao, S. and Bremer, H. (1986) Effect of the bacterial growth rate on replication control of plasmid pBR322 in *Escherichia coli. Mol. Gen. Genet.* **203**, 143–149.
52. Gasunov, V. V. and Brilkov, A. V. (2002) Estimating the instability parameters of plasmid-bearing cells. I. Chemostat culture. *J. Theor. Biol.* **219**, 193–205.
53. Herman, A., Wegrzyn, A., and Wegrzyn, G. (1994) Regulation of replication of plasmid pBR322 in amino acid-starved *Escherichia coli* strains. *Mol. Gen. Genet.* **243**, 374–378.
54. Wrobel, B., and Wegrzyn, G. (1998) Replication regulation of ColE1-like plasmids in amino acid-starved *Escherichia coli. Plasmid* **39**, 48–62.
55. Wang, Z., Le, G., Shi, Y., Wegrzyn, G., and Wrobel, B. (2002) A model for regulation of ColE1-like plasmid replication by uncharged tRNAs in amino acid-starved *Escherichia coli* cell. *Plasmid* **47**, 69–78.
56. Kim, B. G., Good, T. A., Ataai, M. M., and Shuler, M. L. (1987) Growth behaviour and prediction of copy number and retention of ColE1-type plasmids in *E. coli* under slow growth conditions. *Ann. N.Y. Acad. Sci.* **506**, 384–395.
57. Torkel-Nielsen, T. and Boe, L. (1994) A statistical analysis of the formation of plasmid-free cells in populations of *Escherichia coli. J. Bacteriol.* **176**, 4306–4310.
58. Lee, S. B. and Bailey, J. E. (2002) Analysis of growth rate effects on productivity of recombinant *Escherichia coli* populations using molecular mechanism models. *Biotechnol. Bioeng.* **79**, 550–557.
59. Weston-Hafer, K. and Berg, D. E. (1991) Deletions in plasmid pBR322: replication slippage involving leading and lagging strands. *Genetics* **128**, 487.
60. Vilette, D., Ehrlich, S. D. and Michel, B. (1995) Transcription-induced deletions in *Escherichia coli* plasmids. *Mol. Microbiol.* **17**, 493–504.
61. Olavarrieta, L., Hernández, P., Krimer, D. B., and Schvartzman, J. B. (2002) DNA knotting caused by head-on collision of transcription and replication. *J. Mol. Biol.* **322**, 1–6.
62. Sogo, J. M., Stasiak, A., Martínez-Robles, M. L., Krimer, D. B., Hernández, P., and Schvartzman, J. B. (1999) Formation of knots in partially replicated DNA molecules. *J. Mol. Biol.* **286**, 637–643.
63. Benham, C. J. (1993) Sites of predicted stress-induced DNA duplex destabilization occur preferentially at regulatory loci. *Proc. Natl. Acad. Sci. USA* **90**, 2999–3003.
64. Frenkel, L. and Bremer, H. (1986). Incresased concentration of plasmid pBR322 and pBR327 by low concentrations of chloramphenicol. *DNA* **5**, 539–544.
65. Schroetel, A., Riethdorf, S. and Hecker, M. (1988) Amplification of different ColE1 plasmids in *Escherichia coli relA* strain. *J. Basic Microbiol.* **28**, 553–555.
66. Riethdorf, S., Schroeter, A., and Hecker, M. (1989) RelA mutation and pBR322 plasmid amplification in amino acid-starved cells of *Escherichia coli. Genet. Res.* **54**, 167–171.
67. Hofmann, K. H., Neubauer, P., Riethdorf, S., and Hecker, M. (1990) Amplification of pBR322 plasmid DNA in *Escherichia coli relA* strains during batch and fed-batch fermentation. *J. Basic. Microbiol.* **30**, 37–41.
68. Wolfson, J. S., Hooper, D. C., Swartz, M. N., and McHugh, G. L. (1982) Antagonism of the B subunit of DNA gyrase eliminates plasmids pBR322 and pMG110 from *Escherichia coli. J. Bacteriol.* **152**, 338–344.
69. Ishii, S., Murakami, T. and Shishido, K. (1991) Gyrase inhibitors increase the content of knotted DNA species of plasmid pBR322 in *Escherichia coli. J. Bacteriol.* **173**, 5551–5553.

70. Lakshmi, V. V. and Polasa, H. (1991) Curing of pBR322 and pBR329 plasmids in *Escherichia coli* by cis-dichlorodiamine platinum (II) chloride (Cis-DDP). *FEMS Microbiol. Lett.* **62**, 281–284.

71. Lakshmi, V. V., Sridhar, P., Khan, B. T., and Polasa, H. (1988) Mixed-ligand complexes of platinum (II) as curing agents for pBR322 and pBR329 (ColE1) plasmids in *Escherichia coli. J. Gen. Microbiol.* **134**, 1977–1981.

72. Brahati, A. and Polasa, H. (1990) Elimination of ColE1 group (pBR322 and pBR329) plasmids in *Escherichia coli* by alpha-satonin. *FEMS Microbiol. Lett.* **56**, 213–215.

73. Watson, N. (1988) A new revision of the sequence of plasmid pBR322. *Gene* **70**, 399–403.

74. Valenzuela, M. S., Ikpeazu, E. V. and Siddiqui, K. A. (1996) *E. coli* growth inhibition by high copy number derivative of plasmid pBR322. *Biochem. Biophys. Res. Commun.* **27**, 219.

75. Magee, T. R. and Kogoma, T. (1991) Rifampin-resistant replication of pBR322 derivatives in *Escherichia coli* cells induced for the SOS response. *J. Bacteriol.* **173**, 4736–4741.

76. Parada, C. A. and Marians, K. J. (1991) Mechanism of DNA A protein-dependent pBR322 DNA replication. DNA A-mediated trans-strand loading of the DNA B protein at the origin of pBR322 DNA. *J. Biol. Chem.* **266**, 18895–18906.

77. Silberstein, Z. and Cohen, A. (1987) Synthesis of linear multimers of *oriC* and pBR322 derivatives in *Escherichia coli* K-12: role of recombination and replication functions. *J. Bacteriol.* **169**, 3131–3137.

78. Petersen, S. K. and Hansen, F. G. (1991) A missense mutation in the *rpoC* gene affects chromosomal replication control in *E. coli. J. Bacteriol.* **173**, 5200–5206.

79. McNicholas, P., Chopra, I., and Rothstein, D. M. (1992) Genetic analysis of the tetA(C) gene on plasmid pBR322. *J. Bacteriol.* **174**, 7926–7933.

80. Shishido, K., Ishii, S., and Komiyama, N. (1989) The presence of the region on pBR322 that encodes resistance to tetracycline is responsible for high levels of plasmid knotting in *Escherichia coli* DNA topoisomerase mutant. *Nucleic Acids Res.* **17**, 9749–9759.

81. Allard, J. D. and Bertrand, K. P. (1992) Membrane topology of the pBR322 tetracycline resistance protein. *J. Biol. Chem.* **267**, 17809–17819.

82. Lewis, G. S., Jewel, J. E., Phang, T. and Miller, K. W. (2002) Mutational analysis of tetracycline resistance protein transmembrane segment insertion. *Arch. Biochem. Biophys.* **404**, 317–325.

83. McNicholas, P., McGlynn, M., Guay, G. G. and Rothstein, D. M. (1995) Genetic analysis suggests functional interactions between the N- and C-terminal domains of the TetA(C) efflux pump encoded by pBR322. *J. Bacteriol.* **177**, 5355–5357.

84. Valenzuela, M. S., Siddiqui, K. A. and Sarkar, B. L. (1996) High expression of plasmid-encoded tetracycline resistance gene in *E. coli* causes a decrease in membrane-bound ATPase activity. *Plasmid* **36**, 19–25.

85. Chiang, C. S. and Bremer, H. (1988) Stability of pBR322-derived plasmids. *Plasmid.* **20**, 207–220.

86. Pruss, G. J. and Drlica, K. (1986) Topoisomerase I mutants: the gene on pBR322 that encodes resistance to tetracycline affects plasmid DNA supercoiling. *Proc. Natl. Acad. Sci. USA* **83**, 8952–8956.

87. Lodge, J. K., Kazic, T., and Berg, D. E. (1989) Formation of supercoiling domains in plasmid pBR322. *J. Bacteriol.* **171**, 2181–2187.

88. Griffith, J. K., Cuellar, D. H., Fordyce, C. A., Hutchings, K. G., and Mondragón, A. A. (1994) Structure and function of the class C tetracycline/H+ antiporter: three dependent groups of phenotypes are conferred by TetA (C). *Mol. Membr. Biol.* **11**, 271–277.

89. Stavropoulous, T. A. and Strathdee, C. A. (2000) Expression of the TetA (C) tetracycline efflux pump in *Escherichia coli* confers osmotic sensitivity. *FEMS Microbiol. Lett.* **190**, 147–150.

90. Katayama, T. and Nagata, T. (1990) Inhibition of cell growth and stable DNA replication by overexpression of the *bla* gene of plasmid pBR322 in *Escherichia coli*. *Molec. Gen. Genet.* **223**, 353–360.

91. Kuriki, Y. (1987) Requirement of a heat-labile factor(s) for in vivo expression of the *amp* gene of pBR322. *J. Bacteriol.* **169**, 5856–5858.

92. Berg, C. M., Liu, L., Coon, M., Gray, P., Vartak, N. B., Brown, M., et al. (1989) pBR322-derived multicopy plasmids harboring large inserts are often dimmers in *Escherichia coli* K-12. *Plasmid* **21**, 138–141.

93. Boyd, L. A., Woytowich, A. and Selvaraj, G. (1993) Target sequence specificity of transposon *Tn5* in the absence of major hotspots in the plasmid pBR322: identification of a new hotspot. *Plasmid* **30**, 155–158.

94. Lodge, J. K, and Berg, D. E. (1990) Mutations that affect *Tn5* insertion into pBR322: importance of local supercoiling. *J. Bacteriol.* **172**, 5956–5960.

95. Gamas, P., Chandler, M. G., Prentki, P., and Galas, D. J. (1987) *Escherichia coli* integration host factor binds specifically to the ends of the insertion sequence *IS1* and to its major insertion hotspot in pBR322. *J. Mol. Biol.* **195**, 261–272.

96. Hogget, J. G. and Brierley, I. (1992) Kinetics of activation of the P4 promoter by *Escherichia coli* cyclic AMP receptor protein. *Biochem. J.* **287**, 937–941.

97. Brierley, I. and Hogget, J. G. (1992) Binding of the cyclic AMP receptor protein of *Escherichia coli* and DNA bending at the P4 promoter of pBR322. *Biochem. J.* **285**, 91–97.

98. Zhang, P. and Omaye, S. T. (2001) DNA strand breakage and oxygen tension: effects of beta-carotene, alpha-tocopherol and ascorbic acid. *Food Chem. Toxicol.* **39**, 239–246.

99. Rajagopalan, R., Wani, K., Huilgol, N. G., Kagiya, T. V., and Nair, C. K. (2002) Inhibition of gamma-radiation induced DNA damage in plasmid pBR322 by TMG, a water soluble derivative of vitamin E. *J. Radiat. Res.* **43**, 153–159.

100. Melchior, W. B., Jr., Marques, M. M. and Beland, F. A. (1994) Mutations induced by aromatic amine DNA adducts in pBR322. *Carcinogenesis* **15**, 889–899.

101. Kumar, S. S., Chaubey, R. C., Devasagayam, T. P., Priyadarsini, K. I., and Chauhan, P. S. (1999) Inhibition of radiation-induced DNA damage in plasmid pBR322 by chlorophyllin and possible mechanism(s) of action. *Mutat Res.* **425**, 71–79.

102. Kumar, S. S., Devasagayan, T. P., Jayashree, B. And Kesavan, P. C. (2001) Mechanism of protection against radiation-induced DNA damage in plasmid pBR322 by caffeine. *Int. J. Radiat. Biol.* **77**, 617–623.

103. Lourencini da Silva, R., Albano, F., Lopes do Santos, L. R., Tavares, A. D. Jr., and Felzenszwall, I. (2000) The effect of electromagnetic field exposure on the formation of DNA lesions. *Redox Rep.* **5**, 299–301.

104. Levi, B. and Werman, M. J. (2001) Fructose triggers DNA modifications and damage in an *Escherichia coli* plasmid. *J. Nutr. Biochem.* **12**, 235–241.

105. Vanella, A., Russo, A., Acquaviva, R., Campisi, A., Di Giacomo, C., Sorrenti V. et al. (2000) L-propionyl-carnitine as superoxide scavenger, antioxidant, and DNA cleavage protector. *Cell. Biol. Toxicol.* **16**, 99–104

106. Adam, W., Hartung, J., Okamoto, H., Saha-Moller, C. R., and Spehar, K. (2000) N-hydroxy-4-(4-chlorophenyl)thiazole-2(3H)-thione as a photochemical hydroxyl-radical source: photochemistry and oxidative damage of DNA (strand breaks) and 2'-deoxyguanosine (8-oxodG formation). *Photochem Photobiol.* **72**, 619–64.

107. Onoa, G.B. and Moreno, V. (2002) Study of the modifications caused by cisplatin, transplatin, and Pd(II) and Pt(II) mepirizole derivatives on pBR322 DNA by atomic force microscopy. *Int. J. Pharm.* **245**, 55–65.

108. Balbás, P. and Gosset, G. (2001) Chromosomal editing in *Escherichia coli*: vectors for DNA integration and excision. *Mol. Biotechnol.* **19**, 1–12.

109. Balbás, P. and Bolivar, F. (1998) Molecular cloning by plasmid vectors. in *Recombinant DNA*. Principles and Applications (Greene, J.J. and Rao, V.B. eds.), Marcel Dekker, Inc., New York, NY, pp. 383–411.

5

α-Complementation-Enabled T7 Expression Vectors and Their Use for the Expression of Recombinant Polypeptides for Protein Transduction Experiments

Mikhail F. Alexeyev, Viktoriya V. Pastukh, Inna N. Shokolenko, and Glenn L. Wilson

Summary

Over the past few years protein transduction has emerged as a powerful means for the delivery of proteins into cultured cells and into whole mice. This method is based on the ability of proteins containing protein transduction domains (PTDs), short stretches of 9–16 predominantly basic amino acids, to traverse the cytoplasmic membrane and accumulate inside cells in a time- and dose-dependent fashion. The number of PTDs, both natural and synthetic, is constantly expanding, as is the need to test newly discovered PTDs for their ability to mediate the internalization of the corresponding fusion proteins. Here we describe a strategy and methodology that can be used for the construction of vectors for the T7 RNA polymerase-driven expression of PTD fusions. The cloning in these vectors is facilitated by α-complementation. Also, these vectors are small in size (less than 3 kbp) and express influenza virus hemagglutinin tag as well as His tag as part of the fusion for immunological identification and purification respectively of expressed proteins.

Key Words: T7 RNA polymerase; protein transduction; fusion protein; expression; purification; hemagglutinin tag; His tag; α-complementation.

1. Introduction

Over the past few years protein transduction has emerged as a powerful means for the delivery of proteins into cultured cells and into whole mice *(1–7)*. This method is based on the ability of proteins containing protein transduction domains (PTDs), short

From: *Methods in Molecular Biology, vol. 267:*
Recombinant Gene Expression: Reviews and Protocols, Second Edition
Edited by: P. Balbás and A. Lorence © Humana Press Inc., Totowa, NJ

stretches of 9–16 predominantly basic amino acids, to traverse the cytoplasmic membrane and accumulate inside cells in a time- and dose-dependent fashion. The number of PTDs, both natural and synthetic, is constantly expanding *(7–10)*, as is the need to test newly discovered PTDs for their ability to mediate the internalization of the corresponding fusion proteins. The T7 RNA polymerase-promoter system *(11)* is one of the most powerful, reliable, and popular vector systems for protein expression in *E. coli*. However, the identification of recombinant clones in this system remains cumbersome because most vectors do not permit visual identification of recombinant clones through standard methods such as α-complementation of β-galactosidase *(12)*. Here we describe methodology that can be used for the construction of α-complementation-enabled vectors for T7 promoter-driven expression of PTD-containing gene fusions in *E. coli*.

2. Materials

1. Oligonucleotide primers (Integrated DNA Technology [IDT], Coralville, IA). Primer sequences are (*see* **Note 1**):

   #1 TCTAGAAATAATTTTGTTTAAC
   #2 TCTAGAAATAATTTTGTTTAACTTAGAGAAGGAGATATACCAT
        GGGCCACC ATCACCATC
   #3 GGCGCTGGCGGCGTTTTTTGCGGCCATAGCCATGGTGATGGTGA
        TGGTGATGGTGGCCCA
   #4 CAAAAAACGCCGCCAGCGCCGCCGCGGCTATCCGTATGATGT
        GCCGGATTATGCGAGCCT
   #5 CGACAAGCTTGAATTCGAGCTCGGATCCCCCGGCATATGCAGG
        CTCGCATAATCCGGCAC
   #6 GTGCGGCCGCGTCGACAAGCTTGAATTCGAGC

2. T4 DNA ligase, restriction enzymes, Vent DNA polymerase, large fragment of *Escherichia coli* DNA-polymeraseI (Klenow), Taq DNA polymerase and corresponding buffers (New England Biolabs, Beverly, MA).
3. Agarose.
4. Ampicillin.
5. Terrific Broth (TB) *(13)*.
6. EDTA-free protease inhibitor cocktail (PIC, Roche, Indianapolis, IN).
7. Sonifier 250 with ⅛" tapered probe (Branson, Danbury, CT).
8. Qiaquick spin gel extraction kit (Qiagen, Valencia, CA).
9. Plasmids: pUC19 (New England Biolabs), pBSL97 (American Type Culture Collection, [ATCC], Manassas, VA).
10. Deoxynucleotide triphosphates (a 100X mix of all four dNTPs containing 20 mM each).
11. *E. coli* strains DH5α and BL21(DE3).
12. 5-bromo-4-chloro-3-indolyl-β-D-galactoside (X-gal) 50 mg/mL in dimethylformamide.
13. Ni-NTA agarose (Qiagen).
14. Disposable chromatography columns (Pierce, Rockford, IL).
15. PD10 desalting columns (Bio-Rad, Hercules, CA).
16. Resuspension buffer: 20 mM Tris-HCl pH 8.0, 500 mM NaCl, 5 mM imidazole, 1X PIC.

17. Wash buffer: 20 m*M* Tris-HCl pH 8.0, 500 m*M* NaCl, 30 m*M* imidazole, 1X PIC.
18. Elution buffer: 20 m*M* Tris-HCl pH 8.0, 500 m*M* NaCl, 500 m*M* imidazole, 1X PIC.

3. Methods
3.1. Vector Construction

There is one basic problem that confronts enabling α-complementation in T7 expression vectors. The promoter driving the expression of *lacZα* has to transcribe the target gene without elevating the level of its basal expression in the uninduced state. Reconciliation can be achieved only if this promoter transcribes the antisense strand of the target gene. Therefore, in our vector system the promoters driving expression of the target gene (pT7) and *lacZα* (p*lac*) are facing each other. Because the p*lac* promoter in the vectors described below is made constitutive by partially deleting the *lac* operator (*lacO*), this arrangement has an additional advantage: the *lac*-promoter-driven transcription of the complementary strand of the target gene will reduce its expression by the way of antisense inhibition.

3.1.1. Construction of pUC19tet

pUC19tet (**Fig. 1**) was constructed by polymerase chain reaction (PCR)-mediated mutagenesis using an overlap extension strategy *(14)*. First, two PCR reactions were performed using primer pairs lacTetOgene (TCCCTATCAGTGATAGAGATTGTGAGCG-GATAAC) plus NdeF (GGACGTCAGATCTCCCATATGCGGTGTGAAATACC) and lacTetOProm (TATCACTGATAGGGATTCCACACAACATACGAGCC) plus pMB1ori (CACCTCTGACTTGAGCGTCG) using pUC19 plasmid as a template. This manipulation creates two overlapping DNA fragments that cover pUC19 coordinates 2617 to 926, removes part of the *lacO*, and inserts in its place *tetO*. Then, the products of these two PCR reactions were gel-purified, combined and amplified using primers NdeF plus pMB1ori. This round of PCR combined two overlapping fragments into one contiguous piece. Finally, the resulting PCR product was gel-purified, digested with *Afl*III plus *Aat*II, and ligated to similarly digested pUC19, resulting in pUC19tet. Therefore, pUC19tet is a derivative of pUC19 in which *lacO* was partially deleted and replaced with the *tet* operator (*tetO*) (*see* **Note 2**). Also, in this plasmid a deletion in the pUC19 backbone between coordinates 2623 and 183 was introduced and the *Bgl*II site was inserted in place of this deletion. All recombinant DNA manipulations described in this and the following sections were performed using standard molecular biology techniques *(13)*.

3.1.2. Construction of pUC23dME

pUC23dME (**Fig. 2**) is a pUC19tet derivative in which a T7 promoter was introduced between *Bgl*II and *Ehe*I sites. It was constructed by first inserting a chloramphenicol-resistance (Cm-res) gene from pBSL97 (*Ecl*136II-fragment) into the *Xba*I digested, Klenow-treated pET23 plasmid (Novagen, Madison, WI), thus resulting in pET23Cm. Then, the *Blg*II-*Nde*I fragment of pET23Cm encompassing the Cm-res gene and the T7 promoter was ligated to the *Blg*II-*Nde*I digested pUC19tet, thus producing pUC23Cm.

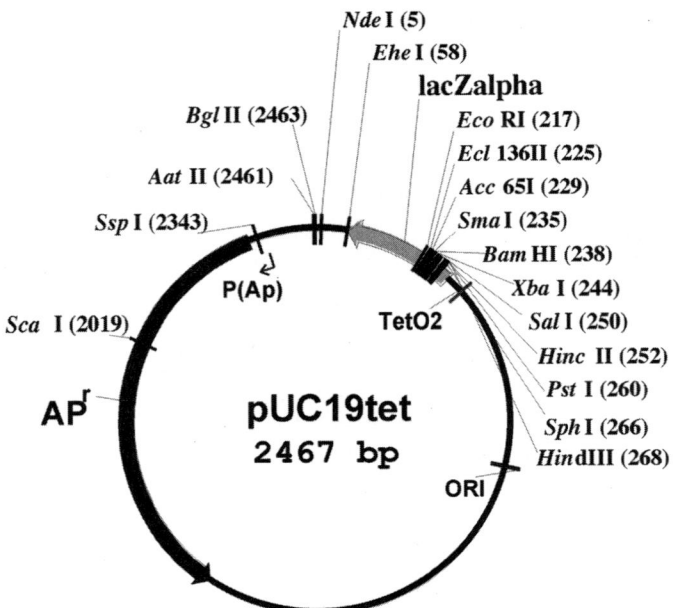

Fig. 1. Physical and genetic maps of pUC19tet. Only relevant unique restriction sites are shown. Ap^r, ampicillin-resistance gene; ORI, pMB1 replication origin; TetO2, the synthetic symmetric operator of tetracycline-resistance gene.

Finally, pUC23Cm was digested with *Mlu*I plus *Ehe*I, treated with Klenow, and circularized. This final manipulation removed the Cm-res gene and part of the pUC19tet backbone from pUC23Cm and generated pUC23dME.

3.1.3. Construction of Expression Vectors

The expression vectors described in this chapter were constructed by replacing the pUC23dME polylinker with a synthetic DNA fragment encoding (a) an N-terminal purification tag consisting of eight consecutive histidine residues, (b) a protein transduction domain, (c) the influenza virus hemagglutinin (HA) epitope tag, and (d) multiple cloning sites (*see* **Note 3**). This procedure will be described using construction of pUC23Tat1 as an example (**Fig. 3A,B**) (*see* **Note 4**).

1. Mix the following components in a 0.5-mL PCR tube on ice:
   a. Primers 1 through 6, 2 pmol each (*see* **Fig. 3A**).
   b. 10X Taq buffer, 10 µL.
   c. dATP, dGTP, dCTP, and dTTP, 0.2 m*M* final concentration.
   d. H$_2$O, to 100 µL.
   e. *Taq* DNA polymerase, 5 units.

2. Program a thermocycler with a heated lid for the following: initial denaturation for 2 min at 94°C, then 15 cycles of 20 s at 94°C, 20 s at 50°C, and 20 s at 72°C.

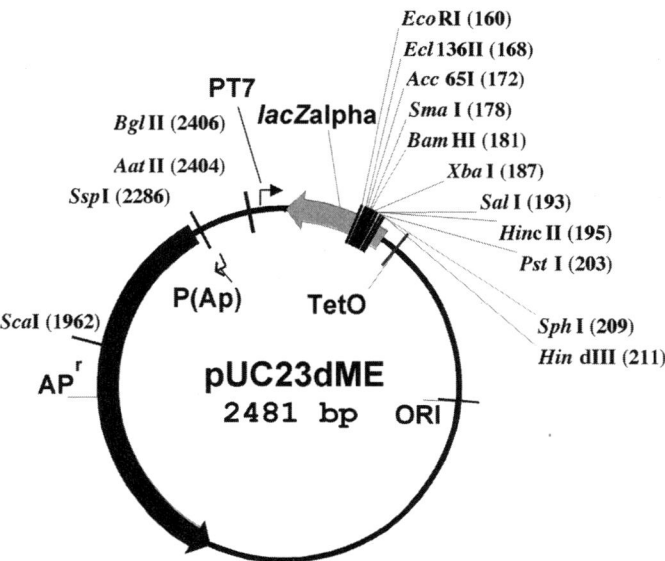

Fig. 2. Physical and genetic maps of pUC23dME. Only relevant unique restriction sites are shown. Apʳ, ampicillin-resistance gene; P_{T7}, T7 promoter; ORI, pMB1 replication origin; TetO2, the synthetic symmetric operator of tetracycline-resistance gene.

3. Start the program and place the tube in the thermocycler as soon as it warms up to 94°C.
4. At the end of the program prepare the following mix:

   a. Primers 1 and 6, 50 pmol each.
   b. 10X Vent buffer, 10 μL.
   c. dATP, dGTP, dCTP, and dTTP, 0.2 μM final concentration.
   d. H₂O, to 100 μL.
   e. The product of the first PCR reaction, 2 μL.
   f. Vent DNA polymerase, 2 units.

5. Program the thermocycler for the initial denaturation 2 min at 94°C, then 25 cycles of 20 s at 94°C, 20 s at 50°C, and 20 s at 72°C, and start the program as above.
6. While the thermocycler is running, digest pUC23dME with EcoRI plus HindIII, precipitate the reaction for 5 min at −70°C with ethanol, pellet DNA in a microcentrifuge for 5 min at 13,000g, dissolve the pellet in 20 μL of water and add 5 μL of 5X Klenow buffer containing 5 mM of each dATP, dGTP, dCTP, and dTTP.
7. Add 5 units of Klenow and incubate 20 min at room temperature.
8. Gel-purify the resulting product using Qiagen's gel extraction kit.
9. In the final volume of 10 μL mix 200 ng of the gel-extracted pUC23dME, 1 μL of 10X T4 DNA-ligase buffer, 1 μL of PstI restriction endonuclease and water (see Note 5).
10. Add 3 μL of the product of the second PCR reaction (step 4) and ligate for 3 h at 25°C in a thermocycler.
11. Transform competent DH5α cells selecting for ampicillin-resistant clones on plates containing 50 μg/mL X-gal.

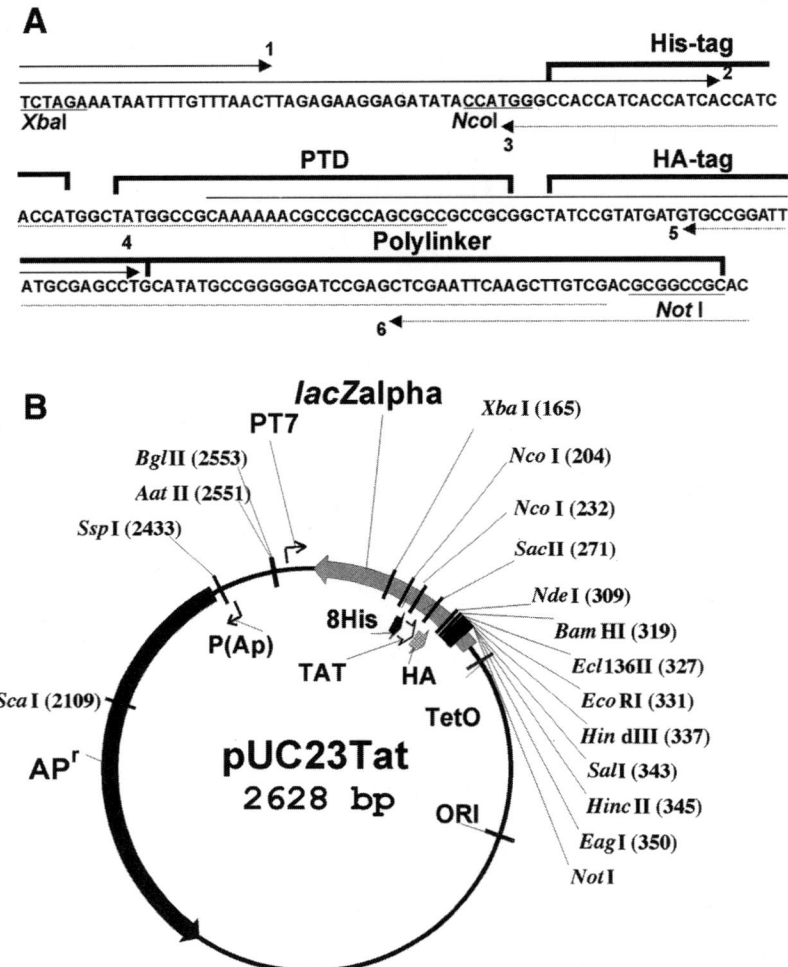

Fig. 3. Construction of pUC23TAT1. (**A**) The strategy for the assembly of the synthetic DNA fragment for the replacement of pUC23dME polylinker. The oligonucleotide primers are denoted with numerals and either solid (sense strand primers) or broken (antisense strand primers) arrows. The heavy solid brackets above the sequence indicate regions encoding functional sites of the vector (His-tag, PTD, HA-tag and polylinker). Positions for the *Xba*I, *Nco*I and *Not*I restriction sites are shown for orientation. (**B**) Physical and genetic maps of pUC23TAT1. Only relevant unique and two *Nco*I restriction sites are shown. Ap^r, ampicillin-resistance gene; P$_{T7}$, T7 promoter; ORI, pMB1 replication origin; TetO2, the synthetic symmetric operator of tetracycline-resistance gene.

12. Isolate the plasmid DNA from blue colonies and screen it for the presence of a *Not*I site.
13. Verify the structure of the vector by sequencing through new polylinker. The resulting plasmid is pUC23Tat1 (**Fig. 3B**).

3.2. Protein Expression and Purification

Cloning in pUC23Tat1 is performed using standard molecular biology techniques (*13*) for the identification of recombinant clones by α-complementation (*see* **Note 6**). Once the desired construct is created, it is introduced into several T7 RNA polymerase-expressing host strains (*see* **Note 7**) using standard transformation techniques, and the best host-vector combination is selected for a large-scale protein purification.

3.2.1. Induction of Protein Expression

1. Inoculate 3–4 colonies of the best host-vector combination into 2-L baffled-bottom flask containing 0.5 L of TB medium supplemented with 200 µg/mL ampicillin and 0.1 mM IPTG.
2. Grow overnight with vigorous shaking (300 rpm) at 37°C.
3. Divide the culture into two 250-mL aliquots and spin at 4000g for 20 min to pellet cells. The cell pellet can either be used immediately or stored at −70°C until needed.

3.2.2. Cell Disruption

1. Resuspend the bacterial pellet obtained from a 250-mL culture in 12 mL of resuspension buffer.
2. Divide the suspension into two 6-mL aliquots in 15-mL plastic conical centrifuge tubes and chill on ice for 15 min.
3. Sonicate each aliquot five times for 20 s using the 50% duty cycle and the output set at 7 (Branson Sonifier 250) on ice alternating the tubes between 20-s bursts.

3.2.3. Protein Purification

1. Transfer the broken cells into two Beckman 331372 14 × 89 mm polyallomer tubes and spin for 20 min at 23°C and 25,000 rpm in SW41 rotor (77.175 g average, 107.170 g maximum).
2. Collect the supernatant into two 15-mL conical centrifuge tubes and add 1 mL of Ni-NTA agarose.
3. Close the lids and allow the protein to bind to the resin for 1 h at room temperature, tumbling the tubes over their long axis.
4. Load two disposable chromatography columns and wash with wash buffer until A_{280} is below 0.05.
5. Elute with elution buffer, collecting 0.5-mL fractions (*see* **Note 8**).
6. Identify fractions containing the eluted protein by measuring A_{280} and pool them.
7. Remove excess salt by dialysis or gel filtration using desalting columns, e.g., PD10 (Bio-Rad).

Figure 4 presents the results of a typical experiment on expression of chloramphenicol acetyltransferase cloned in pUC23TAT1.

4. Notes

1. In our experience, unpurified (desalted) primers produced by IDT can be used in this protocol without compromising the fidelity of the final product.
2. The partial deletion of the *lac*O relieves *lac*I-mediated repression of the *lac* promoter and makes expression of *lac*Zα constitutive in pUC19tet. The introduction of the *tet*O in place of partially deleted *lac*O was necessary to increase the level of *lac*Zα expression. Apparently, destruction of the *lac*O leads to the destabilization of *lac*Zα mRNA and reduces the cellular

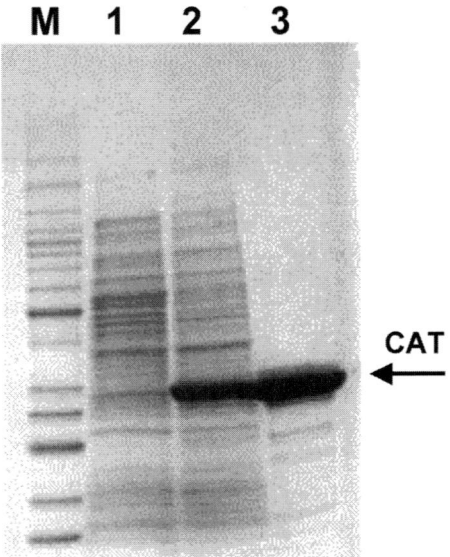

Fig. 4. Expression of the *Tn9*-encoded chloramphenicol acetyltransferase (CAT) in pUC23TAT1. Coomassie-stained SDS-PAGE. Lanes: M-molecular weight markers (BenchMarkÔ, Invitrogen, Carlsbad, CA); 1- lysate of BL21(DE3); 2- lysate of BL21(DE3) containing pUC23TAT1 with CAT cloned between *Nde*I and *Hind*III sites; 8his-TAT-HACAT protein after purification on Ni-NTA agarose.

       levels of *lac*Zα. The stem-loop structures at the 3' end of mRNA have been reported to have a stabilizing effect and to increase the level of protein expression *(15)*.

3. While designing the nucleotide sequence of the new expression vector's polylinker region, it is important to make sure that it does not encode any translation termination codons in frame with *lac*Zα. Usually, this is achieved quite easily by taking advantage of the degeneracy of the genetic code.

4. This protocol describes the construction of the vector containing HIV TAT PTD. However, primers used in this protocol also can be utilized in the construction of vectors containing other PTDs. Indeed, only two of six primers used in this protocol, primers 3 and 4, overlap the region encoding the PTD (**Fig. 3A**). Therefore, to construct a vector containing a new PTD, only primers 3 and 4 need to be redesigned. We have successfully used this approach to construct vectors encoding the following PTDs: PTD4 *(7)*, 9 Arg *(16)*, and s4–13 *(8)*.

5. The addition of *Pst*I into the ligation reaction facilitates the reduction of the background of pUC23dME through biochemical enrichment. This occurs because pUC23dME has one *Pst*I site, whereas pUC23Tat1 has none.

6. Any *E. coli* strain capable of α-complementation can be used in conjunction with pUC23Tat1. However, when using DH5α it is not necessary to include isopropyl-β-D-thiogalactopyranoside (IPTG) in the selection medium, as this strain lacks a functional lac repressor (LacI). Blue-white screening is especially useful when using the *Nde*I site for cloning because treatment of plasmid DNA with *Nde*I often results in incomplete digests, which in turn leads to a

high number of clones that need to be screened before finding a recombinant one.

7. Many host strains for T7 promoter-driven expression have been developed and are marketed by different companies. Oftentimes, the secret to a high-level expression lies in the particular host-vector combination. Therefore, it is advisable to screen several host strains for a high level of target protein expression before initiating purification efforts. The expression construct can be easily introduced into several strains using procedures such as freeze-thawing *(17)*.

8. On occasion we observed aggregation of the eluted protein immediately after elution. This process appears to be protein- and concentration-dependent and can be prevented by inclusion of 300 m*M* arginine into the elution buffer.

Acknowledgments

This work was supported by AHA grant 0255697B to M. F. A., DOD grant DAMD17-00-1-0658 to I. N. S., and NIH grants ES03456 and AG19602 to G. L. W.

References

1. Vocero-Akbani, A. M., Heyden, N. V., Lissy, N. A. Ratner, L., and Dowdy, S. F. (1999). Killing HIV-infected cells by transduction with an HIV protease-activated caspase-3 protein. *Nat. Med.* **5**, 29–33.

2. Vocero-Akbani, A., Lissy, N. A., and Dowdy, S. F. (2000). Transduction of full-length Tat fusion proteins directly into mammalian cells: analysis of T cell receptor activation-induced cell death. *Methods Enzymol.* **322**, 508–521.

3. Schwarze, S. R. and Dowdy, S. F. (2000). In vivo protein transduction: intracellular delivery of biologically active proteins, compounds and DNA. *Trends Pharmacol. Sci.* **21**, 45–48.

4. Schwarze, S. R., Hruska, K. A. and Dowdy, S. F. (2000). Protein transduction: unrestricted delivery into all cells? *Trends Cell. Biol.* **10**, 290–295.

5. Vocero-Akbani, A., Chellaiah, M. A., Hruska, K. A., and Dowdy, S. F. (2001). Protein transduction: delivery of Tat-GTPase fusion proteins into mammalian cells. *Methods Enzymol.* **332**, 36–49.

6. Becker-Hapak, M., McAllister, S. S. and Dowdy, S. F. (2001). TAT-mediated protein transduction into mammalian cells. *Methods* **24**, 247–256.

7. Ho, A., Schwarze, S. R., Mermelstein, S. J., Waksman, G. and Dowdy, S. F. (2001). Synthetic protein transduction domains: enhanced transduction potential in vitro and in vivo. *Cancer Res.* **61**, 474–477.

8. Hariton-Gazal, E., Feder, R., Mor, A., Graessmann, A., Brack-Werner, R., Jans, D., et al. (2002). Targeting of nonkaryophilic cell-permeable peptides into the nuclei of intact cells by covalently attached nuclear localization signals. *Biochemistry* **41**, 9208–9214.

9. Mi, Z., et al. (2000). Characterization of a class of cationic peptides able to facilitate efficient protein transduction in vitro and in vivo. *Mol. Ther.* **2**, 339–347.

10. Futaki, S., Ohash, W., Suzuki, T., Niwa, M., Tanaka, S., Ueda, K., et al. (2001). Stearylated arginine-rich peptides: a new class of transfection systems. *Bioconjug. Chem.* **12**, 1005–1011.

11. Studier, F. W. and Moffatt, B. A. (1986). Use of bacteriophage T7 RNA polymerase to direct selective high-level expression of cloned genes. *J. Mol. Biol.* **189**, 113–130.

12. Welply, J. K., Fowler, A. V., and Zabin, I. (1981). Beta-galactosidase alpha-complementation. *J. Biol. Chem.* **256**, 6804–6810.

13. Sambrook, J., Fritsch, E. F., and Maniatis, T. *Molecular Cloning. A Laboratory Manual.* 2nd

ed. 1989, Cold Spring Harbor Laboratory Press, Cold Spring Harbor, NY.
14. Ho, S. N., Hunt, H. D., Horton, R. M., Pullen, J. K., and Pease, L. R. (1989). Site-directed mutagenesis by overlap extension using the polymerase chain reaction. *Gene* **77**, 51–59.
15. Panayotatos, N. and Truong, K. (1985). Cleavage within an RNase III site can control mRNA stability and protein synthesis in vivo. *Nucleic Acids Res.* **13**, 2227–2240.
16. Han, K., Jeon, M. J., Kim, S. H., Ki, D., Bahn, J. H., Lee, K. S., et al. (2001). Efficient intracellular delivery of an exogenous protein GFP with genetically fused basic oligopeptides. *Mol. Cells* **12**, 267–271.
17. Shokolenko, I. N. and Alexeyev, M. F. (1995). Transformation of *Escherichia coli* TG1 and *Klebsiella oxytoca* VN13 by freezing-thawing procedure. *Biotechniques* **18**, 596–598.

6

Expression of Recombinant Alkaline Phosphatase Conjugates in *Escherichia coli*

Jean-Claude Boulain and Frédéric Ducancel

Summary

The methods described in this article are relative to the use of a positive cloning/screening recombinant system for the generation in *Escherichia coli* of foreign proteins fused to a highly active bacterial alkaline phosphatase (PhoA) variant as reporter enzyme. Appropriate insertion of the DNA encoding the foreign peptides, proteic domains, or proteins between codons +6 and +7 of the *phoa* gene restores the initial frame of the *phoa* gene in the vector. Consequently, only recombinant clones appear as blue colonies when plating onto an agar medium containing a chromogenic substrate for PhoA. The presence of an intact PhoA signal peptide yields to a systematic secretion of the fusion proteins into the periplasm where the PhoA dimerises to its active form, and disulfides can be formed if necessary. The resultant PhoA-tagged proteins are particularly convenient novel tools that can be used in a wide range of applications, including expression, epitope mapping, histochemistry, immunoblotting, mutant analysis, and competition or sandwich ELISAs (*see* **Note 1**). Expression of an scFv antibody fragment derived from an IgG2a/κ immunoglobulin specific for curaremimetic toxins from snake (named M-α2-3), will be used to illustrate the methods utilized for its cloning, expression in *E.coli*, extraction, and functional characterization.

Key Words: Alkaline phosphatase; *E. coli*; recombinant protein; antibody fragments; ELISA; immunoenzymatic conjugates.

1. Introduction

Many prokaryotic expression systems are based on the fusion of a foreign peptide or protein to a bacterial partner. The resulting fusion proteins often display the combined properties of the parent proteins, a situation that often favors their production and purification and increases the solubility and the stability of the foreign moeity, which should

From: *Methods in Molecular Biology, vol. 267:*
Recombinant Gene Expression: Reviews and Protocols, Second Edition
Edited by: P. Balbás and A. Lorence © Humana Press Inc., Totowa, NJ

be finally cleaved off. Alternatively, in many applications it can be necessary to keep the two fused partners associated, and to exploit the bifunctionality of the resulting molecules. Such hybrid molecules are of first importance not only in the field of diagnosis, but also, for instance, in basic research to establish the structural and molecular basis of protein recognition process. However, relatively few efficient recombinant systems are available in that field, and consequently most molecules labeled with reporter enzymes, such as horseradish peroxidase, acetylcholinesterase, or bovine alkaline phosphatase *(1)*, are actually obtained by chemical coupling *(2)*. Recently, an original genetic strategy of coupling *(3,4)* based on the fusion of the DNA encoding various foreign molecules to the genetically engineered gene of a highly active variant of the bacterial alkaline phosphatase of *E. coli (5)* has been developed. That strategy of fusion/expression was successfully applied to produce in *E. coli* various nonprokaryotic molecules, including peptidic hormones *(6,7)*, animal toxins *(1)*, proteic domains *(8,9)*, or antibody fragments *(4,10–13)*. All these fusion proteins were secreted in the periplasm of *E. coli* where they folded correctly, yielding homogeneous, stable, and bifunctional molecules. Among these constructions, those involving monoclonal antibody fragments are especially interesting. Indeed, in the field of diagnosis they constitute the central reagent of immunometric assays, when in research they are used in a large range of applications including ELISA, Western blot, identification of antigenic domains within globular proteins, or mapping of antibody/antigen interfaces. Expression of an scFv antibody fragment derived from an IgG2a/κ immunoglobulin specific for curaremimetic toxins from snake (named M-α2-3), will be used to illustrate the methods utilized for its cloning, expression in *E. coli*, extraction, and functional characterization.

2. Materials

1. pQUANTabody expression system (Qbiogene, Inc).
2. M-α2-3 heavy and light chain cDNAs.
3. *E. coli* strain W3110.
4. Oligonucleotide primers.
5. Restriction enzymes, high-fidelity thermophilic DNA polymerase, and T4 DNA ligase.
6. Agarose, PCR, DNA sequencing, SDS-PAGE, and ELISA equipments.
7. LB (Luria-Bertani) medium.
8. Ampicillin stock solution (100mg/mL): 5g of ampicillin in 50 mL of sterile water.
9. Aliquot in 1-mL fractions, and store at −20°C.
10. XP (5-bromo-4-chloro-3-indolyl-phosphate) 2%: Mix 500 mg of Xp (Sigma Ref. B8503) with 25 mL of DMF (dimethyl formamide; Erba. Ref. 6812-2). Aliquot in 1-mL fractions, and store at −20°C.
11. IPTG (isopropyl-β-δ-thio-galactopyranoside) 0.1 *M*: dissolve 0.48 g IPTG (Eurobio Ref. 018026) into 20 mL of sterile water. Aliquot in 1-mL fractions, and store at −20°C.
12. Lysozyme solution freshly prepared at 10 mg/mL: dissolve 10 mg of lysozyme (Sigma Ref. L-6876) into 1 mL of cold TSE solution: 300 m*M* Tris-HCl, pH 8.0, 20% sucrose, and 1 m*M* EDTA.
13. 0.1 *M* PMSF (phelylmethylsulfonylfluoride): dissolve 0.174 g of PMSF powder (Sigma Ref. P-7626) into 10 mL of absolute ethanol Normapur (Prolabo Ref. 20821.296). Store the solution at −20°C.

14. Hypertonic buffer (TSE): 300 mM Tris-HCl, pH 8.0, 20% sucrose, and 1 mM EDTA.
15. Coating solution: 50 mM Tris-HCl, pH 7.4, and 10 μg/ml toxin solution.
16. Saturation buffer: 100 mM Tris-HCl, pH 7.5, and 5% BSA.
17. Washing buffer: 10 mM Tris-HCl, pH 7.5, and 0.5% Tween.
18. Dilution buffer: 100 mM Tris-HCl, pH 7.5, and 0.1% BSA.
19. DEA buffer, pH 10.0: 1M diethanolamine (DEA), 1 mM MgCl$_2$, 20 mM ZnCl$_2$.
20. Revelation buffer, pH 10.0: 1M DEA, 1 mM MgCl$_2$, 20 mM ZnCl$_2$, and 5 mM pNPP.

3. Methods

The methods described below outline (1) the construction by overlapping polymerase chain reaction (PCR) of the DNA encoding the scFv fragment, (2) the induction of protein expression, (3) the extraction and the calibration of the periplasmic extract, and (4) the functional characterization of the recombinant hybrid protein.

3.1. Expression Plasmid

Since the description of the first PhoA tagged exogen protein *(3)*, we have progressively engineered the initial vector to finally end at the two last versions, named pQUANTabody and pQUANTagen, which have been commercialized by Q.Biogen Company (PhoA*Color System). Whereas pQUANTabody has been especially designed to express scFv antibody fragments, pQUANTagen (*see* **Note 1**) is an expression system devolved to the expression as PhoA hybrids of any other foreign peptide, proteic domain, or protein.

The pQUANTabody (**Fig. 1A**) expression system (Q.Biogen) contains a p15A origin of replication *(14)*. The cloning region is under the transcriptional control of the inducible *tac* promoter and the *Lac*Iq repressor. The signal peptide, together with the transcription terminator, are those of the natural *phoA* gene. Resistance to ampicillin is confered by the β-lactamase gene. Oriented insertion of the DNA coding for the foreign fusion partner is mediated by the rarely found restriction sites *Sfi*I and *Not*I, which should be present at the 5' and 3' extremities of the insert, respectively. In the empty vector, the *phoA* gene is out of frame, resulting in white colonies when transformed bacteria are plated on solid LB/agar/Amp/XP culture medium. Only the cloning of a DNA fragment organized as described in **Fig. 1B** would restore the reading frame of the *phoA* gene. Recombinant bacterial clones could therefore be directly visualized as blue colonies.

The fusion pQUANTabody expression vector contains a mutated version of the bacterial alkaline phosphatase gene, with improved catalytic properties *(5)* (*see* **Note 2**). To ensure a systematic processing of the PhoA signal peptide, the cloning site is located between codons +6 and +7 of the *phoA* gene. The periplasmic localization of the PhoA tagged proteins ensures dimerization of the PhoA moeity into its enzymatically active form, and also the correct folding of the foreign partners whose 3-D structure is reliable to the formation of disulfides, i.e., scFv antibody fragments. Together these characteristics make pQUANTabody and pQUANTagen (*see* **Note 1**) suitable expression systems to produce fully bifunctional PhoA tagged peptides, proteic domains, or proteins, which constitute useful research tools.

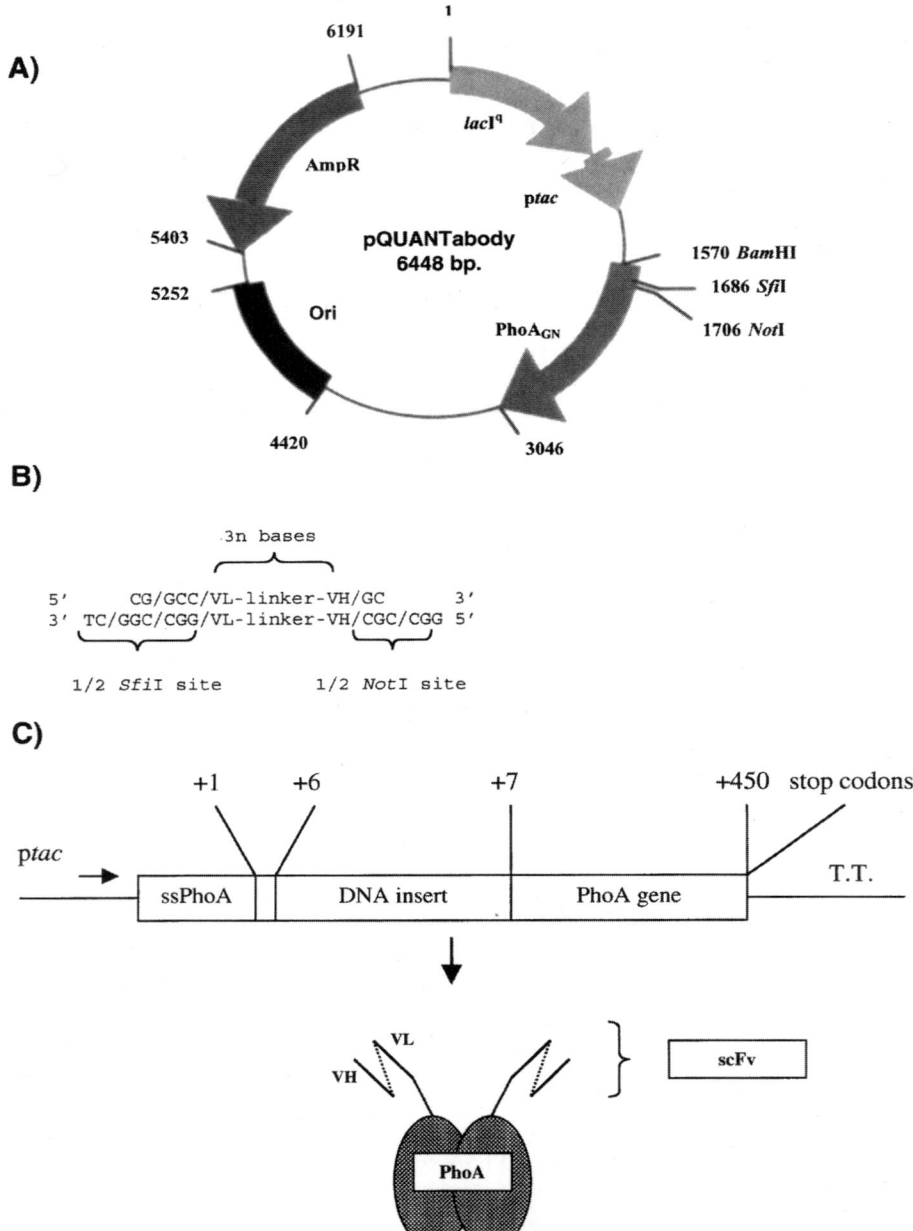

Fig. 1. (**A**) Schematic drawing of pQUANTabody expression plasmid. (**B**) Schematic organization of the DNA fragment necessary to be cloned into pQUANTabody to restore the frame of the PhoA gene. Codons are indicated by <</>> (**C**) Schematic representation of the expression cassette together with its product once secreted into the periplasm. T.T. = transcriptional terminator.

3.2. Construction of the DNA Encoding an scFv Antibody Fragment

Obtaining of the DNA encoding a functional scFv fragment should be described in details since it necessitates to perform standard but also particular DNA manipulations (*see* **Note 3**). The strategy described below can be considered as a general one, since up to now we applied it successfully to more than 10 antibodies having different specificities of recognition against haptens, peptides, or proteins, and characterized by variable isotypes. We will illustrate it in the case of a neutralizing anti-short-chain snake neurotoxins named M-α2-3 *(4,7)*.

3.2.1. Cloning of the Heavy- and Light-Chain Precursors

1. Culture the hybridoma cells producing M-α2-3 in IMDM (Iscove's modified Dulbecco medium) containing 10% fetal calf serum (Gibco-BRL) and antibiotics (100 μg/ml) at 37°C in a 5% CO_2 humidified chamber.
2. Extract and purify poly(A+) RNAs from 1×10^7 hybridoma cells using a mRNA purification kit (Amersham Biosciences) according to the manufacturer's instructions.
3. Perform the reverse transcription of the poly(A+) from 0.1–1 μg mRNA using a RT-PCR kit (Amersham Biosciences) according to the manufacturer's instructions. The backward primers used were deduced from immunoglobulin consensus sequences *(15)*, and correspond to the C-terminal extremities of mouse IgG2a/CH1 and κ/CL domains in the case of M-α2-3.
4. Isolate the cDNAs encoding Fd domain (VH-CH1) and complete light chain (VL-CL) of M-α2-3 antibody using the rapid amplification of cDNA ends (RACE) method as previously described *(16)*.
5. Ligate directly obtained cDNAs into the pCR®2.1 cloning vector using a TA cloning kit (Invitrogen) according to the manufacturer's instructions.
6. Determine the DNA sequences of the cloned inserts.

3.2.2. Assembling of the scFv DNA

Figure 2 illustrates the strategy followed to assemble the DNA encoding the VL-linker-VH scFv fragment of M-α2-3, which combines classical PCR steps and «PCR»-based single overlap extension *(17)*. An 18-amino-acid linker coding sequence, Nter-GSTSGSGKPGSGEGSTKG-Cter, was used to covalently associate the C- and N-terminal extremities of the VL and VH domains, respectively *(18)*. The VL domain was amplified using M-α2-3VLFORWARD primer containing the *Sfi*I sequence, and the six first codons of the VL N-terminus, in combination with M-α2-3VLBACKWARD: the six last codons of the VL C-terminus extremity, followed by the complementary sequence of the two first thirds of the linker (**Fig. 2A**, left part). Amplification of the VH domain was obtained using M-α2-3VHBFORWARD encoding: the two last thirds of the linker, followed by the six first codons of the VH N-terminus extremity, in combination with M-α2-3VHBACKWARD corresponding to the six last codons of the VH domain followed by *Not*I complementary restriction site sequence (**Fig. 2A,** right part). The resulting PCR products are then used to assemble the complete scFv-encoding DNA fragment (**Fig. 2B**).

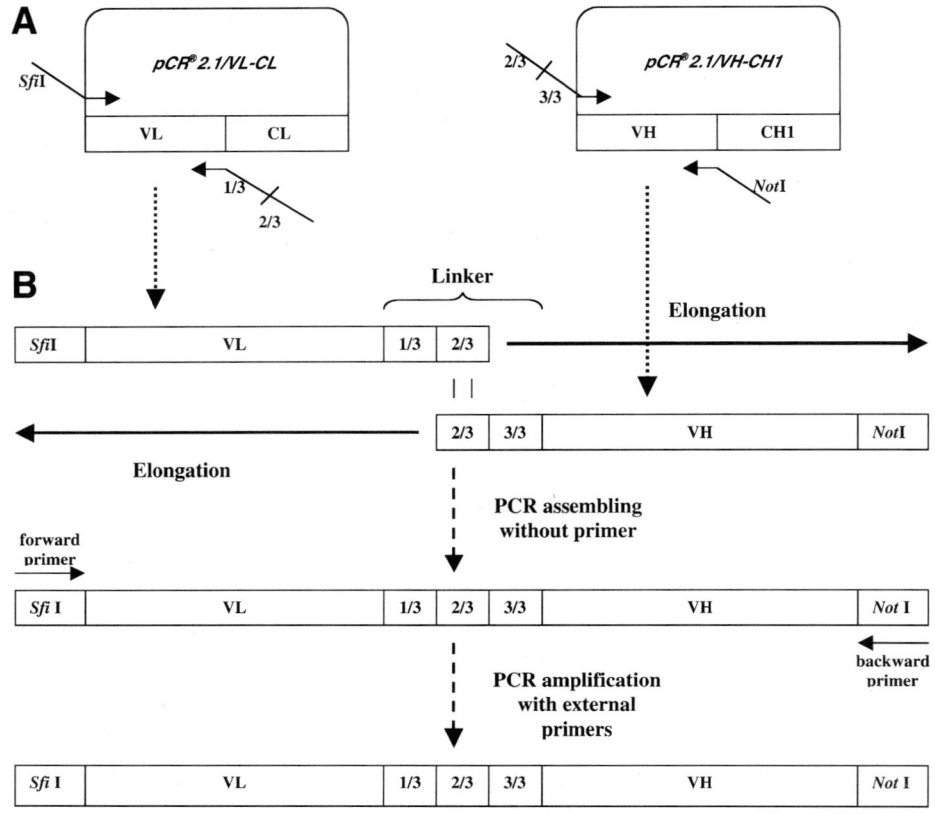

Complete scFv's DNA

Fig. 2. Schematic representation of the strategy followed: (**A**) to amplify the VL and VH domains of M-α2-3, and (**B**) to assemble the DNA encoding a complete scFv antibody fragment.

1. Prepare a solution at 1 μg/mL of the two plasmids containing the DNA encoding the VL/CL and VH/CH1 domains of M-α2-3, using standard molecular biology methods (*19*).
2. Proceed to PCR amplifications using 5 ng of each plasmid DNA solution in combination with the two previous sets of primers, a high-fidelity thermophilic DNA polymerase, according to the following PCR parameters: (a) 1 × [5 min at 95°C], (b) 30 × [1 min at 94°C; 1 min at 55°C; 1 min at 72°C], (c) 1 × [10 min at 72°C].
3. Check on agarose gel the size of the amplified DNA, and proceed to their purification using a QIAEX II Extraction Kit (Qiagen).
4. Proceed to the assembing of the DNA encoding the complete scFv by addition in the same PCR tube of equimolar amounts (50–100 ng) of each extracted PCR products, without primer, in presence of a high-fidelity thermophilic DNA polymerase, according to the following PCR parameters: (a) 1 × [5 min at 95°C], (b) 10 × [1 min at 94°C; 1 min at 50°C; 1 min at 72°C], (c) 1 × [10 min at 72°C].

5. Amplify the DNA encoding the complete scFv fragment using M-alpha2-3VLFORWARD and M-α2-3VHBACKWARD primers according to the following PCR parameters: (a) 1 × [5 min at 95°C], (b) 30 × [1 min at 94°C; 1 min at 50°C; 1 min at 72°C], (c) 1 × [10 min at 72°C].

6. Check on agarose gel the size of the amplified DNA, and proceed to their purification using a QIAEX II Extraction Kit (Qiagen).

7. Perform a classical ethanol precipitation, wash and dry the pellet, resuspend it in sterile water, and determine the concentration of the scFv encoding DNA solution at 260 nm.

3.2.3. Cloning of the scFv Encoding DNA

*Sfi*I and *Not*I restriction sites have been chosen owing to their low abundancy in DNA-encoding IgGs. However, they are charaterized by different cleavage conditions, which necessitates a two-step digestion:

1. Digest 1 µg of scFv DNA in a final volume of 50 µL containing the specific reaction buffer of *Sfi*I (BioLabs), supplemented with BSA, and 20 enzyme units (1 µL). Cover the reaction solution with one drop of mineral oil (Perkin-Elmer), and incubate 2 h at 50°C. If necessary, for instance to increase the cleavage efficacy, add an extra microliter of *Sfi*I enzyme solution into the aqueous phase (beneath the oil surface), and re-incubate for 2 h.

2. Perform an ethanol precipitation, wash and dry the pellet, resuspend it in sterile water, and determine its concentration at 260 nm.

3. According to the manufacturer conditions, proceed in a final volume of 50 µl to the digestion by *Not*I (BioLabs) at 37°C during 2 h.

4. Precipitate the cleaved DNA, determine its concentration at 260 nm, and evaluate on agarose gel (by comparison with undigested DNA) the efficacy of the double digestion.

5. After standard ligation into *Sfi*I and *Not*I restriction sites of pQUANTabody vector (*see* **Note 3**), the DNA is used to transform *E. coli* W3110 (*see* **Note 4**) cells by standard CaCl$_2$ method *(19)*.

6. Plate transformed bacteria on LB/agar/Ampi/Xp Petri dish plates, and incubate overnight at 37°C (100 µL XP/50 mL LB).

3.3. scFv-PhoA Expression

The next step is the production the scFv-PhoA recombinant protein and its functional characterization. The selection of the transformed bacteria is easy and visual, since it is based on restoration of the *phoA* gene reading frame only upon correct insertion of the scFv encoding DNA (*see* **Note 5**). It is noteworthy that appearance of blue colonies indicates not only that the fused foreign molecule does not affect the enzymatic activity of PhoA, which is able to dimerise in its active form, but also that hybrids are correctly secreted within the periplasm.

3.3.1. Protein Expression

1. Select single blue colonies, and grow overnight at 37°C in 5 mL of LB media containing ampicillin. Classically, two or three blue clones are selected, and analyzed in parallel.

2. Inoculate individual flasks containing 50 mL of LB/amp media with 1 mL of of the overnight cultures and allow to grow at 37°C to an optical density of 0.6–0.7 at 600 nm.

3. Induce the cells with IPTG (0.5 m*M*) for 4 h, at 30°C. It should be noted that from one antibody fragment to another, the previous induction conditions can result in variable levels and

quality of expression. Classically, reduction of the IPTG concentration simultaneously with the decrease of the culture temperature, and thus use of a longer time of induction, can increase the yield of expression.

3.3.2. Protein Extraction

The steps described below outline the procedure classically followed for extraction of the crude periplasmic fluid in a manner that gives enough recombinant proteins to carry out their structural and functional characterizations by Western blot and ELISA, respectively.

1. After the 4 h of induction, harvest the cells by centrifugation (5000 rpm; 10 min at 4°C), and discard the supernatant.
2. Resuspend each cell pellet in 5 mL of cold TSE extraction solution, supplemented with 50 μL of a freshly prepared solution of lysozyme (10 mg/mL), and 25 μL of a PMSF solution at 10 mg/mL.
3. Transfer the bacteria solution into a 15 ml falcon tube, and incubate 20 min at 4°C with agitation, then centrifuge the bacteria solution at 13,000 rpm for 30 min at 4°C.
4. Transfer the supernatant, which corresponds to the periplasmic fraction, into a new 15-mL Falcon tube.
5. Filter the supernatant through a 0.45-μM membrane to remove cell debris.
6. To obtain optimal results it is recommended to use the crude extract immediately. However, the extract may be frozen at −20°C for up to 2 wk, without significant breakdown of the hybrid proteins.

3.4. Characterization of scFv-PhoA Hybrids

The periplasmic extracts prepared as described above can be used directly to perform enzymatic assays using the bifunctional characteristics of the generated hybrids. Indeed, only the specific recognition and/or interaction mediated by the foreign molecule, i.e., an scFv fragment should be revealed by the enzymatic capacity of the fused PhoA in the presence of an appropriate chromogenic substrate. These could be XP if a precipitate is required—e.g., Western blot, or pNPP if the reaction requires a soluble chromogen—e.g., ELISA.

3.4.1. Soluble PhoA Activity

Determination of the soluble PhoA activity contained in the periplasmic extracts first allows to follow the overall level of expression of the PhoA hybrids, and second, standardizing the volume of periplasmic extract that should be used in ELISA. Measurement of the soluble PhoA activity is performed in a diethanolamine assay *(20)*. Since 1 mg of mutated bacterial PhoA corresponds to 8×10^5 optical units at 410 nm, one can evaluate the hybrid protein concentration, which classically varies from 0.1–10 mg/L depending on the type of vector and insert used. Note that the amount of enzymatic tracer produced using the pQUANTabody is generally smaller than that produced with the pQUANTagen vectors. This is in general because antibody fragments are less efficiently produced.

1. Perform different dilutions (ranging from ½ to ¹⁄₁₀) of the periplasmic extracts in DEA buffer pH 10.0.
2. Add 10 μL of each dilution in 490 μL of revelation buffer, and incubate 10 min at 37°C.

3. Stop the colorimetric reaction with 0.5 mL of NaOH 1M, and read the optical density at 410 nm against a control tube without periplasmic extract.
4. Evaluation of the hybrid concentration should be done using the linear part of the obtained dilution curve.

3.4.2. Western Blot Characterization

The quality of the scFv-PhoA hybrids contained in the crude extracts is analyzed by Western blot using an anti-*E. coli* alkaline phosphatase monoclonal antibody (VIAP, Caltag Laboratories). This allows to establish the ratio of recombinant scFv-PhoA hybrids versus free PhoA, and thus give an indication of the stability of the hybrids (**Fig. 3**).

1. Apply 20 µL of each periplasmic extract to PAGE-SDS 12% under reducing conditions, then blot on a nitrocellulose membrane.
2. Wash the membrane, and immunoreveal by 1-h incubation with a 1/2500 dilution in PBS-Tween 0.1% of the VIAP antibody.
3. Wash the membrane three times, prior to addition of a 1/2000 dilution of the secondary antibody directed against mouse IgG labeled with horseradish peroxydase (Jackson Immuno-Research Laboratories).

3.4.3. ELISA Characterization

Finally, the ability of each scFv to bind to its natural antigen—i.e., a short-chain neurotoxin from snake—is evaluated by ELISA via the genetically fused PhoA moeity (*see* **Note 6**). Direct ELISA (in the absence of soluble competitor) should be used to establish the overall binding capacity of the scFv fragment, and the volume of periplasmic extract that should be used in competitive ELISA. The specificity of recognition and sensitivity toward coated antigen should be determined by competitive ELISA.

3.4.3.1. DIRECT ELISA

1. Adsorb passively the toxin solution on microtitration weels (Maxisorb, Nunc) at 10 µg/mL in the dilution buffer overnight at room temperature.
2. Wash the plate three times with the washing buffer, then block with the saturation buffer for 1 h at 37°C.
3. Add decreasing concentrations of the standardized crude periplasmic fluid, incubate 1 h at 37°C, then wash three times. Note that "standardized" means an initial dilution of the crude extract giving an OD_{410nm} of 0.3–0.5 in the experimental conditions described in the **Subheading 3.4.1**.
4. Add 200 µL of revelation buffer, and read at 410 nm.

3.4.3.2. COMPETITIVE ELISA

The major difference as compared to the procedure described in **Subheading 3.4.3.1**. concerns **step 3**. Indeed, to ensure a good sensitivity of the competitive assay, the concentration of crude periplasmic extract used, should correspond to the dilution giving 50% of the maximum signal as obtained previously. Once determined, 50 µL of that dilution are added in each microtitration well and incubated together with 50 µL of soluble toxin solution ranging from 1 to 10^4 nM. The resulting inhibition curve obtained

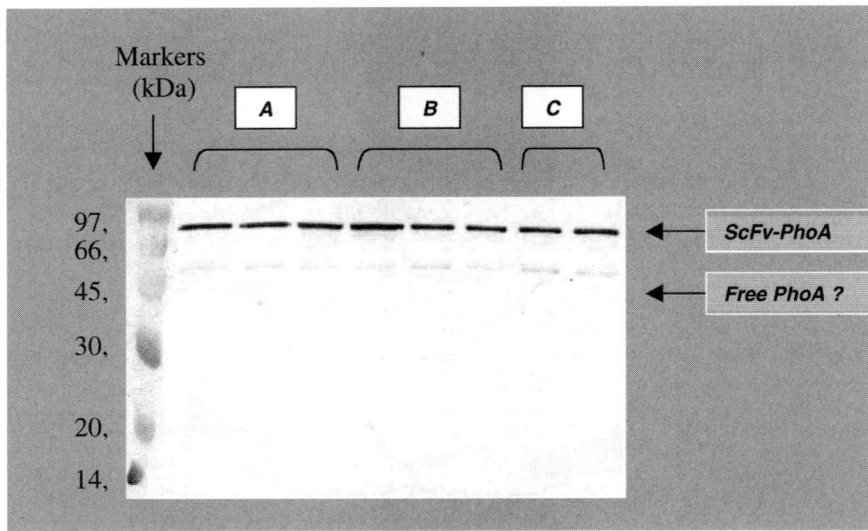

Fig. 3. SDS-PAGE and Western blot analysis of different crude periplasmic fractions. Reve-lation was achieved using an anti-PhoA monoclonal antibody. (**A**) Periplasmic extracts from three different blue colonies obtained from bacteria transformed with the pQUANTabody vector encoding the wild-type scFv$_m$-α2-3/PhoA hybrid. (**B**) and (**C**) are periplasmic extracts of two point mutations in the scFv moiety.

with the recombinant scFv-PhoA hybrid, establishes that the sensitivity of detection of curaremimetic toxins is in the picomole range *(11)*.

4. Notes

1. The plasmid pQUANTagen is an expression vector having the same overall characteristics as pQUANTabody, except that it contains a multiple cloning site (MCS) composed of following cloning sites: *Kpn*I, *Sal*I, *Sna*BI, *Bgl*II, *Nde*I, *Sac*I, *Sma*I/*Xma*I. pQUANTagen (kx) or (xk) have their MCS inverted. To restore the native reading frame of the *phoA* gene, the foreign DNA sequence to be inserted into the pQUANTagen vectors should contain 3n + 1 nucleotides (3 = any complete codon) in addition to the required half-restriction sites and 3n bases in the case of pQUANTabody (*see* **Fig. 1B**). The foreign sequence should start with a complete codon and end with a codon plus one nucleotide. Please note that different evolu-tions of these two fusion/expression systems are planned, i.e., to allow affinity purification of the PhoA hybrids, and to clone directly PCR products. Altogether, these vectors allow the construction of any PhoA tagged proteic molecule, which are useful in a large range of appli-cations in academic research and applied purposes. The most obvious one is the easy design of laboratory-made immunoenzymatic tracers to efficiently and rapidly develop enzyme immunoassays. Another important type of application consists of creating libraries of fusion genes expressing peptides or protein domains derived from whole globular molecules, to map antigenic epitopes or functional domains *(8,9)*.
2. The bacterial *phoA* gene used in the pQUANT vectors contains two point mutations: Asp153Gly and Asp330Asn selected by directed evolution process *(5)*. It results in a highly

active bacterial variant with high thermostability. It is noteworthy that the optimal enzymatic activity of that bacterial variant is obtained under the assay conditions used in the case of the bovine intestinal phosphatase (1 M diethanolamine, pH 10.0, at 37°C).

3. The two external primers used during preparation of the scFv encoding DNA—M-α2-3VLFORWARD and M-α2-3VHBACKWARD—display at their 5' ends an extension of 15 base pairs, which increase the yield of cleavage by the restriction enzymes *Sfi*I and *Not*I. During ligation of the cleaved scFv encoding DNA into pQUANTabody expression vector, different insert/vector ratios (1/1, 5/1, and 10/1) should be tested in parallel.

4. The bacterial strain W3110 has been selected since in general it gives better yields of scFv-PhoA production. Bacterial strains XL1-Blue, JM101, or MC1061 can also be used. The use of a PhoA⁻ strain is not absolutely necessary, as the natural chromosomal PhoA promoter is totally repressed if inorganic phosphate is available, which is the case in any "rich" culture medium such as LB.

5. The presence of the DNA inserted into the *phoA* gene can be confirmed by PCR directly on blue clones using surrounding primers as provided in the PhoA* Color System commercialized by Q.Biogen Company. Furthermore, the presence of transformed blue clones constitutes an indication that the molecule inserted into the N-terminus extremity of the bacterial PhoA does not perturb significantly the enzymatic activity.

6. The example taken in this chapter deals with the characterization of a wild-type scFv antibody fragment. However, we widely applied that strategy of expression to rapidly map the functional paratope of numerous antibodies specific for variable antigens *(11,12,21)*. In these cases, mutated scFv fragments, as obtained by site-directed mutagenesis, were expressed fused to PhoA, and their recognition capacity toward their natural antigen established in a direct ELISA by comparison to the corresponding wild-type antibody fragment. Interestingly, in the case of two anti-steroid antibodies, we recently had a complete confirmation by x-ray crystallography data of the mapping we proposed by site-directed mutagenesis using the PhoA tagged expression system *(22)*.

References

1. Porstmann, T., and Kiessig, S. T. (1992) Enzyme immunoassay techniques: an overview. *J. Immunol. Methods* **150,** 5–21.
2. Lindbladh, C., Mosbach, K., and Bülow, L. (1993) Use of genetically perpared enzyme conjugates in enzyme immunoassay. *TIBS* **18,** 279–283.
3. Gillet, D., Ducancel, F., Pradel, E., Léonetti, M., Ménez, A., and Boulain, J.-C. (1992) Insertion of a disulfide-containing neurotoxin into *E. coli* alkaline phosphatase: the hybrid retains both biological activities. *Protein Eng.* **5,** 273–278.
4. Ducancel, F., Gillet, D., Carrier, A., Lajeunesse, E., Ménez, A., and Boulain, J.-C. (1993) Recombinant colorimetric antibodies: construction and characterization of a bifunctional F(ab)2/alkaline phosphatase conjugate produced in *Escherichia coli. Nature/Biotechnol.* **11,** 601–605.
5. Muller, B. H., Lamoure, C., Le Du, M.-H., Cattolico, L., Lajeunesse, E., Lemaître, F., et al. (2001) Improving *Escherichia coli* alkaline phosphatase efficacy by additional mutations inside and outside the catalytic pocket. *Chembiochem.* **2,** 517–523.
6. Gillet, D., Ezan, E., Ducancel, F., Gaillard, C., Ardouin, T., Istin, M., et al. (1993) Enzyme immunoassay using a rat prolactin-alkaline phosphatase recombinant tracer. *Analytical Chem.* **65,** 1779–1784.

7. Chanussot, C., Bellanger, L., Ligny-Lemaire, C., Seguin, P., Ménez, A., and Boulain, J.-C. (1996) Engineering of a recombinant colorimetric fusion protein for immunodiagnosis of insulin. *J. Immunol. Methods* **197,** 39–41.

8. Schichtholz, B., Legros, Y., Gillet, D., Gaillard, C., Marty, M., Lane, D., et al. (1992) The immune response to p53 in breast cancer patients is directed against immunodominant epitopes unrelated to the mutational hot spot. *Cancer Research* **52,** 6380–6384.

9. Hebrard, E., Drucker, M., Leclerc, D., Hohn, T., Uzest, M., Froissart, R., et al. (2001) Biochemical characterization of the helper component of the *Cauliflower mosaic virus. J. Virology* **75,** 8538–8546.

10. Carrier, A., Ducancel, F., Settiawan, N.B., Cattolico, L., Maillère, B., Léonetti, M., et al. (1995) Recombinant antibody-alkaline phosphatase conjugates for diagnosis of human IgGs: application to an anti-HBsAg detection. *J. Immunol. Methods* **181,** 177–186.

11. Mérienne, K., Germain, N., Zinn-Justin, S., Boulain, J.-C., Ducancel, F., and Ménez, A. (1997) The functional architecture of an acetylcholine receptor-mimicking antibody. *J. Biol. Chem.* **272,** 23775–23783.

12. Bettsworth, F., Monnet, C., Watelet, B., Battail-Poirot, N., Gilquin, B., Jolivet, M., et al. (2001) Functional characterization of two anti-estradiol antibodies as deduced from modelling and site-directed mutagenesis experiments. *J. Mol. Recognition* **14,** 99–109.

13. Muller, B. H., Chevrier, D., Boulain, J.-C., and Guesdon, J.-L. (1999) Recombinant single-chain Fv antibody fragment-alkaline phosphatase conjugate for one-step immunodetection in molecular hybridization. *J. Immunol. Methods* **227,** 177–185.

14. Lazzaroni, J. C., Atlan, D., and Portalier, R. C. (1985) Excretion of alkaline phosphatase by *E. coli* K-12 *pho* constitutive mutants transformed with plasmids carrying the alkaline phosphatase structural gene. *J. Bacteriol.* **164,** 1376–1380.

15. Kabat, E. A., Wu, T. T., Bilofsky, H., Reid-Milner, M., and Perry, H. (1991) *Sequences of immunological interest.* Washington, DC, Public Health Service, N.I.H.

16. Ruberti, F., Cattaneo, A., and Bradbury, A. (1994) The use of the RACE method to clone hybridoma cDNA when V region primers fail. *J. Immunol. Methods* **173,** 33–39.

17. Horton, R. M., Hunt, H. D., Ho, S. N., Pullen, J. K., and Pease, L. R. (1989) Engineering hybrid genes without the use of restriction enzymes: gene splicing by overlap extension. *Gene* **77,** 61–68.

18. Whitlow, M., Bell, B. A., Feng, S. L., Filpula, D., Hardman, K. D., Hubert, S. L., et al. (1993) An improved linker for single-chain Fv with reduced aggregation and enhanced proteolytic stability. *Protein Engin.* **6,** 989–995.

19. Sambrook, J., Fritsch, E. F., and Maniatis, T. (1989) *Molecular Cloning, A Laboratory Manual,* 2 ed. Cold Spring Harbor Laboratory Press, Cold Spring Harbor, NY.

20. Walter, K., and Schutt, C. (1976) *Methods in Enzymatic Analysis,* 2 ed. Academic Press Inc, New York, NY.

21. Germain, N., Mérienne, K., Zinn-Justin, S., Boulain, J.-C., Ducancel, F., and Ménez, A. (2000) Molecular and structural basis of the specificity of a neutralizing acetylcholine receptor-mimicking antibody, using combined mutational and molecular modeling analyses. *J. Biol. Chem.* **275,** 21578–21586.

22. Monnet, C., Bettsworth, F., Stura, E. A., Le Du, M.-H., Ménez, R., Derrien, L., et al. (2002) Highly specific anti-estradiol antibodies: structural characterisation and binding diversity. *J. Mol. Biol.* **315,** 699–712.

7

Overexpression of Chromosomal Genes in *Escherichia coli*

Fernando Valle and Noemí Flores

Summary

Conversion of some carbon sources into desired compounds by a biological system is the goal of many biotechnologists. The understanding of the mechanisms by which an organism does these conversions permits the improved production of specific metabolites. Pathway engineering involves the strategies to modify cells to overproduce desired molecules. We describe here the methodology to modify chromosomal genes by replacing their native regulatory regions with promoter cassettes to increase or deregulate expression of chromosomal genes.

Key words: *E. coli*; chromosomal integration; chromosomal promoter replacement; Cre-*lox*.

1. Introduction

The use of genetically modified microorganisms to obtain valuable commercial products is wide spread. This biotechnological exploitation of microorganisms is based on their tremendous catabolic and biosynthetic capabilities, which enable them to utilize inexpensive substrates for growth and produce a wide range of biomolecules. The purposeful modification of cells to redirect their physiology to overproduce the desired biomolecules has been defined as "pathway engineering" and very often requires extensive remodeling of the microbial genome.

The type and number of genes that need to be modified varies depending on the biomolecule(s) being overproduced. To improve cell performance, it is common that the expression level of several native genes need to be changed. For example, in an *Escherichia coli* strain constructed to overproduce D-lactic acid, the chromosomal genes *pflB, frdBC, adhE,* and *ackA* were deleted (*1*). It can be said that a successful commercial biotechnological process will depend on the ability to balance properly the expression of relevant genes, fulfill the needs of the cells, and sustain the production

From: *Methods in Molecular Biology, vol. 267:*
Recombinant Gene Expression: Reviews and Protocols, Second Edition
Edited by: P. Balbás and A. Lorence © Humana Press Inc., Totowa, NJ

process. Achieving this balance can be time-consuming and in general depends on trial-and-error experiments. The methodology described in this chapter allows the modification of chromosomal genes to quickly replace their native regulatory regions with promoter cassettes and increase or deregulate expression of chromosomal genes. It is based on the use of a strong promoter linked to an excisable antibiotic cassette (*2*), and the λ-red system to facilitate homologous recombination (*3*). The approach is simple and can be used numerous times in the same strain. Furthermore, this approach can be enhanced by using promoter variants that provide different levels of expression or modes of regulation.

2. Materials

1. Bacterial strains: MG1655 (ATCC 47076).
2. pLoxCat2 plasmid (*2*).
3. pTrcCm2 (**Fig. 1**).
4. LB broth: 10 g Bactotryptone, 5 g yeast extract, and 10 g NaCl per liter and pH is adjusted to 7.0–7.2 with NaOH. Solidified LB for plates contains 25 g/L agar (LA plates).
5. Carbenicillin (Carb) and chloramphenicol (Cm) are from Sigma. Strains containing plasmids are selected by using media containing carbenicillin at 50 gmg/mL or chloramphenicol 20 μg/mL.
6. LA medium: LB media solidified with 2.5% BactoAgar containing 10 μgmL of chloramphenicol.
7. SOB medium (per liter): 20 g tryptone, 5 g yeast extract, 0.5 g NaCl, and 2.5 mL 1 M KCl solution. After autoclaving, add 10 mL sterile 1 M MgCl$_2$.
8. SOC medium is prepared as described for SOB medium, but 20 mL of sterile 1 M glucose are added also.
9. Z buffer for β-galactosidase activity: 0.06 M Na$_2$HPO$_4$, 0.04 M NaH$_2$PO$_4$, 0.01 M KCl, 0.001 M MgSO$_4$, 0.05 M β-mercaptoethanol.
10. There is a wide variety of electroporators and Polymerase chain reaction (PCR) machines. Protocols described in this chapter were carried out using a gene pulser from BioRad (Hercules, CA), and a Mastercycler gradient PCR machine from Eppendorf.
11. Synthetic primers were from commercial suppliers.
12. Plasmid DNA was isolated from liquid cultures using a Qiaprep Spin Miniprep kit from Qiagen (Valencia, CA).
13. Agarose gel electrophoresis equipment.
14. PCR products were purified from agarose gels using the Zymoclean kit from Zymo Research.
15. DNA sequencing was done by automated sequencing in a core facility.

3. Methods

A general scheme of the approach is shown in **Fig. 2**. The overall principle is to generate, by PCR, a mutagenic cassette to precisely replace the native regulatory region of a gene, with a promoter linked to an excisable antibiotic marker.

The methods outlined below describe (1) a PCR-based method to obtain a mutagenic cassette to replace regulatory regions of chromosomal genes (2) the integration of the resulting mutagenic cassette into the *E. coli* chromosome in single copy, (3) verification of the genomic modification, and (4) removal of the chloramphenicol antibiotic marker.

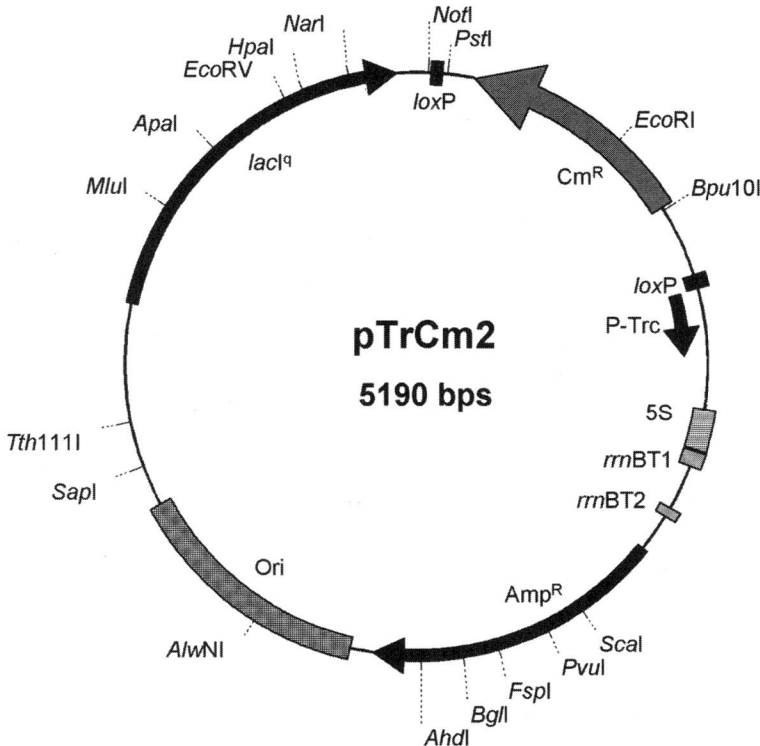

Fig. 1. PCR template plasmid. PTrCm2 is a derivative of pTrc99a, where a chloramphenicol resistance gene (Cm) was introduced upstream of the Trc promoter (P-Trc) at a *Bsp*M 1 site. The Cm gene is flanked by *loxP* sites (black boxes).

3.1. Plasmid pTrcCm2 as a Source of a Strong Promoter Cassette

Plasmid pTrcCm2 is a derivative of pTrc99a where the loxP-cat2 cassette (*2*) was introduced upstream to the Trc promoter at the *Bsp*M 1 site. The general characteristics of pTrcCm2 are shown in **Fig. 1**.

The Trc promoter (P_{Trc}) is a strong promoter that contains the -35 and upstream region of the *E. coli* tryptophan promoter, and the -10 of the *lacZ* promoter (*4*), and it is partially regulated by LacI. To facilitate promoter replacement, we have linked P_{Trc} to an excisable antibiotic marker. This allows the successive modification of multiple genes, without leaving markers in the chromosome (*see* **Note 1**).

3.2. Generation of Mutagenic Cassettes

The method described by Datsenko and Wanner (*3*) utilizes 30–50 nucleotides as regions of homology, to promote homologous recombination between PCR products and the *E. coli* chromosome. Part of the primer sequence must be homologous to the ends of

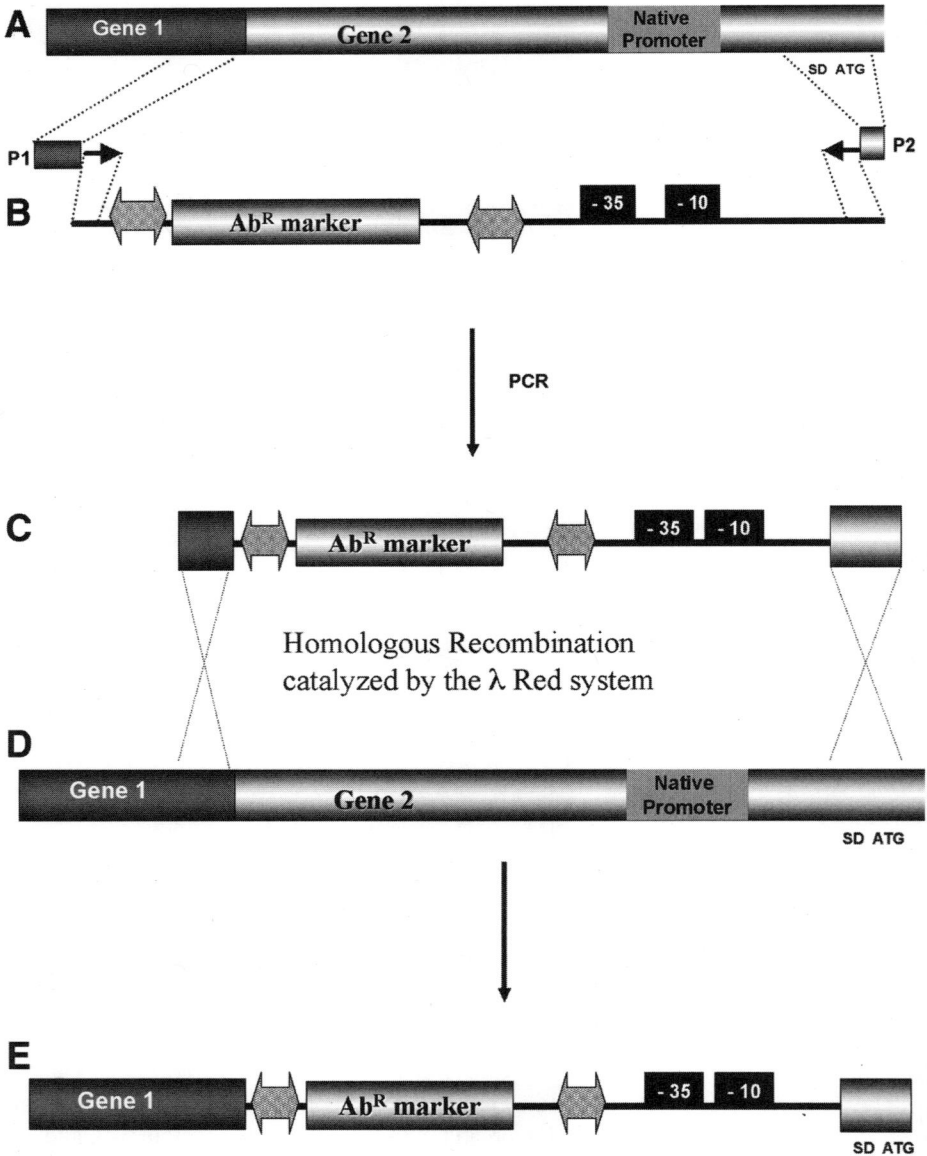

Fig. 2. General strategy to replace a chromosomal regulatory region. The substitution of the native promoter of gene 2 (**A**), is accomplished by first creating a mutagenic cassette by PCR using primers P1 and P2; a cassette that contains a promoter (represented by the two black boxes), and an antibiotic marker (Ab^R). The Ab^R is flanked by *loxP* sites (two-headed arrows) (**B**). Primers P1 and P2 contain sequences complementary to chromosomal regions and to the Ab^R (dotted lines). The PCR product (**C**) is used to replace the original region present in the chromosome (**D**) by homologous recombination catalyzed by the λ Red system. After recombination, the PCR product replaced the original regulatory region (**E**).

the segment to be inserted, and the other part homologous to the chromosomal region to recombine. We used such a method to replace the native regulatory regions of chromosomal genes with P_{Trc}. For such a purpose, the pTrcCm2 was used as templates for PCR reactions. The primers designed to amplify the P_{Trc}-chloramphenicol cassette contained also the regions of homology necessary to allow homologous recombination with the chromosome. To exemplify our approach, we describe here the replacement of the regulatory region of the *lacZYA* operon. **Figure 3A** shows the details of the area where the modifications were made.

Primers lacZ1:

5'-AGCGCAACGCAATTAATGTGAGTTAGCTCACTCATTAGGGATGCATATG-GCGGCCGCA-3'

and lacZ2:

5'-GTCACGACGTTGTAAAACGACGGCCAGTGAATCCGTAATCATGGTCTG-TTTCCTGTGTGAAA-3'

were designed to contain 20 nucleotides complementary to the pTrcCm2 and 39 nucleotides complementary to the *lacZYA* regulatory region (**Fig. 3B**). Using these primers, a 1333-bp DNA fragment was generated by PCR.

3.3. Chromosomal Replacement of lacZ Regulatory Region

1. Competent cells of strain MG1655 containing plasmid pKD46 are prepared accordingly to Datsenko and Wanner (*3*) (*see* Chapter 8).
2. 100 µL of competent cells are transformed by electroporation with 20–100 ng of the 1333-bp PCR product described in **Subheading 3.2.** (*see* **Note 2**).
3. After recovering the cells for 1 h in 1 mL of SOC media, they are plated on four Petri dishes with LA media, containing 10 µg/mL of chloramphenicol.
4. Plates are incubated at 37°C for at least 16 h. Colonies are transferred to a fresh LA medium + 10 µg/ml chloramphenicol plate (*see* **Note 3**).

3.4. Verification of the Chromosomal Modification

Chromosomal DNAs from the MG1655 and some of the transformants obtained by the procedure described in **Subheading 3.3.** are purified and used as substrates for PCR reaction using primers

lacT1 5'-GGCACGACAGGTTTCCCGAC-3' and

lacT2 5'-GAGGGGACGACGACAGTATC-3'

These two primers hybridize with regions outside of where lacZ1 and lacZ2 hybridize (*see* **Note 4**) and should generate PCR products of the following sizes:

MG1655—425 bp.

MG1655:: P_{Trc}-Cm-*lacZ*—1585 bp.

The PCR products were separated in a 2% agarose gel. As can be seen in **Fig. 4**, the expected sizes were obtained for our model system (lanes 1 and 2). To ensure that unexpected mutations were not introduced in the process, the PCR products were sequenced.

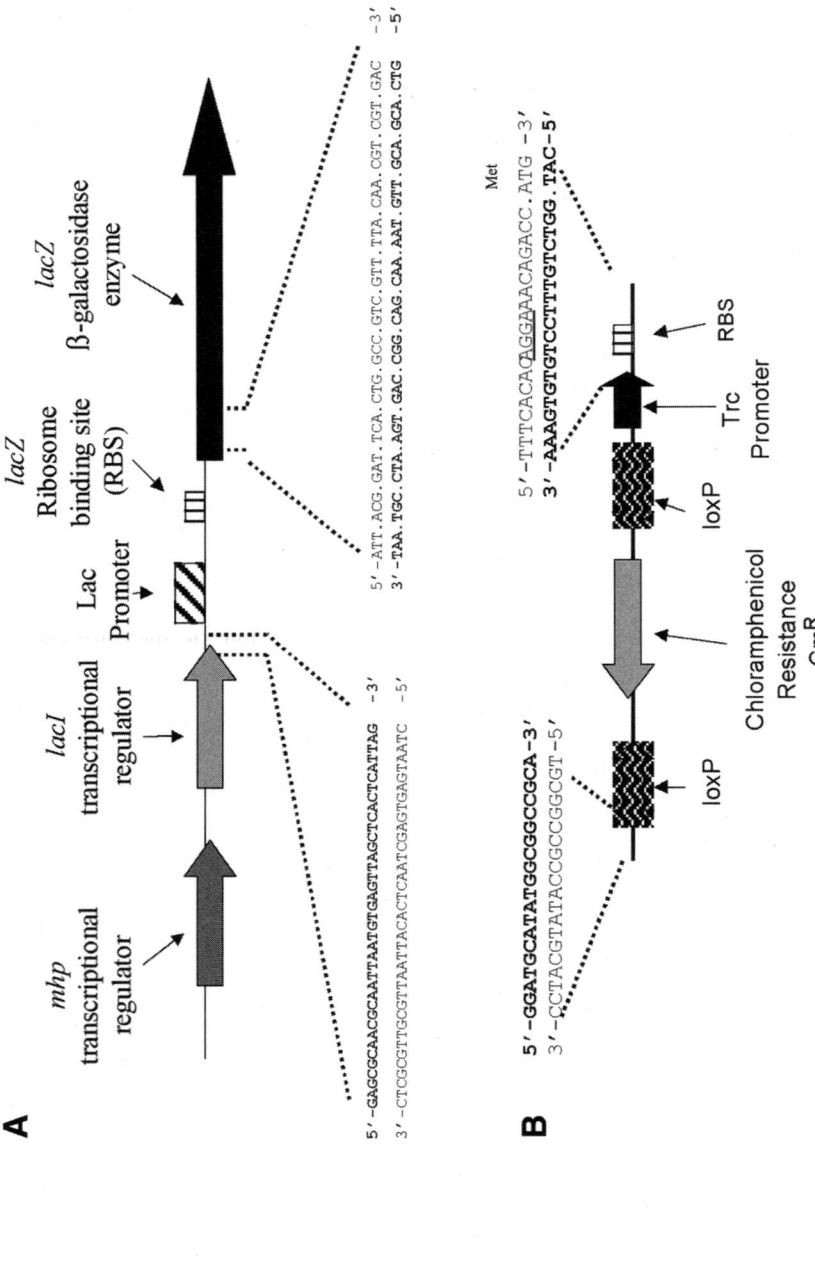

Fig. 3. Details of the *E. coli* chromosomal region upstream the *lacZ* gene. The *lacZ* gene encodes the β-galactosidase enzyme and its regulatory region is located downstream of the *lacI* gene. The nucleotide sequences of the relevant regions included in the PCR primers are shown in bold face (**A**). Details of the pTrCm2 region utilized as a source of the loxP-Cm^R-P_{Trc} region, as well as the nucleotide sequences included in the PCR primers (boldface type) are shown in (**B**). The ribosome binding site (RBS) present in P_{Trc} and the first ATG codon are also indicated.

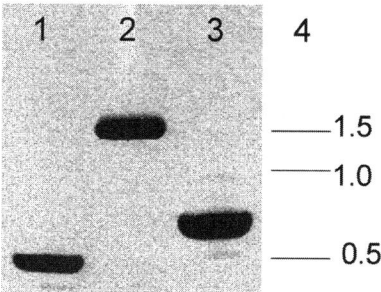

Fig. 4. Corroboration of the modification of the *lacZ* regulatory region. PCR products obtained using primers lacT1 and lacT2 were separated in a 2% agarose gel. Chromosomal DNA isolated from strains MG1655 (lane 1); MG1655:: P_{Trc} -Cm-*lacZ* (lane 2); or MG1655:: P_{Trc} - *lacZ* (lane 3), were used as substrates for PCR reactions. The position of molecular weight markers are indicated also (lane 4).

3.5. Chloramphenicol Antibiotic Marker Removal

Once the proper replacement of the *lacZYA* regulatory region was confirmed, the chloramphenicol antibiotic marker was removed by a variation of the protocol described by Palmeros et al. (*2*).

1. A fresh colony that carries the P_{Trc} -Cm-*lacZ* fusion was resuspended in 1.5 mL of cold water, centrifuged, and washed three times with cold water.
2. After the washing steps, the pellet was resuspended in 100 µL of cold 10% glycerol (*see* Note 5).
3. This suspension was electroporated with plasmid pJW168 (*2*).
4. After electroporation, 1 mL of SOC media was added and the cells were incubated at 30°C in a shaker, and plated on LB plates containing 50 µg of carbenicillin/mL and 20 µM IPTG.
5. Plates were incubated 16 h at 30°C.
6. Colonies were picked and patched into Lb+ carbenicillin (Carb), and Lb + Cm plates. Usually, 100% of the colonies have lost the Cm^R marker.

Chromosomal DNA from a $Carb^R$ Cm^S colony was isolated, subjected to PCR, and analyzed as described (*see* Subheading 3.4.). The expected size of the PCR product after removal of the loxP-Cm cassette was 583 bp. As can be seen in Fig. 4, lane 3, the correct size was obtained.

3.6. Measurement of lacZ Enzymatic Activity

The *lacZ* gene codes for the enzyme β-galactosidase in *E.coli* and has been widely used as a reporter to quantify gene expression. The β-galactosidase activity was measured using the synthetic substrate ONPG (ortho-nitrophenyl-β-D-galactoside) according to the procedure described in detail by Miller (*5*). To quantify the level of expression of the *lacZ*, strains were grown overnight in LB media. These overnights were used to inoculate a 250-mL flask containing 50 mL of LB or LB+ 50 µM IPTG. As a reference point, strain MG1655 was also inoculated. In the case of MG1655, *lacZ* was expressed from its native promoter. To establish a growth curve, aliquots are taken at desired time points and placed on ice. Flasks were incubated in a shaker at 37°C until they reached the middle-exponential phase (approx 0.8 OD at 600 nm) and 1.5-mL samples were collected by centrifugation.

1. Three 1.5-mL aliquots of cells are spun in 1.5-mL tubes, 5 min at 4°C.
2. Supernatant is removed by suction and pellets either frozen or assayed immediately.
3. Resuspend pellet in 1 mL Z buffer (*see* **Note 6**). Determine OD_{600}.
4. An aliquot (*see* **Note 7**) is used and the volume adjusted to a total of 730 µL with Z buffer. A reagent blank containing 730 µL Z buffer is run with the unknowns.
5. 10 µL of 10 mg/mL fresh lysozyme (keep on ice) is added and tubes incubated 5 min in 37°C water bath (*see* **Note 6**).
6. Add 10 µL of 10% triton X100 (*see* **Note 6**).
7. 100 µL of *O*-nitrophenyl-β-D-galactoside (ONPG) at 4.5 mg/mL (*see* **Note 6**) is added (prepare fresh). Start timing the reaction at the addition of ONPG.
8. Tubes are immediately transferred to 28°C water bath for exactly 15 min.
9. 150 µL 1.2 M Na_2CO_3 (*see* **Note 6**) added to stop reaction.
10. Read OD_{420} against the reagent blank (*see* **Notes 7–12**).
11. Once OD_{420} is read, estimate the amount of original cell suspension necessary to result in a spectrophotometric reading between 0.15 and 0.6 OD.
12. Calculate specific activity with Miller formula.

$$\frac{U\ \beta\text{-gal}}{mg\ prot} = \frac{OD_{420} \times 380}{15 \times mg\ prot}$$

OD_{420} = OD from β-gal assay
380 = constant
15 = min at 28°C
mg prot = calculated as follows from OD_{600}

$$\mu g\ prot/mL = OD_{600} \times 83$$

$$mg\ prot = \frac{\mu g\ prot}{mL} = \frac{mL\ used\ in\ \beta\text{-gal assay}}{1000}$$

After correcting for the volumes utilized in the assay, the relative activity of β-galactosidase per unit of optical density (600 nm) was calculated. The results of these measurements are presented in Table 1. As can be seen, the level of β-galactosidase activity produced by the MG1655:: P_{Trc} -*lacZ* strain was almost fivefold higher than the maximal activity obtained with the fully induced P_{lac} chromosomal promoter. Furthermore, the leakiness of P_{Trc} can be seen by the amount of β-galactosidase produced in the absence of IPTG (**Table 1**).

3.7. Perspectives

The approach described in this chapter is very flexible, allowing the quick modification of defined chromosomal regions. If needed, promoter variants can be constructed using the pTrcCm2 plasmid as a starting point. For example, if lower levels of expres-

Table 1
Relative Expression of the *lacZ* Gene

Promoter controlling *lacZ* expression	LB Relative β-galactosidase activity	LB + μ*M* IPTG Relative β-galactosidase activity
wt P$_{lac}$	–	1
P$_{Trc}$	1.46	4.78

Strains were inoculated in LB media containing 50 μ*M* IPTG. Cells were collected at OD$_{600}$ around 0.8. After lysis in buffer Z, ONPG hydrolysis was measured, and the relative activity with respect to the MG1655 + IPTG strain calculated.

sion are needed, point mutations in the −35, −10, or spacer regions can be incorporated. It is known that mutations in these areas will decrease P$_{Trc}$ strength. Furthermore, point mutations can also be used to alter LacI regulation. In this case, the modifications should be made in the LacI binding site (Lac operator) located between the −10 region and the ribosome-binding site (**Fig. 3**).

It is evident that in some cases P$_{Trc}$ may not be the best choice because it cannot be fully repressed and/or other means of induction are needed. It should be very straightforward to construct other Ab^R-promoter cassettes using the approach described here.

The use of the Cre-*loxP* system has proved to be very powerful, and new *loxP* and Cre variants are available. Some of these variants could be used to decrease recombination between *loxP* sites (**6**). Other *loxP* variants are not recognized by wild-type Cre recombinase, but serve as substrates for Cre variants (**7**). The former variants could be used to construct new cassettes susceptible of mobilization by P1 transduction.

4. Notes

1. A drawback of *loxP*-antibiotic cassettes is that they can not be transduced with bacteriophage P1, due to the fact that this phage produces the Cre recombinase, and the marker will be excised when the P1 lysate is prepared. This problem can be solved by using excisable antibiotic markers flanked by FRT sites (**3**).
2. We have noticed that a higher number of transformants can be recovered if we use fresh competent cells instead of frozen ones.
3. It is always a good practice to purify colonies at least once after transformation. This is particularly important after electroporation due to the high cellular density used. This eliminates mixed colonies that contain the native regulatory region.
4. The use of primers that hybridize outside the first set of primers and ensures that the PCR product generated is located in the right place in the chromosome.
5. This quick method to prepare competent cells works when super-coiled plasmid is used and the colonies are fresh. Depending on the size of the colonies, it may be necessary to use more than one colony to get enough cells for electroporation.
6. Prepare solutions with sterile water. Z buffer, lysozyme, and ONPG must be prepared fresh.
7. Usually, a 0.5-mL aliquot will give a representative reading. In cultures with higher densities, less cell suspension can be used.
8. The assay is best performed in two steps. The first part is a pilot test to approximate the amount of β-galactosidase and to determine the amount necessary to assay. Since the OD$_{420}$

is linear between 0.15 and 0.6, the samples are diluted to fall within this range. The second part is to determine exact concentration.

9. If the sample is suspected to have low levels of β-gal, the pellet can be resuspended in 730 μL of Z buffer and assayed directly.

10. If working with fresh culture and the β-gal level is high, up to 30 μL of culture can be brought to a final volume of 730 μL with Z buffer and assayed. This eliminates the pelleting step. It has not been determined whether greater than 30 μL of media will interfere with OD readings.

11. Samples in media can be kept on ice all day (up to 6 h) without a change in β-galactosidase levels. However, once the pellet is resuspended in Z buffer, the activity of β-galactosidase is not stable. After two hours on ice it begins to increase.

12. Repeat assay in duplicate diluting the suspension as predetermined in the pilot assay.

Acknowledgments

N. Flores was partially supported by a fellowship from the Dirección General de Asuntos del Personal Académico (DGAPA), from The National Autonomous University of Mexico (UNAM).

References

1. Zhou, S., Causey, T. B., Hasona, A., Shanmugam, K. T., and Ingram, L. O. (2002) Production of optically pure D-lactic acid in mineral salts medium by metabolic engineered *Escherichia coli* W3110. *Appl. Environ. Microbiol.* **69**, 399–407.

2. Palmeros B., Wild J., Szybalski W., Le Borgne, S., Hernandez-Chavez, G., Gosset, G., et al. (2000) A family of removable cassettes designed to obtain antibiotic-resistance free genomic modifications of *Escherichia coli* and other bacteria. *Gene* **247**, 255–264.

3. Datsenko, K. A. and Wanner, B. L. (2000) One-step inactivation of chromosomal genes in *Escherichia coli* K-12 using PCR products. *Proc. Natl. Acad. Sci. USA* **10**, 6640–6645.

4. Amann, E., Brosius, J., and Ptashne M. (1983) Vectors bearing a hybrid *trp-lac* promoter useful for regulated expression of cloned genes in *Escherichia coli*. *Gene* **25**, 167–178.

5. Miller, J. H. (1992) Procedures for working with lac, in *A Short Course in Bacterial Genetics*. Cold Spring Harbor Laboratory Press. New York, NY, pp. 72–74.

6. Siegel, R. W., Jain, R., and Bradbury A. (2001) Using an in vivo phagemid system to identify non-compatible *loxP* sequences. *FEBS Lett.* 499, 147–153.

7. Santoro, W. S. and Schultz P. G. (2002) Directed evolution of the site specificity of Cre recombinases. *Proc. Natl. Acad. Sci. USA* **99**, 4185–4190.

8

Chromosomal Expression of Foreign and Native Genes From Regulatable Promoters in *Escherichia coli*

Lu Zhou, Ke Zhang, and Barry L. Wanner

Summary

A two-step cloning system for expression of foreign and native genes from heterologous promoters in single copy from the *E. coli* chromosome is described. The system is based on the conditional-replication integration and modular CRIM plasmid technology and new CRIM plasmids described herein. The gene of interest is first synthesized by Polymerase chain reaction (PCR) by using primers with specially designed *Sap*I site extensions and cloned into the *Sap*I CRIM cloning plasmid pKZ20. The gene is then subcloned with *Sap*I into a CRIM expression plasmid, such as pKZ14, pLZ41, or pLZ42, which carry the regulatable promoter *araBp8*, *rhaBp3*, or *lac*UV5, respectively. The system is described for *gfp*, which encodes green fluorescence protein, as an example. The resulting CRIM expression plasmids are then integrated in single copy into a chromosomal phage attachment site by supplying integrase from a helper plasmid. Such integrants can be used for conditional expression of any target gene in single copy on the chromosome, especially in gene-structure-function studies where it is important to avoid copy-number artifacts.

Key Words: CRIM plasmid; *araBp*; *rhaBp*; *lac*UV5 promoter; conditional expression; site-specific recombination.

1. Introduction

Conditional expression is a powerful genetic tool. It was first used to show that a control region is separable from its structural gene in the analysis of a fusion that expressed LacY under purine control (*1*). Shortly afterward conditional (adenine-dependent) expression of this fusion was used to clone the first gene (*lacZ*) into a phage vector by directed transposition (*2*). This approach has been used innumerable times since. For example, it has been used to show that the first *trp-lacZ*, *ara-lacZ*, and *phoA-lacZ* fusions

From: *Methods in Molecular Biology, vol. 267:*
Recombinant Gene Expression: Reviews and Protocols, Second Edition
Edited by: P. Balbás and A. Lorence © Humana Press Inc., Totowa, NJ

were fused to the respective promoters (*3–5*). Likewise, searches for *d*amage-*in*ducible (*din*), *p*hosphate-*s*tarvation-*in*ducible (*psi*), and *in v*ivo *in*ducible (ivi) genes (*6–8*) were logical extensions of this approach.

A number of methods now exist for conditional gene expression. In this chapter, simple methods are described for conditional expression of genes from heterologous promoters in single copy. The methods employ the recently described conditional-replication integration and modular CRIM plasmids (*9*). CRIM plasmids have the replication origin (*oriRγ*) of the natural plasmid R6K, which requires the *trans*-acting Π replication protein (encoded by the *pir* gene) for replication as a plasmid. CRIM plasmids can be maintained at medium (15 copies per cell) or high (250 copies per cell) copy number by using specially engineered *E. coli* hosts with the *pir+* or *pir-116* (high-copy mutant) gene inserted in the chromosome (within the *uidA* gene) (*10*). CRIM plasmids have also a phage attachment (*attP*) site, so they can be easily and efficiently integrated into the respective bacterial attachment (*attB*) site on the chromosome of a normal (non-*pir*) strain by supplying the respective phage inte-grase (*int* gene product) from a helper plasmid. A series of easily curable, low-copy-number helper plasmids encoding different phage Int proteins has been described elsewhere (*9*).

Chromosomal expression of genes in single copy is especially useful for genetic and physiology studies where it is important to avoid plasmid copy number artifacts (Chapters 7, 8, and 11). The protocol described herein involves cloning the gene of interest behind a heterologous promoter in a CRIM plasmid. The resulting CRIM plasmid is then integrated into the chromosome by site-specific recombination. The method can be used for expression of native or foreign genes. Here the *gfp* gene for the green fluorescence protein is used as an example.

2. Materials

1. Bacteria: *E. coli* K-12 strains BW25113 (*lacIq rrnB3 ΔlacZ4787 hsdR514* DE(*araBAD*)*567* DE(*rhaBAD*)*568 rph-1*), BW25141 (*lacIq rrnB3 ΔlacZ4787 ΔphoBR580 hsdR514* DE(*araBAD*)*567* DE(*rhaBAD*)*568 galU95 ΔendA9 ΔuidA3::pir+ rph-1 recA1*), BW25142 (like BW25141, except *ΔuidA4::pir-116*) and BW30383 (like BW25113, except *rph+*). *E. coli* K-12 BW25113 (or BW30383) is used for integration of CRIM plasmids and for direct manipulation of the bacterial chromosome. BW25141 and BW25142 are used for replication of P-dependent replicons at medium and high copy number.
2. *Sap*I CRIM cloning plasmid pKZ20 (AY236520).
3. pKZ22 (AY236521).
4. CRIM expression plasmids: pKZ14 (AY236519), pLZ41 (AY236523), and pLZ42 (AY236524).
5. φ80 Int helper plasmid pAH123 (AY048726).
6. GFP expression plasmids: pKZ28 (AY236522), pLZ43 (AY236525), and pLZ44 (AY236526).
7. LB broth contains 10 g Bacto-tryptone, 5 g yeast extract, and 5 g NaCl per liter and is adjusted to pH 7.2.
8. Tryptone-yeast extract (TYE) agar contains 10 g Bacto-tryptone, 5 g yeast extract, 8 g NaCl, and 15 g Bacto-agar per liter and is adjusted to pH 7.0, as described elsewhere (*11*).

9. Ampicillin and chloramphenicol are from Sigma. Plasmids are maintained with antibiotics by using media containing ampicillin at 100 µg/mL or chloramphenicol at 25 µg/mL. Single-copy integrants are selected on TYE agar containing 6 µg/mL chloramphenicol.

10. SOB (*12*) medium: 2% tryptone, 0.5% yeast extract, 10 mM NaCl, 2.5 m*M* KCl, 10 m*M* MgCl$_2$ and 10 m*M* MgSO$_4$. It is prepared without Mg^{2+} and autoclaved. A 2 *M* Mg^{2+} stock solution (1 *M* MgCl$_2$ and 1 *M* MgSO$_4$) is prepared and filter-sterilized. To prepare SOB, 100 µL of the 2 *M* Mg^2 stock solution is added per 10 mL of the Mg^{2+}-free SOB immediately before use.

11. SOC medium: SOB with 20 m*M* D-glucose. SOC is made by adding 100 µL of the 2 *M* Mg^{2+} stock and 200 µL of filter-sterilized 1 *M* glucose per 10 mL Mg^{2+}-free SOB.

12. A variety of electroporators and PCR machines are suitable. Protocols described in this chapter were carried out using a BRL Cell-Porator with Voltage Booster and an MJ Research PTC-200 Cycler. Temperatures, times, and volumes may require adjustments with equipment of different models or manufacturers.

13. Agarose gel equipment.

14. Oligonucleotide primers are from commercial suppliers.

15. *Sap*I, other restriction enzymes, and DNA ligase are from commercial suppliers.

16. L-arabinose, L-rhamnose, and isopropyl-β-D-thiogalactopyranoside (IPTG) are from Sigma.

17. Plasmid DNAs are isolated using Qiagen kits.

18. DNA sequencing is done by automated sequencing in a core facility.

3. Methods

The method outlined describes (1) a simple two-step protocol for the construction of CRIM plasmids that express *gfp*, as an example, behind heterologous promoters and (2) the integration of the resulting CRIM expression plasmids into the *E. coli* chromosome in single copy.

3.1. CRIM Expression Plasmids
for Cloning Genes Behind Heterologous Promoters

Genes are cloned behind the *araBp8*, *rhaBp3*, or *lac*UV5 promoter in two steps (*see* **Note 1**). First, the gene is synthesized by PCR using a 5' primer about 40 nucleotides (nt) in length with the extension GCAGCTCTTCN juxtaposed to an ATG start site for the 5' end of the gene and a 3' primer of similar length with the extension GCAGCTCTTCN juxtaposed to a TAA stop codon. If the gene has different native start or stop codons (that is, other than an ATG codon or a TAA stop codon), then one or more nucleotides of the ATG and TTA (stop codon complement) must be included within the 5' and 3' extension, respectively, to ensure compatibility with the *Sap*I CRIM cloning plasmid pKZ20 (*see* **Subheading 3.1.1.**). In any case, the PCR primers are designed so that restriction of the PCR product with *Sap*I generates a DNA molecule with an upstream 5' ATG overhang and a downstream 5' TTA overhang (**Fig. 1**). After ligation and transformation, plasmid DNAs are isolated, analyzed by restriction enzyme analyses, and then verified by DNA sequencing. The gene is then subcloned behind the *araBp8*, *rhaBp3*, or *lac*UV5 promoters with *Sap*I and the respective CRIM expression plasmid.

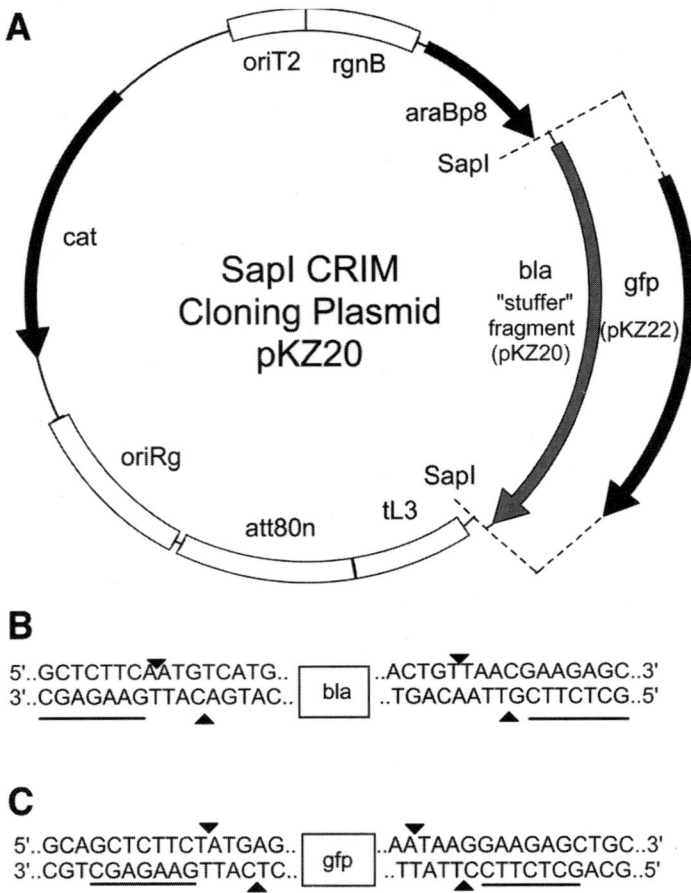

Fig. 1. *Sap*I CRIM cloning plasmid pKZ20. (**A**) Digestion of pKZ20 with *Sap*I releases the *bla* "stuffer" fragment and ends that are compatible with the *gfp* insert of the resulting plasmid pKZ22. (**B**) The *Sap*I sites in pKZ20 are in the plasmid backbone and are therefore retained in hybrid plasmids created with *Sap*I. Digestion with *Sap*I leaves a 5' CAT overhang upstream and 5' TAA overhang downstream in the backbone. (**C**) *Sap*I sites in the *gfp* PCR fragment lie outside the respective cut sites. *Sap*I recognition sequences are underlined. Arrows mark the cut sites on each strand.

3.1.1. Sapl CRIM Cloning Plasmid pKZ20

The CRIM plasmid pKZ20 permits cloning and subcloning of DNA molecules with *Sap*I (**Fig. 1A**). *Sap*I recognizes a seven-base-pair sequence and cuts the DNA asymmetrically one and four nucleotides from the *Sap*I site (**Fig. 1B**). pKZ20 contains the *bla* gene as a "stuffer" fragment that is released upon digestion with *Sap*I so the "backbone" DNA is easily separable from uncut or partially cut vector DNA on an agarose gel. In pKZ20, the *Sap*I sites lie within the backbone so they are retained in the initial clone, e.g., pKZ22 (**Fig. 1A**). A gene cloned into pKZ20 is behind *araB8p*; however, the gene is weakly expressed, even in the presence of the inducer arabinose. This is because placement of the *Sap*I site upstream of an ATG start codon interferes with creation of an efficient ribosome-binding site.

3.1.2. First-Step SapI Cloning

*Sap*I cuts infrequently because it has a seven-base-pair recognition sequence. Therefore many genes of interest are clonable with *Sap*I. In the example described, *gfp* was PCR-amplified, and the PCR product was purified and digested with *Sap*I. The gel-purified *Sap*I-digested *gfp* DNA molecule was ligated to *Sap*I-digested pKZ20. The ligation mixture was transformed into BW25141. Chloramphenicol-resistant colonies were selected on TYE agar containing chloramphenicol and arabinose. Colonies showing green fluorescence under UV light exposure were candidates for being correct and were further verified by restriction enzyme analysis. The strain BW25142 can be used for preparing larger amounts of plasmid DNA, provided that the high-copy plasmid is not deleterious.

3.1.3. CRIM Expression Plasmids

The CRIM expression plasmids pKZ14, pLZ41, and pLZ42 are similar to pKZ20, except for the arrangement of the upstream *Sap*I site (**Fig. 2**). While the upstream *Sap*I site lies within the backbone of pKZ20 (**Fig. 1B**), the upstream *Sap*I sites in the CRIM expression plasmids lie within the *bla* stuffer fragments of pKZ14, pLZ41, and pLZ42 (**Fig. 2B**). Accordingly, a gene of interest, in this case *gfp*, was first "captured" in pKZ20 and verified. It was then subcloned with *Sap*I into one or more CRIM expression plasmids.

3.1.4. Second-Step SapI Subcloning

The *gfp* gene was subcloned with *Sap*I from pKZ22 (**Fig. 1**) into the CRIM expression plasmids pKZ14, pLZ41, and pLZ42 (**Fig. 2**) to generate the plasmids pKZ28, pLZ43, and pLZ44 (**Fig. 3**), respectively. Plasmids were verified by standard restriction enzyme analyses. They were also tested for inducer-dependent phenotypes (in this case, GFP synthesis). As expected, BW25141 transformants harboring these plasmids show L-arabinose, L-rhamnose, and IPTG-dependent GFP synthesis, respectively, when propagated on TYE agar containing 1 m*M* inducer and 25 mg/mL chloramphenicol.

3.2. Chromosomal Integration of CRIM Plasmids

Integration is carried out by transformation of a normal (non-*pir*) host such as BW25113 (or BW30383; *see* **Note 2**) carrying a helper plasmid synthesizing the respective Int. The plasmid pAH123 was used here because it encodes φ80 Int (**Fig. 4**). Replication of the Int helper plasmids is temperature-sensitive. Also, Int synthesis is inducible by a temperature-shift. Therefore, cells are first propagated at 30°C in media with ampicillin to maintain the plasmid. They are then shifted to an elevated temperature (42°C), which induces Int synthesis and simultaneously arrests plasmid replication.

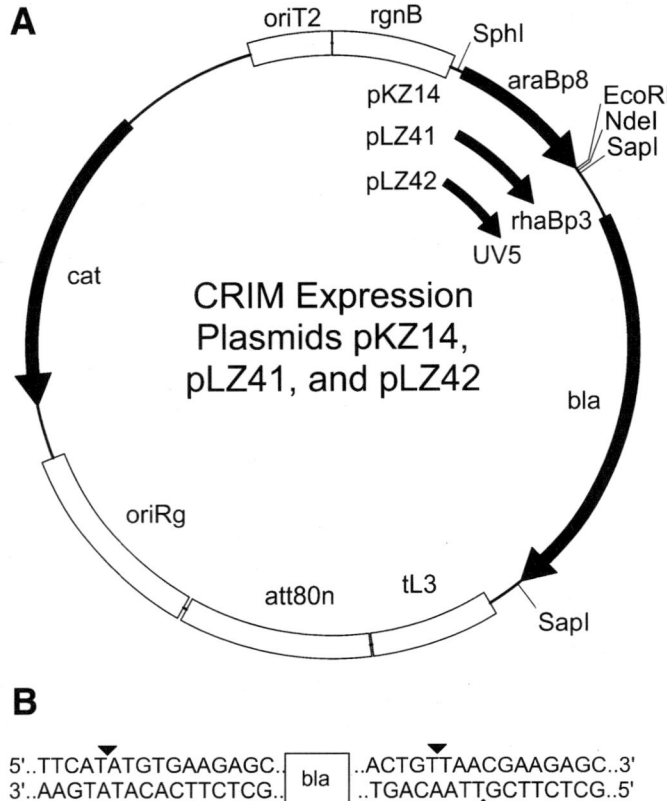

B

```
5'..TTCATATGTGAAGAGC.┌─────┐..ACTGTTAACGAAGAGC..3'
3'..AAGTATACACTTCTCG..│ bla │..TGACAATTGCTTCTCG..5'
                      └─────┘
```

Fig. 2. CRIM expression plasmids. (**A**) shows the structures of the CRIM expression plasmids pKZ14, pLZ41, pLZ42 which permit cloning genes with *Sap*I behind any of the three promoters: *araBp8*, *rhaBp3*, and *lac*UV5, respectively. As shown, pKZ14 has *Eco*R1, *Nde*I, and *Sap*I sites that can be used for cloning inserts behind *araBp8*. pLZ41 and pLZ42 are similar, except they have only *Nde*I and *Sap*I (and no *Eco*R1) sites for cloning inserts behind *rhaBp3* and *lac*UV5, respectively. Accordingly, inserts can be cloned into pKZ14 by using an upstream *Eco*R1, *Nde*I, or *Sap*I site and a downstream *Sap*I site. Inserts can be cloned into pLZ41 or pLZ42 by using an upstream *Nde*I or *Sap*I site and a downstream *Sap*I site. Because the upstream *Sap*I site in these plasmids lies within the *bla* stuffer fragment (**B**), the upstream *Sap*I site is lost upon cloning with *Sap*I.

3.2.1. Transformation With Int Helper Plasmid

1. To make cells electrocompetent, 1.5-mL Mg^{2+}-free SOB cultures of BW25113 (or its *rph*$^+$ derivative, BW30383) are grown to mid-log phase (A$_{600}$ approx 0.5) in 18-mm tubes in a tube roller.
2. Cultures are transferred to standard microfuge tubes and placed on ice for 2 min. The cells are then collected by centrifugation for 1 min at maximum speed in a high-speed microfuge at 4°C. Supernatants are discarded and the cells are washed three times. This is done by resuspending the cells with 1 mL ice-cold 10% glycerol, followed by 30-s centrifugations.

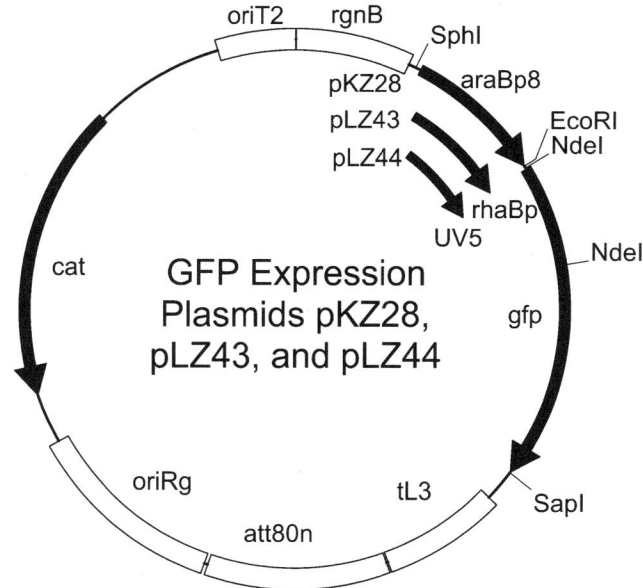

Fig. 3. GFP expression plasmids. pKZ28, pKZ43, pKZ44 are derivatives of pKZ14, pKZ43, and pKZ44, respectively, which were made by subcloning *gfp* from pKZ22 with *Sap*I.

After the final wash, the cells are resuspended with 50 µL ice-cold 10% glycerol and kept in an ice-water bath until use.

3. For electroporation, 20 µL electrocompetent cells and 1 µL pAH123 DNA (50 ng) are combined and placed into an electroporation cuvet. Electroporation is done as recommended by the manufacturer.

4. Following electroporation, the cell-DNA mixture is added to 1 mL SOC (no antibiotic) in an 18-mm tube, and incubated at 30°C for 1.5.

5. Portions are spread on TYE agar containing ampicillin and then incubated overnight at 30°C.

3.2.2. Integration of the CRIM Plasmid

CRIM plasmid integration is carried out as described below and as shown for pKZ28 in **Fig. 5**.

1. An isolated and freshly grown colony of *E. coli* K-12 BW25113 (or BW30383) carrying pAH123 is chosen from a TYE ampicillin agar plate that had been incubated at 30°C for 18 to 20 h. The colony is inoculated into a small flask containing 1.5-mL Mg^{2+}-free SOB with ampicillin and incubated in a 30°C shaking water bath.

2. Cells are grown to an A_{600} of approx 0.5 to 0.6 (approx 5 to 6 h), after which the flask is moved to a 42°C shaking water bath, incubated about 20 min longer, and then placed on ice. Rapid temperature shift is important for efficient induction, which is afforded by use of shaking water baths.

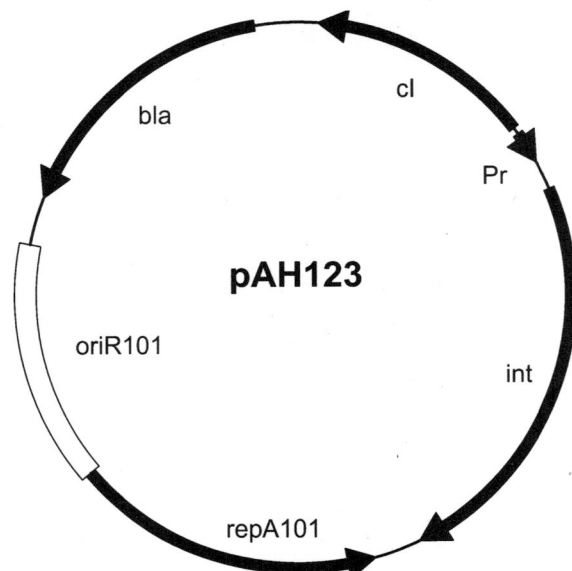

Fig. 4. Int helper plasmid pAH123. Like other CRIM helper plasmids (*9*), pAH123 is a low-copy-number plasmid that is temperature-sensitive for replication and hence easily curable by a temperature shift. Int is synthesized from λp_R under *cI857* control, which is also encoded by pAH123.

3. Cells are collected by centrifugation and made electrocompetent as described above.
4. After electroporation, the cell-DNA mixture is added to 1 mL SOC (without ampicillin) in an 18-mm culture tube. The tube culture is at 42°C for 30 min, and then at 37°C for 1 to 1.5 longer.
5. Portions are spread onto TYE agar containing chloramphenicol, but without ampicillin, to select for the integrated CRIM plasmid and permit loss of the Int helper plasmid. The agar plates are incubated at 37°C for 16 to 20 h. Cells are colony-purified once without an antibiotic and tested for loss of the helper plasmid, as verified by its ampicillin sensitivity (*see* **Note 3**).

3.2.3. PCR Test of Integrant Copy Number

This protocol pertains to CRIM plasmids described herein. Primers P1 and P4 are specific for the phage *att* site. The annealing temperatures depend on the primer sequences. Conditions used for other CRIM plasmids are described elsewhere (*9*).

1. Isolated colonies are picked up with a plastic tip or glass capillary (but not with a wooden toothpick) and suspended in 20 μL water in a microfuge tube.
2. 5 μL of the cell suspension, 10 pmol of primers P1 to P4 (**Fig. 5**), and 0.5 U of *Taq* DNA polymerase (New England Biolabs) are combined in 1X PCR buffer-2.5 m*M* MgCl$_2$ with deoxynucleoside triphosphates in a final volume of 20 μL. PCR is carried out for 25 cycles (denaturing for 1 min at 94°C, annealing for 1 min at 63°C, and extending for 1 min at 72°C). Single PCR reactions are run with all four primers in the same tube. Different anneal-

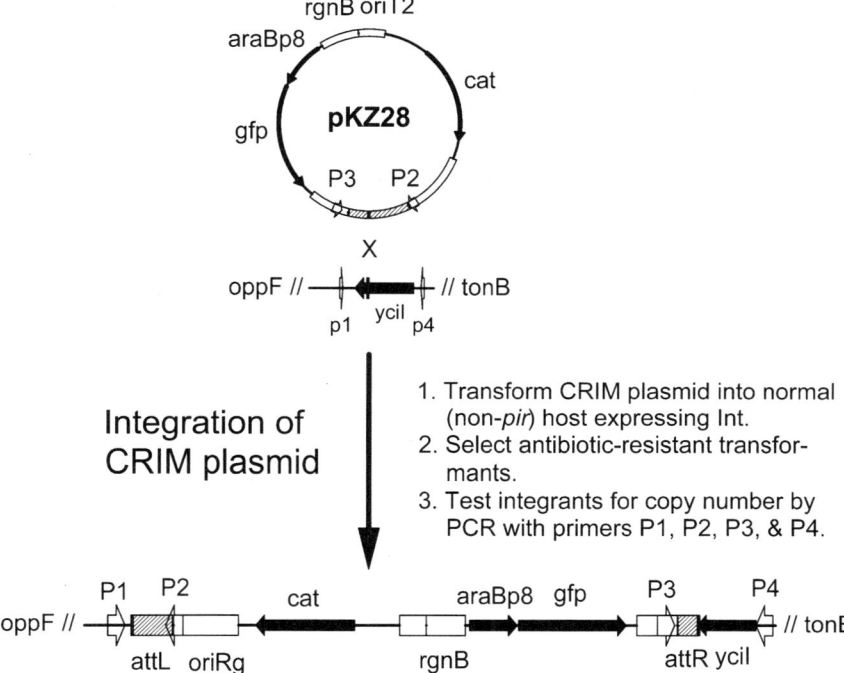

Fig. 5. Integration of CRIM plasmid pKZ28. CRIM plasmids are integrated into a normal (non-*pir*) host under conditions of Int synthesis. Integration of an *att\phi80n* CRIM plasmid occurs by site-specific recombination at *attB*, bacterial *att\phi80*, which lies inside the *ycil* gene. P1, P2, P3, and P4 mark priming sites used for PCR verification of integrant copy number. PCR tests are carried out with all four primers in a single reaction tube. When using these primers, a control strain with an empty *attB* site yields a single *attB* (546-nt, P1/P4) PCR product and those with a single integrated CRIM plasmid yield both *attL* (409-nt, P1/P2) and *attR* (657-nt, P3/P4) PCR products. Ones with multiple CRIM plasmids integrated in tandem yield three products: (1) an *attL* (409-nt, P1/P2) PCR product, (2) an *attR* (657-nt, P3/P4) PCR product, and (3) a CRIM plasmid-specific (520-nt, P2/P3) PCR product. It should be noted that the CRIM plasmid pKZ28 has an *att80* sequence, denoted *att80*n, which differs in length from those described elsewhere (*9*). Therefore, the P3/P4 fragment is 657-nt, instead of 732-nt, and the P2/P3 fragment (for multiple integrants) is 520-nt, instead of 595-nt. Primer sequences are: P1, CTGCTTGTGGTGGTGAAT; P2, ACTTAACGGCTGACATGG; P3, ACGAGTATCGAGATGGCA; P4, TAAGGCAAGACGATCAGG.

ing temperatures and P1 and P4 primers are used for CRIM plasmids that integrate at different *attB* sites, as described in elsewhere (*9*).

3. Usually three or four colonies are picked directly from the selection plate and initially tested by PCR for single-copy ones (refer to the expected sizes for PCR products in **Fig. 5**). Two or three single-copy candidates are then colony-purified nonselectively once or twice and retested by PCR to be sure that they are pure and stable (*see* **Note 4**).

4. Integrants are tested for inducer-dependent GFP syntheis by growing them in the presence and absence of L-arabinose, L-rhamnose, or IPTG (*see* **Note 5**).

4. Notes

1. Methods are described for conditional expression of a target gene behind alternative promoters. Which promoter is most suitable for a particular application can vary. For example, the *lac*UV5 promoter is most leaky; however it is insensitive to catabolite repression. Also, in the case of a gene requiring high-level expression, the leakiness of the *lac*UV5 promoter is not problematic. The arabinose-inducible *araBp* promoter (also called P_{araB} or P_{BAD}) is less leaky than the *lac*UV5 promoter, and the rhamnose-inducible *rhaBp* promoter (also P_{rhaB}) is even less leaky than the *araBp* promoter. However, the *araBp* and *rhaBp* promoters are both subject to catabolite repression (***13***).
2. The strains BW25113 and BW30383 are deleted of *lacZ*, *araBAD*, and *rhaBAD* and carries the *lacI* promoter up mutation *lacI*^{q}, for elevated synthesis of the LacI protein. The *lacZ* deletion permits use of *lacZ* fusions. The Δ*araBAD* and Δ*rhaBAD* mutations eliminate genes for arabinose and rhamnose catabolism, thus allowing use of low concentrations of arabinose and rhamnose as inducers.
3. Because the Int helper plasmid is unstable at 37°C and ampicillin is omitted from the selection medium, the helper plasmid is usually lost following electroporation and selection of integrants. If it is not lost, then a few colonies can be colony-purified nonselectively once at 43°C and retested. Once integrants are chosen from the initial selective media, it is preferable not to maintain them on an antibiotic medium to prevent inadvertent selection of multiple-copy integrants. Since integration of CRIM plasmids occurs by site-specific recombination, the integrants are extremely stable and there is no reason to maintain antibiotic selection.
4. Single-copy integrants are usually the majority. The most frequent undesirable event is the formation of multicopy (tandem) integrants at the respective *attB* site. These can predominate if too high an antibiotic concentration is used in the selective medium. It is for this reason that chloramphenicol is used at 6 µg/mL. The recommended concentrations for other antibiotics are described elsewhere (***9***). These concentrations can be varied to reduce background growth or to find single-copy integrants. A second undesirable event results if the CRIM plasmid recombines elsewhere on the chromosome and not at the respective *attB* site. These are recognizable by the standard PCR test for integrant copy number. They usually occur via homologous recombination of the CRIM plasmid with short homologies on the bacterial chromosome. For example, a CRIM plasmid carrying a native *E. coli* gene can recombine via the gene sequence in common with the chromosome. Homologous recombination can also occur with the resident CRIM plasmid if the cell already contains an integrated CRIM plasmid at a different *attB* site. This is because all CRIM plasmids have sequences in common (tL3, *oriR*g, *rgnB*). Importantly, these events are seldom problematic, due to the high efficiency of site-specific recombination. When they occur, they are caused by inadequate Int synthesis. The simplest remedy is to repeat the integration with newly prepared electrocompetent cells. Sometimes, a new transformant carrying the CRIM helper plasmid is used. Changing to another CRIM helper plasmid that synthesizes Int under a different control (e.g., induction by temperature shift or by arabinose or IPTG addition) can also be used.
5. It is often useful to vary gene expression levels by varying inducer concentrations. Unfortunately, inducible systems encoding an associated transporter can display an autocatalytic behavior and suffer from the "all or none" effect (***14,15***). This can be avoided by using mutants lacking the associated transporter. For example, the all-or-none effect is eliminated in a *lacY* mutant (***14***). Since the Δ*lacZ* mutation in BW25113 (and BW30383) removes *lacP*, this strain is phenotypically LacY negative and therefore should not show the all-or-none effect with IPTG induction. An alternative way to modulate gene expression levels in a

homogeneous manner is to vary the expression levels by catabolite repression by using different carbon sources (e.g., fructose, mannitol, or glycerol) in the presence of saturating inducer concentrations (*13*). Alternatively, the respective transporter gene itself can be expressed from a heterologous (*16*) or constitutive promoter (*17*), which has now been shown to eliminate the all-or-none effect. Importantly, fusions that express arabinose (*17*) and rhamnose (K. A. Datsenko and B. L. Wanner, unpublished data) transporters from constitutive promoters are available. These include chromosomal fusions that are marked with FRT-flanked antibiotic resistance cassettes, which facilitates transferring them to other strains. It has recently been shown that a plasmid encoding a mutant *lacY* protein is also useful for homogeneous expression of arabinose-inducible promoters (*18*).

Acknowledgments

Research was supported by grants from the National Institutes of Health and the National Science Foundation.

References

1. Jacob, F., Ullmann, A., and Monod, J. (1965) Délétions fusionnant l'opéron lactose et un opéron purine chez *Escherichia coli. J. Mol. Biol.* **31**, 704–719.
2. Beckwith, J. R. and Signer, E. R. (1966) Transposition of the *lac* region of *Escherichia coli*. I. Inversion of the *Lac* operon and transduction of *Lac* by ϕ80. *J. Mol. Biol.* **19**, 254–265.
3. Reznikoff, W. S., Miller, J. H., Scaife, J. G., and Beckwith, J. R. (1969) A mechanism for repressor action. *J. Mol. Biol.* **43**, 201–213.
4. Casadaban, M. J. (1975) Fusion of the *Escherichia coli lac* genes to the *ara* promoter: A general technique using bacterial Mu-1 insertions. *Proc. Natl. Acad. Sci. USA* **72**, 809–813.
5. Sarthy, A., Fowler, A., Zabin, I., and Beckwith, J. (1979) Use of gene fusions to determine a partial signal sequence of alkaline phosphatase. *J. Bacteriol.* **139**, 932–939.
6. Kenyon, C. J. and Walker, G. C. (1980) DNA-damaging agents stimulate gene expression at specific loci in *Escherichia coli. Proc. Natl. Acad. Sci. USA* **77**, 2819–2823.
7. Wanner, B. L., Wieder, S., and McSharry, R. (1981) Use of bacteriophage transposon Mu *d1* to determine the orientation for three *proC*-linked phosphate-starvation-inducible (*psi*) genes in *Escherichia coli* K-12. *J. Bacteriol.* **146**, 93–101.
8. Mahan, M. J., Slauch, J. M., and Mekalanos, J. J. (1993) Selection of bacterial virulence genes that are specifically induced in host tissues. *Science* **259**, 686–688.
9. Haldimann, A. and Wanner, B. L. (2001) Conditional-replication, integration, excision, and retrieval plasmid-host systems for gene structure-function studies in bacteria. *J. Bacteriol.* **183**, 6384–6393.
10. Metcalf, W. W., Jiang, W., and Wanner, B. L. (1994) Use of the *rep* technique for allele replacement to construct new *Escherichia coli* hosts for maintenance of R6Kg origin plasmids at different copy numbers. *Gene* **138**, 1–7.
11. Wanner, B. L. (1994) Gene expression in bacteria using TnphoA and TnphoA' elements to make and switch phoA gene, lacZ (op), and lacZ (pr) fusions. In Methods in Molecular Genetics, vol. 3 (Adolph, K. W., ed.). Academic Press, Orlando, FL, pp. 291–310.
12. Hanahan, D. (1983) Studies on transformation of *Escherichia coli* with plasmids. *J. Mol. Biol.* **166**, 557–580.
13. Haldimann, A., Daniels, L. L., and Wanner, B. L. (1998) Use of new methods for construction of tightly regulated arabinose and rhamnose promoter fusions in studies of the *Escherichia coli* phosphate regulon. *J. Bacteriol.* **180**, 1277–1286.

14. Novick, A. and Weiner, M. (1957) Enzyme induction as an all-or-none phenomenon. *Proc. Natl. Acad. Sci. USA* **43**, 553–556.
15. Siegele, D. A. and Hu, J. C. (1997) Gene expression from plasmids containing the *araBAD* promoter at subsaturating inducer concentrations represents mixed populations. *Proc. Natl. Acad. Sci. USA* **94**, 8168–8172.
16. Khlebnikov, A., Risa, O., Skaug, T., Carrier, T. A., and Keasling, J. D. (2000) Regulatable arabinose-inducible gene expression system with consistent control in all cells of a culture. *J. Bacteriol.* **182**, 7029–7034.
17. Khlebnikov, A., Datsenko, K. A., Skaug, T., Wanner, B. L., and Keasling, J. D. (2001) Homogeneous expression of the P_{BAD} promoter in *Escherichia coli* by constitutive expression of the low-affinity high-capacity AraE transporter. *Microbiol.* **147**, 3241–3247.
18. Morgan-Kiss, R. M., Wadler, C., and Cronan, J. E., Jr. (2002) Long-term and homogeneous regulation of the *Escherichia coli araBAD* promoter by use of a lactose transporter of relaxed specificity. *Proc. Natl. Acad. Sci. USA* **99**, 7373–7377.

9

Plasmid Vectors for Marker-Free Chromosomal Insertion of Genetic Material in *Escherichia coli*

Sylvie Le Borgne, Francisco Bolívar, and Guillermo Gosset

Summary

A method to achieve the insertion of genetic material into the chromosome of *Escherichia coli* is described. The method is based on the use of integration vectors from the pBRINTs-rAnb^R family. These vectors offer the choice of using the antibiotics chloramphenicol, gentamycin, or kanamycin to select for chromosomal integration events. In addition, it is possible to eliminate these chromosomal antibiotic resistance markers, after integration has taken place. The overall insertion strategy is as follows: a fragment containing the gene(s) to be integrated in the chromosome is inserted into the multiple cloning site of a pBRINTs-rAnb^R vector and the resulting plasmid is used to transform *E. coli* cells. The plasmid is first allowed to replicate in the cell at the permissive temperature of 30°C. Next, the temperature of the culture is raised to 44°C to inhibit plasmid replication and to select for the integrants in the presence of the appropriate antibiotic. Chromosomal excision of the Anb^R gene can then be catalyzed by the Cre recombinase that is transiently expressed in the cell from the temperature-sensitive pJW168 plasmid. This plasmid is finally eliminated from the cells by increasing the temperature of the culture to 44°C.

Key Words: Recombinant DNA tools; chromosomal integration; homologous recombination; Cre recombinase; *loxP*; unmarked strains.

1. Introduction

Insertion of genetic material into the chromosome of *Escherichia coli* provides a method to avoid the inherent instability of most plasmid vectors. A gene inserted into the chromosome is completely stable, even in the absence of a selection scheme. This approach is also useful when a single copy per cell of a cloned gene is desired to study its function under closer to natural conditions.

From: *Methods in Molecular Biology, vol. 267:*
Recombinant Gene Expression: Reviews and Protocols, Second Edition
Edited by: P. Balbás and A. Lorence © Humana Press Inc., Totowa, NJ

One approach for achieving chromosomal integration is based on allele replacement, which results from the homologous recombination activities of the host cell *(1)*. To use this method, the gene of interest should be flanked by 5' and 3' sequences with homology to a region of the chromosome. This genetic construct is introduced in *E. coli* usually as part of a plasmid that cannot replicate. Once inside the cell, recombination activities promote single- and double-crossover events that result in the chromosomal insertion of the foreign gene. Any chromosomal region could be the site used for recombination. However, a gene target can be selected so that its inactivation does not have a negative effect on the cell's physiology and the resulting strain displays an easily selectable phenotype. Plasmids designed with these principles have been reported *(2–6)*. An example of this type of vectors is the pBRINTs-rAnb^R family *(6)*. By following a simple and efficient temperature-based procedure, this vector family allows the insertion of foreign DNA into the chromosome of *E. coli*. The procedure to achieve chromosomal insertion and the subsequent excision of the antibiotic resistance (Anb^R) gene will be illustrated by using the pBRINT$_s$-Cat2 insertion vector.

2. Materials

1. *Escherichia coli* strain W3110.
2. pBRINTs-Cat2 and pJW168 plasmids *(5,6)*.
3. Luria-Bertani (LB) liquid and solid media.
4. Carbenicillin (Cb) at a final concentration of 50 μg/mL in all cases.
5. Chloramphenicol (Cm).
6. Sterile toothpicks for replica plating.
7. 30°C and 44°C petri dish incubators.
8. 30°C and 44°C orbital shakers/incubators.
9. XGal (5-bromo-4-chloro-3-indolyl galactopyranoside) at a final concentration of 33 μg/mL in all cases. Dissolve in *N,N*-dimethylformamide, cover with aluminum foil, and store at −20°C.
10. IPTG (isopropyl-β-D-thiogalactopyranoside) at a final concentration of 0.1 m*M* in all cases. Dissolve in water, filter-sterilize, and store at 4°C.
11. Elongase Enzyme Mix (Invitrogen-Life Technologies, Carlsbad, CA) and its optimized reaction buffer containing 1.6 m*M* MgCl$_2$ provided by the manufacturer.
12. dNTPs (desoxyribonucleoside triphosphates).
13. Oligonucleotide primers lac1 (5'-GAAATTATCGATGAGCGTGGTGGTTATGCC-3') and lac4 (5'-CATTGGCACCATGCCGTGGGTTTCAATATT-3').
14. Agarose gel electrophoresis equipment.
15. Thermocycler for PCR.

3. Methods

The methods described include (1) the integration of the pBRINTs-Cat2 plasmid into the *E. coli* chromosome (*see* **Note 1**), (2) the excision of the chloramphenicol resistance gene cassette, (3) the elimination of the Cre-expression plasmid pJW168, and (4) the PCR analysis of the chromosomal insertion and excision events.

3.1. pBRINTs-rAnb^R Chromosomal Insertion Vectors

The pBRINTs-rAnb^R chromosomal insertion vector family *(6)* is composed of three plasmids with the following main characteristics (*see* **Note 2**) (**Fig. 1**):

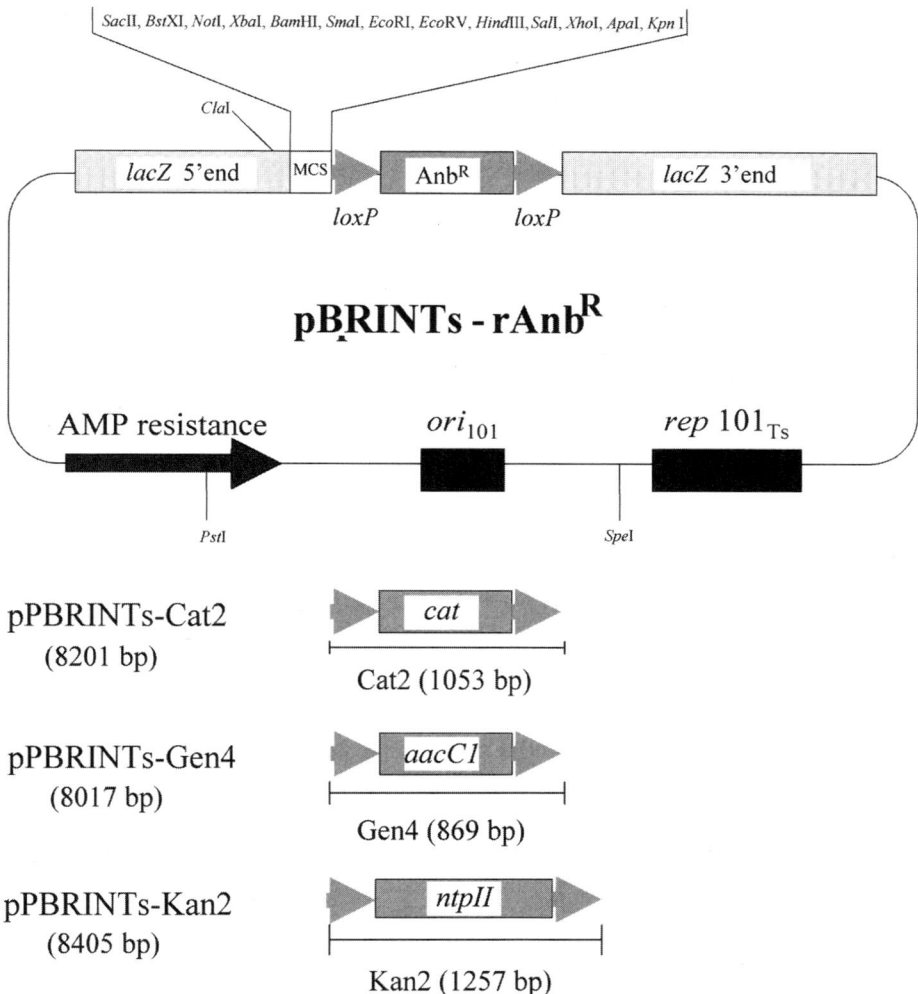

Fig. 1. Schematic presentation of the pBRINTs-rAnb^R integration vectors. Unique restriction sites are indicated. The *Eco*RI/*Xba*I and *Eco*RI/*Sal*I restriction sites are not unique in plasmids pBRINT$_S$-Cat2 and pBRINT$_S$-Gen4, respectively.

1. A temperature-sensitive replicon derived from plasmid pMAK705 (ori_{101}, *rep* 101_{Ts}) (*7*).
2. A multiple cloning site (MCS) next to either of the three Anb^R genes (*cat*, chloramphenicol; *aacC1*, gentamycin; or *ntpII*, kanamycin), flanked by regions of homology to the 5' and 3' ends of the chromosomal *E. coli lacZ* gene.
3. Anb^R genes are flanked by *loxP* sites to allow its Cre recombinase-dependent chromosomal excision.
4. A gene encoding ampicillin (AMP) resistance.

Any of these plasmids can be used to insert genetic material into the *lacZ* locus of the chromosome of any RecA⁺, *lacZ*⁺ *E. coli* strain (*see* **Note 3**) following the strategy illustrated in **Fig. 2**.

3.2. Integration of the pBRINTs-Cat2 Plasmid into the E. coli Chromosome

The steps described here involve the transformation of *E. coli* with the pBRINTs-Cat2 integration plasmid followed by a simple and rapid temperature-based procedure that allows selection of cells where chromosomal integration occurred. Here, the wild-type *E. coli* W3110 strain was used as a model host.

1. Transform *E. coli* W3110 with covalently closed circular pBRINTs-Cat2 (*see* **Note 4**).
2. Plate the transformants on LB plates containing Cb (*see* **Note 5**) and 30 μg/mL of Cm and incubate 60 h at 30°C.
3. Pick a single colony and grow it for 4 h at 30°C in a revolving shaker at 200 rpm in 0.5 mL of LB medium without any antibiotic.
4. Take 0.5 mL of the previous culture to inoculate 10 mL of fresh LB medium without any antibiotic and incubate the culture for 6 additional hours at 30°C in a revolving shaker at 200 rpm. Next, increase the incubation temperature to 37°C and maintain the culture overnight in the orbital shaker.
5. The next day, plate 150 μL of several serial dilutions of the previous culture (from 10^{-3} to 10^{-6}) on LB plates containing XGal, IPTG and 15 μg/mL Cm (*see* **Note 6**).
6. Incubate the plates 24 h at 44°C. Cells where inactivation of *lacZ* took place generate white colonies. This is an indication that pBRINTs-Cat2 integrated into the *lacZ* chromosomal gene (*see* **Note 7**). The observed frequency of white, Anb^R-resistant cells is approximately 0.1% of total viable cells.
7. Screen 100 to 200 single white colonies (*see* **Note 8**) from these plates and select those resistant to Cm and sensitivity to Cb by replica plating at 44°C on LB plates containing XGal, IPTG and 15 μg/mL Cm and LB plates containing XGal, IPTG, and Cb. The observed frequency of Cm-resistant (Cm^R) and Cb-sensitive (Cb^S) cells is in the range of 2–20%.
8. To confirm the absence of cytoplasmic pBRINTs-Cat2, replicate all the white Cm^R Cb^S colonies selected in step 7 at 44°C on LB plates containing XGal, IPTG and 15 μg/mL Cm. Then replica plate again the white Cm^R Cb^S colonies at 30°C on LB plates containing Cb. Select colonies resistant to Cm and sensitive to Cb.
9. PCR analysis of some of these strains is performed to verify expected chromosomal recombination event. The W3110 *lacZ::*MCSCat2 strain is obtained.

3.3. Excision of the Cat2 Chloramphenicol-Resistance Gene Cassette

The steps described here allow the excision of the Cm^R gene marker from the chromosome of the W3110 *lacZ::*MCSCat2 strain to obtain a marker-free W3110 *lacZ::*MCS*loxP* strain (*see* **Note 9**). This is performed by transforming the W3110 *lacZ::*MCSCat2 strain with the Cre-producing plasmid pJW168 *(5)* (**Fig. 3**) and expressing the *cre* gene.

1. Transform *E. coli* W3110 *lacZ::MCSCat2* with covalently closed circular DNA of pJW168 (*see* **Note 4**).
2. Plate the transformants on LB plates containing 50 μg/mL of Cb and 15 μg/mL of Cm and incubate 60 h at 30°C.

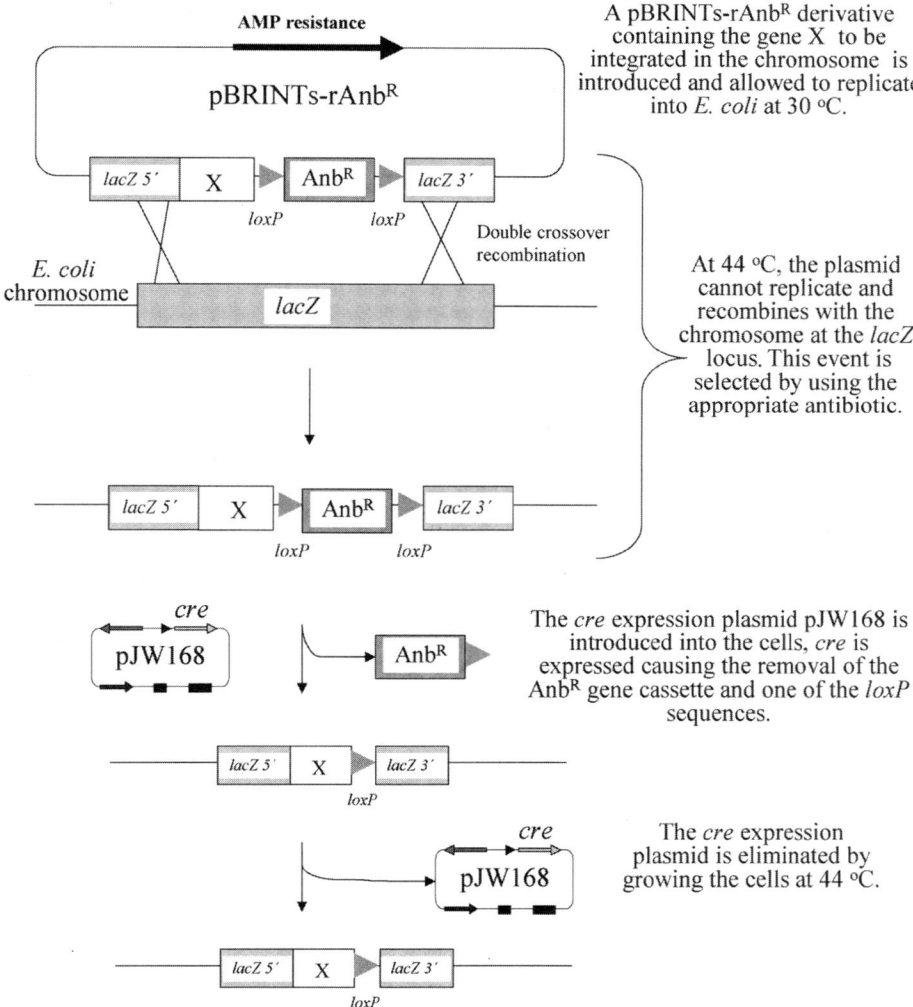

Fig. 2. Overall strategy for marker-free chromosomal insertion using the pBRINT$_S$-rAnb^R vector family and the pJW168 vector expressing Cre recombinase.

3. Pick 100 isolated colonies on LB plates containing IPTG and incubate overnight at 30°C to induce *cre* expression.

4. Screen these 100 colonies for resistance to Cb and sensitivity to Cm by replica plating at 30°C on LB plates containing Xgal, IPTG, and Cb and LB plates containing XGal, IPTG, and 15 µg/mL Cm, respectively. The appearance of Cb^R Cm^S white colonies indicate that these cells have lost the Cm^R gene marker but still remain *lacZ*$^-$ and contain the pJW168 plasmid. The Cre-mediated excision process when using plasmid pJW168 is very efficient, as it shows a frequency of nearly 100%.

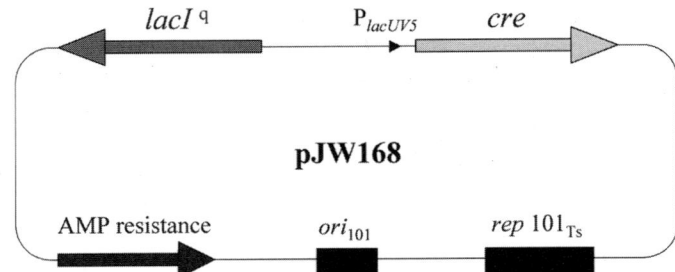

Fig. 3. Schematic presentation of the pJW168 vector. The *cre* gene is transcribed from the IPTG-inducible *lac*UV5 promoter. This plasmid replicates in *E. coli* at culture temperatures of 30°C or lower.

5. PCR analysis of these strains is performed to verify expected chromosomal excision event. The W3110 *lacZ::*MCS*loxP*[pJW168] strain is obtained.

3.4. Elimination of the cre-Expression Plasmid pJW168

The following steps describe how to cure the cells from the pJW168 plasmid after elimination of the Anb^R marker. Plasmid pJW168 carries a temperature-sensitive replicon, which allows replication of this vector at 30°C but not at temperatures above 37°C. So, a temperature-based procedure similar to that described in **Subheading 3.2.** can be applied to eliminate this plasmid from the cells.

1. Grow one single colony of *E. coli* W3110 *lacZ::*MCS*loxP*[pJW168] 4 h at 30°C in an orbital shaker at 200 rpm in 0.5 mL of LB medium without any antibiotic.
2. Reinoculate the 0.5 mL of the resulting culture in 10 mL of fresh LB medium without any antibiotic and allow to grow for 6 additional hours at 30°C in an orbital shaker at 200 rpm to obtain more biomass.
3. Incubate this culture overnight at 37°C in an orbital shaker at 200 rpm.
4. Plate 150 μL of several dilutions of the resulting culture (from 10^{-3} to 10^{-6}) on LB plates.
5. Incubate the plates 24 h at 44°C.
6. Replica plate 10 colonies at 30°C on LB plates containing Xgal, IPTG, and Cb and LB plates containing Xgal and IPTG, respectively, to screen for the colonies that lost the pJW168 plasmid. The W3110 *lacZ::*MCS*loxP* strain is obtained.

3.5. PCR Analysis of the Chromosomal Integration and Excision Events

The chromosomal insertion and excision events can be further confirmed by a PCR-based analysis strategy consisting of amplifying the *lacZ* region of genomic DNA isolated from the previously constructed strains (**Fig. 4**) (*see* **Note 10**). Only two primers, lac1 and lac4, are required to perform these analyses.

1. Extract genomic DNA from 5 mL of an overnight culture of strains W3110, W3110 *lacZ::*MCS*Cat2* and W3110 *lacZ::*MCS*loxP* using cetyltrimethylammonium bromide as previously described *(8)* or any other suitable method.
2. Use the genomic DNA as templates for PCR amplification.

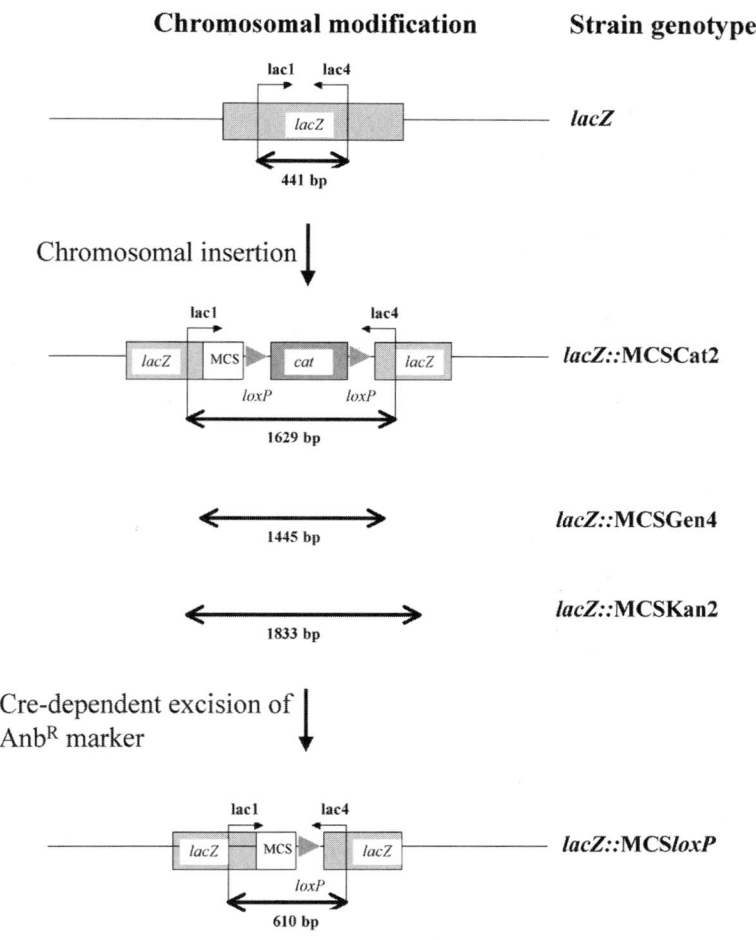

Fig. 4. Schematic presentation showing the expected chromosomal modifications, genotypes, and PCR amplification products using the lac1 and lac4 primers and chromosomal DNA from wild-type and strains modified with the pBRINT$_S$-Cat2 vector. Thick horizontal arrows show the size of the expected PCR amplification products.

The 50 μL reaction mix contains:

    a. 100 ng of genomic DNA.
    b. 50 pmol lac1 primer.
    c. 50 pmol lac4 primer.
    d. 0.25 mM dNTPs.
    d. Optimized reaction buffer.
    f. 1 μL of Elongase Enzyme Mix.

3. Program and run the automated thermocycler as follows: Pre-amplification denaturation: 94°C for 3 min (1 cycle);

Thermal cycling (30 cycles);
Denaturation: 95°C for 30 s;
Annealing: 60°C for 45 s;
Extension: 68°C for 2.5 min; and
Final extension: 68°C for 10 min (1 cycle)
4. Analyze 10 μL of each reaction by electrophoresis on a 1.2% (w/v) agarose gel in 1X TBE buffer.

4. Notes

1. It is important to point out that, for clarity, a nonmodified pBRINTs-Cat2 vector was chosen to illustrate the integration-excision and PCR analysis procedures. In this case, the end result is the inactivation of the *lacZ* gene in the chromosome by the insertion of a *loxP* site and the MCS from pBRINTs-Cat2. However, under normal usage conditions, cloned DNA would be inserted into the MCS of any of these vectors and the resulting plasmid derivative will be used to transfer this DNA into the *lacZ* gene in the chromosome. Using vectors of the pBRINTs-rAnb^R family, it has been possible to insert up to 7 kb of cloned DNA into the chromosome of *E. coli* W3110.

2. The full sequence of each of the three members of the pBRINTs-rAnb^R family is known. The accession numbers for the sequences of pBRINT$_S$-Cat2, pBRINT$_S$-Gen4, and pBRINT$_S$-Kan2 in the EMBL Nucleotide Sequence Database are AJ278278, AJ278279, and AJ278280, respectively.

3. For chromosomal integration to take place and be easily selected using the pBRINT$_S$-rAnb^R vector family, the *E. coli* strain used must have a complete *lacZ* gene. A simple way to determine if an *E. coli* strain has a functional *lacZ* gene is to streak it on LB plates containing 33 μg/per mL of Xgal and 0.1 mM IPTG and incubate it overnight at 37°C. Blue colonies are an indication of a *lacZ*$^+$ phenotype. It should be pointed out that chromosomal integration can also be achieved in *E. coli* strains like JM101, which does not have a functional chromosomal *lacZ* gene. In this case, integration can occur at the Δ(*lacZ*)M15 allele present in the F' episome of this strain, which contains an incomplete *lacZ* gene encoding an inactive beta-galactosidase enzyme. In this case, the white/blue selection scheme cannot be used. Cells where chromosomal integration occurred form large colonies (*see* **Note 8**), and can be selected only for resistance to Cm and sensitivity to Cb at 44°C. Many *E. coli* strains commonly used in recombinant DNA experiments have the Δ(*lacZ*)M15 allele in their F' episome and no *lacZ* gene in their chromosome.

4. Bacterial transformations can be carried out either using CaCl$_2$-treated or electrocompetent cells. It is important to remember that the incubation time to recuperate the cells after heat shock or electroporation must be carried out at 30°C.

5. Cb was used instead of ampicillin because it is a more stable antibiotic.

6. For vectors pBRINT$_S$-Gen4 and pBRINT$_S$-Kan2, 5 μg/mL of gentamycin and 15 μg/mL of kanamycin are used for selection of chromosomal integration, respectively. If either of these two plasmids are used with this method, the respective antibiotic should replace Cm.

7. On very rare occasions a strain obtained from a white colony does not have the expected pBRINT$_S$-rAnb^R chromosomal insertion. This is the result of random mutations that inactivate *lacZ*. This is why it is necessary to verify by PCR that the expected recombination event took place at the *lacZ* locus.

8. The colonies in which the plasmid had integrated into the *lacZ* gene of the chromosome are typically larger, in addition to being white and Cm^R. Colonies presenting these characteristics are good candidates for replica plating.

9. As can be seen in **Fig. 1** and **Fig. 2**, two *loxP* sites flank the Anb^R genes. After the site-specific recombination event promoted by Cre recombinase, the Anb^R and one *loxP* site are removed and a single *loxP* site and any DNA sequences inserted into the MCS remain in *lacZ*.
10. To estimate the size of the expected PCR amplification products in each particular integration experiment, it is necessary to add the size of the DNA cloned into the MCS of the pBRINTs-rAnb^R vector to the sizes shown in **Fig. 4**. It is also important to point out that depending on which pBRINTs-rAnb^R vector is used, the size of the PCR product will be different. This is due to the size difference of each Anb^R marker (**Fig. 1**).

Acknowledgments

This work was supported by grant NC-230 from the Consejo Nacional de Ciencia y Tecnología, México.

References

1. Balbás, P. and Gosset, G. (2001) Chromosomal editing in *E. coli*: vectors for DNA integration and excision. *Mol. Biotechnol.* **19**, 1–12.
2. Balbás, P., Alvarado, X., Bolívar, F., and Valle, F. (1993) Plasmid pBRINT: a vector for chromosomal integration of cloned DNA. *Gene* **136**, 211–213.
3. Balbás, P., Alexeyev, M., Shokolenko, I., Bolívar, F., and Valle, F. (1996) A pBRINT family of plasmids for integration of cloned DNA into the *Escherichia coli* chromosome. *Gene* **172**, 65–69.
4. Le Borgne, S., Palmeros, B., Valle, F., Bolívar, F. and Gosset, G. (1998) pBRINT-Ts: a plasmid family with a temperature-sensitive replicon, designed for chromosomal integration into the *lacZ* gene of *Escherichia coli*. *Gene* **223**, 213–219.
5. Palmeros, B., Wild, J., Szybalski, W., Le Borgne, S., Hernández-Chávez, G., Gosset, G., et al. (2000) A family of removable cassettes designed to obtain antibiotic-resistance-free genomic modifications of *Escherichia coli* and other bacteria. *Gene* **247**, 255–264.
6. Le Borgne, S., Palmeros, B., Bolívar, F. and Gosset, G. (2001) Improvement of the pBRINT-Ts plasmid family to obtain marker-free chromosomal insertion of cloned DNA in *Escherichia coli*. *BioTechniques* **30**, 252–256.
7. Hamilton, C. M., Aldea, M., Washburn, B. K., Babitzke, P. and Kushner, S. R. (1989) New method for generating deletions and gene replacements in *Escherichia coli*. *J. Bacteriol.* **171**, 4617–4622.
8. Ausubel, F. M., Brent, R., Kingston, R. E., Moore, D. D., Seidman, J. G., Smith, J. A., et al. (1995) Short protocols in molecular biology, 3rd ed. John Wiley and Sons Inc., New York, NY.

10

Copy-Control pBAC/oriV Vectors for Genomic Cloning

Jadwiga Wild and Waclaw Szybalski

Summary

The use of the improved BAC system for cloning genomic DNA and library constructions is described. This system retains all the advantages of the original BACs but, in addition, permits, on command, amplification of the BAC plasmids and cloned DNA. This system consists of (1) plasmid pBAC/oriV containing an additional replication origin, *oriV*, and (2) a host carrying the up-mutants of the *trfA* replicator gene expressed from the L-arabinose-inducible *P*ara promoter. The pBAC/oriV clones are always maintained in the single-copy state, but if more DNA is required, they could be amplified up to 100-fold, depending on the size of the cloned insert.

Key Words: BAC libraries; amplifiable copy number; F plasmid *oriS*/RepE replicon; *oriV*/TrfA replication.

1. Introduction

Bacterial artificial chromosome (BAC) low-copy plasmids, developed in Dr. Mel Simon's laboratory at CalTech (*1*), are the vectors of choice for construction of stable genomic DNA libraries with large DNA inserts. These BAC vectors, which are based on the F plasmid *oriS*/RepE replicon, allow stable maintenance of large genomic DNA fragments as very-low-copy or single-copy (SC) plasmids. The best known of these is vector pBeloBAC11. Genomic libraries constructed in this vector were shown to be stable (*2*). However, original BAC vectors also have disadvantages; they produce a low yield of both the vector and cloned DNA. Consequently, this low DNA yield results in costs higher than those for high-copy (HC) vectors, both for vectors and shotgun libraries to be used for end sequencing and for restriction fingerprinting. Furthermore,

From: *Methods in Molecular Biology, vol. 267:*
Recombinant Gene Expression: Reviews and Protocols, Second Edition
Edited by: P. Balbás and A. Lorence © Humana Press Inc., Totowa, NJ

the very low yield of DNA affects its purity by increasing contamination with the host DNA.

To overcome these limitations, Dr. W. Szybalski's laboratory at the University of Wisconsin has re-engineered pBeloBAC11 and its host to allow conditional production of multiple copies of the cloned DNA, while retaining all advantages of the SC vectors (*3–5*). This amplifiable pBAC/oriV system corresponds, in its maintenance phase, to the original SC BAC; when required to amplify DNA, however, it switches to the high-copy replication mode. As a result of such an SC-to-HC switch (1) up to 50 times more vector and cloned DNA can be prepared from the same culture volume, and (2) the resulting DNA is up to 50-fold less contaminated with the host DNA.

An amplifiable pBAC/oriV system consists of the pBAC/oriV vector and a cognate host strain. The vector contains a new, additional origin of replication (*oriV*) derived from a broad-host-range plasmid RK2 (*5,6*). The host strain, a modified *E. coli* DH10B, upon induction with the L-arabinose supplies *in trans* the *oriV*-specific amplification function (TrfA). When not induced, the *oriV*-TrfA replicon is silent and therefore the pBAC/oriV vector behaves as the original SC BAC. This is the maintenance mode, which depends on the *oriS* system. Only when required, the amplification function (TrfA) is provided by induction with L-arabinose, and replication of DNA is then taken over by the *oriV*-TrfA replicon. Such a temporal switch from the SC to HC plasmid copy number results in the amplification and high yield of DNA (*4,5*).

2. Materials

1. Bacterial strains: JW366 (DH10B, with *trfA*203 integrated at the *attB* site); JW463 (DH10B, with *trfA*250 integrated at the *attB* site); JW371 (JW366 carrying plasmid pBAC/oriV); and JW389 (JW366 carrying plasmid pBAC/oriV+40-kb insert). All strains are described in **ref. 5**.

   Plasmids: Vector pBAC/oriV (pJW360; *see* **Fig. 1**) and recombinant plasmid pBAC/oriV+ 40-kb insert are described in Ref. *5*. Additional vectors are described in **Subheadings 4.1., 4.2.**

2. LB medium: 1% Bacto-tryptone, 0.5% yeast extract, 1% NaCl sterilized by autoclaving.
3. LB plates: LB medium supplemented before autoclaving with 1.5% agar.
4. Solutions: 2% L-arabinose (Sigma), 20% D-glucose both filter-sterilized, X-gal (dissolved in DMSO, final concentration 40 µg/mL), 0.1 M CaCl$_2$.
5. Antibiotic: stock solution of chloramphenicol (Cm: 12.5 mg/mL)
6. Restriction enzymes of choice, appropriate buffers for digestion and 10 × BSA (bovine serum albumine). In **Subheading 3.2.**, the use of NcoI* and buffer 4 (NEB) to digest pBAC/oriV is described. Alkaline phosphatase (Roche), T4 DNA ligase (Epicentre, Madison, WI, 10 U/µL).
7. Phenol-chloroform, ethanol, 70% ethanol.
8. Wizard *Plus* SV Minipreps DNA Purification System (Promega) and PSIχClone BAC DNA Kit (Princeton Separations).
9. Agarose, ethidium bromide (EtdBr, 500 µg/mL), gel-loading buffer, and TBE electrophoresis buffer (*7*).

---

*New nomenclature (no italics) according to Roberts et al. (*8*).

10. Required laboratory equipment includes gel boxes and power supplies for agarose gel electrophoresis, apparatus for electroporation, a microfuge, water bath, incubators, and autoclave.

3. Methods

3.1. Preparation of Large Quantities of pBAC/oriV Vectors

To construct genomic libraries, it is important to have sufficient quantity of pure vector DNA, a rather difficult task with the SC BAC vectors (*1,2,9,10*). Therefore, the SC/HC pBAC/oriV vectors are ideally suited for that purpose. The amplification and preparation of pBAC/oriV DNA is described below:

1. Using a sterile wooden applicator or inoculation loop transfer a lump of frozen culture from a freezer stock to inoculate 5 mL of LB medium supplemented with 12.5 µg Cm/mL. Grow culture overnight at 37°C with shaking.
2. To induce amplification of the pBAC/oriV plasmid, inoculate 10 mL of fresh LB + Cm medium with 50 µL of overnight culture. Grow culture at 37°C with shaking to A_{590nm} = 0.2–0.3 (usually 30 min). At that cell density add 50 µL of 2% L-arabinose (0.01% final concentration) and grow culture(s) for additional 4–5 h before harvesting cells for extracting DNA.
3. Routinely, DNA is prepared from 4.5 mL of induced culture. To make DNA minipreps, follow the protocol for Wizard *Plus* SV Minipreps DNA Purification System (Promega). The DNA is eluted from the column with 50 µL of nuclease-free water. When larger quantities of DNA are needed, grow 100 mL of culture and induce DNA amplification by adding 500 µL of 2% L-arabinose. Centrifuge 4.5 mL of culture in 1.5-ml Eppendorf tubes, using 10 tubes. Extract DNA as described above and after elution with 50 µL of nuclease-free water, combine all samples (to a total volume of 0.5 mL). When 96-well plates are used to grow cultures (*see*, e.g., *11*), it helps to supplement LB with 0.05%–0.1% glucose to improve yield of bacterial mass. However, one should not use higher than 0.1% concentration of glucose, since 0.2% D-glucose will suppress DNA amplification and thus would interfere with induction by L-arabinose.

3.2. Quantity and Quality of Amplified Vector DNA

To evaluate the amount of DNA purified from the induced culture, a digestion with a restriction enzyme (which cuts once or twice within the vector DNA (e.g., NcoI) should be performed prior to the gel electrophoresis. In the total volume of 20 µL, combine 3–5 µL of DNA purified from induced culture, 2 µL 10 × BSA, 2 µL of a buffer appropriate for the chosen restriction enzyme, 1–2 µL of restriction endonuclease, and water. Mix well and incubate at appropriate temperature (consult the Promega or NEB catalogs for buffers and temperatures of digestion with the chosen enzymes; for NcoI use 37°C and buffer 4) for 1–2 h. Add dye solution, transfer the sample into the well in 0.8% agarose gel, electrophorese, and stain DNA with EtdBr, either by adding it to the melted agarose or by incubating gel in the EtdBr solution for 5–10 min. EtdBr-stained DNA is visualized under UV light. Digestion of the pBAC/oriV DNA with NcoI results in two bands (1.8 and 6.1 kb), as shown in **Fig. 2A** lanes 1, 2, 3. By comparing lanes 1 and 2 (DNA from uninduced cultures) with lane 3

(DNA from induced culture) one can estimate fold of DNA amplification (approx 50-fold in **Fig. 2**). (For glucose effect in lane 2, *see* **Subheading 4.3.**)

3.3. Cloning Genomic Fragments into pBAC/oriV Vectors

3.3.1. Vector DNA

Once a sufficient quantity of pure vector DNA is prepared, it can be used for cloning the various DNA fragments and for construction of genomic libraries (*9–11*). Initial preparation (amplification and purification) of vector DNA is described in **Subheading 3.1.** DNA fragments can be cloned into any one or between any two of the several unique restriction sites of the vector's MCS that are within the *lacZα*, allowing for insert screening by α-complementation (*12*) or between two NotI sites (*see* legend to **Fig. 1**; **ref. *13***). Upon choosing the appropriate restriction endonuclease, the digestion reactions are set up similarly as described in **Subheading 3.2.**, using 3–5 µL of amplified pBAC/oriV DNA. In the next step, dephosphorylation of 5' ends of DNA is carried out as described in **ref. *7***.

3.3.2. Genomic Fragments

Various methods are used for preparation of genomic fragments for construction of libraries. Target DNA could be fragmented by the partial digestion with the enzyme of choice (most often HindIII; *see* **refs. *1*** and ***2***), by digestion with NotI (*13*) or by random mechanical shear. After enzymatic or mechanical fragmentation, DNA is usually fractionated according to its size to prepare uniform-size libraries. Detailed methods for preparing genomic fragments are as given in **ref. *7***. Methods for obtaining cDNA, addition of synthetic DNA linkers and adaptors, as well as fractionation of cDNA are described in **ref. *7***.

3.3.3. Cloning

1. To clone genomic fragments, amplified, digested, and dephosphorylated DNA of pBAC/oriV vector is ligated with the appropriately prepared insert DNA. Details on ligation between a vector and an insert are as described in **ref. *7***.
2. Perform ligations for 2 h at room temperature using highly concentrated (10 U/µL) ligase (Epicentre).
3. After 2 h ligation, precipitate DNA with ethanol, dry, and resuspend in 20 µL of water.
4. Electroporate 10-µL portions of the ligation reactions into electrocompetent cells. Electrocompetent cells are prepared according to the standard protocol (*7*) from the culture of strain JW366 (suitable for amplification of DNA up to 80 kb in size) or JW463 (suitable for amplification of DNA fragments larger than 80 kb). Both strains are described in **Subheading 2** and in Epicentre catalog, EPI300.
5. After electroporation, cells are supplemented with 1 mL of LB and incubated for 1 h at 37°C. Cells are spread on LB plates containing 12.5 µg Cm/mL. To allow for white/blue screening of transformants carrying pBAC/oriV clones, the X-gal indicator dye (40 µg/mL) has to be included in the medium (*12*).
6. Using a sterile wooden applicator or inoculation loop, pick up from the plate a single colony of transformant (white, or preferably use lacZα, like here when using *lacZα*-complementation), inoculate 5 mL of LB + Cm medium, and grow culture overnight at 37°C with shaking. Each white colony will represent a different clone.

    Transfer 1.2 mL of overnight culture into a cryogenic vial (Nalge Co.), add 0.3 mL of sterile 50% glycerol, mix, and freeze at −70°C.

3.3.4. Maintenance

To allow stable maintenance of large genomic DNA fragments, clones must be kept at the single-copy (SC) state. This is discussed in **Subheading 4.** As an alternative option, the available original BAC clones can be retrofitted to allow their amplification, as described in **Subheading 4.**

3.4. Amplification of pBAC/oriV Clones

Using a sterile wooden applicator or inoculation loop transfer a lump of frozen culture from a freezer stock to inoculate 5 mL of LB medium supplemented with 12.5 µg Cm/mL. Grow culture overnight at 37°C with shaking. Use this overnight inoculum to start cultures for DNA amplification, as described for the vector in **Subheading 3.1.2.** Amplification could be modulated as described in **Subheading 4.**

3.5. Extraction of DNA from Induced pBAC/oriV Clones

1. After 4–5 h of growth in LB + Cm medium supplemented with 0.01% L-arabinose to induce DNA amplification, collect cells from the 4.5-mL volume of the culture and resuspend in 100 µL of Solution I.
2. Add 200 µL of Solution II, mix the contents by inverting the tube several times, and incubate for 10 min.
3. After the cells have lysed, add 150 µL of Solution III and keep the samples on ice for 5 min (Solutions I, II, and III are from a kit, Wizard *Plus* SV Minipreps DNA Purification System, Promega).
4. Centrifuge samples at 12,000g for 10 min.
5. Transfer supernatant to a fresh tube and add equal volume (450 µL) of phenol-chloroform.
6. Mix samples and centrifuge 2 min at 12,000g to separate phases.
7. Transfer the top aqueous phase to a fresh tube and precipitate with 2 volumes of ethanol.
8. Centrifuge samples for 10 min at 12,000g to pellet DNA, then wash with 70% ethanol and dry.
9. Resuspend the dried DNA pellet in 50–150µL (for inserts over 100 kb) of nuclease-free water. Alternatively, for extraction of very large DNA fragments (100 kb and over), the PSIχClone BAC DNA Kit (Princeton Separations) could be used.
10. Resuspend precipitated DNA in 50–150µL of nuclease-free water.

3.6. Evaluation of Cloned, Amplified DNA and Application

To evaluate the amount of DNA extracted from the induced culture carrying pBAC/oriV plasmid with cloned insert, digest with a restriction enzyme, which cuts insert DNA only very infrequently. If the distribution of restriction sites in cloned DNA is not known, use restriction enzymes with 8-bp recognition site, such as AscI, FseI, NotI, PacI, PmeI, SbfI or SwaI (*see* e.g., NEB Catalog). To prepare a digest, follow the protocol in **Subheading 3.2.** To determine the size of DNA fragments, perform electrophoresis on 0.6% agarose gel or use pulsed-field gel electrophoresis.

Cloned genomic DNA is mainly used for sequencing. For such a purpose, this DNA has to be free of any contamination with other DNAs, principally that of the host. It must represent the original genome, which has not mutated because of all the cloning manipulations and/or improper maintenance. The currently available methods permit precise sequencing and sequence assembly for most of the genomic DNA, with the exception of highly repetitive DNA. Therefore, sequencing of the latter kind of DNA is described in **Subheading 4.6.**

4. Notes

1. Additional vectors based on pBAC/oriV plasmid. We have constructed four derivatives of the pBAC/oriV vector with additional useful features (**5**) that are described below. The pBAC/oriV vector contains the *lacZα* fragment encoding the α-peptide of β-galactosidase (**Fig. 1**) that has been widely used for color screening of recombinant clones based on insertional inactivation (between ATG and codon 7). However, this screening method often leads to false results, both positive (white colonies that do not contain the insert) and negative (blue colonies that contain the insert). Vectors that offer very high accuracy in blue/white screening and provide a much darker blue color of colonies have been constructed and are available commercially. These are pIndigoBAC-5 (Epicentre) and TrueBlue-BAC2 (Genomics One). To obtain their amplifiable derivatives we cloned into each vector the modified *oriV*, creating (*1*) pIndigoBAC-5/oriV and (*2*) pTrueBlue-BAC2/oriV.

   As optical mapping became a tool for constructing of restriction mega-maps (**16**), it became desirable to have a very rare and reliable restriction site on the vector for efficient and convenient linearization of clones. Thus we added to pBAC/oriV the I-SceI restriction site for the intron homing endonuclease (**17**), obtaining (*3*) pBAC/oriV/SceI and (*4*) pTrueBlue-BAC2/oriV/SceI (**5**).

2. Commercially available pBAC/oriV derivatives. The biotechnology company Epicentre (Madison, WI) has been distributing our pBAC/oriV vector and its cognate host as essential components of several CopyControl (CC) kits. The Introduction to the CopyControl Cloning Systems (Epicentre 2003 Catalog, p. 162) outlines the principle of "on command" switch from a SC to HC number plasmid. Epicentre is supplying: (*1*) CopyControl(blunt cloning-ready) Cloning kits for construction of the BAC libraries (Epicentre 2003 Catalog, p. 164), (2) Copy-Control Fosmid Library Production Kit for constructing libraries of cosmid-sized (approx 40 kb) clones (Epicentre 2003 Catalog, p. 168) and (*3*) CopyControl PCR Cloning Kits for cloning any PCR or RT-PCR product or for construction of cDNA libraries (Epicentre 2003 Catalog, p. 174). The CopyControl BAC Cloning kits contain BamHI, EcoRI, or HindIII-linearized, dephosphorylated and highly purified DNA of pCC1 (Blunt cloning-ready) vector ready for cloning and construction of libraries (*see*, e.g., **ref. 11**). The CopyControl PCR Cloning Kits (Epicenter 2003 Catalog, p. 174) consist of pCC1 (blunt cloning-ready) vector and either electrocompetent or chemically competent EPI300 host cells (Epicenter 2003 Catalog, p. 184). Furthermore, elements of our *oriV*-based amplification system are included in the EZ::TN<oriV/KAN-2> Insertion Kit that is designed to simplify and speed up both DNA purification and sequencing of existing SC BAC and fosmid clones (Epicentre 2003 Catalog, p. 112) (*see also* **Subheading 4.6.**). Along a similar line, another biotechnology company, Invitrogen, developed the GeneJumper oriV Transposon Kit (Invitrogen 2003 Catalog, p. 458), which allows for obtaining high-BAC DNA yields as well as the insertion of primer-complementary sequences for complete sequencing.

3. Stability of the pBAC/oriV clones; effect of glucose. When testing various media for optimal DNA amplification, we have discovered that glucose, at 0.2%, reduces the number of BAC copies (**5**) as shown in **Fig. 2A**. This makes the 0.2% glucose an important novel tool for the maintenance of BACs at a SC state, since such a state prevents any undesirable rearrangements that might occur as a result of recombination between two or more plasmids present within a single host cell.

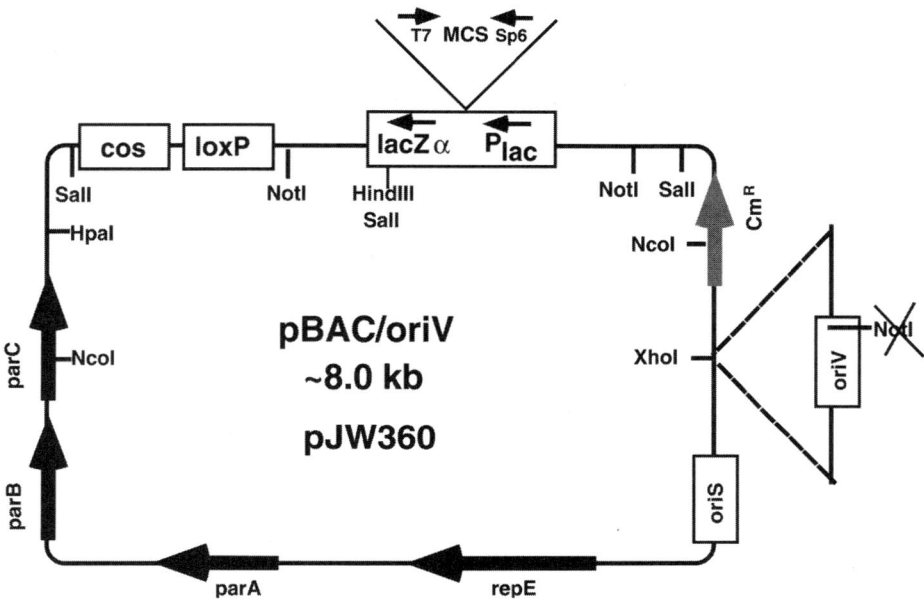

Fig. 1. The pBAC/oriV vector (pJW360) permitting its SC maintenance and alternatively, its conditional, tightly regulated DNA amplification. This one, among several new *oriV*-based derivatives of the pBeloBAC11 vector, preserves most of its original specific features, including the plasmid F-derived SC maintenance system based on the *oriS-repE-parABC* genes (*2*), but is equipped with an HC origin of DNA replication, *oriV*, from the broad-host-range plasmid RK2 (*6*). The MCS contains unique ApaLI, BsaHI-KasI-NarI-SfoI, BamHI, SphI, and HindIII restriction sites that are the same as in pBeloBAC11 (NEB 2002-03 Catalog, p. 294). The MCS is within the *lacZα*, allowing detection of cloned inserts by α-complementation (white/blue screen). The MCS is flanked by two NotI sites, permitting easy excision of cloned inserts. The T7 and SP6 phage promoters contain sequences complementary to generally used sequencing primers. The *cosN* site of phage λ allows (*1*) packaging into phage λ particles, (*2*) specific labeling of λ cohesive ends used for restriction mapping and (*3*) in vitro linearization by λ terminase. The *loxP* site permits linearization and/or introduction of additional DNA fragments via the Cre-*loxP* system of phage P1. The inactivated NotI site within *oriV* is indicated.

4. Easy conversion of SC BAC clones into copy-control pBAC/oriV clones. We have developed a system that permits in vivo retrofitting of SC BAC clones with the conditionally inducible *oriV* (*18,19*). To introduce *oriV*, we employed the *lox* site present on BAC vectors and Cre recombinase. Retrofitting is carried out using a host that provides inducible TrfA function and carries two plasmids. One plasmid contains *oriV*, *loxP* and Km[R] elements (to introduce *oriV* by recombination between two *lox* sites, while using the Km[R] selection). The second plasmid features an ability to transiently supply the Cre protein synthesized under control of inducible *lac* promoter. Both retrofitting plasmids have specific features permitting their easy elimination (*18,19*). The retrofitted BAC/oriV clones permit production of copious amounts of large genomic fragments in the same manner as for originally constructed pBAC/oriV clones (*5*).

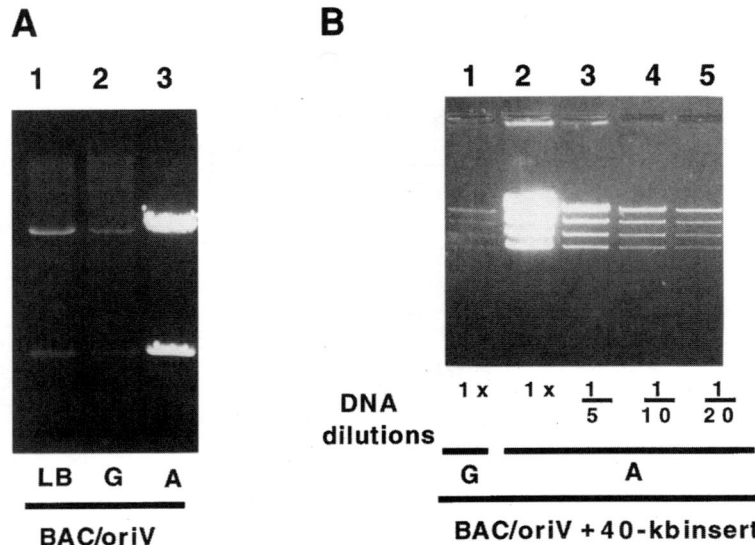

Fig. 2. Maintenance and amplification of pBAC/oriV vector and clones carrying 40-kb insert. (**A**) Amplification of the pBAC/oriV vector. The DNA was digested with NcoI and run on a 0.8% agarose gel. Lanes 1–3, strain JW371 (carrying pBAC/oriV) grown in the LB medium (lane LB), LB +0.2% D-glucose (lane G) or LB + 0.01% L-arabinose (lane A). (**B**) Assessment of amplification by diluting the amplified DNA of pBAC/oriV clones containing 40-kb inserts. The DNA analysis after SalI digestion and electrophoresis is shown. The G and A lane designations are as for panel A. Lane 1, uninduced strain JW389 (carrying pBAC/oriV with the 40-kb insert) grown in LB+G; lanes 2–5, induced strain JW389 grown in LB+A; DNA in lanes 1 and 2 is undiluted. In lanes 3, 4, and 5, DNA was diluted, as specified below the lanes, prior to SalI digestion.

Retrofitting of SC BAC clones could also be performed in conjunction with the transposon-mediated sequencing (or independent from it) using our *oriV*-carrying transposon Tn5 insertion kit, EZ::TN™<*oriV*/KAN-2> (Cat No. EZI02VK) that is distributed by Epicentre and is described at: http://www.epicentre.com/item.asp?ID=386. For some applications, the position of the Tn5 element insertion and the efficiency of amplification have to be assessed.

5. Regulation of amplification of the pBAC/oriV clones by L-arabinose, D-fucose, and D-glucose. The extent of DNA amplification could be regulated either by varying the concentration of L-arabinose or by modulating the DNA amplification by other sugars (5). The optimal plasmid amplification is achieved at the L-arabinose concentration of 0.01%. Lowering the concentration to 0.001% reduces the yield of DNA only slightly. In the presence of 0.0002% L-arabinose, amplification is about 10 times lower and at concentrations below 0.0001% there is very little DNA amplification. The D-glucose and D-fucose interfere with L-arabinose induction of DNA amplification. Inhibition by glucose exhibits a very sharp transition between noninhibitory (0.1%) and very inhibitory (0.18–0.2%) concentrations. It is of practical significance that glucose concentrations just below 0.1% enhance the growth of the host and the final yield of the amplified DNA, while the glucose is metabolized

(diauxic growth). Fucose blocks quite effectively the induction of DNA replication at concentrations above 0.01%. Modulation of DNA replication is mainly applicable for the new class of expression vectors that are based on the pBAC/oriV SC/HC plasmids (*14,15*).
6. Novel method for the sequencing of the repetitive DNA in the pBAC/oriV clones. Our method (developed in cooperation with M. Mendez Lago and A. Villasante) combines (*1*) the sequencing primed at two sequences inserted with each Tn*5* element with (*2*) the molecular mapping of the SceI-marked primer sites in relation to the SceI-marked sites on our pBAC/oriV vectors. To this end, we have reconstructed our *oriV*-carrying transposon Tn*5* element, EZ::TN™<*oriV*/KAN-2>, to incorporate the very rare I-SceI and PI-SceI restriction sites (*17*). When these new Tn*5*-based elements are used to insert the primer-complementary site into DNA clones in BACs that carry the very rare I–SceI or PI-SceI restriction site (e.g., our pBAC/oriV/SceI and pTrueBlue-BAC2/oriV/SceI; see **Subheading 4.1**), one could determine the exact location of the restriction cuts corresponding to each twin primer-complementary site. This permits mapping of the relative positions of each of the 1000–2000-nt sequence runs on the DNA clone and thus assembling the sequence of the entire clone, even though the individual 1000–2000-nt sequence repeats might be identical or nearly identical (*see* **20**).

References

1. Shizuya, H., Birren, B., Kim, U-J., Mancino, V., Slepak, T., Tachiiri, Y., et al. (1992) Cloning and stable maintenance of 300-kilobase-pair fragments of human DNA in *Escherichia coli* using an F-factor-based vector. *Proc. Natl. Acad. Sci. USA* **89,** 8794–8797.
2. Kim, U-J., Birren, B. W., Slepak, T., Mancino, V., Boysen, C., Kang, H-L., et al. (1996) Construction and characterization of a human bacterial artificial chromosome library. *Genomics* **34,** 213–218.
3. Hradecná, Z., Wild, J., and Szybalski, W. (1998) Conditionally amplifiable inserts in pBAC vectors. *Microbial & Comp. Genomics* **3,** 58.
4. Szybalski, W., Wild, J., and Hradecna, Z. (1999) Conditionally amplifiable BAC vector. US Patent No. 5,874,259 [revised].
5. Wild, J., Hradecná, Z., and Szybalski, W. (2002) Conditionally amplifiable BACs: switching from single-copy to high-copy vectors and genomic clones. *Genome Res.* **12,** 1434–1444.
6. Stalker, D. M., Thomas, C. M., and Helinski, D. R. (1981) Nucleotide sequence of the region of the origin of replication of the broad host range plasmid RK2. *Mol. Gen. Genet.* **181,** 8–12.
7. Sambrook, J., Fritsch, E. F., and Maniatis, T. (1989) *Molecular Cloning. A Laboratory Manual*, 2nd ed. Cold Spring Harbor Laboratory Press, Cold Spring Harbor, NY.
8. Roberts, R. J., Belfort, M., Bestor, T., Bhagwat, A. S., Bickle, T. A., Bitinaite, J., et al. (2003) A nomenclature for restriction enzymes, DNA methyltransferases, homing endonucleases and their genes. *Nucl. Acids Res.* **31,** 1805–1812.
9. Osoegawa, K., Woon, P. Y., Zhao, B., Frengen, E., Tateno, M., Catanase, J. J., et al. (1998) An improved approach for construction of bacterial artificial chromosome libraries. *Genomics* **52,** 1–8.
10. Schibler, L., Vaiman, D., Oustry, A., Guinec, N., Dangy-Caye, A. L., Billault, A., et al. (1998) Construction and extensive characterization of a goat bacterial artificial chromosome library with threefold genome coverage. *Mamm. Genome* **9,** 119–124.
11. Piffanelli, P. (2002) Construction of four Copy Control BAC libraries by BACTROP—a BAC-based platform to study tropical plant species. *Epicentre Forum* **9, No. 3,** 1–2.
12. Ullmann, A. (1992) Complementation in β-galactosidase: from protein structure to genetic engineering. *BioEssays* **14,** 201–205.

13. Mozo, T., Fischer, S., Shizuya, H., and Altmann, T. (1998) Construction and characterization of the IGF *Arabidopsis* BAC library. *Mol. Gen. Genet.* **258,** 562–570.
14. Wild, J., Hradecná, Z., and Szybalski, W. (2001) Single-copy/high-copy (SC/HC) pBAC/oriV novel vectors for genomics and gene expression. *Plasmid* **45,** 142–143.
15. Szybalski, W., Wild, J., and Hradecná, Z. (2002) Expression vector with dual control of replication and transcription. US Patent No. 6,472,177.
16. Giacalone, J., Delobette, S., Gibaja, V., Ni, L., Skiadas, Y., Qi, R., et al. (2000) Optical mapping of BAC clones from the human Y chromosome *DAZ* locus. *Genome Res.* **10,** 1421–1429.
17. Monteilhet, G., Perrin, A., Thierry, A., Colleaux, L., and Dujon, B. (1990) Purification and characterization of the *in vitro* activity of I-*Sce*I, a novel and highly specific endonuclease encoded by a group I intron. *Nucleic Acids Res.* **18,** 1407–1413.
18. Wild, J., Hradecná, Z., and Szybalski, W. (2001) Easy conversion of single-copy BAC libraries into conditionally high-copy pBAC/oriV clones. The 2001 Molecular Genetics of Bacteria & Phages Meeting. Madison, WI, July 31–August 5, 2001, p. 163.
19. Szybalski, W., Wild, J., and Hradecná, Z. (2003) Method for converting single-copy BAC vectors to conditional high-copy pBAC/oriV vectors. U.S. Patent Publ. No. US-2003-0049665, March 13, 2003.
20. Mendez Lago, M., Wild, J., Abad, J. P., Martin-Gallardo, A., Villasante, A., and Szybalski, W. (2004, abstract). Transposon-based innovative method for sequencing highly repetitive heterochromatic DNA in BAC/oriV clones. Abstracts, The 45th Annual Drosophilia Research Conference, Washington, DC, March 24–28, 2004.

11

Copy-Control Tightly Regulated Expression Vectors Based on pBAC/oriV

Jadwiga Wild and Waclaw Szybalski

Summary

A novel type of expression vectors with a dual regulation of both the plasmid copy number and gene expression, is described. The most important and beneficial feature of these vectors is that when they are not induced, they are maintained as a single-copy plasmid, and therefore, any residual expression is much more tightly regulated than for the conventional multicopy expression vectors. The simplest version of these copy-control expression vectors is based on the pBAC/oriV plasmid that carries the *trfA* up-mutant gene under control of the L-arabinose-inducible *P*ara promoter (*araC-P*$_{BAD}$). The same promoter controls expression of a gene cloned into MCS. Thus, addition of the inducer (L-arabinose) simultaneously turns on amplification of the plasmid and expression of the cloned gene. Net result is about a 50,000-fold increase in the cloned gene expression. However, when not induced, background expression level is very low, which is important for the maintenance of any "toxic" genes. This vector could be used in most *E. coli* hosts. Similar versions of the described vector employ the rhamnose-inducible *P*$_{rha}$ promoter (*rhaS-P*$_{rha}$). Other expression systems allow independent regulation of the plasmid amplification and of the cloned gene expression, and some also use the $P_{\text{LtetO-1}}$ promoter. Copy-control expression vector pETcoco, based on the *pT7lacO* promoter, is commercially available.

Key Words: Dual regulation; P_{ara} promoter; P_{rha} promoter; *oriV*; plasmid copy number.

1. Introduction

To maximize gene expression, most of the currently available expression vectors are based on high-copy (HC) plasmids that carry a strong and inducible promoter. Such HC vectors, even when noninduced, usually exhibit a high level of background expression. When a "toxic" gene is cloned in such vectors, such a gene is expressed at background

From: *Methods in Molecular Biology, vol. 267:*
Recombinant Gene Expression: Reviews and Protocols, Second Edition
Edited by: P. Balbás and A. Lorence © Humana Press Inc., Totowa, NJ

levels which often results in poor growth of the cells due to toxicity that is proportional to the vector copy number. Therefore, to minimize the background expression, we reengineered our copy-control pBAC/oriV vector (*1,2*), which, when not induced, is maintained as a single-copy (SC) plasmid. The most important and beneficial feature of the pBAC/oriV vector is its tightly regulated switch from SC to HC replication mode. Vector pBAC/oriV contains two origins (*ori*'s) for DNA replication: (1) SC maintenance *ori* of the *oriS*-RepE system together with the *parABC* partition determinants from BAC vectors (*3*), and (2) inducible HC *ori* of the *oriV*-TrfA system of the broad-host-range plasmid RK2 (*1,2,4,5*). The HC replication originating at *oriV* is driven by the copy-up variant(s) of TrfA protein encoded by gene *trfA* mutant(s) that is (are) under control of an inducible promoter. The newly synthesized TrfA protein initiates the HC replication at *oriV* only when the inducer is added. We reengineered this conditional SC/HC pBAC/oriV vector to generate a family of expression vectors by adding strong, controllable promoters to drive expression of any target gene cloned in a convenient multiple cloning site (MCS) (*6*). Two classes of expression vectors were constructed: (1) vectors carrying the same controllable promoter to express both the target gene and the copy-controlling *trfA* gene, and (2) vectors having one promoter to express the target gene and a separate promoter for expression of the copy-controlling gene *trfA*. Both promoters are independently and very tightly regulated. The SC mode of replication and maintenance of our novel expression vectors assures high stability of clones, because the background expression level is very low. This is especially important when the leaking product is toxic to the host. It is well known that "toxic" genes are often not "clonable" into the HC plasmids and vectors with a cloned toxic insert are not stably maintained. Such clones are counter-selected, even when the expression of a cloned gene is not induced, and they might undergo progressive "evolution" by acquiring undesirable spontaneous modifications affecting the integrity of the product or its final expression.

For target gene expression we employed four different promoters: $araC$-P_{BAD} (*7*), $rhaS$-P_{rha}(*8*), synthetic $P_{LtetO-1}$ (*9*), and phage $p_{T7}lacO$ promoter (*10*). Downstream of each specific promoter, there is a MCS for cloning the gene to be expressed.

The major advantage of our copy-control expression vectors is their unprecedented tightness, as any leakage from only copy (SC) is obviously 50 times lower than that from the 50-copy (HC) vector, whereas upon induction, the expression from our vectors is as high as that from the HC expression vectors. Thus, using these SC/HC expression vectors one can obtain very high ratios of induced to noninduced levels of expression.

The additional advantage of our copy-control expression vectors, as compared to various currently available HC vectors, is the enhanced viability of the host-vector system when maintained in a noninduced state as well as after the properly chosen induction procedure.

The activity of each promoter in these novel expression vectors can be modulated by adjusting concentrations of the inducer. The expression level can also be modulated by D-glucose ($araC$-P_{BAD}, $rhaS$-P_{rha}, $P_{T7}lacO$) (*1,8,11*) or D-fucose (*1*), when added together with inducers, enhancing the versatility of these SC/HC copy-control expression systems. Varying concentrations of an inducer and glucose or fucose allows attainment of an intermediate level of the cloned gene expression.

2. Materials

2.1. Bacterial Strains

All strains are derivatives of *Escherichia coli* K-12. Strains DH10B and JW366 (DH10B, with *trfA203* integrated at the *attB* site) are described in refs. *1* and *2*. The JW413 strain is a derivative of DH5α (New England BioLab [Beverly, MA] Catalog) with the *tetR* gene integrated into the *attB* site (*9*) and *araC-P$_{BAD}$-trfA203* integrated into the *lacZ* gene (*12*). Strain *E. coli* Tuner(DE3) is described in the Novagen catalog (2003). The biotechnology company Epicentre of Madison, WI, USA, has been distributing TransforMax EPI300 electrocompetent and chemically competent cells of a host suitable for the use with copy-control pJW493 expression vector (**Fig. 1**).

2.2. Expression Vectors

All expression vectors described in this chapter are based on the copy-control vector pBAC/oriV (pJW360) that is discussed in detail in **refs. *1*** and *2*.

2.2.1. Vectors with Expression Unit and Copy Number Simultaneously Controlled by L-Arabinose Regulatory System

1. Vector pJW493 is shown in **Fig. 1**. It contains an *araC-P$_{BAD}$*RBS unit (*7*) followed by the MCS (BamHI, SphI, HindIII).
2. Expression tester plasmid pJW532 (**Fig. 1**) has gene *lacZ* cloned into the SphI site of pJW493 MCS.
3. Vector pJW544, shown in **Fig. 2**, contains the L-arabinose-regulated expression unit consisting of *araC-P$_{BAD}$*RBS (*7*) followed by the *trfA203* gene (*1*) and the unique HindIII site.
4. Expression tester plasmid pJW546 (**Fig. 2**) has gene *lacZ* cloned into the filled-in HindIII site of pJW544 MCS. Unique sites in MCS are BamHI, SphI, and HindIII.

2.2.2. Vectors That Have Expression Unit and Copy Number Simultaneously Controlled by L-Rhamnose Regulatory System

1. Vector pJW572, shown in **Fig. 3**, contains the L-rhamnose-regulated expression unit consisting of *rhaS-P$_{rha}$*RBS (*8*) followed by the *trfA250* gene and the MCS. Unique site in MCS is SacI.
2. Plasmid pJW570 (**Fig. 3**) has the *rhaS-P$_{rha}$*RBS-*trfA250-lacZ* module cloned into the HpaI site of pBAC/oriV (*1,2*).

2.2.3. Vectors That Provide Independent Control of Target-Gene Expression and of Plasmid Copy Number

1. Vector pJW383, shown in **Fig. 4**, carries synthetic promoter $P_{LtetO-1}$ (*9*), whose expression is regulated by repressor (TetR) and inducer anhydrotetracycline (aTc). The promoter is followed by a restriction site MluI. This vector requires a specific host JW413 described in **Subheading 2.1**.

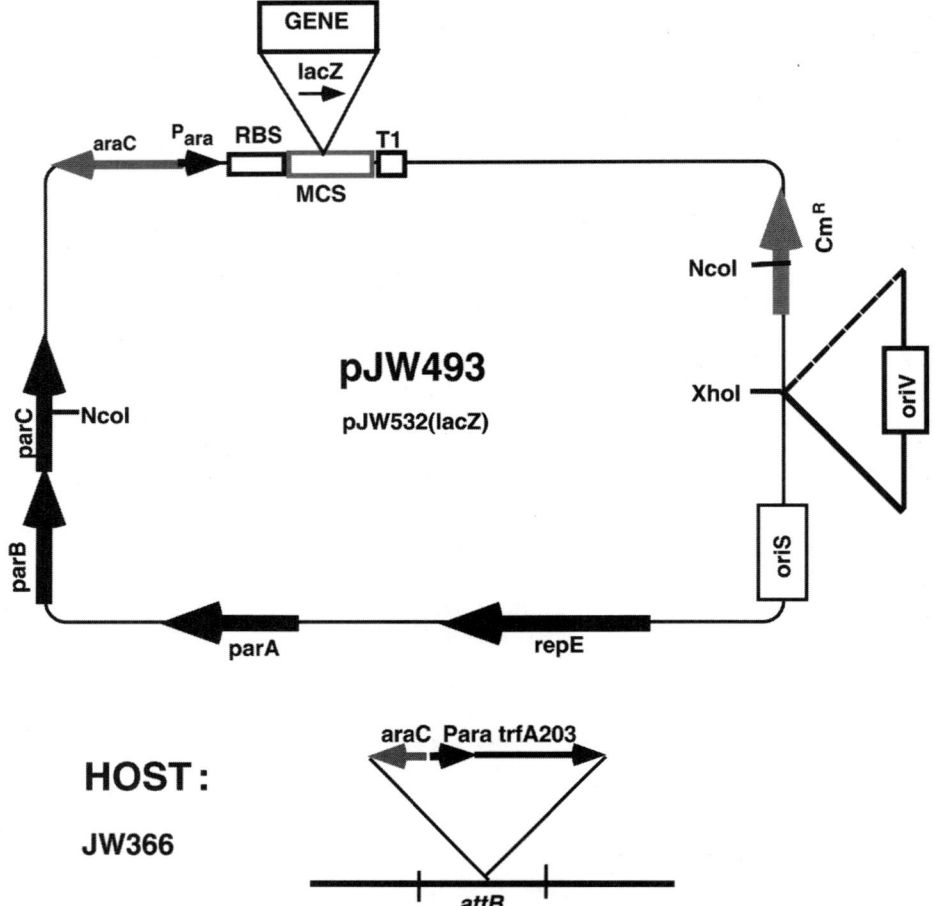

Fig. 1. The pBAC/oriV-based expression vector, pJW493, which permits the SC maintenance of any cloned gene (GENE) and its HC expression upon induction. It differs from pJW544 (**Fig. 2**) by not containing the *trfA* gene in its backbone. Therefore, for a switch from SC to HC mode of replication, it requires a host (JW366) that supplies the TrfA function under control of the *araC-P_{BAD}*RBS promoter system. The pJW532(lacZ) tester plasmid is the same as pJW493, but carries the *lacZ* reporter gene (**Table 1**). Both plasmids are described in **Subheading 2.2.** For the JW366 host, the switch from SC to HC mode of replication is achieved by induction with L-arabinose. Host JW366 is described in **Subheading 2.1.** and is represented by a drawing below the plasmid. Unique sites in MCS are BamHI, SphI, and HindIII.

2. Expression tester plasmid pJW396 (**Fig. 4**) has *lacZ* gene cloned into the BamHI site of pJW383.
3. Vector pETcoco-1, shown in **Fig. 5**, contains expression unit *lacI-P_{T7}/lacO-MCS-t_{T7}* from Novagen's pET-24(+) plasmid followed by the *trfA203* gene (*11*). The latter can also be independently controlled by the *araC-P_{BAD}*RBS unit (as in vector pJW493).

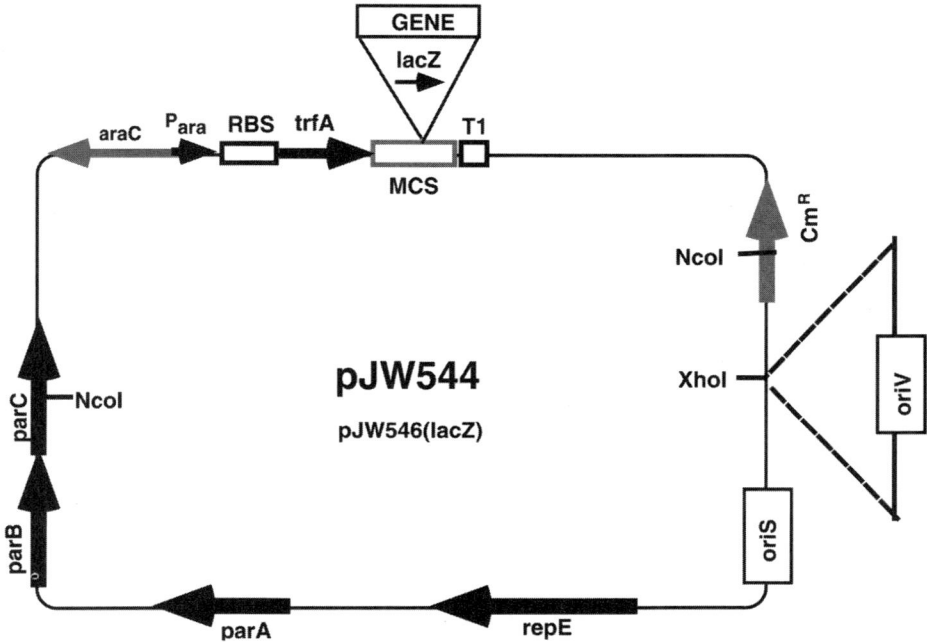

Fig. 2. The pBAC/oriV-based expression vector, pJW544, which permits the SC maintenance of any cloned gene (GENE) and its expression at HC upon induction with L-arabinose. The copy-controlling TrfA protein, as well as the cloned gene, are under control of the $araC$-P_{BAD}RBS promoter system. The pJW546(lacZ) tester plasmid is the same as plasmid pJW544, but carries the $lacZ$ reporter gene (**Table 1**). Both plasmids are described in **Subheading 2.2.1.** The switch from SC to HC mode of replication plasmid is achieved by induction with L-arabinose in most *E. coli* hosts, because this vector contains the L-arabinose-inducible $trfA$ gene. Arrowheads indicate orientation of genes and promoters. Unique site in MCS is HindIII.

4. Expression tester plasmid pETcoco-1.lacZ which carries gene *lacZ* cloned in MCS downstream of $P_{T7}lacO$, but upstream of P_{BAD} promoter.

2.3. Media, Solutions, and Enzymes

1. LB medium: 1% Bacto-tryptone, 0.5% yeast extract, 1% NaCl, sterilized by autoclaving.
2. LB plates: LB medium supplemented before autoclaving with 1.5% agar.
3. Solutions: 2% L-arabinose (Sigma), 2% L-rhamnose, 10 µg/mL anhydrotetracycline (aTc), 100 mM IPTG, 20% D-glucose, X–gal (dissolved in DMSO, final concentration 40 µg/mL), 0.1 M CaCl$_2$, 4 mg/mL ONPG, 0.1% SDS, 1 M Na$_2$CO$_3$, buffer Z (**13**), chloramphenicol (Cm: 12.5 µg/mL), chloroform.
4. Restriction enzymes of choice, appropriate buffers for digestion and 10 × BSA (bovine serum albumin), Klenow fragment of RNA polymerase, alkaline phosphatase (Roche), T4 DNA ligase (Epicentre, 10 U/µL).
5. Wizard *Plus* SV Minipreps DNA Purification System (Promega), agarose, ethidium bromide (EtdBr, 500 µg/mL), gel loading buffer, and TBE electrophoresis buffer.

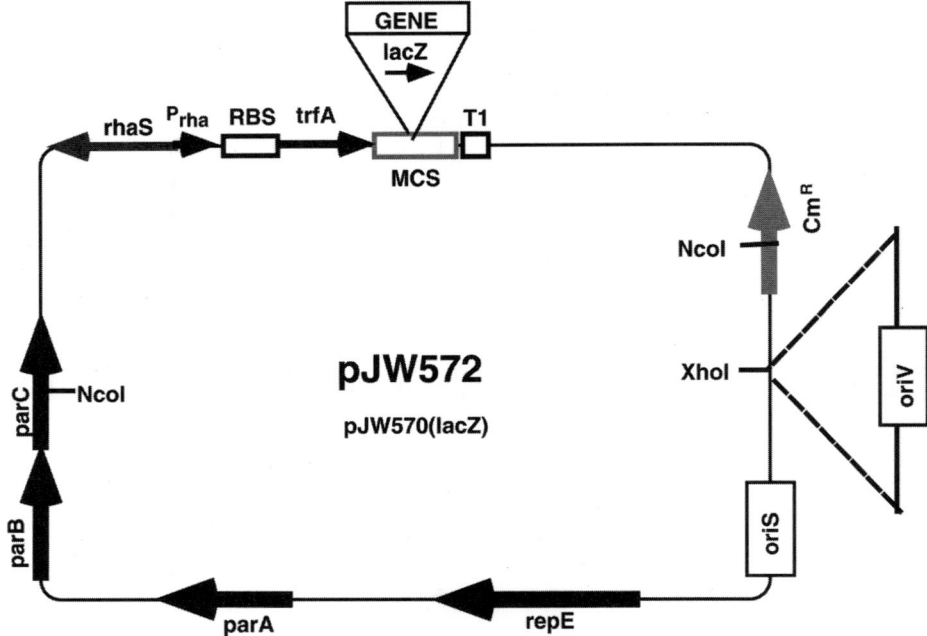

Fig. 3. The pBAC/oriV-based expression vector, pJW572, which permits the SC maintenance of any cloned gene (GENE) and its expression at HC number upon induction with L-rhamnose. This vector is analogous to pJW544 (**Fig. 1**) but it uses the *rhaS-P$_{rha}$*RBS promoter system to control expression from the *trfA* as well as from the cloned gene. The pJW570 (lacZ) tester plasmid is the same as pJW572 but carries the *lacZ* reporter gene. Both plasmids are described in **Subheading 2.2.2.**

The switch from the SC to HC mode of replication is achieved by induction with L-rhamnose in almost any *E. coli* host, because this vector carries the L-rhamnose-inducible *trfA* gene. Unique site in MCS is SacI.

2.4. Required Laboratory Equipment

Spectrophotometer, gel boxes, and power supplies for agarose gel electrophoresis, a microfuge, water bath, incubators, and an autoclave.

3. Methods

3.1. Preparation of Vectors for Cloning Target Genes

The backbone of all the expression vectors described in **Subheading 2.2.** is that of copy-control plasmid pBAC/oriV (*1,2*). Therefore, these expression vectors can be easily switched to the HC replication mode, as to obtain ample amount of DNA for cloning. Depending on the regulatory system employed to synthesize TrfA, amplification can be induced either by L-arabinose (pJW544, pJW493, pJW383) or by L-rhamnose (pJW572). In the case of pETcoco-1, addition of L-arabinose induces only

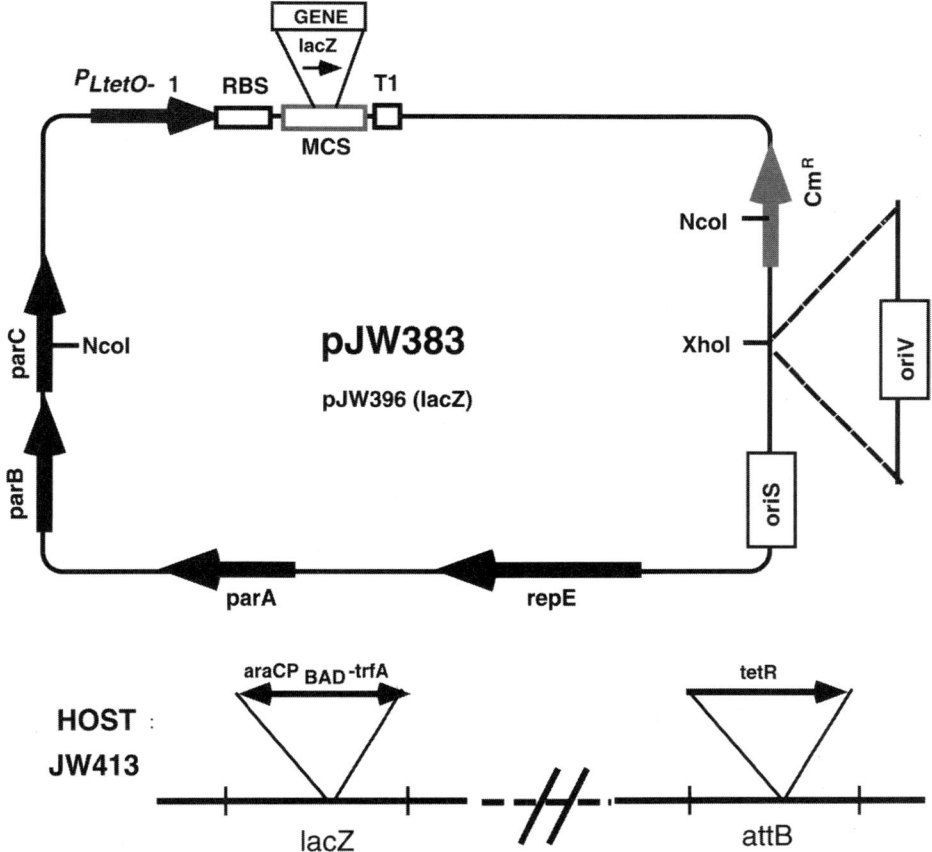

Fig. 4. The pBAC/oriV-based dual-control expression vector pJW383. This vector contains $P_{LtetO-1}$ promoter for the inducible expression of any cloned gene and does not contain the *trfA* gene. In host JW413, its copy number is controlled by the *araC-P*$_{BAD}$ regulatory system, independently from the gene-expression unit. Expression from $P_{LtetO-1}$ is inducible by aTc. When the host *trfA* is not induced by L-arabinose, this plasmid is maintained as an SC. The pJW396(lacZ) tester plasmid is the same as pJW383 but it carries the *lacZ* reporter gene. Both plasmids are described in **Subheading 2.2.3.** The host, JW413, is described in **Subheading 2.1,** and is represented by a drawing below the plasmid. Unique site in MCS is MluI.

DNA amplification, whereas IPTG induces both the DNA amplification and expression of the target gene.

3.1.1. Induction of TrfA Synthesis With L-Arabinose to Amplify the Vector

Cultures inoculated with overnight-grown cells are propagated in LB+Cm (12.5 µg/mL) medium at 37°C with shaking to A_{590nm} = 0.2–0.3 (usually 30 min). At that cell density L-arabinose is added (0.01% final concentration) and cultures are grown for additional 4–5 h before harvesting cells to obtain DNA.

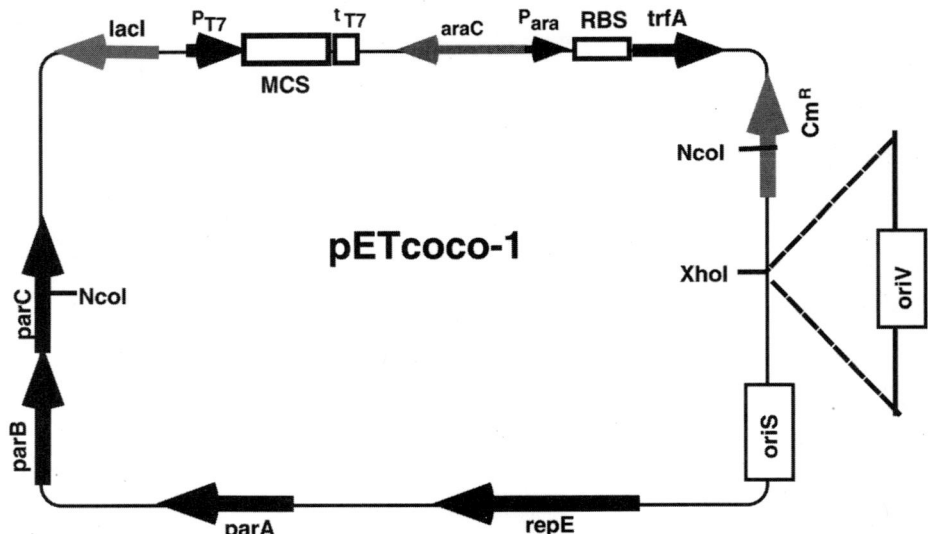

Fig. 5. The pBAC/oriV-based expression vector, pETcoco-1, which permits the SC maintenance of any target gene and its very high expression upon induction with IPTG. Expression of a
target gene is controlled by the T7 promoter (P_{T7}) and the T7 RNA polymerase, the latter supplied
by the IPTG-induced host under the control of *lac* promoter/operator. Therefore, expression from
pETcoco-1 is a multistep process: (1) induction by IPTG of the T7 RNA polymerase synthesis
from P_{lac} in the host strain *E. coli* Tuner(DE3), (2) target gene expression from the T7 promoter
of pETcoco-1, (3) expression of the *trfA* gene also from the T7 promoter, (4) amplification of the
pETcoco-1 or its clones by the *trfA* gene product, followed by (5) enhanced expression of the
cloned target gene, with all steps being inducible by IPTG. Since induction with L-arabinose
results only in plasmid amplification, it is used to amplify the vector's DNA prior to the target
gene cloning and also to amplify the clones. The pETcoco-1.lacZ tester plasmid is the same
pETcoco-1 plasmid, but it carries the *lacZ* reporter gene in MCS. Both plasmids are described in
Subheading 2.2.3. and primarily in **ref. *11***.

3.1.2. Induction of TrfA Synthesis With L-Rhamnose to Amplify the Vector

Amplification of vector DNA by using L-rhamnose (0.1% final concentration) is as
described above for induction with L-arabinose.

3.1.3. Evaluation of Quantity and Quality of Amplified Vector's DNA

Routinely, DNA is purified from 4.5 mL of the induced culture using Wizard Plus SV
Minipreps columns (Promega) by eluting DNA with 50 µL of nuclease-free water.

To evaluate the amount of DNA purified from the induced culture, a digestion with a
restriction enzyme (used for cloning) should be performed.

1. In the total volume of 20 µL, combine 3–5 µL of DNA purified from induced culture, 2 µL
   10X BSA, 2 µL of a buffer appropriate for the chosen restriction enzyme, 1–2 µL of restriction endonuclease, and water.

2. Mix well and incubate at appropriate temperature (consult the Promega or NEB catalogs for temperatures of digestion with the chosen enzymes). When the digestion is completed, add dye solution, transfer the sample into a well in the 0.8% agarose gel, perform electrophoresis.
3. Stain DNA with EtdBr, either by adding it to the melted agarose before pouring gel or by keeping gel in the EtdBr solution for 5–10 min. EtdBr-stained DNA is visualized under UV light.

3.2. Cloning Target Genes into Expression Vectors

Fragments of DNA to be expressed are cloned into copy-control expression vectors using standard DNA recombinant methods as described in ref. *13*. Each expression vector (*see* **Figs. 1–5**) contains specific MCS that is described in **Subheading 2.2.** and in the appropriate figure caption. These unique restriction sites serve for convenient cloning of inserts. The DNA fragments with cohesive or blunt ends can be cloned as well as PCR-derived fragments carrying ends compatible with the digested expression vector.

3.3. Regulation of Target Gene Expression

3.3.1. Induction With L-Arabinose

Vectors carrying L-arabinose-controlled expression units are listed in **Subheading 2.2.1.** and are shown as in **Figs. 1** and **2**. For induction of expression of a target gene cloned into pJW544 and/or pJW493 expression vectors, 0.01% L-arabinose is used and induction is performed as described in **Subheading 3.1.1.**

1. Vector pJW544 carries the *araC-P$_{BAD}$-trfA203* cassette with MCS for cloning a target gene downstream of the *trfA* gene. When using this vector, addition of L-arabinose induces expression from *ara* promoter resulting in simultaneous expression of a target gene and plasmid amplification. The pJW544 vector can be used in any *E. coli* strain. To test expression from this vector, the *lacZ* reporter gene was cloned into MCS of pJW544, resulting in pJW546. The residual levels of the *lacZ* gene product (βGal) in noninduced cells were low (39 Miller units). Upon induction with 0.01% L-arabinose, βGal synthesis has reached almost 50,000 units (*see* **Table 1**), which corresponds to 1200-fold induction.
2. Vector pJW493 carries the *araC-P$_{BAD}$*RBS expression unit followed by MCS for cloning the target gene. Depending on the bacterial host, this vector can be either permanently maintained at SC or can be maintain as a conditional SC/HC vector. When pJW493 resides in any wild-type *E. coli* strain, e.g., DH10B, it remains as a SC, since HC replicator protein, TrfA, is not available. However, when pJW493 is in strain JW366 carrying in its chromosome L-arabinose-regulated *araC-P$_{BAD}$-trfA203* cassette, its replication mode depends on the availability of TrfA protein. Upon induction of TrfA synthesis with L-arabinose, the HC replication at *oriV* resumes and pJW493 becomes an HC vector. Simultaneously, L-arabinose induces expression of the target gene controlled by the *araC-P$_{BAD}$* promoter. Thus, in the presence of L-arabinose, both expression of the target gene and its amplification are induced. To test expression from this vector, the *lacZ* reporter gene was cloned into MCS of pJW493, resulting in the pJW532 tester plasmid. Levels of βGal (**Table 1**) confirm that the expression from this vector is very tightly controlled in the JW366 host (7 βGal units without induction). Upon addition of L-arabinose, induced level of βGal is very high (40, 000 units) with the induction ratios close to 6000-fold (*see* **Table 1**). When βGal synthesis from pJW532 was assayed in DH10B host, the induced βGal level was moderate and reached 6000 units. This host-vector system is very versatile, allowing attainment of varying levels of the target gene expression.

Table 1
Levels of β-Galactosidase (βGal) Measured at SC and HC Replication Mode of pBAC/oriV Expression Vectors

Expression vector [a]	Host [b]	Inducer [c]	βGal Level [d] Noninduced	βGal Level [d] Induced	Fold induction
pJW546	DH10B	L-ara	39	47,000	~1200
pJW532	DH10B	L-ara	33	6000	~180
	JW366	L-ara	7	40,000	~5700
pJW570	DH10B	L-rha	35	33,500	~1000
pJW396	JW413	L-ara	8	87	~10
	JW413	aTc	8	850	~100
	JW413	L-ara + aTc	8	16,200	~2000
pETcoco-1.lacZ	Tuner(DE3)	L-ara	9[e]	120	~13
	Tuner(DE3)	IPTG	9[e]	48,000	15,300

[a] Vectors are described in **Subheading 2.2.**
[b] Hosts are described in **Subheading 2.1.**
[c] Inducers and induction processes are described in **Subheading 3.3.**
L-ara, L-arabinose; L-rha, L-rhamnose; aTc, anhydrotetracycline; IPTG, isopropyl-β-D-thiogalactoside.
[d] Assay of βGal is described in **Subheading 3.4.1.** and in ref. *14*.
[e] We performed βGal assays using an analogous commercial strain, BL26(DE3)[pET24-*lacZ*], which carries the multi-copy pET24-*lacZ* plasmid. When not induced with IPTG, it produces 498 units of βGal, which is about 50 times more than our 9 βGal units. Even the standard noninduced *E. coli* strain MG1655(*lacZ*[+]), which has a single copy of *lacZ* in its genome, produces 401 units of βGal.

3.3.2. Induction With L-Rhamnose

Vector pJW572, which carries the L-rhamnose-regulated expression unit, is described in **Subheading 2.2.2.** and shown in **Fig. 3**. For induction of expression of a target gene cloned into pJW572 expression vector, 0.1% L-rhamnose is used and induction is performed as described in **Subheading 3.1.2.**

Vector pJW572 carries the *rhaS*-P_{rha}RBS expression cassette (*8*) with MCS for cloning a target gene downstream of the *trfA* gene. When using this vector, addition of L-rhamnose induces expression from the *rha* promoter, resulting in expression of a target gene and plasmid amplification. The pJW572 vector can be used in any *E. coli* strain. To test expression from L-rhamnose-controlled promoter, vector pJW570 carrying the *rhaS*-P_{rha}RBS-*trfA250-lacZ* module was constructed and tested. Noninduced levels of βGal (*see* **Table 1**), reflecting the background level of expression, were low (35 units), whereas upon induction the expression increased by about 1000-fold, up to 33,000 units. The pattern of expression regulation in this vector is similar to that of pJW544, the latter regulated by L-arabinose and described in **Subheadings 2.2.1.** and **3.1.1.**

3.3.3. Induction With Anhydrotetracycline (aTc)

Vector pJW383 (described in **Subheading 2.2.3.** and shown in **Fig. 4**) is the dual-control expression vector. In this vector, expression of a target gene is driven by the synthetic promoter, $P_{\text{LtetO-1}}$ (**9**), regulated by a repressor (TetR) and an inducer (aTc). The copy number of the pJW383 vector is controlled by L-arabinose, which induces synthesis of TrfA from *araC-P$_{BAD}$-trfA*203 cassette integrated into the bacterial chromosome. Therefore, for regulated expression and amplification, plasmid pJW383 requires a special host, JW413 (*see* **Subheading 2.1.**), that expresses both regulators, TetR and TrfA. The protocol for L-arabinose induction is the same as that described in **Subheading 3.1.1.** and for pJW493 in **Subheading 3.3.1.** To induce expression from $P_{\text{LtetO-1}}$ promoter, the overnight-grown culture is diluted 1:100 in B+Cm (12.5 µg/mL) medium, aTc is added to a final concentration of 0.1 µg/mL, and cultures are grown at 37°C with shaking for 3 h to $A_{600nm} = 0.6$–0.8. At that absorbance, cells are harvested for βGal assay and/or to obtain DNA.

To evaluate the regulation of expression and amplification for this dual-control expression vector, the *lacZ* reporter gene was cloned into MCS of pJW383 and the levels of βGal in the resulting pJW396 tester plasmid were measured (**Table 1**). Upon induction with L-arabinose alone, plasmid copy number increased, resulting in expression level 10-fold above the background. With aTc alone, the target-gene expression increased 100-fold, reflecting the induced expression from the SC vector. When both inducers (L-arabinose and aTc) were present, expression increased 2000-fold, reaching 16,000 units. Since the level of expression from the pJW396 tester plasmid can be increased by 10-, 100-and 2000-fold, depending on the inducers used, vector pJW382 is very helpful when various levels of expression are to be obtained.

3.3.4. Induction With IPTG

The pETcoco-1 vector, described in **Subheading 2.2.3.3.** and shown in **Fig. 5**, is probably the most tightly controlled while being among the most highly expressed vectors currently available. It carries expression unit *lacI-P$_{T7}$lacO-MCS-t$_{T7}$* (**11**), which is indirectly inducible by IPTG, via induction of synthesis of the T7 RNA polymerase in the host. The T7 RNA polymerase, in turn, activates the P_{T7}*lacO* promoter. Therefore, to achieve maximal levels of expression, strain *E. coli* Tuner(DE3) (Novagen, Madison, WI) must be used; upon induction, it produces the T7 RNA polymerase. To induce expression from *lacI-P$_{T7}$lacO-MCS-t$_{T7}$* unit, the overnight-grown culture is diluted 1:200 in LB+Cm (12.5 µg/mL) medium and incubated at 37°C with shaking for 1.5 h. At that time IPTG is added (0.5 m*M* final concentration) and cultures are grown for additional 3 h before harvesting cells for performing an enzyme assay and/or obtaining DNA. To measure expression controlled by the T7 promoter, the *lacZ* reporter gene was cloned into pETcoco-1, resulting in the pETcoco-1.lacZ tester plasmid. Background expression from pETcoco-1.lacZ is very low (**Table 1**), being reduced to as little as 1/40 of the levels obtained with Novagen's HC pET vectors (**Table 1**, footnote e). Induced levels of *lacZ* expression from pETcoco-1.lacZ are as high as those for pJW546 and other pET vectors (**11**). When the pETcoco-1 DNA is to be amplified, one uses only L-arabinose, in the same manner as described in **Subheading 3.1.1.**

3.4. Measuring Level of Target Gene Expression

3.4.1. Assay of β-Galactosidase (βGal)

To measure βGal activity, ONPG assays were performed according to **ref. *14*** using 50 or 100 μL of a culture.

1. To lyse bacteria, add 50 μL of chloroform and 25 μL of 0.1% SDS.
2. Vortex thoroughly, when cells are lysed add 0.2 mL of 4 mg/mL ONPG.
3. When a yellow color develops, stop the reaction by adding 0.5 mL of 1 M Na_2CO_3.
4. Read the A_{420} of the reaction samples. Also determine the A_{600} of the cultures. Calculate units of βGal activity (as in **ref. *14***).

3.4.2. Assay by SDS-PAGE Gel Electrophoresis and Other Methods

The level of target gene expression can be monitored by denaturing polyacrylamide gel electrophoresis (SDS-PAGE) and nondenaturing PAGE *(15)*. Evaluation of the amount of synthesized protein is usually done by Coomassie staining, Western blotting, and other immunological assays *(13,15)*. Furthermore, expression product can be direct measured by spectrophotometric, colorimetric, and/or isotopic assays, as per each specific gene product.

References

1. Wild, J., Hradecná, Z., and Szybalski, W. (2002) Conditionally amplifiable BACs: Switching from single-copy to high-copy vectors and genomic clones. *Genome Res.* **12**, 434–1444.
2. Wild, J. and Szybalski, W. (2004) Copy-control pBAC/oriV vectors for genomic cloning, in *Protocols for Recombinant Gene Expression* (Balbás, P. and Lorence, A., ed.), Humana Press, pp. 145–154.
3. Kim, U.-J., Birren, B. W., Slepak, T., Mancino, V., Boysen, C., Kang, H.-L., et al. (1996) Construction and characterization of a human bacterial artificial chromosome library. *Genomics* **34**, 213–218.
4. Stalker, D. M., Thomas, C. M., and Helinski, D. R. (1981) Nucleotide sequence of the region of the origin of replication of the broad host range plasmid RK2. *Mol. Gen. Genet.* **181**, 8–12.
5. Durland, R. H., Toukdarian, A., Fang, F., and Helinski, D. R. (1990) Mutations in the *trfA* replication gene of the broad-host-range plasmid RK2 result in elevated plasmid copy numbers. *J. Bacteriol.* **172**, 3859–3867.
6. Szybalski, W., Wild, J., and Hradecná, Z. (2002) Expression vector with dual control of replication and transcription. US Patent No. 6, 472, 177.
7. Guzman, L.-M., Belin, D., Carson, M. J., and Beckwith, J. (1995) Tight regulation, modulation, and high-level expression by vectors containing the arabinose P_{BAD} promoter. *J. Bacteriol.* **177**, 4121–4130.
8. Haldimann, A., Daniels, L. L., and Wanner, B. L. (1998) Use of new methods for construction of tightly regulated arabinose and rhamnose promoter fusions in studies of the *Escherichia coli* phosphate regulon. *J. Bacteriol.* **180**, 1277–1286.

9. Lutz, R. and Bujard, H. (1997) Independent and tight regulation of transcriptional units *in Escherichia coli* via the LacR/O, the TetR/O and AraC/I_1-I_2 regulatory elements. *Nucleic Acids Res.* **25,** 1203–1210.
10. Studier, F. W., Rosenberg, A. H., Dunn, J. J., and Dubendorff, J. W. (1990) Use of T7 polymerase to direct expression of cloned genes. *Methods Enzymol.* **185,** 60–89.
11. Sektas, M. and Szybalski, W. (2002) Novel single-copy pETcoco vector with dual controls for amplification and expression. *In**Nova**tions* **14,** 6–8.
12. Palmeros, B., Wild, J., Szybalski, W., Le Borgne, S., Hernández-Chávez, G., Gosset, G., et al. F. (2000) A family of removable cassettes designed to obtain antibiotic-resistance-free genomic modifications of *Escherichia coli* and other bacteria. *Gene* **247,** 255–264.
13. Sambrook, J., Fritsch, E. F., and Maniatis, T. (1989) *Molecular Cloning. A Laboratory Manual*, 2nd ed. Cold Spring Harbor Laboratory Press, Cold Spring Harbor, NY.
14. Miller, J. H. (1972) in *Experiments in Molecular Genetics*. Cold Spring Harbor Laboratory Press, Cold Spring Harbor, NY, pp. 352–355.
15. Tuan, R. S. (ed.) (1997) *Methods in Molecular Biology, Vol. 63. Recombinant Protein Protocols: Detection and Isolation*. Humana Press, Totowa, NJ.

12

Cell-Free Protein Synthesis
With Prokaryotic Combined Transcription-Translation

James R. Swartz, Michael C. Jewett, and Kim A. Woodrow

Summary

Cell-free biology exploits and studies complex biological processes in a controlled environment without intact cells. One model system is prokaryotic cell-free protein synthesis. This technology offers an attractive and convenient approach to produce properly folded recombinant DNA (rDNA) proteins on a laboratory scale, screen PCR fragment libraries in a high-throughput format, express pharmaceutical proteins, incorporate labeled or unnatural amino acids into proteins, and activate microbial physiology to allow for investigation of biological systems. We describe the preparation of materials necessary for the expression, quantification, and purification of rDNA proteins from active *Escherichia coli* extracts.

Key Words: Prokaryotic combined transcription-translation; S30 extract; cell-free protein synthesis; in vitro; chloramphenicol acetyl transferase; T7 bacteriophage RNA polymerase; protein purification.

1. Introduction

Prokaryotic cell-free protein synthesis harnesses the catalytic machinery of the *Escherichia coli* cell to produce a desired protein in vitro. The capability of these systems to produce a target protein has increased several orders of magnitude over the past decade (*1–10*). This accomplishment has made in vitro translation a practical technique for laboratory-scale research, and has also provided a platform technology for high-throughput protein expression. The necessary enzymes and factors for translation are present in the cell lysate prepared from *E. coli* cells harvested in exponential phase. Additional substrates are added for protein synthesis to occur; these include amino acids, nucleotides, salts, an energy-regenerating source, an exogenous RNA polymerase, and

From: *Methods in Molecular Biology, vol. 267:*
Recombinant Gene Expression: Reviews and Protocols, Second Edition
Edited by: P. Balbás and A. Lorence © Humana Press Inc., Totowa, NJ

the DNA template for the target protein. After protein production, the open environment provided by cell-free systems is advantageous for the direct recovery of purified and properly folded protein products (*11*). We describe the preparation of materials necessary for the production, characterization, and purification of DNA gene products using the cell extract prepared from *E. coli* cells. The cell-free expression of chloramphenicol acetyl transferase is used for illustration.

2. Materials

1. *E. coli* strains A19 and BL21(DE3)/pAR1219.
2. LB (Luria-Bertani) media (amount per 1 L): 10.0 g tryptone, 5.0 g yeast extract, 5.0 g sodium chloride, 1.0 mL 1 *N* sodium hydroxide.
3. 2X YT media (amount per 1 L): 16.0 g tryptone, 10.0 g yeast extract, 5.0 g sodium chloride; adjust pH to 7.0 with sodium hydroxide.
4. S30 buffer: 10 m*M* tris-acetate pH 8.2, 14 m*M* magnesium acetate, 60 m*M* potassium acetate, 1 m*M* dithiothreitol, DTT at 4°C.
5. Pre-incubation mixture: 0.3 *M* Tris-acetate, pH 8.2, 13.2 m*M* magnesium acetate, 13.2 m*M* adenosine triphosphate, ATP, 4.4 m*M* DTT, 0.04 m*M* 20 amino acids, 6.7 U/mL pyruvate kinase (Sigma), 84.0 m*M* phosphoenolpyruvate (PEP) (Roche).
6. Spectra/Por MWCO 6-8000 dialysis tubing (Spectrum, Rancho Dominguez, CA).
7. Dialysis membrane buffer: 10 m*M* ethylenediaminetetraacetic acid (EDTA), pH 8.0, 100 m*M* sodium bicarbonate.
8. Ampicillin.
9. IPTG (isopropyl-β-D-thio-galactopyranoside) (Invitrogen).
10. Talon Superflow metal-affinity resin (Clontech).
11. T7 Equilibration/wash buffer: 50 m*M* sodium phosphate, pH 7.0, 300 m*M* sodium chloride.
12. T7 Elution buffer: 50 m*M* sodium phosphate, pH 7.0, 300 m*M* sodium chloride, 300 m*M* imidazole.
13. T7 storage buffer: 5 m*M* Tris-HCl, pH 8.2, 7 m*M* magnesium acetate, 30 m*M* potassium acetate, 5 m*M* DTT.
14. *E.coli* total tRNA mixture (Roche Molecular Biochemicals).
15. 5X Master Mix: 286 m*M* HEPES-KOH, pH7.5, 6 m*M* ATP, 4.3 m*M* each of GTP, UTP, and CTP, 333 μ*M* folinic acid, 853 μg/mL *E. coli* total tRNA.
16. 10X Salt solution: 2 M potassium glutamate, 0.8 M ammonium acetate, and 0.16 M magnesium acetate.
17. Nicotinamide adenine dinucleotide (NAD).
18. Poly (ethylene glycol) 8000.
19. L- [U-^{14}C]-leucine (Amersham Pharmacia Biotechnology).
20. [^{3}H]-uridine triphosphate (Amersham Pharmacia Biotechnology).
21. Trichloroacetic acid (TCA).
22. Whatman chromatography paper: 3 MM Chr.
23. Staining buffer: 45% methanol, 10% acetic acid, 0.5% Coomassie blue, 44.5% water.
24. Destaining buffer: 45% methanol, 10% acetic acid, 45% water.
25. 4X-NuPAGE LDS sample buffer.
26. NuPAGE 10X-reducing agent (0.5 *M* DTT in stabilizing buffer).
27. Acetyl–coenzyme A.
28. DTNB (5,5'-dithiobis-2-nitrobenzoic acid).
29. Chloramphenicol.

30. Chloramphenicol caproate resin (Sigma).
31. CAT load buffer: 50 m*M* Tris-HCl, pH 7.8.
32. CAT wash buffer: 50 m*M* Tris-HCl, pH 7.8, 300 m*M* sodium chloride.
33. CAT elution buffer: 50 m*M* Tris-HCl, pH 7.8, 300 m*M* sodium chloride, 8 m*M* chloramphenicol.
34. CAT storage buffer: 5 m*M* Tris-HCl, pH 7.8, 1 m*M* DTT.
35. CAT column regeneration buffer: 50 m*M* Tris-HCl, pH 7.8, 1 *M* sodium chloride.
36. Braun C-10-2 No.153–10L fermentor (B. Braun Biotech Inc., Allentown, PA).
37. Bioflow–5L fermentor (New Brunswick Scientific, Edison, NJ).
38. Emulsiflex C-50 homogenizer (Avestin, Ottawa, Canada).
39. Fisher Scientific 700 homogenizer.
40. Bradford assay reagents (Bio-Rad).
41. Beckman LS3801 Scintillation Counter.
42. Centriprep30 (Millipore, Billerica, MA).
43. Qiagen Plasmid Maxi Kit.

3. Methods

The methods described below outline (1) the construction of the expression plasmid, (2) preparation of T7 RNA polymerase, (3) cell extract preparation, (4) protein expression in an *E. coli*-based coupled transcription-translation system, (5) evaluation of protein production in the cell-free reaction, and (6) product purification.

3.1. Expression Plasmid

All oligonucleotides were synthesized by the Protein and Nucleic Acid (PAN) facility at Stanford University, Stanford, CA. Both forward and reverse primers were modified to contain functional 5'-*Nde*I and *Sal*I restriction sites, respectively. Forward primer FwdCAT: 5'-TGCTAACTGTCATAT GGAGAAAAAAATCACTGGATATA-3', and reverse primer RevCAT: 5'-TGTCAATCGTGTCGA CTTACCAATGTTACGCCC-CGCCCTGCCAC-3' were used for PCR amplification of the chloramphenicol acetyl transferase gene from *E. coli* genomic DNA using standard procedures (*12*). Underlined regions denote nucleotides complementary to the coding sequence of CAT.

The PCR product and pK7 expression vector (**Fig. 1**) were digested at 37°C for approx 24 h with 20 U each of *Nde*I and *Sal*I in a reaction buffer comprised of 150 m*M* NaCl, 10 m*M* Tris-HCl (pH 7.9, 25°C), 10 m*M* MgCl$_2$, 1 m*M* DTT, and 100 µg/mL BSA. The digested vector was treated with alkaline phosphatase (NEB) and the desired fragment was isolated by gel electrophoresis and then purified using a gel extraction kit (Qiagen). The PCR product was combined in approximately three molar excess with the vector DNA and ligated using standard techniques (*12*). The ligation products were used to transform competent *E. coli* strain TG1 and transformants were selected by plating onto agar plates containing 20 µg/mL kanamycin. Restriction digest analysis of small-scale plasmid preparations was used to identify colonies that incorporated the pK7CAT expression plasmid and was further verified by DNA sequencing. Plasmids used for in vitro combined transcription/translation were isolated and purified using a Qiagen Plasmid Maxi kit.

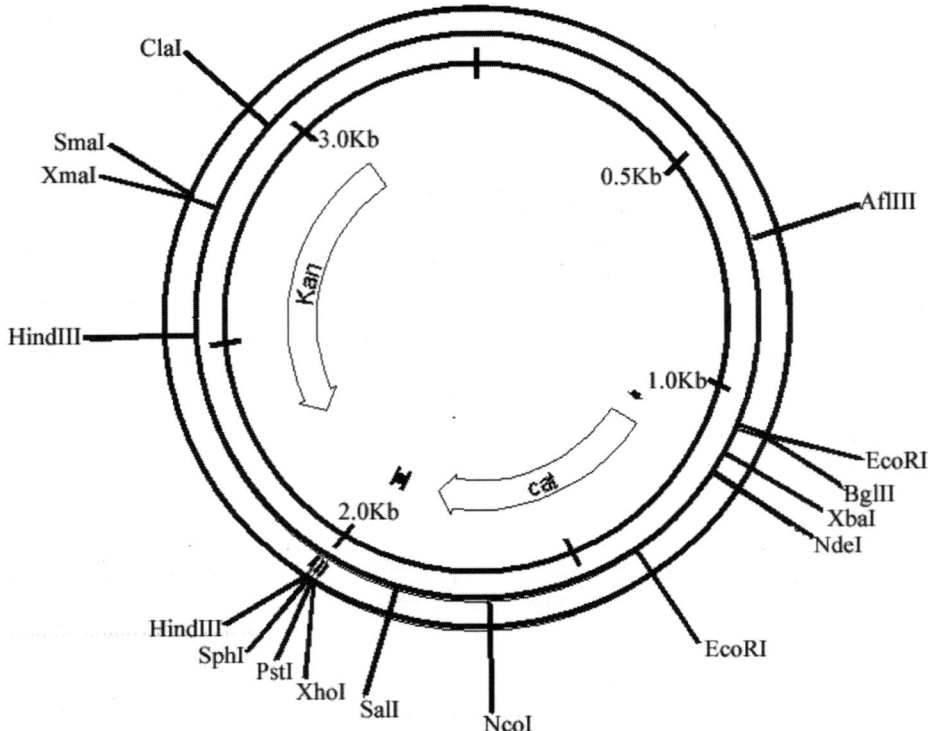

Fig. 1. pk7CAT expression vector.

3.2. Preparation of T7 RNA Polymerase (RNAP)

T7 RNA Polymerase (RNAP) was prepared according to a modified procedure from Davanloo et al. (**13**).

3.2.1. In Vivo Expression and Purification of T7His₆ RNAP

1. Inoculate 5 mL of LB media supplemented with ampicillin (100 μg/mL, LBAmp100) with *E. coli* strain BL21(DE3) harboring the pAR1219 plasmid for expression of an N-terminal histidine-tagged T7 RNAP. Grow for 9 h at 37°C on a rotary shaker (280 rpm).
2. Take entire preculture volume and transfer to 50 mL LBAmp100 and grow for 15 h on a rotary shaker (280 rpm).
3. Use 40 mL of the overnight culture to inoculate a Bioflow–5 L fermentor containing 4 L of LBAmp100 growth media under sterile conditions. The following parameters are used during the fermentation: 37°C, 4 standard liters per minute of air, and 300 rpm agitation speed. Ammonium hydroxide (1 *M*) is used to maintain a pH of 7.1.
4. Cell growth is monitored at 600 nm and the culture is induced with 100 m*M* IPTG when the optical density reaches 0.4. Growth is sustained for an additional 4 h before harvest.
5. Harvest cells by centrifugation at 7000*g* for 20 min at 4°C.
6. Decant the supernatant and resuspend the cell pellet to homogeneity in 80 mL of T7 equilibration/wash buffer using a Fisher Scientific 700 homogenizer.

7. The cells are lysed by a single pass through an Emulsiflex C-50 homogenizer set at 17,500 PSI and a flow rate of approx 1 mL/min.
8. Centrifuge the cell lysate at 30,000*g* for 30 min at 4°C to remove cellular debris.
9. Collect the supernatant and dilute with 5 volumes of equilibration buffer. Filter the lysate through a 0.2-μm filter before applying to the affinity column.
10. Pack a 100-mL bed volume column with the Talon Superflow metal-affinity resin according to procedures provided by the manufacturer (Clontech). The column purification is conducted using a peristaltic pump at a flow rate of 5 mL/min.
11. Charge the resin with half a column volume of cobalt chloride (50 m*M*) and equilibrate with 5 column volumes of T7 equilibration/wash buffer.
12. Apply the filtered (0.2-μm) cell lysate to the column, wash with 10 column volumes of equilibration/wash buffer, and elute using a linear gradient of imidazole (0 m*M* to 300 m*M*) in a total of 5 column volumes. Collect the eluent in 5-mL fractions and determine the fractions containing T7His$_6$ RNAP by Coomassie blue staining of an SDS-PAGE gel (Invitrogen) (*see* **Note 1**).
13. Pool fractions containing the purified enzyme and dialyze (MWCO 6-8000) against 2 L S30 Buffer for 4 h at 4°C.
14. Centrifuge at 25,000*g* for 20 min and determine the protein concentration using a Bradford assay (Bio-Rad).
15. Transfer the supernatant into a Centriprep30 ultrafiltration module and concentrate according to the manufacturer's recommendations to a final protein concentration of 5 mg/mL
16. Dialyze against T7 storage buffer as in **step 13** and supplement with an equal volume of 80% (v/v) glycerol. Store at −20°C.

3.3. Cell Extract Preparation

Prokaryotic cell-free protein synthesis is performed using a crude extract prepared from *Escherichia coli* K-12. The following procedure describes the preparation of S30 extract from *E. coli* strain A19 (*see* **Note 2**) according to a modified protocol of Pratt (*14*).

3.3.1. Preparation of S30 Extract

1. Establish a preculture by inoculating a culture tube containing 5 mL of LB growth media with the desired strain and incubating for 7 h at 37°C on a rotary shaker (280 rpm).
2. Use the entire pre-culture volume to inoculate 200 mL of 2X YT growth media in a 2-L baffled flask and grow overnight for 15 h at 37°C on a rotary shaker (280 rpm).
3. Use the 200-mL overnight culture to inoculate a Braun C-10-2 No.153–10L fermentor containing 8 L of 2x YT growth media under sterile conditions. The following parameters are used during the fermentation: 37°C, 8 standard liters per minute of air, and 750 rpm agitation speed. Ammonium hydroxide (1 *M*) is used to maintain a pH of 7.1.
4. Harvest the culture in mid-exponential phase when the optical density at 600 nm reaches 3.0. Typical growth rates are 1.0–1.2 h^{-1}. The culture is chilled immediately from the fermentor by passage through a metal coil immersed in ice and collected into 4X 4-L flasks on ice.
5. Harvest the cells immediately by centrifugation at 7000*g* for 20 min at 4°C.
6. Decant the supernatant and wash the pellet by resuspending the cells in 400 mL of S30 buffer using a Fisher Scientific 700 homogenizer.
7. Transfer the cell suspension to preweighed centrifuge bottles and centrifuge at 9000*g* for 10 min at 4°C. Repeat the wash step twice.
8. Weigh the cell pellet and store overnight at −80°C.

9. Add 1 mL of S30 buffer per gram of wet cell mass and allow the mixture to thaw on ice.
10. Use a Fisher Scientific 700 homogenizer to resuspend the cells to homogeneity.
11. The cells are lysed by a single pass through an Emulsiflex C-50 homogenizer at 17,500 PSI and a flow rate of approx 1 mL/min.
12. DTT is added to the lysate as it comes out of the homogenizer to a final concentration of 1 mM.
13. The lysate is centrifuged twice at 30,000g for 30 min at 4°C to remove cell debris and genomic DNA (see **Note 3**).
14. The supernatant is collected in sterile conical tubes and 3 mL of pre-incubation mixture is added to every 10 mL of cell extract. The mixture is wrapped in aluminum foil and incubated at 37°C on a rotary shaker (120 rpm) for 80 min (see **Note 4**).
15. Following the pre-incubation step, the S30 extract is dialyzed against four exchanges of 20 volumes of S30 buffer for 45 min each.
16. The dialyzed extract is centrifuged at 4000g for 10 min at 4°C and immediately aliquoted into 50-, 100- and 200-µL samples and 1 and 2-mL stocks as desired. The aliquots are frozen rapidly in liquid nitrogen and stored at −80°C.

3.4. Protein Expression in an
E. coli-Based Coupled Transcription-Translation System

The general approach for protein production using cell-free synthesis utilizes a coupled transcription-translation scheme in a batch configuration (see **Note 5**). This involves incubating the necessary substrates with cell extract in a closed system. The standard reaction mixture for a coupled transcription-translation protocol, which we have named the PANOx procedure (2), contains the following components: 1X salt solution, 1X master mix (see **Note 6**), 2 mM each of the 20 unlabeled amino acids (see **Note 7**), 33 mM PEP (see **Note 8**), 0.33 mM NAD, 0.26 mM CoenzymeA, 2.7 mM sodium oxalate, 2% poly (ethylene glycol) 8000, 11 µM L-[^{14}C]-leucine, 100 µg/mL T7His$_6$ RNAP, 13.3 µg/mL plasmid (see **Note 9**) and 0.24 volume of S30 extract (see **Note 10**). Laboratory scale reactions are generally 15 to 100 µL in volume (see **Note 11**).

1. A "pre-mix" is formed by combining the reaction components in an Eppendorf tube at 4°C in the order listed above but excluding T7His$_6$ RNAP, plasmid, and cell extract. Upon addition of each component, the reagents are mixed by pipetting up and down approx 5–10 times.
2. Autoclaved, milliQ water (4°C) is added to the pre-mix to bring the reagents to the final concentrations as specified above.
3. The pre-mix components are mixed by vortexing. One may centrifuge (approx 3–5 s) at 4°C in order to bring down any residual solution on the tube walls. Alternatively, one can mix thoroughly by pipetting up and down.
4. The appropriate volume of cell extract (3.6 µL for a 15-µL reaction) is aliquoted into the bottom of a separate Eppendorf tube (the batch reactor) that has been prechilled at 4°C. Avoid introducing air bubbles.
5. T7His$_6$ RNAP and the plasmid are added sequentially and carefully to the pre-mix to avoid introducing air bubbles. Upon each addition, pipet up and down approx 5–10× (avoid vortexing these components). Once all the reagents have been combined, mix 10–15× by pipet-

ting 50% of the total reaction volume (*see* **Note 12**). The reagent mixture now contains all the necessary substrates and salts for the reaction. A separate reagent mixture is also prepared without the plasmid to assess any background protein expression levels.

6. Next, 11.4 µL of the reagent mixture is added to the extract to obtain a final 15 µL reaction volume and this is mixed by pipetting approx 5–10× (*see* **Note 13**).
7. The reaction mixture is brought to 37°C and is incubated for 3 h.

3.5. Evaluating Protein Production in the Cell-Free Reaction

3.5.1. Determination of Total and Soluble Protein Yields

Total and soluble protein yields are quantified after the combined transcription and translation system by determining the incorporation of [^{14}C]-leucine into TCA-precipitable counts using a liquid scintillation counter (*15*). We first describe a method for determining the total amount of protein produced in the system (*see* **Note 14**).

1. Ten to 15 µL of the coupled transcription-translation reaction is treated with 100 µL of 0.1 *N* sodium hydroxide.
2. The sample is then incubated at 37°C for 20 min.
3. For each sample, two tabs of Whatman 3 MM chromatography paper are cut to fit into a scintillation vial and labeled (in pencil) with an appropriate designation. For example, one may write "1-wash" on one piece and "1-unwashed" on the other for sample 1. After labeling, they are suspended above styrofoam using straight pins. Fifty microliters of the total sample volume from **step 2** is aliquoted onto each piece of filter paper and then dried under a heat lamp for 1 h.
4. The set of filter papers (one from each pair) designated with "wash" are placed into a beaker on ice without removing the pins. The filter papers are immersed in cold 5% (v/v) trichloroacetic acid (TCA) to precipitate the proteins. The other filter papers, labeled "unwashed," remain on the piece of styrofoam.
5. After 15 min, the TCA solution is poured off and fresh TCA (4°C) is added as in **step 4**. This is repeated for a total of three exchanges with 5% (v/v) TCA.
6. Following the third precipitation, the filter papers are immersed once in 100% ethanol for 10 min at room temperature.
7. The wet filter papers are mounted on the styrofoam and dried under a heat lamp for approx 1 h.
8. Radioactivity of both the TCA-precipitated (labeled "wash") and non TCA-precipitated (labeled "unwashed") samples is measured using a liquid scintillation counter (Beckman LS3801). The straight pins are removed prior to placing the filter papers into the scintillation vials.
9. The fraction of incorporated leucine (the counts of the "washed" sample divided by the counts of the "unwashed" sample) is divided by the number of leucine residues in the expressed protein and multiplied by the total concentration of leucine in the reaction to determine the amount of protein produced. This value can be manipulated from a molar to gram basis using the molecular weight of the expressed protein. For example, consider the calculation of the amount of CAT produced if the washed count was 3000 cpm, the unwashed count was 17,000 cpm, and the background count from an experiment without plasmid DNA was 150 cpm. In order to determine the concentration of CAT produced in the reaction we must also know that there was a total of 2000 µ*M* leucine in the

reaction, that CAT contains 13 leucine residues and that CAT has a molecular weight of 25 kDA. The result follows:

[((3000–150)/17000) * 2000 µ*M* * (25,000 µg/µmol)] / [(13) * (1000 mL/L)] = 645 µg/mL.

10. To determine the soluble protein yield, the reaction mixture is centrifuged at 14,000 rpm for 15 min at 4°C prior to step 1. Following centrifugation, 10–15 µL of the supernatant is removed and combined with 100 µL of 0.1 *N* NaOH. After this, **steps 2–9** are completed as outlined above.

3.5.2. Estimation of Messenger RNA Concentration and Stability

To estimate mRNA concentrations in a cell-free protein synthesis reaction, [³H]-uridine triphosphate (5 µ*M*) is substituted for L-[¹⁴C]-leucine in the standard reaction mixture for coupled transcription-translation as described in **Subheading 3.5.1.**

1. Two tabs of Whatman 3 MM chromatography paper are prepared as in **step 3** of **Subheading 3.5.1**.
2. The reaction is quenched by directly spotting 7 µL of the cell-free reaction mixture onto each piece of filter paper pair and drying under a heat lamp for one hour. TCA-precipitable counts are determined by following **steps 4–8** of **Subheading 3.5.1**.
3. The fraction of incorporated uridine (the counts of the "washed" sample divided by the counts of the "unwashed" sample) is divided by the number of uridine residues in the expressed message and multiplied by the total concentration of uridine in the reaction to determine the amount of message produced. This value can be manipulated from a molar to gram basis using the molecular weight of the expressed message. The calculation is similar to that described in **step 9** of **Subheading 3.5.1**.

3.5.3. SDS Polyacrylamide Gel Electrophoresis and Autoradiography

1. Assemble the Mini-cell according to the manufacturer's instructions (Invitrogen).
2. Prepare protein samples (≤ 30 µg/mL) by addition of 1X LDS sample loading buffer and 1X NuPAGE reducing agent (if running under reduced conditions). Heat samples for 10 min at 70°C.
3. Pour 1X NuPAGE MES SDS running buffer (supplement with reductant, e.g., 50 m*M* PTT if running under reduced conditions) into the inner chamber and ensure that there are no leaks. Fill the lower chamber with 600 mL of 1X NuPAGE running buffer.
4. Use a pipet to rinse the sample wells with running buffer.
5. Load up to 20 µL of each sample into the bottom of the wells.
6. Run at a constant current of 50 mA for approx 1 h or as suggested by the manufacturer.
7. Pry apart the gel cassette using a spatula or other device, such as a wedge, and gently remove the gel from the cassette.
8. Immerse the gels in a sufficient volume of staining buffer to permit easy gel mobility. Incubate at room temperature on a rotary shaker for 3–6 h or overnight.
9. Decant the staining buffer and rinse the gel with two exchanges of MilliQ water for 10 min each.
10. Immerse the gels in a sufficient volume of destain to permit easy gel mobility and destain for at least 7 h at room temperature on a rotary shaker with two exchanges of destain.
11. Rinse gel with several exchanges of MilliQ water.

12. Mount gel onto Whatman filter paper and dry using a gel dryer at 80°C for 1 h according to the manufacturer's instructions.
13. Use the gel to expose photographic paper (Kodak) for autoradiographic analysis.

3.5.4. Protein Activity Assay

Analytical methods for characterizing the active protein expressed will depend on the properties of the target protein. We describe here a method for measuring the active amount of chloramphenicol acetyl transferase (CAT), the model protein often expressed in bacterial cell-free systems. The enzymatic activity of synthesized CAT is determined by the spectrophotometric procedure described by Shaw (*16*) (*see* Chapter 5). This experiment measures the formation of reduced coenzymeA (CoA). Reduced CoA reacts with 5,5'-dithiobis-2-nitrobenzoic acid to form a colorometric product, 5-thio-2-nitrobenzoate. The reactions follow:

Chloramphenicol + acetyl-S-CoA → chloramphenicol 3-acetate + HS-CoA

HS-CoA + 5,5'-dithiobis-2-nitorbenzoic acid → CoA-thionitrobenzoic acid + 5-thio-2-nitrobenzoate

Samples from a cell-free reaction mixture are centrifuged at 14,000g for 15 min at 4°C. The supernatant is collected and diluted 40 times with water at 4°C. The assay mixture is prepared and prewarmed to 37°C. The final concentrations of the components in the mixture are 100 mM Tris-HCl (pH 7.8), 0.1 mM Acetyl—Coenzyme A, 0.4 mg/mL DTNB, and 0.1 mM chloramphenicol. Ten µL of the diluted sample is added to 990 µL of the assay mixture in a 1 mL cuvet. The rate of absorbance increase is measured at 412 nm and 37°C. The change of absorbance of a negative control (cell-free reaction that has been incubated without a DNA plasmid) is subtracted from this rule of absorbance increase at the sample. This value is divided by the extinction coefficient of free 5-thio-2-nitrobenzoate (13.6 mM^{-1} cm) to yield the units of CAT activity, which is defined as the number of mmoles of chloramphenicol acetylated per minute within the cuvet.

3.6. Protein Purification

Analytical methods for protein purification from cell-free systems will depend on the specific physicochemical properties of the target protein. We describe here a method for purifying chloramphenicol acetyl transferase (CAT), the model protein often expressed in bacterial cell-free systems. The following results are from a 10-mL PANOx, fed-batch reaction where 33 mM phosphoenolpyruvate, 6 mM magnesium acetate, 2 mM of each unlabeled amino acid, and 11 µM ^{14}C-leucine were added at 1, 2, and 3 h. **Figure 2** shows the time course of CAT accumulation. **Figure 3** shows the results from the purification after a 4-h reaction (*17*).

1. Centrifuge the reaction volume at 14,000g for 20 min at 4°C and collect the supernatant.
2. Dilute supernatant with an equal volume of CAT load buffer and filter through a 0.2-µm filter.
3. Pack a 10-mm i.d. by 10-cm-long column with 3 mL of chloramphenicol caproate resin (Sigma) according to procedures provided by the manufacturer. Use a column flow adapter

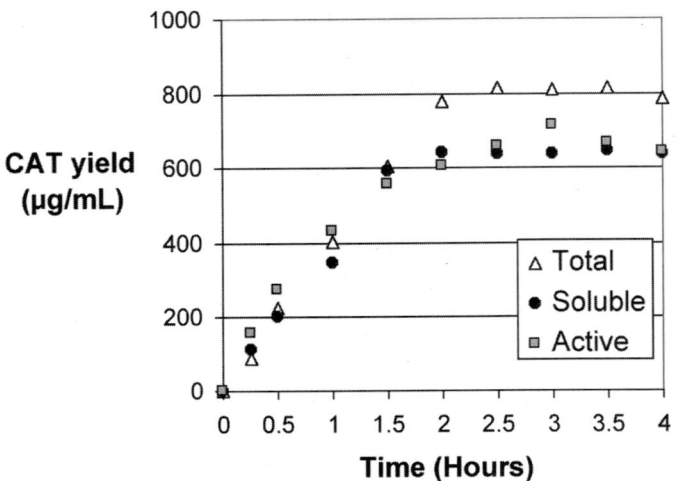

Fig. 2. Time course for the production of CAT in a 10-mL reaction. The amount of total and soluble protein was determined by ^{14}C-leucine incorporation. The activity of CAT was determined using the assay described in the text. Approximately 80% of the total protein produced was soluble and active.

    to ensure even flow and negligible dead-volume. Column purification is conducted at 4°C using a peristaltic pump at a flow rate of 0.5 mL/min.

4. Equilibrate the column with 5 column volumes (15 mL) of CAT load buffer.
5. Load the diluted supernatant from **step 2** onto the column, wash away proteins that bind nonspecifically to the resin with 20 mL of CAT wash buffer, elute CAT from the resin using a linear gradient of chloramphenicol (0 m*M* to 8 m*M*) in a total volume of 20 mL of CAT wash buffer. Collect the eluent in 1.5-mL fractions. Regenerate the column using 10 column volumes (30 mL) CAT regeneration buffer and store the column according to the manufacturer's recommendations.
6. Determine the fractions containing CAT by Coomassie blue staining of a SDS-PAGE or assay the fractions for chloramphenicol acetyltransferase activity.
7. Pool fractions containing the purified enzyme and dialyze (MWCO 6-8000) against 2 L S30 buffer for 1 h at 4°C. Exchange the buffer and repeat once.
8. Centrifuge at 25,000*g* for 20 min and determine the protein concentration using a Bradford assay (Bio-Rad).
9. Transfer the supernatant into a Centriprep 10 ultrafiltration module and concentrate according to the manufacturers recommendations to the desired protein concentration.
10. Dialyze against storage buffer and supplement with equal volume of 80% (v/v) glycerol. Store at −20°C.

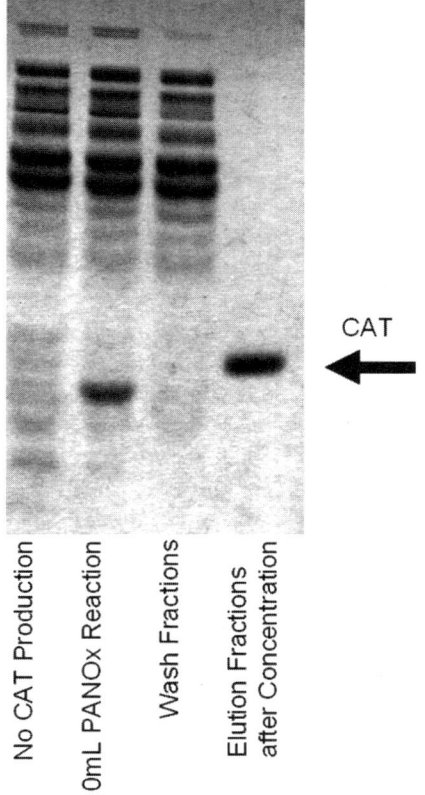

Fig. 3. Affinity purification of CAT. Lane 1, Reaction mixture from control reaction without plasmid addition. Lane 2, Reaction mixture with CAT after production in a 10-mL PANOx reaction. Lane 3, Wash fractions before the elution with CAT extraction buffer. Lane 4 Elution fractions after concentration to 1 mg/mL using a centrifugal filter unit with a 10,000 molecular weight cutoff. The samples were run on an SDS-PAGE gel and stained with Coomassie blue. The arrow denotes the purified protein product. There was a 69% recovery of CAT.

4. Notes

1. During the gel analysis of the purified T7His$_6$ RNAP, two proteolytic forms are identified. These fragments appear to be 80K and 20K in molecular weight. This mixture has been shown to be active (native T7 RNAP is approx 100K) (*17*).

2. S30 extract is prepared from a genetically engineered strain of A19. The strain is of genotype A19 $\Delta speA$, $\Delta tnaA$, $\Delta tonA$, $\Delta endA$ and has had a methionine auxotrophy reverted into methionine protrophy by P1 phage transduction from an *E. coli* prototroph. Cell extract from this modified organism has similar performance relative to the original A19 strain (Michel-Reydellet, N. and Swartz J. R., submitted).

3. The pellet after the first centrifugation tends to be noncompact. It is therefore important to immediately transfer the supernatant of the first centrifugation into another centrifuge bottle for the second centrifugation. Experiments have shown that two centrifugations yield more active extracts as compared to extracts performed with one centrifugation.

4. The "run-off reaction" is thought to be necessary for endogenous mRNA to be released from the ribosome and degraded. Schindler et al. (*20*) reported that the performance of the extract can vary by altering the protocol at this step.

5. Cell-free protein synthesis can be performed in several types of reactors. Typically, the batch configuration, which involves incubating the necessary substrates with cell extract in a closed system, has been used due to numerous operational advantages. In addition to this traditional approach, continuous and semicontinuous systems, initially designed through the pioneering work of Spirin and coworkers (*21*), have shown tremendous success. These reactor configurations involve the continuous supply of substrates and removal of inhibitory byproducts to and from the reaction mixture. Continuous and semicontinuous systems typically increase the duration of the translation reaction and the protein yield as compared to the batch system (*1*).

6. Optimal protein expression in vitro depends on the pH of the reaction mixture. Our laboratory has seen that the most favorable pH for protein expression can range between 6.6 and 8.2, depending on the particular protein produced. It is important to note that the pH is not homeostatic in reactions containing a high-energy phosphate donor, such as phosphoenolpyruvate. In these reactions, the pH typically drops over the first 30 min of the reaction. The pH of the HEPES solution used in the master mix may be altered in order to adjust pH. (Note that the pH of the HEPES solution is 7.5 but this is not necessarily the pH of the master mix, nor the pH of the reaction mixture.) Alternatively, the pH of the amino acid or salt solution can be adjusted to investigate the pH effect on protein synthesis.

7. An amino acid stock solution is prepared by combining each of the 20 amino acids (use amino acids with greater than 98% purity) to a final concentration of 50 mM. A white precipitate forms and it is important to mix the solution thoroughly before addition to the cell-free reaction.

8. The secondary energy source does not have to be phosphoenolpyruvate. Other high-energy phosphate compounds and glycolytic intermediates have been used. These include acetyl-phosphate, creatine phosphate, glucose-6-phosphate, and pyruvate. While most systems use high-energy phosphate compounds, it is preferable that the secondary energy source be homeostatic with respect to phosphate accumulation. This is because phosphate accumulation is inhibitory for the coupled transcription-translation process (*see* **ref.** *1* for review).

9. The plasmid can encode for any desired gene product.

10. The cell-free system offers a flexible format for protein expression. This flexibility allows for numerous modifications to the component concentrations without adversely affecting the system. One could potentially reevaluate every component for each batch of extract. However, when the concentration of one particular component of the reaction mixture is changed another component also may need to be changed accordingly for optimal protein synthesis yields. In the energy system described (PANOx), for example, changing the magnesium concentrations affects the optimal concentrations of poly (ethylene glycol) and phosphoenolpyruvate within the reaction mixture. While looking at the entire composition of reaction components may lead to improved protein synthesis, it seems vital to investigate only the magnesium concentration for each batch of extract. Synthesis with a model protein should be performed over a range of magnesium concentrations, typically between 12 and 20 mM, to determine the optimal magnesium concentration for every batch of extract.

11. Cell-free reactions are typically somewhat less productive when the scale is increased. Similar results are attainable for protein production within the 15-to-100-µL volume range. However, a 500-µL PANOx batch reaction generally produces less protein, in terms of final concentration, than a 15-µL PANOx reaction. For different reactor configurations, this may not be the case.

12. It is important that the reaction components are evenly mixed. Cell-free protein synthesis has been plagued by variable results. One cause for this variability in our laboratory has been the lack of complete mixing of the reaction components. Specifically, this can be a problem when setting up numerous reactions from the same pre-mix. As an example, rather than setting up a pre-mix for 20 reactions with the T7His$_6$ RNAP and DNA template, separate a mixture containing all the reaction components for 20 reactions (without the T7His$_6$ RNAP, DNA template, and extract) into two to four smaller pre-mixes. To each of these smaller pre-mixes, add the T7His$_6$ RNAP and DNA template for the appropriate number of reactions. We have found that by adding the T7His$_6$ RNAP and DNA template for less than or equal to 10 reactions (≤150 µL), the variability among experiments is reduced. This is most likely due to better mixing of these necessary substrates.

13. The extract is always added last. Enzymes active in the extract will nonproductively degrade the energy sources and other substrates that are required for protein synthesis if the extract is added to the reaction components without the DNA template *(20)*.

14. Radioactive amino acid incorporation measurements determine the amount of protein that is precipitated onto the filter paper. Thus, all the measured protein does not have to be full-length. However, our experience is that the majority of the protein produced is full-length.

References

1. Jewett, M. C., Voloshin, A., and Swartz, J. R. (2002) Prokaryotic systems for in vitro expression, in *Gene Cloning and Expression Technologies* (Weiner, M. P., and Lu, Q., eds.), Eaton Publishing, Westborough, MA, pp. 391–411.

2. Kim, D. M., and Swartz, J. R. (2001) Regeneration of ATP from glycolytic intermediates for cell-free protein synthesis. *Biotechnol. Bioeng.* **74**, 309–316.

3. Jermutus, L., Ryabova, L. A., and Pluckthun, A. (1998) Recent advances in producing and selecting functional proteins by using cell-free translation. *Curr. Opin. Biotechnol.* **9**, 534–548.

4. Kigawa, T., Yabuki, T., Yoshida, Y., Tsutsui, M., Ito, Y., Shibata, T., et al. (1999) Cell-free production and stable-isotope labeling of milligram quantities of proteins. *FEBS Lett.* **442**, 15–19.

5. Kim, R. G. and Choi, C. Y. (2000) Expression-independent consumption of substrates in cell-free expression system from *Escherichia coli*. *J. Biotechnol.* **84**, 27–32.

6. Kim, D. M. and Swartz, J. R. (1999) Prolonging cell-free protein synthesis with a novel ATP regeneration system. *Biotechnol. Bioeng.* **66**, 180–188.

7. Nakano, H. and Yamane, T. (1998) Cell-free protein synthesis systems. *Biotechnol. Adv.* **16**, 367–384.

8. Shimizu, Y., Inoue, A., Tomari, Y., Suzuki, T., Yokogawa, T., Nishikawa, K., and Ueda, T. (2001) Cell-free translation reconstituted with purified components. *Nat. Biotechnol.* **19**, 751–755.

9. Stiege, W. and Erdmann, V.A. (1995) The potentials of the *in vitro* protein biosynthesis system. *J. Biotechnol.* **41**, 81–90.

10. Yokoyama, S., Matsuo, Y., Hirota, H., Kigawa, T., Shirouzu, M., Kuroda, Y., et al. (2000) Structural genomics projects in Japan. *Pro. Biophys. Mol. Biol.* **72**, 363–376.

11. Alimov, A. P., Khmelnitsky, A. Y., Simonenko, P. N., Spirin, A. S., and Chetverin, A. B. (2000) Cell-free synthesis and affinity isolation of proteins on a nanomole scale. *BioTechniques* **28,** 338–344.

12. Sambrook, J., Fritsch, E. F., and Maniatis, T. (1989) *Molecular Cloning, A Laboratory Manual*, 2 ed. Cold Spring Harbor Laboratory Press, Cold Spring Harbor, New York.

13. Davanloo, P., Rosenberg, A. H., Dunn, J. J., and Studier, F. W. (1984) Cloning and expression of the gene for bacteriophage T7 RNA polymerase. *Proc. Natl. Acad. Sci. USA* **81,** 2035–2039.

14. Pratt, J. M. (1984) Coupled transcription-translation in prokaryotic cell-free systems, in *Transcription and Translation: A Practical Approach* (Hames, B. D. and Higgins, S. J., eds.), IRL Press, New York, NY, pp. 179–209.

15. Kim, D. M., Kigawa, T., Choi, C. Y., and Yokoyama, S. (1996) A highly efficient cell-free protein synthesis system from *Escherichia coli. Eur. J. Biochem.* **239,** 881–886.

16. Shaw, W. V. (1975) Chloramphenicol acetyltransferase from chloramphenicol-resistant bacteria. *Methods Enzymol.* **43,** 737–755.

17. Jewett, M. C. and Swartz, J. R. (2004) Rapid expression and purification of 100 nmol quantities of active protein using cell-free protein synthesis. *Biotechnol. Prog.* In press.

18. Muller, D. K., Martin, C. T., and Coleman, J. E. (1988) Processivity of proteolytically modified forms of T7 RNA polymerase. *Biochemistry* **27,** 5763–5771.

19. Michel-Reydellet, N., Calhoun, K. A., and Swartz, J. R. (2004) Amino acid stabilization for cell-free protein synthesis by modification of the *E. coli* genome. *Metabolic Engineering.* In press.

20. Schindler, P. T., Baumann, S., Reuss, M., and Siemann, M. (2000) *In vitro* transcription translation: effects of modification in lysate preparation on protein composition and biosynthesis activity. *Electrophoresis* **21,** 2606–2609.

21. Spirin, A. S., Baranov, V. I., Ryabova, L. A., Ovodov, S. Y., and Alakhov, Y. B. (1988) A continuous cell-free translation system capable of producing polypeptides in high yield. *Science* **242,** 1162–1164.

22. Kim, D. M. and Swartz, J. R. (2000) Prolonging cell free protein synthesis by selective reagent additions. *Biotechnol. Prog.* **16,** 385–390.

13

Genetic Tools for the Manipulation of Moderately Halophilic Bacteria of the Family *Halomonadaceae*

Carmen Vargas and Joaquín J. Nieto

Summary

Moderately halophilic bacteria of the family *Halomonadaceae* (*Halomonas, Chromohalobacter*, and *Zymobacter*) have promising applications in biotechnology as a source of compatible solutes (stabilizers of biomolecules and cells), salt-tolerant enzymes, biosurfactants, and extracellular polysaccharides, among other products. In addition, they offer a number of advantages to be used as cell factories, alternative to conventional prokaryotic hosts like *E. coli* or *Bacillus*, for the production of recombinant proteins: (1) their high salt tolerance decreases to a minimum the necessity for aseptic conditions, resulting in cost-reducing conditions; (2) they are very easy to grow and maintain in the laboratory, and their nutritional requirements are simple; and (3) the majority can use a large range of compounds as a sole carbon and energy source. In this decade, the efforts of our group and others have made possible the genetic manipulation of this bacterial group. In this review, the most relevant tools are described, with emphasis given to cloning vectors, genetic exchange mechanisms, mutagenesis approaches, and reporter genes. Due to its relevance for genetic studies, complementary sections describing the influence of salinity on the susceptibility of moderately halophilic bacteria to antimicrobials, as well as the growth media most routinely used, culture conditions, and nucleic acid isolation procedures for these microorganisms, are included.

Key Words: Halophilic bacteria; genetic exchange; reporter genes; mutagenesis; *Zymobacter; Chromohalobacter, Halomonas*

1. Introduction

Moderately halophilic bacteria are defined as prokaryotes that grow optimally in media containing 3 to 15% NaCl, although they can also grow above and below this range of salt concentrations (*1*). Besides extremely halophilic archaea, they constitute

From: *Methods in Molecular Biology, vol. 267:*
Recombinant Gene Expression: Reviews and Protocols, Second Edition
Edited by: P. Balbás and A. Lorence © Humana Press Inc., Totowa, NJ

the most important group of microorganisms adapted to live and thrive in hypersaline niches. Due to their abundance in habitats such as solar salterns, brines, hypersaline soils and lakes, salty foods, and the like, they play an important role in the ecology of such extreme environments, representing an excellent example of adaptation to frequent changes in extracellular osmolarity. This heterogeneous physiological group, which includes a great variety of Gram-positive and Gram-negative bacteria, has recently been reviewed concerning its ecology, taxonomy, physiology, and genetics, among other aspects of this fascinating group of microorganisms (*2*). Despite some of the species belong to genera that also include halotolerant (tolerate high salt concentrations without requiring salt) and nonhalophilic (grow best in media with less than 1% NaCl) representatives such as *Pseudomonas*, *Bacillus*, or *Halomonas*, most of them are encompassed in genera that include only moderately halophilic species such as *Salinicoccus*, *Salibacillus*, and *Chromohalobacter* (*2*).

Among moderate halophiles, members of the family *Halomonadaceae* show a remarkable versatility with respect to their salt tolerance, showing suitable features to study the molecular basis of bacterial osmoadaptation (*3*). In fact, some species, such us *Halomonas elongata* and *Chromohalobacter salexigens*, have been used in recent years for the study of osmoregulatory mechanisms in halophilic bacteria (*2*). Very recently, the phylogenetic status of the family *Halomonadaceae* has been examined and reevaluated after comparative 23S and 16S analyses (*4*). As a consequence, although only three genera were included (*Halomonas*, *Chromohalobacter*, and *Zymobacter*), a clear distinction of the genus *Halomonas* into two different phylogenetic groups as well as the existence of six species that did not fall into either of the two former groups, was found. These findings confirm the high phenotypic heterogeneity reported for members of the genus *Halomonas* (*4*).

2. Biotechnological Perspectives of Moderately Halophilic Bacteria

Apart of their ecological interest, these extremophiles are currently receiving great attention because of their potential for exciting and promising applications in biotechnology. For instance, they are good sources for salt-tolerant enzymes (i.e., amylases, proteases, and nucleases) useful in a variety of biotechnological processes (*2*). Recently, we reported the production of one α-amylase from *H. meridiana* by cloning and molecularly characterizing the encoding gene (*amyH*), constituting the first gene encoding one extracellular amylase isolated from moderately halophilic bacteria (*5*). Other organisms from this group produce salt-resistant biosurfactants or extracellular polysaccharides, which have a great potential as enhancers of oil recovery processes and for the pharmaceutical or food industry. Some of them are able to degrade toxic compounds from such as phenol, organophosphorus compounds and others that are generated during manufacture—e.g., that of pesticides and herbicides—from hypersaline wastewaters, constituting an interesting potential for the decontamination of such polluted waters (*6,7*).

One of the most interesting current and potential applications of these organisms is the commercial use of the osmolytes that accumulate in their cytoplasm as a response to external salinity, the compatible solutes. Although there is a relative diversity of compatible solutes in moderate halophiles, they are mainly poliols (glucosylglycerol), aminoacids (proline) and diaminoacids, quaternary amines (betaines), and ectoines, which have been demonstrated to be the predominant osmolytes accumulated by this group of halophiles (**8–10**). Whereas betaines are very widespread in nature, ectoines can only be produced biotechnologically from some *Halomonas elongata* strains using a "bacterial milking" process (**11**) that has been subsequently upscaled by bitop GmbH (Witten, Germany). Since compatible solutes display additional biological properties in the cell, such as the stabilization and protection of enzymes, nucleic acids, and organelles against different stress agents, they have received great interest as general protecting agents with a wide range of biotechnological applications. Thus, some of them—such as betaines, and especially ectoines—can be used as powerful stabilizers of enzymes, membranes, nucleic acids, antibodies, and whole cells against a variety of denaturating stresses caused by urea, salt, heating, freezing, and desiccation. They have also a promising potential in cosmetics; for example, ectoine is a component of some beauty creams that help to protect skin from dehydration (**9,12,13**). Other applications, such as a highly efficient expression of functional recombinant proteins and optimization of PCR, have also been claimed (**14**).

One of the most interesting potentials of compatible solutes is the use of the genes responsible for their synthesis in the generation of transgenic organisms and crop plants that have an in-built tolerance to osmotic stress conditions such as drought or salinity. These transgenic approaches to the enhancement of stress tolerance in plants have been mainly done in model plants such as *Arabidopsis thaliana,* rice, and tobacco (**15,16**). In these studies, metabolic engineering of glycine betaine, mannitol, or trehalose synthesis has been mostly employed (**17–19**). However, experiments conducted with ectoine synthesis genes are extremely scarce and performed only with cultured tobacco cells (**20**). Very recently we have initiated several studies granted by the European Commission aimed at transferring genes for some compatible-solute synthesis from several microorganisms and plants to generate stress-resistant legumes and host-specific *Rhizobium* strains.

Finally, an emerging field is the use of the *Halomonadaceae* for the production of recombinant proteins, as cell factories alternative to conventional prokaryotic hosts like *E. coli* or *Bacillus*. Moderate halophiles offer a number of advantages for their use as host cells. First, their high salt tolerance enables their cultivation under nonsterile, and thus cost-reducing, conditions. Second, they are very easy to grow, and their nutritional requirements are simple. The majority can use a large range of compounds as their sole carbon and energy source. Because of the highly corrosive nature of salt-rich media, these organisms cannot be cultivated in stainless-steel-containing bioreactors. This constraint can be solved by replacing the parts of the bioreactor that are in contact with the medium by alternative materials. This approach has been successfully applied for the optimization of polymer production by two extreme halophiles by using a corrosion-resistant bioreactor composed of polyetherether ketone (**21**).

In this decade, the efforts of our group and others have made possible the genetic manipulation of this bacterial group. This manipulation has been revealed to be relatively simple, very similar to that of members of the closely related *Pseudomonadaceae*. Cloning and expression vectors are available, as well as the methodology to transfer recombinant clones to *Halomonas* and *Chromohalobacter* (*see* Chapter 14 for a description of the protocol). Both native and heterologous promoters have proved to be functional in these halophiles to express a number of heterologous proteins of commercial interest (*see* Chapter 14), such as the ice-nucleation protein from *P. syringae* (*22,23*) or some amylases from hyperthermophiles (*Pyrococcus woesei*, **24**) or nonhalophiles (*B. licheniformis*, **5**). Although few recombinant proteins have been produced as yet, the results show that the *Halomonadaceae* have great potential as alternative cell factories. Bacterial ice nuclei (formed by the aggregation of ice nucleation, InaZ, and proteins) can be useful in various biotechnological applications, such as artificial snow making, frozen food industry, or cryo-concentration. However, the plant pathogenic character of the ice nucleation bacteria hampers the commercialization of most of these applications. In this respect, the expression of ice nucleation genes in nonpathogenic bacterial hosts (e.g., *Halomonadaceae*) is of great importance. Notably, recombinant *Halomonas* strains expressing the *inaZ* gene released ice nuclei in their growth medium, whereas in their natural hosts they usually remain membrane-associated. In addition, these ice nuclei were more heat-resistant than the ice nuclei from all other bacterial sources tested so far (*23*). Therefore, *Halomonadaceae* are a good source of ice nucleation protein.

Due to the inherent difficulty of cultivation of hyperthermophiles, proteins of hyperthermophilic origin need to be expressed in mesophilic hosts for biotechnological purposes. An extracellular α-amylase gene from the hyperthermophilic archaea *Pyrococcus woesei* has been cloned under the control of a native promoter and expressed in *H. elongata*. The recombinant pyrococcal enzyme purified from *H. elongata* displayed catalytic properties similar to the native enzyme purified from *E. coli* with respect to thermal stability, pH and optimal temperature, and effect of various reagents on activity (*24*). Thus, *H. elongata* could be used at least as well as *E. coli* for the production of functional α-amylase in recombinant form.

In this review, the most relevant tools for the manipulation of moderate halophiles of the family *Halomonadaceae* are described. Emphasis has been given to cloning vectors, genetic exchange mechanisms, mutagenesis approaches, and reporter genes. Due to its relevance for genetic studies, an additional section devoted to the influence of salinity on the susceptibility of moderate halophiles to antimicrobials has been included. Complementary sections describe the most important growth media for these microorganisms, as well as culture conditions and nucleic acid isolation procedures.

3. Genetic Manipulation

3.1. Media and Growth Conditions

Members of the *Halomonadaceae* grow best in media containing 3 to 15% NaCl (*2*). Apart from this, they have simple growth requirements and complex and defined media are available for all species. In our laboratory, the pH of all media is adjusted to 7.2 with a solution of 1 *M* KOH. Solid media contain 2% (w/v) Bacto-Agar (Difco). Unless otherwise stated, cultures are incubated at 37°C; liquid cultures are incubated in an orbital shaker at 200 rpm.

3.1.1. Complex Media

Strains of *Halomonas* and *Chromohalobacter* are routinely grown in complex SW-10 medium, which contains 10% (w/v) total salts and 0.5% (w/v) yeast extract (*25, see* **Note 1**). Modified versions of this medium containing a lower percentage of total salts can also be used for specific purposes, such as when cells are grown for conjugation. Many other rich media have been reported in the literature, such as the MY medium supplemented with sea salt (*26*), the casaminoacid (CAS) medium supplemented with NaCl described by Vreeland et al. (*27*), the Artificial Organic Lake Peptone (AOLP) medium supplemented with a salt solution and phosphate reported by James et al. (*28*), or the K10 medium described by Severin et al. (*29*), to cite some examples.

We have also tested the growth of some *Halomonas* and *Chromohalobacter* strains in the *Escherichia coli* LB medium, which contains 1% NaCl, and LB containing 2% NaCl (**Table 1**). Whereas some of the strains grew well in LB + 2% NaCl, the original LB medium was growth-inhibitory for all but three strains of *Halomonas,* indicating again the salt requirement of the *Halomonadaceae* (**Table 1**).

3.1.2. Defined Media

When supplemented with NaCl, M63 (a defined medium used for *Enterobacteria, 30*) supports the growth of all *Halomonas* and *Chromohalobacter* strains tested in our laboratory and is usually used as the minimal medium of election. Glucose at a final concentration of 20 m*M* is used as the carbon source. We have also found that members of the *Halomonadaceae* grow well in the defined medium CDMM described by Kamekura et al. (*31*) for *Salinivibrio costicola*. Other defined media have also been described, as the minimal medium used by Vreeland et al. (*27*), the MBM medium reported by Cummings and Gilmour (*32*), or the glucose-mineral medium G10 described by Severin et al. (*29*). It is worth noticing that the defined medium described by Vreeland et al. (*27*) and the CDMM medium contain sodium glutamate. Thus, the amount of NaCl has to be adjusted to allow for the sodium contributed by the sodium glutamate.

3.2. Effect of Salinity
on Antimicrobial Susceptibility of Moderate Halophiles

Antimicrobials are used extensively as selective agents in genetic exchange experiments. The early work reported by our group (*33*) demonstrated that when moderate halophiles are grown at their optimal salt concentration, usually 10%, they tolerate very high concentrations of the majority of antimicrobials. This fact makes difficult not only the use of antimicrobials as genetic markers, but also the design of suitable selection media when these halophiles are used as recipients for plasmids or transposons. Other studies on the susceptibility of *H. elongata* to some antibiotics indicated that a decrease in salinity resulted in an enhanced sensitivity to these antimicrobials (*3,34*).

Table 1
**Growth of Some Strains[a] of *Halomonas* and *Chromohalobacter* on Five Complex Media
After 48 h of Incubation at 37°C[b]**

	SW-10[c]	SW-2	SW-1	LB[d]	LB-2[e]
Halomonas elongata ATCC 33173[Tb]	++	++	+	+	++
Halomonas meridiana DSM 5425[T]	++	++	++	−	++
Halomonas eurihalina ATCC 49336[T]	++	+	−	−	+
Halomonas halophila DSM 4770[T]	++	++	−	−	−
Halomonas halmophila ATCC 19717[T]	++	−	−	−	−
Halomonas halodurans ATCC 29686[T]	++	++	++	+	++
Halomonas subglaciescola DSM 4683[T]	++	++	++	++	++
Chromohalobacter marismortui ATCC 17056[T]	++	+	−	−	−
Chromohalobacter salexigens DSM 3043[T]	++	++	++	−	+
Chromahalobacter salexigens ATCC 33174	++	++	++	−	+
Chromahalobacter canadensis ATCC 43984[T]	++	−	−	−	−
Chromohalobacter israelensis ATCC 43985[T]	++	++	+	−	−

[a] Cells were grown on plates with solid media.
[b] T, type strain.
[c] Media SW-10, SW-2 and SW-1 contain 10%, 2%, and 1% (w/v) total salts, respectively, and 0.5% (w/v) yeast extract.
[d] The original recipe of medium LB contains 1% (w/v) NaCl, 1% (w/v) tryptone, and 0.5% (w/v) yeast extract.
[e] LB-2. LB with 2% NaCl.

Based on these findings, a deeper analysis was performed on the influence of salinity on the susceptibility of moderately halophilic bacteria to antimicrobials (*35*). For comparative purposes, the MICs for *E. coli* strains DH5α and S17-1, widely used in genetic transfer studies, were also determined at different salt concentrations. Three different patterns of tolerance were found for moderate halophiles when salinity was varied from 10 to 1% (w/v) total salts in the testing media. The first group included the responses to ampicillin and rifampicin, where only minimal effects on the susceptibility were found. All moderate halophiles showed a high sensitivity to rifampicin regardless of the salt concentration. On the other hand, many of the tested strains showed a high resistance to ampicillin, indicating the presence of β-lactamases that inactivated the antibiotic regardless of the salt concentration.

In the second group, including the response to aminoglycosides (gentamycin, kanamycin, neomycin, and streptomycin), a remarkable and gradual increase of the toxicity was detected at lower salinities. Thus, by using media with low salinity, genes encoding aminoglycoside resistance become suitable as genetic markers for plasmids or transposons to be transferred to moderate halophiles. The two *E. coli* strains tested also showed a gradually increased tolerance to aminoglycosides at high salt concentration (*see* **Note 2**), indicating that this effect is not exclusive for moderate halophiles. In fact, salt dependence for the MICs of tetracycline, kanamycin, and streptomycin has been reported very recently in photosynthetic moderately halophilic bacteria of the *Rhodospirillaceae* group (*36*).

The third group included the rest of antimicrobials assayed (trimethoprim, nalidixic acid, spectinomycin, and tetracycline), where the effect of salinity was moderate and dependent on both the individual strain and the antimicrobial tested. All these data greatly facilitated further genetic studies on moderate halophiles. Thus, they simplified the design of selection media for genetic exchange experiments and enabled the use of the genes encoding resistance to some antimicrobials, especially aminoglycosides, as genetic markers for this bacterial group. **Table 2** summarizes the final concentrations of the antimicrobials used in our laboratory to grow *Halomonas* and *Chromohalobacter* on solid SW media with different osmolarity.

3.3. Nucleic Acid Isolation

Plasmid and total DNA can be isolated from *Halomonas* and *Chromohalobacter* by using standard methods. Small- to medium-sized plasmids can be isolated by using the alkaline lysis method (*37, see* **Note 3**), or the Wizard SV plus kit (Promega). However, the "boiling" method (*38*), widely used for minipreps from *E. coli*, is not a suitable procedure to isolate plasmid DNA from *Halomonadaceae*. For large-scale preparation, alkaline lysis-isolated plasmid DNA can be further purified through a Qiagen column. The presence of megaplasmids can be tested by using a modification of the procedure of the Eckhardt method (*39*) described by Plazinski et al. (*40*). The CTBA-NaCl method described by Ausubel et al. (*41*) or the procedure described by Marmur (*42*) are both good choices to isolate total DNA from *Halomonas* and *Chromohalobacter*.

3.4. Cloning Vectors

Undoubtedly, cloning vectors are indispensable tools for the genetic manipulation of any bacterial group. A number of native plasmids have been isolated from Gram-negative moderate halophiles (**Table 3**), and cloning vectors based on two of the best-characterized replicons have been constructed (**Table 4**). In addition, some broad-host range vectors for Gram negative bacteria, especially those of the IncQ and IncP incompatibility groups, have shown their utility as cloning and expression vectors for members of the *Halomonadaceae* (**Table 4**).

3.4.1. Native Plasmids Isolated
From Gram-Negative, Moderately Halophilic Bacteria

Our group reported in 1992 the first plasmid isolated from moderately halophilic bacteria, pMH1 (*43*). This plasmid was found to be harbored by four strains, including three strains of *Halomonas* and one strain of *Salinivibrio costicola* (**Table 3**). Despite the fact that *S. costicola* does not belong to the *Halomonadaceae*, we have found that this species is present in the same ecological niches as *Halomonas* and *Chromohalobacter*. Therefore, it might be possible that pMH1 encodes functions that are common for these microorganisms. The occurrence of pMH1 in the original halophilic strains was confirmed by Southern hybridization. However, the plasmid could not be detected on agarose gels by ethidium bromide staining of standard plasmid DNA preparations from the parental strains, but could be observed when these preparations were introduced into *E. coli* by transformation. The question remains why the yield of pMH1

Table 2
Final Concentration of Antimicrobials Used for SW Media of Different Salinities

Antimicrobial	Final concentration[a] (µG/mL)					
	SW-10[b]	SW-7.5	SW-5	SW-3	SW-2	SW-1
Ampicillin (Amp)	1000	500	500	200	100	100
Rifampicin (Rif)	25	25	25	25	25	25
Gentamycin (Gm)	500	500	200	50	20	20
Kanamycin (Km)	1000	500	200	100	75	50
Neomycin (Nm)	1000	1000	300	50	50	50
Streptomycin (Sm)	1000	1000	500	200	200	200
Trimethoprim (Tp)	500	500	200	150	150	150
Nalidixic acid (Nd)	1000	500	200	150	150	150
Spectinomycin (Sp)	1000	1000	500	500	200	200
Tetracycline (Tc)	1000	1000	500	200	125	125

[a] Concentrations referred to solid media.
[b] Media SW-10 to SW-1 contain 10% (w/v) to 1% (w/v) total salts. This table is a general guideline to select the appropriate antimicrobial concentration for members of the *Halomonadaceae*, which may vary depending on the particular strain. For a more precise information on a particular strain of *Halomonas* or *Chromohalobacter*, as well as on two strains of *E. coli, see* **ref. 35**.

Table 3
Native Plasmids Isolated From Gram-Negative Moderately Halophilic Bacteria

Plasmid	Natural host strain	Features	References
pMH1	H. elongata ATCC 33173 H. halmophila ATCC 19717 H. halophila CCM 3662 Salinivibrio costicola NCMB 701	11.5 kb, Km^r, Nm^r, Tc^r; unique EcoRI, EcoRV and ClaI sites; able to replicate in E. coli.	43
pHE1	C. salexigens ATCC 33174	4.1 kb; unique BglII, EcoRI, PstI, XbaI, and XhoI sites; unable to replicate in E. coli; Mob$^+$; complete sequence available (EMBL accession No. AJ132759 and AJ243735)	44–46
pHI1	H. israelensis ATCC 43985	ca. 48 kb	44
pHS1	H. subglaciescola UQM 2927	ca. 70 kb	44
pCM1	C. marismortui ATCC 17056	17.5 kb; unique BamHI, BglII and XbaI sites; unable to replicate in E. coli; minimal replicon sequenced (GenBank accession No. X86092)	47
pVE1 and pVE2	H. eurihalina strain F2-7	pVE1: 8.2 kb; unique SmaI site pVE2: 5.8 kb; unique BglII site	48
pVC1, pVC2 and pVC3	S. costicola strain E-367	pVC1: 2.95 kb; unique BamHI site pVC2: 19 kb pVC3: 21 kb	49

Table 4
Cloning and Expression Vectors Useful for Gram-Negative Moderately Halophilic Bacteria

Vector	Features[a]	Cloning sites	Insertional inactivation	Host range				References
				E. coli	Halomonas	Chromo-halobacter	Halomonas Halomonas	
Native-plasmid-based								
pHS15	12.25 kb, pHE1-derivative, Amp^r, Sm^r, Sp^r, ColE1 ori, RK2 oriT, Tra^-, Mob+	BamHI, EcoRI, KpnI, NotI, PstI, SalI, SmaI, SpeI	-	+	+	+	+	44
pHS134	9 kb, pHE1-derivative, Amp^r, Sm^r, Sp^r, ColE1 ori, Tra^-, Mob+	BamHI, BglII, SalI, EcoRV	-	+	ND^b	+	ND	46
pEE3	6.6 kb, pCM1-derivative, Amp^r, Tp^r, ColE1 ori, Tra^-, Mob+	NotI, SacI, Sma	-	+	+	+	-	50
pEE5	7 kb, pCM1-derivative, Amp^r, Tp^r, ColE1 ori, lacZ, Tra^-, Mob+	(BamHI, SacI, SalI, SmaI, XbaI	LacZ (BamHI, SacI, SalI, SmaI, XbaI)	+	+	+	-	50

Host range

Vector	Features[a]	Cloning sites	Insertional inactivation	E. coli	Halomonas	Chromo-halobacter	Halomonas Halomonas	References
Broad-host-range								
pVK102	23 kb, Tc[r], Km[r], Tra[−], Mob[+], IncP, cosmid vector	SalI, HindIII, XhoI	Tc (SalI) Km (HindIII, XhoI)	+	+	+	+	51,52
pMP220	10.5 Kb, Tra[−], Mob[+], IncP, promoterless lacZ, promoter-probe vector	BglII, EcoRI, PstI, SphI, XbaI	−	+	+	+	+	53
pKT230	11.9 kb, Km[r], Sm[r], Tra[−], Mob[+], IncQ	BamHI, EcoRI, HindIII, SacI, SmaI, SstI, SstII, XhoI	Km (HindIII, SmaI,XhoI) Sm (EcoRI, SacI, SstI, SstII)	+	+	+	+	51,54
pML123	12.5 kb, Gm[r], Nm[r], Tra[−], Mob[+], IncQ, expression vector	BamHI,ClaI, EcoRI, HindIII, SacI XbaI, XhoI	−	+	+	+	+	55
pGV1124	10.8 kb, Sm[r], Cm[r], Tra[−], Mob[+], IncW	BamHI, EcoR, HindIII, SalI	Cm (EcoRI)	+	+[c]	+[c]	−	51,56

a See **Table 2** for abbreviations of antimicrobials.

b ND, not determined.

c Only H. elongata and C. salexigens, and with very low transfer frequencies.

was so low as to escape detection by conventional procedures. Some possible explanations are the presence of nucleases that rapidly degrade DNA after cellular lysis, or a close association of pMH1 with the cytoplasmic membrane or with the chromosome.

The aforementioned difficulties prompted us to look for other native plasmids that could be more easily manipulated. In a second screening performed on species of the family *Halomonadaceae*, three autochthonous cryptic plasmids were isolated from *C. salexigens* (formerly *H. elongata*) ATCC 33174 (pHE1), *H. israelensis* (pHI1), and *H. subglaciescola* (pHS1) by using a slightly modified alkaline lysis procedure (**Table 3**). When a variation of the Eckhardt method (**40**) was utilized, no megaplasmids were detected in any of the strains tested (**44**).

Because of its small size (4185 bp), pHE1 was selected for further characterization and construction of a shuttle vector for moderate halophiles. Analysis of the complete sequence of pHE1 revealed that the plasmid is basically composed of two functional parts: the replication (*rep*) and the mobilization (*mob*) regions (45,46). The basic replicon consists of a 1.77-kb *Bgl*II-*Eco*RI fragment, which contains the genetic information required for autonomous replication and stable maintenance. Analysis of its sequence revealed the presence of two genes, *repAB*, probably organized in one operon. *repA* encodes the replication initiator protein (RepA), which appeared to have a high degree of homology to the θ-replicase proteins of ColE2-related plasmids. The *repB*-encoded product showed certain similarity with the RepB proteins of the same family of replicons. Deletion analysis suggests that pHE1 origin of replication (*ori*) is located in an 800-bp region upstream of *repA*. A third putative gene, *incA*, was found in the complementary strand to the leader region of *repA* mRNA. This, together with the presence in the 5' end of the *repA* mRNA of inverted repeats that could form stable stem-loops structures, suggests that the *IncA* gene encodes a small antisense RNA. A possible control mechanism of pHE1 replication involving an RNA molecule, which sequesters the translational initiation region of the replication protein RepA, was proposed (45). The *mob* region, located in the 2.4-kb *Eco*RI-*Bgl*II remaining half of pHE1, contains four genes (*mobCABD*), which showed a complex organization with two of them (*mobB* and *mobD*) entirely overlapped by a third (*mobA*). The deduced proteins appeared to have a high degree of homology to Mob proteins of ColE1 and closely related plasmids. Upstream of the *mob* genes, an *oriT* region with a putative nick sequence highly homologous to that of ColE1 plasmids was identified. pHE1was found to be mobilizable from *E. coli* to *H. elongata* assisted by the helper plasmid pRK600 (46). To our knowledge, this is the first mobilizable plasmid found in moderate halophiles.

In a different screening, Mellado et al. (**47**) isolated pCM1, a narrow-host-range plasmid from *C. marismortui* (**Table 3**). Plasmid regions competent for self-replication were identified by cloning into suicide vectors. The minimal replicon of pCM1 was localized in a 1.6-kb fragment and sequenced. The replicon was found to contain two functionally discrete regions, the *oriV* region and the *repA* gene. *oriV*, located on a 0.7-kb fragment, contains four iterons 20 bp in length adjacent to a DnaA box that is dispensable but required for efficient replication of pCM1, and requires *trans*-acting functions. The *repA* gene encodes a replication protein of 289 residues, which is similar to the replication proteins of other Gram-negative bacteria such as that of plasmid pFA3 from *Neisseria gonorrhoeae*. However, despite pCM1 and pHE1 being isolated from two species of *Chromohalobacter*, their Rep proteins do not show significant homology to each other (*45*). This is not surprising, since although both plasmids have been proposed to have a θ-type replication mechanism, the pHE1 replicon seems to be regulated via antisense RNA (*45*), whereas the pCM1 replicon shares common features of iteron-regulated plasmids (*47*). In agreement with this, pHE1 and pMC1 have been shown to be compatible, allowing simultaneous introduction into the same host of vectors derived from their replicons for complementation studies or strain improvement.

H. eurihalina strain F2-7 produces large amounts of an anionic exopolysaccharide called polymer V2-7, with a number of promising applications in industry. The strain was found to contain two plasmids, pVE1 and pVE2 (**Table 3**), which were characterized with respect to their restriction maps and analysis of homologous sequences in other moderately halophilic strains. After treatment at high temperature, cured derivative strains lacking one or two plasmids were constructed. However, they produced the same mucoid phenotype as the parent strain, indicating that none of the plasmids is involved in polysaccharide production. Attempts to correlate pVE1 and pVE2 with other phenotypes—i.e., morphological, physiological, biochemical and nutritional aspects—and antimicrobial and heavy-metal resistance were unsuccessful. Due to their small size, the plamids pVE1 and pVE2 are good candidates to develop cloning and expression vectors for Gram-negative moderate halophiles, including the *Halomonadaceae* (*48*).

An other potential candidate to be used for the construction of cloning vectors is pCV1, a very small plasmid isolated from *S. costicola* strain E-367 (**Table 3**). Mellado et al. (*49*) investigated for the first time the genome organization of a moderate halophile by using pulsed-field electrophoresis. The genome size of 32 strains of *S. costicola* ranged from 2100 to 2600 kb, about half the size reported for *E. coli* but similar to that determined for *Vibrio cholerae*. In addition, plasmid content was analyzed in all strains by using the same technique, but only strain E-367 was found to harbor extrachromosomal DNA. This consisted of three small plasmids, which were designated pCV1, pCV2, and pCV3, and one megaplasmid. Hybridization experiments by using the three small plasmid species as probes showed that pCV2 and pCV3 are unique since they do not probe to one another, and none of these is related to the megaplasmid. In addition, no DNA homology among plasmid pCV1 and plasmids pMH1, pHE1, and pCM1 could be detected.

3.4.2. Cloning and Expression Vectors

Two of the best characterized native plasmids have been used as a base to create cloning vehicles; pHE1 from *C. salexigens* (formerly *H. elongata*) ATCC 33174 and pCM1 from *C. marismortui*.

The first shuttle vector for Gram-negative moderate halophiles, pHS15 (**Table 4**), was constructed from the *C. salexigens* plasmid pHE1 (*44*). Since pHE1 appeared to be unable to replicate into *E. coli* cells, a number of mobilizable pHE1-derived hybrid plasmids were constructed that could be selected and maintained both in *E. coli* and *C. salexigens*. For this purpose, the *E. coli* pKS plasmid, the omega cassette (encoding resistance to streptomycin and spectinomycin), the *oriT* region (encoding the mobilization functions of plasmid RK2), and the entire pHE1 plasmid individually digested with *Pst*I, *Eco*RI, or *Bgl*II were assembled to give pHS7, pHS9, and pHS13, respectively. Plasmid pHS13 was improved by deletion of some repeated restriction sites to create pHS15. This vector contains a number of restriction sites commonly used for cloning and encodes resistance to streptomycin, spectinomycin, and ampicillin. It should be noted that, with some exceptions, the ampicillin-resistance gene is not very useful for the *Halomonadaceae*, as most of the strains have been shown to be resistant to this antibiotic (*35*).

It was fortunate that pHS13 was selected as the precursor to construct the shuttle vector pHS15. In pHS13, the native plasmid pHE1 was inserted as one *Bgl*II fragment. Analysis of the pHE1 sequence, subsequent to pHS15 construction, revealed that the *Bgl*II recognition site is located in the intergenic region between the *repAB* and *mobCABD* genes of pHE1, and therefore this site does not interrupt plasmid replication or mobilization functions. At the moment of pHS15 construction it was unknown that pHE1 was mobilizable itself, and the mobilization functions were provided by cloning the *oriT* of plasmid RK2. Thus, pHS15 contains two *mob* regions, one from RK2 and one from pHE1. A simplified version of pHS15, vector pHS134 (**Table 4**), has been constructed by joining the *E. coli* plasmid pKS, the omega cassette, and the plasmid pHE1 (*44*). This vector can be used directly for cloning or may constitute the basis for specialized pHE1-derived mobilizable vectors.

Two useful vectors have been constructed that are based on the minimal replicon of the *C. marismortui* plasmid pCM1, pEE3, and pEE5 (**Table 4**). To generate pEE3, the trimethoprim resistance gene from plasmid pAS396 and the *oriV* of pCM1 were cloned in the plasmid pUC18mob, a derivative of pUC18 with the RK2 mob DNA (*50*). pEE3 carries unique *Not*I, *Sac*I, and *Sma*I sites, useful for cloning purposes. The characteristics of pEE3 as a cloning vector were improved by removing superfluous restriction sites, and introducing the α-peptide of *lacZ* and a multiple cloning site. This new vector, pEE5 (*50*; **Table 4**), offers the possibility of white/blue colony selection for cloning experiments. Both pEE3 and pEE5 carry a selection marker different from that of pHS15, the trimethoprim resistance gene. Trimethoprim has been shown to be highly toxic for most species of the *Halomonadaceae*, regardless of the salt concentration of the culture media. This makes it a useful genetic marker for this bacterial group, especially in experiments in which high salinities need to be used.

Besides shuttle vectors, which are generally high-copy-number and restricted in host range, broad-host-range for Gram-negative bacteria (usually low-copy-number, apart from IncQ plasmids) also represent an important alternative for genetic studies. Our results indicate that IncQ and IncP plamids can be particularly useful as cloning and expression vectors for the *Halomonadaceae* (*see* **Subheading 3.5**). Some of these plasmids have been included in **Table 4**.

3.5. Genetic Transfer Mechanisms

So far, conjugation is the only genetic transfer mechanism described for Gram-negative moderate halophiles. Natural transformation has not been reported, and approaches such as electroporation or CaCl$_2$ treatment have either been unsuccessful or given nonreproducible results. Although some bacteriophages have been described for moderate halophiles, transduction methods have not yet been developed.

Conjugation has been successfully used to transfer both vectors derived from native plasmids (*44,50,51*), and broad-host range vectors (*34,51*) from *E. coli* to *Chromohalobacter, Halomonas,* and *Salinivibrio.* Several factors affecting the efficiency of conjugation (cell growth phase, mating time, donor-recipient ratio, and composition and salinity of the mating medium) were evaluated to optimize the conditions for conjugation between *E. coli* and moderate halophiles (*see* Chapter 14 for a detailed protocol). In addition, comparative studies on plasmid host range, stability, and compatibility in species of the *Halomonadaceae* (*Halomonas* and *Chromohalobacter*) and *S. costicola,* were done. **Table 4** summarizes the host range of the main cloning vectors reported. Whereas pHS15 (*44*), the IncP plasmids pVK102 (*52*) and pMP220 (*53*), and the IncQ plasmids pKT230 (*54*) and pML123 (*55*) were able to replicate in all strains tested, pEE3 and pEE5 (*50*) did not replicate in *S. costicola,* pGV1124 (IncW, *56*) was maintained only in *H elongata* ATCC 33173 and *C. salexigens* (formerly *H. elongata*) ATCC33174, and pCU1 (IncN, *57*) could not be established in any of the hosts tested (*51*).

On the other hand, conjugation between moderate halophiles was demonstrated by using the self-transmissible IncP plasmid RK2 (*51*). Transfer of this plasmid between *Halomonas* species in SW-2 (2% total salts) medium was found at frequencies up to 1.2 $\times 10^{-3}$ to 2.8×10^{-4}, which were the highest conjugation frequencies achieved. When the salinity of the mating medium was raised up to 5% or 7%, conjugation was observed but the transfer frequencies were lower. No genetic transfer of RK2 was detected at 10% total salts, which are actually the salinity conditions that predominate in the natural habitats where moderate halophiles grow. The reason for this is unknown, but it might be related to a failure of the conjugation machinery encoded by RK2 to function (i.e., assemble or retrieve the sexual *pilus*) at high salt. At this respect, it would be interesting to test whether any of the relatively large plasmids isolated in our laboratory, such as pHI1 or pHS1 (**Table 3**) are conjugative and, if so, whether or not they are able to mediate genetic exchange in hypersaline environments.

For optimal use of the cloning vector it is desirable that, once introduced, it is stably maintained in the host cell, even in a nonselective medium. The stability of the vectors pEE5 (*50*), and pHS15, pVK102, pKT230, and pGV1124 (*51*) has been assayed and compared up to approximately 80 generations. Out of the host strains tested, pEE5 was stable in *H. subglaciescola* and *H. halophila* for 80 generations, whereas it was demonstrated to be highly unstable in *C. marismortui* and, to a lower extent, in *H. eurihalina*. The hybrid plasmid pHS15 (*44*) could not be maintained, in the absence of selective pressure, for a long time in any of the moderate halophiles tested, including its parental strain. In any case, the highest stability of pHS15 occurred in the parental strain (60% loss after 80 generations) and the lowest in *S. costicola* (complete loss after 40 generations) and *H. halophila* (complete loss after 60 generations). The IncQ plasmid pKT230 was found to be stably inherited in all strains tested, except for *S. costicola,* where it was completely lost after 20 generations. Conversely, pVK102 (IncP) was highly unstable in most moderate halophiles assayed, with the exception of *C. marismortui*. Maintenance of pGV1124 in *H. elongata* was observed to be very unstable under nonselective conditions. Shuttle vector pHS15 was found to be compatible with pEE5 and with each of the broad-host-rage vectors assayed.

3.6. Mutagenesis

Mutagenesis and complementation analysis are powerful techniques in the characterization of operons, genes, and proteins. The creation of mutants is an essential step in determining gene function. In general, two main approaches are available to generate bacterial mutants: (1) in vivo random mutagenesis using chemical agents or transposons or (2) gene replacement of previously cloned DNA that has been manipulated in vitro by introducing insertions, deletions or base replacements.

3.6.1. Chemical Mutagenesis With Hydroxylamine

To our knowledge, hydroxylamine (HA) is by far the only mutagen used in a chemical mutagenesis approach in moderate halophiles of the family *Halomonadaceae* (**58**). It was chosen as a DNA-damaging agent because it induces a high mutagenicity. In addition, it is among the few agents that can directly cause a change in base pairing by favoring a tautomeric shift. The effect of different concentrations of HA on survival of *C. salexigens* DSM 3043 (formerly *H. elongata* DSM 3043) and *H. meridiana* as a function of the time exposure was investigated. A concentration of 0.2 *M* (for *C. salexigens*) or 0.1 *M* (for *H. meridiana*) of mutagen was selected for further experiments because it offered a sufficient number of survivors for reliable detection of mutants. In addition, a clear effect of decreasing salinity on the enhancing of the killing action of HA in both species was found. This can be due to changes on membrane permeability due to the hyposmotic shock that result in a higher availability of HA for the cells. From these experiments, SW-5 (5% total salts) was chosen as the treating medium for the two strains. Subsequently, the induced mutagenicity caused by treatment with HA was measured as the frequency of streptomycin-resistant mutants isolated when cells of *C. salexigens* and *H. meridiana* were exposed to 0.2 *M* or 0.1 *M*, respectively, at different times of exposure. As a consequence of these results, the optimal conditions of HA mutagenesis for the two species were established (*see* **Note 4**). Finally, these conditions were used to isolate a number of auxotrophic mutants of *C. salexigens* as well as different salt-sensitive mutants of *C. salexigens* and *H. meridiana*. Some of these latter mutants appeared to be affected in the synthesis of compatible solutes (**58**).

3.6.2. Transposon Mutagenesis

3.6.2.1. TRANSPOSON TN*1732*

Tn*1732*(**59**), a derivative of Tn*1721* (**60,61**), which in turn is a member of the Tn*21* subgroup of the Tn*3* family, was used for the first time in 1995 by Kunte and Galinski (**34**) for transposon mutagenesis in *H. elongata*. Tn*1732* consists of the "basic transposon" of Tn*1721* (Tn*1722*) that encodes the transposition functions (transposase, *tnpA*, and resolvase, *tnpR*, genes), and a kanamycin resistance marker for positive selection of mutants. In view of the problems found with the use of Tn*5* in this species (see below),

Tn*1732* was assayed as an alternative transposon since it was successfully used to overcome similar problems with Tn*5* mutagenesis of *Xanthomonas*. The suicide plasmid pSUP102-Gm (Cm^r) (*see* **Note 5**) was used to introduce Tn*1732* into *H. elongata* via *E. coli* SM10-mediated conjugation (*34*). Southern hybridization analysis of a number of transconjugants indicated the presence of single, randomly distributed insertion sites within the chromosome. Furthermore, the possible co-integration of the vector pSUP102 in *Halomonas* DNA was excluded, as shown by hybridization experiments and the finding that Cm-resistant transconjugants were not observed. The phenotypic analysis of more than 3000 Tn*1732* transconjugants proved the usefulness of the system for insertion mutagenesis in *Halomonas*. Thus, several auxotrophic mutants, as well as different salt-sensitive mutants, were isolated. These latter mutant strains occurred with a very high frequency (2%), confirming that Tn*1732* is a useful tool for transposon mutagenesis in this species. Since then, Tn*1732* has been used successfully for generalized insertion mutagenesis in different strains of *Halomonas* and *Chromohalobacter* (*see* **Note 6**), yielding a number of useful mutants that were used in reverse genetics experiments to isolate the corresponding genes. These included *H. elongata* (*62*) and *C. salexigens* (*63*) mutants affected in the synthesis of the compatible solute ectoine, *H. elongata* mutants impaired in ectoine and hydroxyectoine uptake (*64*), and *H. meridiana* mutants defective in extracellular amylolytic activity (*5*).

3.6.2.2. Other Transposons

Tn*5*, which has been widely used for mutagenesis in Gram-negative bacteria, was tested by Kunte and Galinski (*34*) for insertional mutagenesis in *H. elongata*. The transposon was successfully transferred from *E. coli* SM10 containing the suicide vector pSUP101::Tn*5* to *H. elongata* by biparental mating. However, analysis of Tn*5* transconjugants showed that transposon and vector DNA co-integrated in a nonrandom manner into the chromosome. This was suggested to be due to the presence of hot spots for Tn*5* integration in *H. elongata* DNA, rather than the result of homologous recombination via vector DNA (*34*). We have attempted to use mini-Tn*5* for insertional mutagenesis in *C. salexigens* and also found evidence for the presence of hot spots into the chromosome. From these data, it can be concluded that Tn*5* is not a suitable tool for transposon mutagenesis in members of the *Halomonadaceae*. Attempts made by our group to use transposons belonging to the Tn*7* and Tn*10* families for insertional mutagenesis in *C. salexigens* have likewise been unsuccessful.

These findings cause severe limitations to the use of transposons in *Halomonadaceae*, with Tn*1732* being so far the only suitable tool for transposon mutagenesis in this bacterial family. One disadvantage of Tn*1732*, related to transposons of the Tn*3* family, is its predilection for plasmid vs chromosomal DNA. That is, if one (or more) plasmid is present in the strain to be mutagenized, a high frequency of plasmid insertions will occur to the detriment of chromosome insertions. On the other hand, new Tn*1721*-based transposons with distinct selection markers would be desirable, for instance, if multiple insertions in the same strains are required. Indeed, there are a number of Tn*1721*-derived transposons carrying antibiotic resistance genes other than kanamycin (*59,65*). However, all these transposons contain the cognate transposase gene between their inverted repeated sequences. This fact implies that a recipient cell already containing one of these Tn*1732*-derived transposons becomes immune to further transposition rounds. New systems to deliver Tn*1721*-derived transposons in *Halomonadaceae* are therefore necessary, as the equivalent to the mini-Tn*5* transposons described by De Lorenzo et al. (*66*), which carry the transposase gene into the suicide vector (*in cis*), but external to the mobile element. In this way, a single recipient cell can be used for repeated insertion events with differentially marked minitransposons.

3.6.3. In Vitro Site-Directed Mutagenesis Followed by Gene Replacement

In vitro mutagenesis is an invaluable technique for studying protein structure-function relationships and for identifying intramolecular regions or amino acids, both of which may mediate gene expression and vector modification. Conventional DNA manipulation techniques (e.g., deletion of a specific DNA region by digestion and re-ligation, insertion of a cassette carrying one antibiotic resistance marker or a reporter gene, and others) or PCR-based site-directed mutagenesis approaches can be used to generate the mutation of interest. Following in vitro manipulation of the DNA, the mutated region must be transferred to the host parental strain and gene replacement must occur by homologous (double) recombination.

Very recently, Grammann et al. (*64*) reported the used of the splicing-by-overlap-extension (SOE) PCR technique (*67*) to generate a number of deletion mutants of *H. elongata* affected in the synthesis or transport of the compatible solute ectoine. For this purpose, the DNA regions upstream and downstream of the gene to be deleted were joined together by applying the SOE-PCR methodology. The resulting PCR fragments containing the deleted gene were cloned into the suicide vector pKS18*mobsac* (*68*). This is a derivative of the *E. coli* plasmid pK18 carrying the *sacB* gene from *Bacillus subtilis*, which is inducible by sucrose and is lethal when expressed in Gram-negative bacteria. The incorporation of this conditionally lethal gene greatly simplifies the identification of double recombination events (those leading to gene replacement) among the vast majority of single crossovers (those that still contain the vector). This avoids time-consuming replicaplating or toothpicking. Constructs carrying the deletions were transferred into *H. elongata* by *E. coli* S17-1-mediated conjugation. Deletion mutants arising after double crossover were selected on complex medium LBG containing 22% (w/v) sucrose at 37°C. The deletion sites were verified by PCR and DNA sequencing.

In our laboratory, a similar methodology has been used to generate deletion mutants affected in the region encoding ectoine synthesis in *C. salexigens*. In this case, deletion of the *ectABC* genes was achieved by using the QuikChange site-directed mutagenesis kit developed by Stratagene. This is a widely used PCR-based system that eliminates the necessity to subclone the amplified fragment. Following in vitro mutagenesis, a fragment containing a promoterless *lacZ* gene and a streptomycin-resistant marker (the Ω cassette, *69*) was inserted at the deletion point. The reporter *lacZ* was included to monitor the activity of the of the *ectABC* promoter. The Ω cassette served to facilitate the selection of mutants. The assembled Δ*ectABClacZSm* fragment was subcloned into the vector pJQ200SK (*70*), a suicide mobilizable vector similar to pKS18*mobsac* that also carries the conditional lethal *sacB* gene. Gene replacement was achieved after transferring the deleted fragments to a spontaneous rifampicin resistant mutant of *C. salexigens* by triparental matings assisted by the helper plasmid pRK600 (*71*). Selection was on SW-2 medium with streptomycin (100 μG/mL) and sucrose (10% w/v). Deletion was confirmed by Southern hybridization using an internal *ectB* probe, as well as by the absence of the compatible solute ectoine in the mutants.

3.7. Reporter Genes

Reporter genes are indispensable tools to study gene regulation, protein processing, or protein export. When reporter genes are used, the presence of cellular intrinsic activity leads to interferences in the assay interpretation. In this respect, many species of the family *Halomonadaceae* have intrinsic β-galactosidase or alkaline phosphatase activities, two of the most widely used reporter systems (**Table 5**). For Lac⁻ strains, such as *C. salexigens* or *H. meridiana*, a chromogenic assay with X-gal (*see* **Note 7**) can be used on plates to qualitatively monitor gene expression. We have made transcriptional fusions of osmoregulated promoters from *C. salexigens* and the *E. coli lacZ* gene, encoding β-galactosidase. Our results indicate that salinities as high as 10% (1.72 *M* NaCl) do not interfere with the in vivo assay on plates, probably because the substrate is taken up by the cells and the *lacZ*-encoded enzymatic reaction occurs within cell cytoplasm. However, in our hands, the quantitative in vitro assay for β-galactosidase activity, which requires cell lysis previous to substrate (*o*-nitrophenyl-β-D-galactopyranoside) addition, is unviable at salinities higher than 4.3 % (0.75 *M*) NaCl. Several variations of the standard method, with additional washing steps before cell lysis, or spheroplast preparation (to remove NaCl that might have bound to the outer membrane) have been tested in our laboratory, but yielded no reproducible results when cells are grown at salinities above 0.75 *M* NaCl. This, which might be related to a rapid inactivation of the enzyme by salt, is an important constraint to the use of the *E. coli lacZ* gene as a reporter for members of the *Halomonadaceae*, where salinities up to 2.5 *M* (14.5 %) NaCl are used to assay gene expression.

The above mentioned difficulties prompted us to look for alternative reporter systems for moderate halophiles. The ice nucleation (*inaZ*) and green fluorescent protein (*gfp*) genes are totally absent from halophilic microorganisms. Therefore, they were tested as reporter systems for members of the *Halomonadaceae*. The signal generated from the product of these two genes is due not to enzymatic catalysis but to a physical phenomenon. Both assays are in vivo tests. It was shown that both reporters were efficiently expressed in moderate halophiles of the genera *Halomonas* and *Chromohalobacter* under the control of either heterologous or native promoters (*22,72*)

The product of the *inaZ* gene is the ice nucleation protein (about 185 kDa). Expression of the gene can be easily detected in cultures of cells growing at temperatures below 24°C by a freezing-droplet test using a supercool bath at temperatures ranging from −3°C to −10°C (*73*; *see* Chapter 14 for a detailed description of the method). The *inaZ* gene from the plant pathogen *Pseudomonas syringae* was the first reporter system described in Gram-negative moderately halophilic bacteria (*22*). A promoterless version of *inaZ* was introduced into two different restriction sites (*Eco*RI and *Pst*I) of the pHE1 part of vector pHS15. One orientation of both recombinant constructs expressed high

Table 5
Intrinsic β-Galactosidase and Alkaline Phosphatase Activity[a] of Some Strains
of *Halomonas* and *Chromohalobacter*

Strain	β-galactosidase[c]	Alkaline phosphatase
Halomonas elongata ATCC 33173[Tb]	++[d]	++
Halomonas meridiana DSM 5425[T]	−	+
Halomonas eurihalina ATCC 49336[T]	−	+
Halomonas halophila DSM 4770[T]	−	++
Halomonas halmophila ATCC 19717[T]	+	−
Halomonas halodurans ATCC 29686[T]	+	+
Halomonas subglaciescola DSM 4683[T]	++	++
Chromohalobacter marismortui ATCC 17056[T]	+	+
Chromohalobacter salexigens DSM 3043[T]	−	−
Chromahalobacter salexigens ATCC 33174	++	++
Chromahalobacter canadensis ATCC 43984[T]	(+)	+
Chromohalobacter israelensis ATCC 43985[T]	(+)	++

[a]Activity was observed after 48 h of incubation at 37°C on plates of minimal medium M63 with 10% NaCl and 40 μ/mL of the substrate.
[b]T, type strain.
[c]The chromogenic substrates used were 5-bromo-4-chloro-3-indolyl-β-D-galactopyranoside to test β-galactosidase activity and 5-bromo-4-chloro-3-indolyl-phosphate to test alkaline phosphatase activity.
[d]++, strong positive; +, positive; (+) weak positive; −, negative.

levels of ice nucleation activity in *Halomonas*, indicating that *inaZ* was probably introduced in the correct orientation downstream of putative native promoters. A recombinant construct carrying a tandem duplication of *inaZ* in the same orientation yielded significantly higher ice nucleation activity, showing that *inaZ* is appropriate for gene dosage studies. i*naZ* was also expressed in *Halomonas* under the control of heterologous promoters, such as the β-lactamase (Pbla) promoter of *E. coli* and the piruvate decarboxylase promoter (Ppdc) of *Zymomonas mobilis* (*22*). Remarkably, the salt concentration of cultures growing on SW-10 medium (10% w/v total salts) did not cause any apparent inhibition of the ice nucleation activity. For this reason, SW-10 was the choice medium to make the *inaZ* assays.

The existence of a promoter located upstream of the *Pst*I site of native plasmid pHE1 was further confirmed by the construction of a transcriptional fusion of the 1.3-kb *Eco*RI-*Pst*I fragment of pHE1 (containing the putative promoter) and *inaZ* (*74*). Subsequent analysis of the complete sequence of pHE1 (*45,46*) revealed that this region contained the promoter of the pHE1 *mob* operon (*mobCABD*) encoding mobilization functions (*46*). The exact localization of Pmob, which is functional in *E. coli* (*74*), needs experimental confirmation. However, the first 90 pb of the 1.3-kb region must be necessary for promoter function, as judged by the fact that deletion of this DNA sequence leads to a complete loss of *inaZ* activity (*74*).

The green fluorescent protein (GFP), a 27-kDa protein from the marine bioluminescent jellyfish *Aequorea victoria* (*75*), is a unique marker that can be detected by noninvasive methods. No substrates, other enzymes, or cofactors are required for GFPassay. To monitor gene expression in *H. elongata*, Douka et al. (*72*) used a mutated version of this gene (*76*), which allows the establishment of GFP as a convenient expression marker for bacteria, because it can fold correctly and remain soluble in the cells, resulting in 100-times increased fluorescence as compared to the wild-type GFP when expressed in *E. coli*. To investigate *gfp* expression in *H. elongata*, the shuttle vector pHS15 was used in which the *gfp* gene was placed under a native (Pmob) and an heterologous (Ppdc) promoter, giving the constructs pHS15G1 and pHS15G2, respectively. Both recombinant constructs were transferred to *H. elongata* by conjugation and GFP activity was measured at various concentrations of NaCl (2%, 5%, and 10%). Although the *gfp* gene was expressed in all transconjugants, as shown by RT-PCR analysis, expression from Pmob raised as the salt concentration in the growth medium increased from 2% to 10%, suggesting that the mobilization functions of the native plasmid pHE1 might be osmoregulated (*72*).

4. Notes

1. SW-10 is a saline medium containing 10% (w/v) total salts to which 0.5% (w/v) yeast extract is added. The composition of the 10% total salt solution is as follows: 81 g/L, (NaCl), 7 g/L, $MgCl_2$; 9.6 g/L, $MgSO_4$; 0.36 g/L, $CaCl_2$; 2 g/L, KCl; 0.06 g/L, $NaHCO_3$; and NaBr, 0.0026 g/L (*25*). The pH of the medium is adjusted to 7.2 with KOH.

2. Although *E. coli* is a nonhalophilic microorganism, it has the ability to adapt rapidly to moderate salinities provided that compatible solutes are supplied with the grown medium. Therefore it should be considered as a slightly halotolerant bacterium. We were able to grow *E. coli* in the same media as moderate halophiles do (SW complex medium) up to 5% total salts (*35*).

3. Prior to the alkaline lysis method, cells can be washed with 0.1% SDS in TE buffer (10 m*M* Tris-HCl, 1 m*M* EDTA, pH 8.0). This washing step greatly improves the isolation method, yielding cleaner plasmid DNA preparations, suitable for digestion with restriction enzymes.

4. Optimal experimental condition of HA mutagenesis: (1) for *C. salexigens*: exposure of cellular suspensions containing about 10^8 log-phase cells/mL to 0.2 *M* HA in SW-5 medium, pH 5.2, for 90 min at 37°C; (2) for *H. meridiana*: exposure of a suspension of about 10^8 log-phase cells/mL to 0.1 *M* HA in SW-5 medium, pH 5.2, for 120 min at 37°C.

5. The use of *E. coli* vectors carrying a conjugal transfer (*mob*) sequence is a standard way to introduce transposons into a variety of bacteria, where such vectors cannot replicate. For this reason they are termed "suicide" vectors. We have tried to use other suicide Tn*1732* donors based on unstable (thermosensitive) replicons (*59*) but failed to obtain Tn*1732*-induced mutants. Therefore, the choice of the appropriate delivery vector for Tn*1732* into the cell is crucial for the success of mutagenesis.

6. In our laboratory, Tn*1732* induced mutagenesis is performed by conjugal transfer of pSUP102-Gm::Tn*1732* from *E. coli* SM10 to a spontaneous Rif^r mutant of the wild type strain to be mutagenized as described by Cánovas et al (*63*). Matings are carried out by mixing the donor and recipient cultures at a ratio of 1:4 (100 μL of donor, 400 μL of recipient). The mixed cultures are washed with sterile SW-2 medium to eliminate the antibiotics. The pellet is then suspended in 100 μL of SW-2 and placed on a 0.45- μM pore filter onto SW-2 solid medium (which allows the growth of *E. coli*). After overnight incubation at 30°C, cells are resuspended in 20% (v/v) sterile glycerol and, after appropriate dilutions, inoculated onto

SW-2 + Rif + Km plates at a density resulting in about 100–200 colonies per plate. Colonies from these master plates are subsequently transferred with sterile toothpicks to the appropriate media for selection of mutants.

7. X-gal (5-bromo-4-chloro-3-indolyl-β-D-galactopyranoside) is added to the plates at a concentration of 40 μg/mL.

Acknowledgments

Research in the authors' lab was financially supported by grants from the European Commission INCO-Dev Program (grant ICA4-CT-2000-30041), the Spanish Ministerio de Ciencia y Tecnología (grant BIO2001-2663), and Junta de Andalucía.

References

1. Kushner, D. J. and Kamekura, M. (1988) Physiology of halophilic eubacteria, in *Halophilic Bacteria* (Rodríguez-Valera, F., ed.), CRC Press, Boca Raton, FL. pp. 109–140.
2. Ventosa, A., Nieto, J. J., and Oren, A. (1998) Biology of moderately halophilic aerobic bacteria. *Microbiol. Mol. Biol. Rev.* **62**, 504–544.
3. Vreeland, R. H. (1992) The family *Halomonadaceae*, in *The Prokaryotes*, 2nd ed. (Balows, A., Trüper, H. G., Dworkin, M., Harder, W., and Schleifer, K. H., eds.), Springer-Verlag, New York, NY, pp. 3181–3188.
4. Arahal, D. R., Ludwig, W., Schleifer, K. H., and Ventosa, A. (2002) Phylogeny of the family *Halomonadaceae* based on 23S and 16S rDNA sequence analyses. *Int. J. Syst. Evol. Microbiol.* **52**, 241–249.
5. Coronado, M. J., Vargas, C., Mellado, E., Tegos, G., Drainas, C., Nieto, J. J., et al. (2000) The α-amylase gene of the moderate halophile *Halomonas meridiana*: cloning and molecular characterization. *Microbiology* 146, 861–868.
6. Ventosa, A. and Nieto, J. J. (1995) Biotechnological applications and potentialities of halophilic microorganisms. *World J. Microbiol. Biotechnol.* **11**, 85–94.
7. Margesin, R. and Schinner, F. (2001) Potential of halotolerant and halophilic microorganisms for biotechnology. *Extremophiles* **5**, 73–83.
8. Galinski, E. A. (1995) Osmoadaptation in bacteria. *Adv. Microb. Physiol.* **37**, 273–328.
9. Da Costa, M. S., Santos, H., and Galinski, E. A. (1998) An overview of the role and diversity of compatible solutes in *Bacteria* and *Archaea*. *Adv. Biochem. Eng. Biotechnol.* **61**, 117–153.
10. Bremer, E. and Kramer, R. (2000). Coping with osmotic challenges: osmoregulation through accumulation and release of compatible solutes in bacteria, in *Bacterial Stress Responses* (Storz, G., and Hengge-Aronis, R., eds.), ASM Press, Washington, DC, pp. 77–97.
11. Sauer, T. and Galinski, E. A. (1998) Bacterial milking: a novel bioprocess for production of compatible solutes. *Biotechnol. Bioeng.* **59**, 128.
12. Knapp, S., Landstein, R., and Galinski, E. A. (1999) Extrinsic protein stabilization by the naturally occurring osmolytes b-hydroxyectoine and betaine. *Extremophiles* **3**, 191–198.
13. Borges, N., Ramos, A., Raven, N. D., Sharp, R. J., and Santos, H. (2002) Comparative study of the thermostabilizing properties of mannosylglycerate and other compatible solutes on model enzymes. *Extremophiles* **6**, 209–216.
14. Barth, S., Huhn, M., Matthey, B., Klimka, A., Galinski, E. A., and Engert, A. (2000) Compatible-solute supported periplasmic expression of functional recombinant proteins under stress conditions. *Appl. Environ. Microbiol.* **66**, 1572–1579.
15. Nuccio, M. L., Rodhes, D., McNeil, S. D., and Hanson, A. D. (1999) Metabolic engineering of plants for osmotic stress resistance. *Curr. Opin. Plant Biol.* **2**, 128–134.

16. Zhu, J. K. (2000) Genetic analysis of plant salt tolerance using *Arabidopsis*. *Plant Physiol.* **124**, 941–948.

17. Holmberg, N. and Bülow, L. (1998) Improving stress tolerance in plants by gene transfer. *Trends Plant Sci.* **3**, 61–66.

18. Sakamoto, A. and Murata, N. (2000) Genetic engineering of glycinebetaine synthesis in plants: current status and implications for enhancement of stress tolerance. *J. Exp. Bot.* **51**, 81–88.

19. Chen, T. H. and Murata, N. (2002) Enhancement of tolerance of abiotic stress by metabolic engineering of betaines and other compatible solutes. *Curr. Opin. Plant Biol.* **5**, 250–257.

20. Nakayama, H., Yoshida, K., Ono, H., Murooka, Y., and Shinmyo, A. (2000) Ectoine, the compatible solute of *Halomonas elongata*, confers hyperosmotic tolerance in cultured tobacco cells. *Plant Physiol.* **122**, 1239–1247.

21. Hezayen, F. F., Rehm, B. H. A., Eberhardt, R., and Steinbüchel, A. (2000) Polymer production by two newly isolated extremely halopilic archaea: application of a novel corrosion-resistant bioreactor. *Appl. Microbiol. Biotechnol.* **54**, 319–325.

22. Arvanitis, N., Vargas, C., Tegos, G., Perysinakis, A., Nieto, J. J., Ventosa, A., et al. (1995) Development of a gene reporter system in moderately halophilic bacteria by employing the ice nucleation gene of *Pseudomonas syringae*. *Appl. Environ. Microbiol.* **61**, 3821–3825.

23. Tegos, G., Vargas, C., Perysinakis, A., Koukkou, A. I., Nieto, J. J., Ventosa, A., et al. (2000) Release of cell-free ice nuclei from *Halomonas elongata* expressing the ice nucleation gene *inaZ* of *Pseudomonas syringae*. *J. Appl. Microbiol.* **89**, 785–792.

24. Frillingos, S., Linden, A., Niehaus, F., Vargas, C., Nieto, J. J., Ventosa, A., et al. (2000) Cloning and expression of a-amylase from the hyperthermophilic archaeon *Pyrococcus woesei* in the moderately halophilic bacterium *Halomonas elongata*. *J. Appl. Microbiol.* **88**, 495–503.

25. Nieto, J. J., Fernández-Castillo, R., Márquez, M. C., Ventosa A., and Ruiz-Berraquero, F. (1989) A survey of metal tolerance in moderately halophilic eubacteria. *Appl. Environ. Microbiol.* **55**, 2385–2390.

26. Rodríguez-Valera, F., Ruiz-Berraquero, F., and Ramos-Cormenzana, A. (1981) Characteristics of the heterotrophic bacterial populations in hypersaline environments of different salt concentrations. *Microb. Ecol.* **7**, 235–243.

27. Vreeland, R. H., Anderson, R., and Murray, R. G. E. (1984) Cell wall and phospholipid composition and their contribution to the salt tolerance of *Halomonas elongata*. *J. Bacteriol.* **160**, 879–883.

28. James, S. R., Dobson, S. J., Franzmann, P. D., and McMeekin, T. A. (1990) *Halomonas meridiana*, a new species of extremely halotolerant bacteria isolated from Antarctic saline lakes. *Syst. Appl. Microbiol.* **13**, 270–278.

29. Severin, J., Wohlfarth, A., and Galinski, E. A. (1992) The predominant role of recently discovered tetrahydropyrimidines for the osmoadaptation of halophilic eubacteria. *J. Gen. Microbiol.* **138**, 1629–1638.

30. Cohen, G. N. and Rickenberg, R. H. (1956). Concentration specifique reversible des amino-acides chez *E. coli*. *Ann. Ins. Pasteur. Paris* **91**, 693–720.

31. Kamekura, M., Wallace, R., Hipkiss, A. R., and Kushner, D. J. (1985) Growth of *Vibrio costicola* and other moderate halophiles in a chemically defined minimal medium. *Can. J. Microbiol.* **31**, 870–872.

32. Cummings, E. P. and Gilmour, D. J. (1995) The effect of NaCl on the growth of a *Halomonas* species: accumulation and utilization of compatible solutes. *Microbiology* **141**, 1413–1418.

33. Nieto, J. J., Fernández-Castillo, R., García, M. T., Mellado, E., and Ventosa, A. (1993) Survey of antimicrobial susceptibility of moderately halophilic eubacteria and extremely

halophilic aerobic archaeobacteria: utilization of antimicrobial resistance as a genetic marker. *Syst. Appl. Microbiol.* **16,** 352–360.

34. Kunte, H. J. and Galinski, E. A. (1995) Transposon mutagenesis in halophilic eubacteria: conjugal transfer and insertion on transposon Tn*5* and Tn*1732* in *Halomonas elongata. FEMS Microbiol. Let.* **128,** 293–299.

35. Coronado, M. J., Vargas, C., Kunte, H. J., Galinski, E., Ventosa, A., and Nieto, J. J. (1995) Influence of salt concentration on the susceptibility of moderately halophilic bacteria to antimicrobials and its potential use for genetic transfer studies. *Curr. Microbiol.* **31,** 365–371.

36. Borghese, R., Zagnoli, A., and Zannoni, D. (2001) Plasmid transfer and susceptibility to antibiotics in the halophilic phototrophs *Rhodovibrio salinarum* and *Rhodothalassium salexigens. FEMS Microbiol. Lett.* **197,** 117–121.

37. Morelle, G. (1989) A plasmid extraction procedure on a miniprep scale. *BRL Focus* **11,** 7–8.

38. Sambrook, J. and Russell, D. W. (2001) *Molecular Cloning: A Laboratory Manual, 3rd ed.* Cold Spring Harbor Laboratory Press, Cold Spring Harbor, NY.

39. Eckhardt, T. (1978) A rapid method for the identification of plasmid DNA in bacteria. *Plasmid* **1,** 584–588.

40. Plazinski, J., Cen, Y. H., and Rolfe, B. G. (1985) General method for the identification of plasmid species in fast-growing soil microorganisms. *Appl. Environ. Microbiol.* **48,** 1001–1003.

41. Ausubel, F. M., Brent, R., Kingston, R. E., Moore, D. D., Seidman, J. G., Smith, J. A., et al. (1989). *Current protocols in molecular biology.* Greene Publishing Associates, John Wiley & Sons, New York.

42. Marmur, J. (1961) A procedure for the isolation of deoxyribonucleic acid from microorganisms. *J. Mol. Biol.* **3,** 208–218.

43. Fernández-Castillo, R., Vargas, C., Nieto, J. J, Ventosa, A., and Ruiz-Berraquero, F. (1992) Characterization of a plasmid from moderately halophilic eubacteria. *J. Gen. Microbiol.* **138,** 1133–1137.

44. Vargas, C., Férnandez-Castillo, R., Cánovas, D., Ventosa, A., and Nieto, J. J. (1995) Isolation of cryptic plasmids from moderately halophilic eubacteria of the genus *Halomonas*. Characterization of a small plasmid from *H. elongata* and its use for shuttle vector construction. *Mol. Gen. Genet.* **246,** 411–418.

45. Vargas, C., Tegos, G., Drainas, C., Ventosa, A., and Nieto, J. J. (1999) Analysis of the replication region of the cryptic plasmid pHE1 from the moderate halophile *Halomonas elongata. Mol. Gen. Genet.* **261,** 851–861.

46. Vargas, C., Tegos, G., Vartholomatos, G., Drainas, C., Ventosa, A., and Nieto, J. J. (1999) Genetic organization of the mobilization region of the plasmid pHE1 from *Halomonas elongata. Syst. Appl. Microbiol.* **22,** 520–529.

47. Mellado, E., Asturias, M. A., Nieto, J. J., Timmis, K. N., and Ventosa, A. (1995) Characterization of the basic replicon of pCM1, a narrow-host-range plasmid from the moderate halophile *Chromohalobacter marismortui. J. Bacteriol.* **177,** 3433–3445.

48. Llamas, I., del Moral, A., Béjar, V., Girón, M. D., Salto, R., and Quesada, E. (1997) Plasmids from *Halomonas eurihalina*, a microorganism which produces an exopolysaccharide of biotechnological interest. *FEMS Microbiol. Lett.* **156,** 251–257.

49. Mellado, E., García, M. T., Nieto, J. J., Kaplan, S., and Ventosa, A. (1997) Analysis of the genome of *Vibrio costicola*: pulsed-field gel electrophoresis analysis of genome size and plasmid content. *Syst. Appl. Microbiol.* **20,** 20–26.

50. Mellado, E., Nieto, J. J., and Ventosa, A. (1995) Construction of novel shuttle vectors for use between moderately halophilic bacteria and *Escherichia coli. Plasmid* **34,** 157–164.

51. Vargas, C., Coronado, M. J., Ventosa, A., and Nieto, J. J. (1997) Host range, stability and compatibility of broad-host-range plasmids and a shuttle vector in moderately halophilic bacteria. Evidence of intragenic and intergenic conjugation in moderate halophiles. *Syst. Appl. Microbiol.* **20,** 173–181.

52. Knauf, V. C. and Nester, E. W. (1982) Wide host range cloning vectors: a cosmid clone bank of an *Agrobacterium* Ti plasmid. *Plasmid* **8,** 45–54.

53. Spaink, H. P., Okker, R. J. H., Wiffelman, C. A., Pees, E., and Lutenberg, B. J. J. (1987). Promoters in the nodulation region of the *Rhizobium leguminosarum* Sym plasmid pRLJI. *Plant Mol. Biol.* **9,** 27–39.

54. Bagdasariam, M., Lurz, R., Rückert, B., Franklin, F. C. H., Bagdasariam, M. M., Frey, J., et al. (1981) Specific purpose plasmid cloning vectors, II. Broad-host-range, high-copy-number RSF 1010-derived vectors and host-vector system for gene cloning in *Pseudomonas*. *Gene* **16,** 237–247.

55. Labes, M., Pühler, A., and Simon, R. (1990). A new family of RSF1010-derived expression and *lac*-fusion broad-host-range vectors for Gram-negative bacteria. *Gene* **89,** 37–46.

56. Leemans, J., Langenakens, J., de Greve, H., Deblaere, R., van Monatgu, M., and Schell, J. (1982). Broad-host-range cloning vectors derived from the W-plasmid Sa. *Gene* **19,** 361–364.

57. Konarska-Kozlowska, M., Thatte, V., and Iyer, V. N. (1983). Inverted repeats in the DNA of plasmid pCU1. *J. Bacteriol.* **153,** 1502–1512.

58. Cánovas, D., Vargas, C., Ventosa, A., and Nieto, J. J. (1997) Salt-sensitive and auxotrophic mutants of *Halomonas elongata* and *H. meridiana* by use of hydroxylamine mutagenesis. *Curr. Microbiol.* **34,** 85–90.

59. Ubben, D. and Schmitt, R. (1986) Tn*1721* derivatives for transposon mutagenesis, restriction mapping and nucleotide sequence analysis. *Gene* **41,** 145–152.

60. Schmitt, R., Bernhard, E., and Mattes, R. (1979) Characterization of Tn*1721*, a new transposon containing a tetracycline resistance gene capable of amplification. *Mol. Gen. Genet.* **172,** 53–65.

61. Allmeier, H., Cresnar, B., Greck, M., and Schmitt, R. (1992) Complete nucleotide sequence of Tn*1721*: gene organization and a novel gene product with features of a chemotaxis protein. *Gene* **111,** 11–20.

62. Göller, K., Ofer, A., and Galinski, E. A. (1998) Construction and characterization of an NaCl-sensitive mutant of *Halomonas elongata* impaired in ectoine biosynthesis. *FEMS Microbiol. Let.* **161,** 293–300.

63. Cánovas, D., Vargas, C., Iglesias-Guerra, F., Csonka, L. N., Rhodes, D., Ventosa, A., et al. (1997) Isolation and characterization of salt-sensitive mutants of the moderate halophile *Halomonas elongata* and cloning of the ectoine synthesis genes. *J. Biol. Chem.* **272,** 25794–25801.

64. Grammann, K., Volke, A., and Kunte, H. J. (2002) New type of osmoregulated solute transporter identified in halophilic members of the *Bacteria* domain: TRAP transporter TeaABC mediates uptake of ectoine and hydroxyectoine in *Halomonas elongata* DSM 2581§. *J. Bacteriol.* **184,** 3078–3085.

65. Ubben, D. and Schmitt, R. (1987) A transposable promoter and transposable promoter probes derived from Tn*1721*. *Gene* **53,** 127–134.

66. De Lorenzo, V., Herrero, M., Jakubzik, U., and Timmis, K. (1990) Mini-Tn*5* transposon derivatives for insertion mutagenesis, promoter probing, and chromosomal insertion of cloned DNA in Gram-negative bacteria. *J. Bacteriol.* **172,** 6568–6572.

67. Horton, R. M., Hunt, H. D., Ho, S. N., Pullen, J. K., and Pease, L. R. (1989) Engineering hybrid genes without the use of restriction enzymes: gene splicing by overlap extension. *Gene* **77,** 61–68.

68. Schäfer, A., Tauch, A., Jäger, W, Kalinowski, J., Thierbach, G., and Pühler, A. (1994) Small mobilizable multipurpose cloning vectors derived from the *Escherichia coli* plasmids pK18 and pK19: selection of defined deletions in the chromosome of *Corynebacterium glutamicum*. *Gene* **145,** 69–73.

69. Prentki, P. and Krisch, H. M. (1984) In vitro insertional mutagenesis with a selectable DNA fragment. *Gene* **29,** 303–313.

70. Quandt, J. and Hynes, M. (1993) Versatile suicide vectors which allow direct selection for gene replacement in Gram-negative bacteria. *Gene* **127,** 15–21.

71. Kessler, B., de Lorenzo, V., and Timmis, N. K. (1992) A general system to integrate *lacZ* fusions into the chromosome of Gram negative bacteria: regulation of the Pm promoter of the TOL plasmid studied with all controlling elements in monocopy. *Mol. Gen. Genet.* **233,** 293–301.

72. Douka, E., Christogianni, A., Koukkou, A. I., Afendra, A., and Drainas, C. (2001) Use of a green fluorescent protein as a reporter in *Zymomonas mobilis* and *Halomonas elongata*. *FEMS Microbiol. Lett.* **201,** 221–227.

73. Lindgren, P. B., Frederick, R., Govindarajan, A. G., Panopoulos, N. J., Staskawicz, B. J., and Lindow, S. E. (1989) An ice nucleation reporter gene system: identification of inducible pathogenicity genes in *Pseudomonas syringae* pv. *Phaseolicola*. *EMBO J.* **8,** 2990–3001.

74. Tegos, C., Vargas, C., Vartholomatos, G., Perysinakis, A., Nieto, J. J., Ventosa, A., et al. (1997) Identification of a promoter region on the *Halomonas elongata* plasmid pHE1 employing the *inaZ* reporter gene of *Pseudomonas syringae*. *FEMS Microbiol. Lett.* **154,** 45–51.

75. Chalfie, M., Tu, Y., Euskirchen, G., Ward, W. W., and Prasher, D. C. (1994) Green fluorescent protein as a marker for gene expression. *Science* **263,** 802–805.

76. Cormark, B. P., Valdivia, R. H., and Falkow, S. (1996) FACS-optimized mutants of the green fluorescent protein (GFP). *Gene* **173,** 33–38.

14

Gene Transfer and Expression of Recombinant Proteins in Moderately Halophilic Bacteria

Amalia S. Afendra, Carmen Vargas, Joaquín J. Nieto, and Constantin Drainas

Summary

Moderately halophilic bacteria (MHB) of the genera *Halomonas* and *Chromohalobacter* have been used as hosts for the expression of heterologous proteins of biotechnological interest, thus expanding their potential to be used as cell factories for various applications. This chapter deals with the methodology for the construction of recombinant plasmids, their transfer to a number of MHB, and the assaying of the corresponding heterologous proteins activity. The transferred genes include (1) *inaZ*, encoding the ice nucleation protein of the plant pathogen *Pseudomonas syringae*, (2) *gfp*, encoding a green fluorescent protein from the marine biolumi-nescent jellyfish *Aequorea victoria*, and (3) the α-amylase gene from the hyperthermophilic archeon *Pyrococcus woesei*. Vector pHS15, which was designed for expression of heterologous proteins in both *E. coli* and MHB, was used for the subcloning and transfer of the above genes. The recombinant constructs were introduced to MHB by assisted conjugal transfer from *E. coli* donors. The expression and function of the recombinant proteins in the MHB transconjugants is described.

Key Words: Halophilic bacteria; *Halomonadaceae*; *Chromohalobacter*; GFP; α-amylase; *inaZ*.

1. Introduction

Besides halobacteria, moderately halophilic bacteria are the most important group of microorganisms adapted to hypersaline habitats. Various Gram-positive and Gram-negative bacteria are included in this heterogeneous and versatile physiological group (*1*). Apart from their ecological interest, these extremophiles are currently receiving

From: *Methods in Molecular Biology, vol. 267:*
Recombinant Gene Expression: Reviews and Protocols, Second Edition
Edited by: P. Balbás and A. Lorence © Humana Press Inc., Totowa, NJ

considerable attention because of their potential for exciting and promising applications in biotechnology (*see* Chapter 13, for a review). On one hand, they are a source of interesting products for industry, such as extracellular enzymes, polysaccharides, or osmolytes. On the other, they possess useful physiological properties that can facilitate their use as cell factories alternative to traditional hosts such as *E. coli* for the production of heterologous proteins. For instance, most of them can grow at high salinity, minimizing the risk of contamination, and they are easy to grow with simple nutritional requirements. Among moderate halophiles, members of the *Halomonadaceae* (*Halomonas* and *Chromohalobacter*) possess the additional advantage of relatively easy genetic manipulation (*1*). Among others, genetic tools such as transposons, plasmid vectors, and gene transfer technology have been developed for this group of microorganisms (*see* Chapter 13, for a review). In this chapter, the methodology to transfer recombinant plasmids from *E. coli* to members of the *Halomonadaceae* is described. The generation of recombinant plasmids carrying heterologous proteins to be expressed in *Halomonas* and *Chromohalobacter*, as well as assaying the activity of these proteins, are included as examples to support the potentiality of some members of *Halomonadaceae* as alternative cell factories for the production of recombinant proteins.

2. Materials

2.1. Bacterial Strains and Plasmids

1. Spontaneous rifampicin-resistant mutants of the following moderately halophilic strains: *Chromohalobacter salexigens* (formerly *Halomonas elongata*) ATCC 33174, *Halomonas meridiana* DSM 5425, *H. subglaciescola* UQM 2927, *H. halodurans* ATCC 29696, *H. eurihalina* ATCC 49336, and *H. halophila* CCM 3662.
2. *E. coli* strain DH5α is used as host for subcloning and maintaining of recombinant plasmids.
3. *E. coli* DH5α containing the helper plasmid pRK2013 for triparental matings.
4. Expression vector pHS15 (*2*).
5. Source of heterologous proteins: pUZ119 (*3*), pDS3154-*inaZ* (*4*) and pPTZ3-*inaZ* (containing the *Pseudomonas syringae inaZ* gene, *4,5*), pGreenTIR (containing the *Aequorea victoria* green fluorescent protein *gfp* gene, *6*), and pET15-b/$P_{wo}amy$ (containing the *Pyrococcus woesei* α-amylase gene, *7*).

2.2. DNA Manipulation

1. Wizard SV Plus mini-prep isolation kit (Promega).
2. GENECLEAN II (BIO 101, La Jolla, CA).
3. Oligonucleotide primers.
4. Restriction enzymes, T4 DNA ligase, Taq polymerase.
5. Agarose and DNA electrophoresis equipment.
6. Thermocycler.

2.3. Growth Media, Antibiotics, and Conjugation

1. Luria-Bertani (LB) medium for *E. coli*: 1% (w/v) tryptone, 0.5% (w/v) yeast extract, 1% (w/v) NaCl.
2. SW-2, SW-5, and SW-10 media for moderately halophilic strains (*see* **Note 1**).

3. Nitrocellulose filters, 0.45-μm pore diameter (Millipore).
4. Antibiotics for genetic selection or plasmid maintenance:

	Kanamycin (Km)	Rifampicin (Rif, *see* **Note 2**)	Streptomycin (Sm)	Ampicillin (Amp)
Stock solution	50 mg/mL in sterile distilled water	50 mg/mL in methanol	100 mg/mL in sterile distilled water	100 mg/mL in sterile distilled water
Final concentration (*see* **Note 3**)	50 μg/mL	25 μg/mL	200 μg/mL	100 μg/mL

2.4. Protein Activity Assays

1. Refrigerated circulating bath.
2. 2% paraffin solution in xylene.
3. Fluorimeter (Perkin-Elmer LS-3).
4. Washing buffer for GFP activity: 10 mM Tris-HCl pH 8, 600 mM NaCl.
5. Amylase reaction buffer: 50mM sodium acetate, pH 5.5.
6. Mini Beadbeater cell disrupter (Biospec Products, Bartlesville, OK).
7. Zirconium beads.
8. 50 mM sodium acetate, pH 5.5, containing 0.72% (w/v) soluble starch.
9. Iodine solution: 1.5% (w/v) KI, 0.5% (w/v) I$_2$.

3. Methods

The methodology described below includes (1) the construction of the recombinat plasmids, (2) the introduction of the recombinant plasmids into *Chromohalobacter* and *Halomonas* strains, and (3) the measurement of the activity of recombinant proteins.

3.1. Recombinant Plasmids

Plasmid preparations from *E. coli*, restriction enzyme digestions, ligations, *E. coli* transformations, PCR, and DNA electrophoresis were performed according to standard protocols *(8)*. Plasmid DNA from *C. salexigens* was isolated by using the protocol of Wizard SV Plus mini prep isolation kit (Promega). DNA bands from agarose gels were purified according to the protocol of GENECLEAN II (BIO 101).

The construction of all recombinant plasmids described in this chapter involved the use of the vector pHS15, which was designed for expression of genes in both *E. coli* and moderately halophilic bacteria *(2)*. pHS15 was generated by assembling a streptomycin-resistant gene (for positive selection in moderate halophiles), the mobilization functions of the broad-host-range IncP plasmid RK2, and the plasmid pHE1 from *C. salexigens,* into the conventional *E. coli* vector pKS. The nucleotide sequence of the native plasmid pHE1 has been determined. It contains a replication region, consisting of the *repAB* genes, and a mobilization region, consisting of four ORFs (*mobCABD*) that show a complex organization with two of them (*mobB* and *mobD*) entirely overlapped by a third (*mobA*). Vector pHS15 contains single *Pst*I (which lies within the *mobB* gene) and *Eco*RI (which lies downstream of the *repB* gene) sites suitable for cloning, which are located in the pHE1 part. Cloning a heterologous gene in these sites in the appropriate orientation leads to the expression of this gene due to readthrough from the corresponding *mob* or *rep* native promoters (*9,10*; *see* **Fig. 1**).

3.1.1. Construction of Recombinant Plasmids Containing the inaZ Gene of the Plant Pathogen Pseudomonas syringae

Ice nucleation proteins are products of single genes of plant pathogenic or epiphytic bacteria of the genera *Pseudomonas, Erwinia,* or *Xanthomonas,* which are responsible for frost damage of crops *(11)*. They are glycolipoproteins located at the outer membrane, where they form aggregates of about 40 monomers, which act as an ice nucleus. Apart from their deleterious effects in agriculture, ice nucleation proteins may also have several biotechnological applications in artificial snow-making, frozen food industry, diagnostic kits, and the like *(12)*. Furthermore, their genes can serve as efficient nonconventional reporters for studies of gene expression *(13)*. Therefore, transfer and expression of the ice nucleation gene in nonpathogenic microorganisms such as moderately halophilic bacteria may attract considerable biotechnological interest. A great advantage is that these bacterial species have no background ice nucleation activity that may affect the ice nucleation assay.

For the expression of the ice nucleation protein InaZ in moderately halophilic bacteria, the promoterless *inaZ* gene was excised from plasmid pUZ119 as a 3.7-kb *Pst*I fragment and fused in the *Pst*I site of the pHE1 sequence of pHS15, yielding recombinant plasmid pHS18 (*14,* **Fig. 1**). Furthermore, the *inaZ* gene under the control of the promoter of β-lactamase of *E. coli* (P_{bla}) and pyruvate decarboxylase of *Zymomonas mobilis* (P_{pdc}) was isolated from plasmids pDS3154-*inaZ* and pPTZ3-*inaZ* as 4.5-kb and 4.0-kb *Eco*RI fragments, respectively. These fragments were subcloned into the *Eco*RI site of pHS15, giving the recombinant plasmids pHS22 and pHS23, respectively (*14,* **Fig. 1**).

3.1.2. Construction of Recombinant Plasmids Containing the gfp Gene From the Marine Bioluminescent Jellyfish Aequorea victoria

The green fluorescent protein (GFP), a 27-KDa protein from *Aequorea victoria* (*15*), is a unique marker that, as in the case of ice nucleation protein, can be detected by noninvasive methods, without the use of substrates, other enzymes, or cofactors. To express GFP in *C. salexigens*, a mutated version of the *gfp* gene was used (*16*), which facilitates the expression of GFP as an active form in bacteria. By this form GFP can fold correctly and remain soluble in the cells, resulting in 100-times increased fluorescence as compared to the wild-type GFP when expressed in *E. coli*.

For the expression of GFP in *C. salexigens* the shuttle vector pHS15 was used (*2*), in which the mutated *gfp* gene was placed under a native and a heterologous promoter, respectively. In the first case, the *gfp* gene was excised as a 0.8-kb *Pst*I fragment from pGreenTIR and subcloned in the *Pst*I site of pHS15 at the same as in the case of *inaZ* expression (*14*), producing plasmid pHS15G1 (*17,* **Fig. 2**). In the case of the heterologous promoter, *gfp* was subcloned under the control of the pyruvate decarboxylase gene

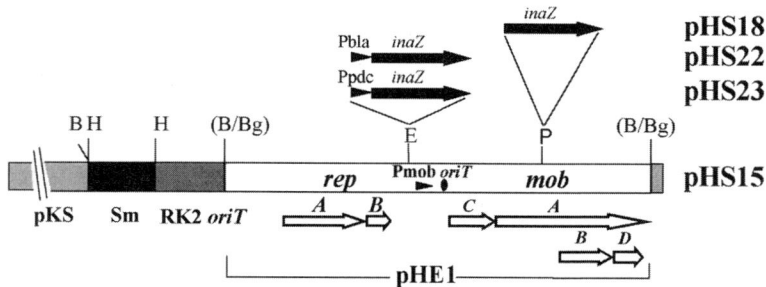

Fig. 1. pHS15-derived recombinant plasmids carrying the *inaZ* gene from *P. syringae*. Maps are not drawn to scale.

promoter of *Z. mobilis* (P$_{pdc}$) as a 1-kb partial *Eco*RI fragment in the *Eco*RI site of pHS15. The outcome was recombinant plasmid pHS15G2 (*17*, **Fig. 2**).

3.1.3. Construction of Recombinant Plasmids Containing the α-Amylase Gene From the Hyperthermophilic Archeon Pyrococcus woesei

The *P. woesei* α-amylase is an extremely thermostable enzyme classified in family 13 of the glucosyl hydrolase superfamily (*18–21*). Enzymes isolated from hyperthermophilic microorganisms present a great biochemical and biological interest for the structural basis of their thermostability. On the other hand, the use of such thermostable enzymes has considerable biotechnological interest. However, due to the practical difficulties in the cultivation of hyperthermophiles, these proteins need to be expressed in mesophilic hosts. The *P. woesei* α-amylase gene was subcloned in the overexpression vector pET15-b, resulting in pET15-b/P$_{wo}$*amy*, and expressed in *E. coli* BL21 (DE3), where it conferred thermostable amylolytic activity (*7*). This plasmid was used as template DNA for the experiments described below.

The DNA sequence encoding the *P. woesei* α-amylase gene was amplified by PCR from plasmid pET15-b/P$_{wo}$*amy* and subcloned in pHS15. Two different constructs were obtained. In the first case, the α-amylase coding region (P$_{wo}$*amy*) was amplified as a *Pst*I fragment and subcloned in the *Pst*I site of pHS15 at the same orientation as in the case of the expression of *inaZ* (*14*) and *gfp* (*17*), giving the expression plasmid pHS15/P$_{wo}$*amy* (*7*, **Fig. 3**). In the second case, the above coding region along with an upstream 58-bp noncoding region from pET15b containing its ribosome binding site (more appropriate for transcription in *E. coli*), was amplified as a *Pst*I fragment and subcloned in *Pst*I-digested pHS15, resulting in plasmid pHS15/P$_{wo}$*amy* (SD) (*7*).

3.2. Introduction of the Recombinant Plasmids Into MHB

The recombinant constructs were conjugally transferred from *Escherichia coli* DH5α donors to the moderate halophiles by triparental matings (*2*) with pRK2013 as a helper plasmid, as follows:

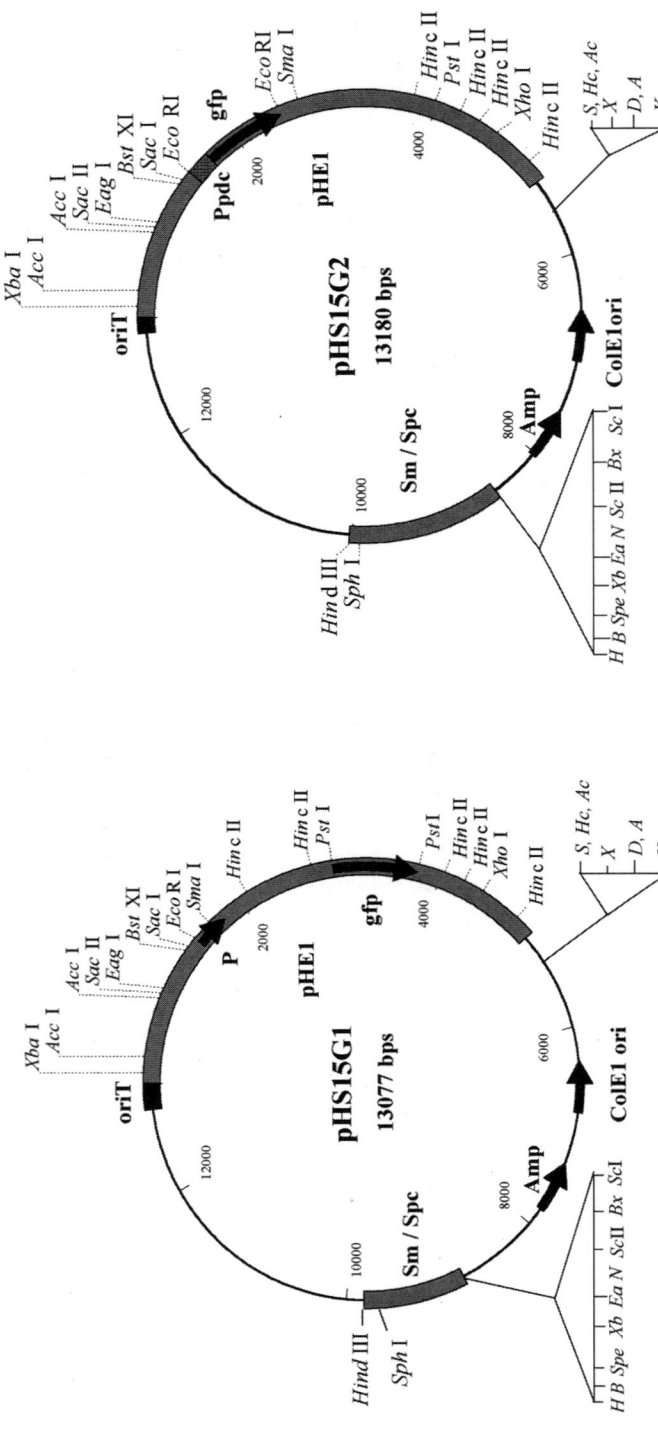

Fig. 2. Plasmids pHS15G1 and pHS15G2. P: pHE1 native promoter. (The above plasmids are reprinted from Douka et al., Use of a green fluorescent protein gene as a reporter in *Zymomonas mobilis* and *Halomonas elongata*, FEMS Microbiology Letters, Vol. 201, No 2, pp. 221–227 (2001), with permission from Elsevier.)

1. Grow 5 mL of liquid cultures of donor (in LB with Sm), helper (in LB with Km), and recipient cells (in SW-2 with Rif) at 37°C with shaking to the middle logarithmic phase (OD_{600}~0.4, *see* **Note 4**).
2. Mix 100-µL aliquots of each donor, helper, and recipient strain into a 1.5-mL Eppendorf tube.
3. Centrifuge at 2500*g* for 5 min (*see* **Note 5**), discard the supernatant, corresponding to the culture media, and wash the cell pellet with 1 mL of SW-2 medium. Centrifuge at 2500*g* for 5 min and discard the supernatant (*see* **Note 6**).
4. Resuspend the mating mixture in 100 µL of SW-2 conjugation medium (*see* **Note 7**). Place a nitrocellulose filter (0.45µm pore diameter) onto a SW-2 agar plate and place the mating mixture on the filter cell surface side up (*see* **Note 8**).
5. Incubate the filter overnight at 30°C.
6. The next day, place the filter in a 10-mL sterile tube and resuspend the mating mixture in 2 mL of SW-2 medium containing 20% glycerol (*see* **Note 9**).
7. Make serial dilutions and plate aliquots of 100 µL of them on SW-2 agar plates with rifampicin (for counterselection of *E. coli* donor and helper) and streptomycin (to select for transconjugants carrying the pHS15-derived recombinant plasmids). Plate 100 µL of each donor, helper, and recipient strains on the same selection media as a control for spontaneous mutations.
8. To estimate the total number of recipient cells, plate appropriate dilutions of the mating mixture onto SW-2 medium with rifampicin. The efficiency of conjugation (expressed as transfer frequency) is given as the number of transconjugants per final number of surviving recipient cells.

3.3. Measurement of Recombinant Protein Activity of the Cells

3.3.1. Measurement of the Ice Nucleation Activity

1. Grow SW-10 (*see* **Note 10**) liquid cultures of *Chromohalobacter* and *Halomonas* recombinant strains carrying *inaZ* at 24°C for 20 h to obtain maximum ice nucleation activity (*see* **Notes 11** and **12**). Grow the corresponding strains carrying the vector alone as negative controls.
2. Make serial 10-fold dilutions up to 10^{-8} from the overnight cultures in sterile distilled water. Use different tip for each dilution.
3. Plate aliquots of 100 µL of dilutions 10^{-5}, 10^{-6}, and 10^{-7} on SW-2 plates containing streptomycin to estimate the number of cells. Incubate plates at 37°C (*see* **Note 13**) overnight. The next day, determine the number of CFU/mL for the calculation of ice nucleation activity.
4. Use the dilutions for the freezing droplet test.

   a. Set a refrigerated circulating bath of ethanol or ethylene glycol (*see* **Note 14**) at −9°C.
   b. Place aluminum foil coated with paraffin on the surface of the bath liquid (*see* **Note 15**).
   c. Take 200 µL of the highest (10^{-8}) dilution. Place this aliquot divided in 20 droplets of about equal volume (10 µL each as possible) on the aluminum foil sheet. Count the number of freezing drops in a time interval not longer than 2 min.
   d. Follow the same procedure for the rest of the dilutions, placing drops from the highest to the lowest dilution.
   e. InaZ activity is quantified by using the equation of Vali (*22*), which takes into account the number of freezing droplets in each dilution, the dilution, and the total number of cells (*see* **Note 16**). Alternatively, the ICE software program (William G. Miller, USDA ARS,

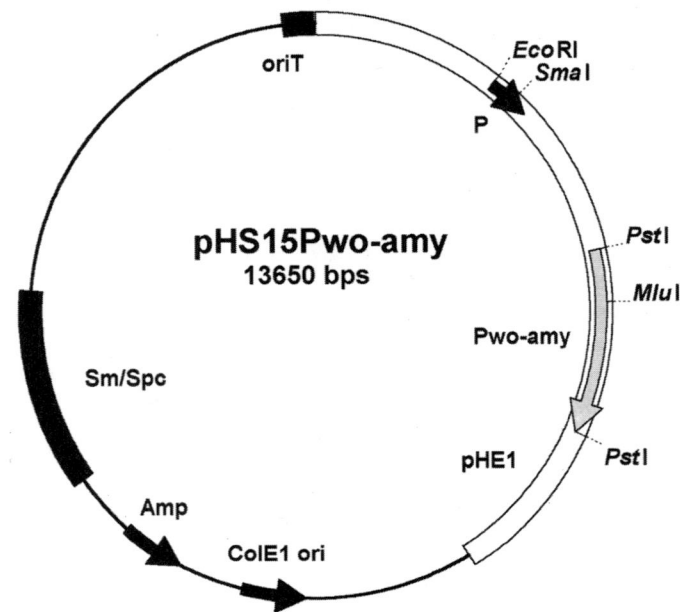

Fig. 3. Plasmid pHS15/*Pwoamy*. P: pHE1 native promoter.

personal communication) can be used to estimate the ice nucleation activity. InaZ activity is usually expressed as the logarithm of ice nuclei per CFU [log(ice nuclei/cell)] (*see* **Note 17**).

Strains of *Halomonas* and *Chromohalobacter* carrying pHS15 have no native ice nucleation activity. High levels of ice nucleation activity were obtained with all recombinant moderately halophilic transconjugants, especially those under the control of the heterologous promoters P_{bla} and P_{pdc}, which could express in moderately halophilic bacteria. In this respect nearly every recombinant cell of the population could act as an active ice nucleus (log [ice nuclei/cell] ≈ zero). An interesting finding here was that the recombinant ice nuclei from moderately halophilic bacteria possessed higher thermostability compared with other bacterial sources (*23*). In conclusion, the high activity of the recombinant ice nucleation protein produced from the recombinant moderately halophilic bacteria makes them an attractive alternative source of biologically "safe" ice nuclei.

3.3.2. Measurement of the GFP Fluorescence

The GFP fluorescence emitted by the transconjugant cells is measured fluorometrically (excitation at 488 nm and emission at 511 nm) and assayed as follows:
Liquid cultures (*24*):

1. Grow 30 mL liquid SW-2, SW-5, or SW-10 cultures of the *Halomonas* and *Chromohalobacter* recombinant strains at 37°C under agitation (*see* **Note 18**) to the end of the exponential

phase of growth (*see* **Note 12**). Grow the corresponding strains carrying the vector alone as negative control.

2. Harvest cells by centrifugation (6000*g*, 10 min).
3. Wash in 5 mL of 10 m*M* Tris-HCl pH 8, 600 m*M* NaCl buffer.
4. Resuspend in 3 mL of the above buffer.
5. Immediately measure the fluorescence in a fluorimeter (Perkin-Elmer LS-3) set to excite the cells at 488 nm and detect emission at 511 nm (*see* **Note 19**).
6. In parallel, use a small aliquot of 30 µL of each culture for protein determination by the Lowry method (*25*).
7. Express the fluorescence value as fluorescence per mg of protein for each sample.
8. Subtract the fluorescence value emitted by the negative control from the fluorescence values of the *gfp* recombinant strains to obtain the fluorescence emitted by the GFP protein.

Colonies grown on agar plates (*26*):

1. Spread appropriate dilutions of *Halomonas* and *Chromohalobacter* recombinant strains liquid cultures at the end of the exponential phase of growth (*see* **Note 12**) on solid SW-2, SW-5, or SW-10 and incubate at 37°C overnight to obtain single colonies. Grow similarly the corresponding strains carrying the vector alone as negative control.
2. After the colonies appear, store the plates with the colonies at 4°C for 28 h (*see* **Note 20**).
3. Scrape off 50–70 colonies from an agar plate for each strain.
4. Resuspend the colonies in 3 mL of the same buffer as above.
5. Follow **steps 5–8** of the above protocol to determinate the fluorescence emitted by the GFP protein.

Transconjugants carrying the *gfp* plasmids showed a high emission at 511 nm, while the fluorescence of those containing pHS15 was undetectable. The expression of *gfp* under the control of the native promoter of plasmid pHS15 appeared to be significantly higher as compared with the expression under the control of the heterologous promoter, P_{pdc} (*17*). It was also shown that the expression of *gfp* from the *mob* promoter of plasmid pHE1 was higher as the concentration of NaCl in the growth medium increased from 2% to 10% in both liquid and solid media. This result suggests that the mobilization functions of pHE1 are osmoregulated at the transcriptional level. The *gfp* plasmids were found to be reasonably stable in the *C. salexigens* transconjugants for at least 100 cell cycles grown under nonselective conditions and retained a similar copy number regardless of the salt concentration. The expression of GFP in the transconjugant cells was also verified by RT-PCR analysis. These results demonstrate that, like *inaZ*, the *gfp* gene can also be used as a reporter in *C. salexigens* in liquid, as well as in solid cultures.

3.3.3. Measurement of α-Amylase Activity

1. Grow *E. coli* and/or *C. salexigens* harboring the α-amylase gene (as well as the strains carrying the vector pHS15 as negative control) up to the late exponential phase (*see* **Note 12**) in 100–150-mL cultures of LB (for *E. coli*) or SW-2 (for *C. salexigens*) medium.
2. Harvest cells by centrifugation (6000*g*, 10 min). Keep the pellet on ice. Wash once with 5 mL of amylase reaction buffer.
3. Resuspend cells in 1 mL of the above buffer.
4. Mix cell suspensions with equal volume of zirconium beads of 0.1 mm diameter and homogenize rapidly in a Mini-BeadBeater cell disrupter.

5. Make a quick spin (10,000g for 1 min) to remove the zirconium beads (*see* **Note 21**).
6. Centrifuge the supernatant at 25000g for 15 min at 4°C to spin out cell debris (*see* **Note 22**).
7. Incubate 20–50 μl of the supernatant in 350 μL of 50 mM sodium acetate pH 5.5 containing 0.72% (w/v) soluble starch at 95°C for 10 min.
8. Terminate the reaction by cooling the mixture in ice water (*see* **Note 23**).
9. Add 50 μL iodine solution (*see* **Note 24**).
10. Measure developed color at 600 nm.
11. Prepare a standard curve of absorbance at 600 nm vs various starch concentrations mixed with iodine solution. The standard curve should consist of readings from tubes containing starch solution at various concentrations from 0 up to 0.72% (w/v) mixed with iodine solution. If the readings are of high absorbance, dilute the starch solution with distilled water.
12. One unit of α-amylase is defined as the enzyme activity required to hydrolyze 1 mg starch in one minute, as determined by the absorbance decrease in the starch-iodine reaction (*see* **Note 25**).

As expected, pHS15/P$_{wo}$*amy* (SD) conferred increased levels of thermostable amylolytic activity in comparison to pHS15/P$_{wo}$*amy* in *E. coli*, but not in *C. salexigens*, where both constructs gave similar levels of activity (*7*). *C. salexigens* cells harboring either recombinant plasmid produced α-amylase that was active in a broad temperature and pH range with maximal values at 90–100°C and pH 5.5–7.0, and a very high thermal stability at temperatures up to 125°C. These properties are comparable to those of the native α-amylase purified from *P. woesei* culture supernatants (*28*). The amylase activity measured at optimal conditions (95°C, pH 5.5) in the recombinant *C. salexigens* strains (0.60–0.65 U/mg protein) was 4-fold lower than that measured in *E. coli* DH5α harboring pHS15/P$_{wo}$*amy* (SD). In the same conditions, relatively low amylase activity (0.74 U/mg protein) is measured with *E. coli* DH5a or BL21(DE3) harboring pHS15/P$_{wo}$*amy*, a construct lacking the typical *E. coli* RBS upstream of the α-amylase sequence (*7*).

Despite the fact that the native enzyme in *P. woesei* is extracellular, the recombinant α-amylase from both *E. coli* and *C. salexigens* expressed by either plasmid was recovered completely in the crude membrane fraction of the cell homogenates. This finding implied that the heterologously expressed amylase remains associated with the cell envelope in an insoluble form and it cannot be processed for secretion to the culture medium. Even treatment of the pellet with an effective detergent (Triton X-100) did not release the enzyme into a soluble form, indicating the formation of protein aggregates (*7,19*). It is possible that factors necessary for the secretion of the pyrococcal α-amylase may be absent from the new hosts. Alternatively, the heterologously expressed enzyme may fold improperly and become trapped in the crude membrane pellet. In any case, the amylolytic activity of either crude preparation or purified enzyme isolated and assayed from this membrane fraction showed catalytic properties almost identical to those of the native enzyme. Nevertheless, it was demonstrated that a moderate halophile can be used at least as well as *E. coli* for the production of functional hyperthermophilic pyrococcal α-amylase in a recombinant form. The transfer and expression of a hyperthermophilic enzyme in *C. salexigens* opens new horizons for the industrial applications of moderately halophilic bacteria.

4. Notes

1. *Halomonas* and *Chromohalobacter* strains are routinely grown in saline media such as SW-10, SW-5, or SW-2, containing 10%, 5%, or 2% (w/v) total salts, respectively. SW media are prepared from SW-30, a salt stock solution made with: 234 g/L NaCl; 39 g/L MgCl$_2$ · 6H$_2$O; 61 g/L · MgSO$_4$ · 7H$_2$O; 1 g/L CaCl$_2$; 6 g/L KCl; 0.2 g/L NaHCO$_3$; and 0.7 g/L NaBr (optional). To prepare SW-30 salt stock, dissolve the components in distilled water with continuous stirring. MgCl$_2$ · 6H$_2$O and CaCl$_2$ should be dissolved apart; otherwise they will precipitate. Once the salts are dissolved, adjust the volume up to 1 L and filter the SW-30 stock solution. To make the SW media, dilute down the SW-30 solution to achieve the selected salinity (for example, to make SW-10, SW-5, or SW-2, dilute 33.33, 16.66, or 6.7 mL, respectively, of SW-30 in 100 mL of distilled water). After this, add 0.5% yeast extract and adjust the pH to 7.2 with KOH. Solid media contain 2% (w/v) agar.
2. Rifampicin is dissolved completely in methanol. The cell viability is not affected by this low amount of methanol added to the culture media. Alternatively, rifampicin can be dissolved in ethanol, but not at concentrations higher than 2 mg/mL.
3. These final concentrations are defined for LB or SW-2 (2% total salts). The susceptibility of moderate halophiles to the aminoglycosides (Km and Sm), but not to Rif or Amp, is greatly affected by the salinity of the medium (*29*). Therefore, the concentration of Km and Sm should be increased accordingly if salinities higher than 2% are used. In contrast, all strains tested are very sensitive to low concentrations of rifampicin. This fact was exploited to isolate and use spontaneous rifampicin mutants of the moderately halophilic strains as recipients for genetic transfer experiments. Most of the strains tested (*C. salexigens, H. subglaciescola, H. eurihalina*, and *H. halophila*) are highly resistant to ampicillin, probably due to the presence of β-lactamases that inactivate the antibiotic regardless of the salt concentration of the culture medium (*29*). Thus, in general, ampicillin is not an appropriate antibiotic to be used as genetic marker for plasmids or transposons to be transferred to these group of moderate halophiles.
4. To achieve the appropriate cell density, it is advisable to inoculate 0.5 mL of *E. coli* donor and helper strains, and recipient *Halomonas* or *Chromohalobacter* overnight cultures, into 5 ml of fresh media added with the corresponding antibiotics and incubate at 37°C under agitation until OD$_{600}$ ~0.4. Alternatively, the growth of *E. coli* cultures to the middle logarithmic phase will be achieved, if liquid cultures of 30 mL are inoculated overnight at 37°C without agitation. The next morning they will be at OD$_{600}$ ~0.4 and ready to use.
5. Low spin is necessary to avoid breakdown of conjugation pili.
6. This wash step is necessary to eliminate the antibiotics.
7. SW-2 (2% total salts) is used as conjugation medium to allow growth of *E. coli*. One alternative conjugation medium is LB supplemented with 2% (instead of the usual 1%) NaCl. Salinities between 2 and 5% NaCl greatly reduce the growth of *E. coli*. Salinities higher than 5% NaCl are inhibitory for *E. coli*.
8. The use of a nitrocellulose filter helps to concentrate the conjugation mix, increasing the conjugation frequency. In addition, it facilitates the recovery of the conjugation mix in step 6. If the conjugation frequency is low, place the filter under vacuum, so that the cells are even more concentrated.
9. The addition of 20% glycerol to the SW-2 medium allows the storage of the conjugation mix at −20°C without loss of cell viability. This can be useful if conjugation frequency is unknown and several dilutions have to be assayed, or just to keep the mix for further experiments.

Alternatively, conjugation mix can be resuspended and stored just in 20% (v/v) sterile glycerol.

10. The salinity of the medium that may potentially interfere with freezing is not affecting nucleation, because it is gradually reduced below the confounding levels by the serial dilutions. Therefore, moderately halophilic strains can be grown in the usual SW-10 medium.

11. In general, temperature plays a key role for the ice nucleation activity and gives a maximum value at growth temperatures $\leq 24°C$. In particular, recombinant *E. coli* carrying *inaZ* does not show any activity when cultured above 24°C. However, growth temperature is not so critical for the ice nucleation activity of the ice$^+$ recombinant moderately halophilic strains. Activity can be detected even in cultures growing above the permissive 24°C and up to 37°C, although at 37°C the activity is slightly reduced.

12. It has been shown for both bacteria that the ice nucleation and α-amylase activity, as well as GFP fluorescence, are higher, when the cells reach the end of the exponential phase. This corresponds to OD_{600} ~0.8 of 1:1 diluted liquid culture of *C. salexigens* ATCC 33174 measured in a Bausch & Lomb spectrophotometer. At this stage the number of cells is ~3 × 10^8/mL of culture.

13. The optimal temperature for growth of *Halomonas* and *Chromohalobacter* strains is 37°C.

14. Both coolants work well. Ethylene glycol is more viscous, but is much cheaper. We recommend concentrated commercial car antifreeze liquid diluted up to 1:2 with distilled water.

15. The aluminium foil surface is prepared as follows: a few racks of a laboratory oven are covered with a medium-weight foil, placed in a hood, and sprayed on the side to be coated with 2% paraffin solution in xylene. After spraying, the foil-covered racks are placed in the oven and heated at 80°C for 1 h, so that the solvent is evaporated and the paraffin is melted on the surface of the sheet. After cooling, the foil sheets are cut to the desired size and stored at room temperature.

16. Equation of Vali (3): $N = \ln[1/(1-f)]/(V_{dr} D_s)$. *N*: concentration of active ice nuclei in the initial solution (i.e., nuclei per cell). $\ln[1/(1-f)]$: calculated nucleation frequency based on a Poisson distribution of freezing events among the collection of droplets. *f*: fraction of frozen droplets (number of droplets that froze / total number of droplets). *N* should be calculated from dilutions in which the fraction of droplets that froze is greater than 0 and less than 1). V_{dr}: volume of each droplet (mL). D_s: factor describing the dilution of the solution from which droplets were tested from the initial solution. The values obtained by this equation are negative. Theoretically, the maximum that can be detected is 1 nucleus per cell. In some cases, however, more than one nucleus per cell is observed, because the cells can retain INA for a short time after they lose viability, or because some bacteria release ice nuclei as membrane blebs in the extracellular medium. The droplets containing these particles will also freeze, even if they do not contain viable cells. This can give positive ice nucleation activity values.

17. Statistics for InaZ. The values were estimated from five independent experiments (mean ± standard deviation). InaZ activity can also be expressed as the logarithm of ice nuclei per mL of culture or per mg of total protein.

18. Measurement of GFP activity requires a posttranslational oxidation of the protein. Therefore, the exposure to aerobic conditions is required (*24*).

19. Statistics for GFP. The values were estimated from four independent experiments (mean ± standard deviation).

20. By this treatment the fluorescence intensity increased significantly (*26*).

21. Zirconium beads are reusable following washing with water and soap and thorough rinsing with deionized water.

22. For amylase assays in the membrane fraction, dilute the cell homogenate tenfold, pellet down membranes by ultracentrifugation at 150,000g for 3 h and resuspend in 1 mL of reaction buffer. Assay as described in **steps 5–10**.
23. By assaying at 95°C, the aqueous solution evaporates. It is necessary to re-estimate the volume after cooling the reaction mixture, by adding distilled water. A good alternative is to use Eppendorf tubes, poking a hole in the cap with a needle.
24. Prepare iodine solution as follows: first, dissolve 1.5 g KI in approx 95 mL of distilled water. Add 0.5 g iodine, dissolve, and adjust the final volume to 100 mL.
25. Statistics for α-amylase. The values were estimated from six independent experiments (mean ± standard deviation).

Acknowledgments

The authors wish to thank W. G. Miller for providing the software program for the calculation of the ice nucleation activity, as well as plasmid pGreenTIR. Research in the laboratory of Joaquín J. Nieto and Carmen Vargas was financially supported by grants from the European Commission INCO-Dev Program (grant ICA4-CT-2000-30041), the Spanish Ministerio de Ciencia y Tecnología (grant BIO2001-2663), and Junta de Andalucía. Research in the laboratory of Constantin Drainas was financially supported by the E. C. Biotechnology Programme "Extremophiles as Cell Factories" (BIO4-CT96-0488) and the Greek Secretariat for Research and Technology (Programs EPET II, 323; PENED 99ED67).

References

1. Ventosa, A., Nieto, J. J., and Oren, A. (1998) Biology of moderately halophilic aerobic bacteria. *Microbiol. Mol. Biol. Rev.* **62,** 504–544.
2. Vargas, C., Fernández-Castillo, R., Cánovas, D., Ventosa, A., and Nieto, J. J. (1995) Isolation of cryptic plasmids from moderately halophilic eubacteria of the genus *Halomonas*. Characterization of a small plasmid from *H. elongata* and its use for shuttle vector construction. *Mol. Gen. Genet.* **246,** 411–418.
3. Baertlein, D. A., Lindow, S. E., Panopoulos, N. J., Lee, S. P., Mindrinos, M. N., and Chen, T. H. H. (1992) Expression of a bacterial ice nucleation gene in plants. *Plant Physiol.* **100,** 1730–1736.
4. Drainas, C., Vartholomatos, G., and Panopoulos, N. J. (1992) The ice nucleation gene from *Pseudomonas syringae* as a sensitive gene reporter for promoter analysis in *Zymomonas mobilis*. *Appl. Environm. Microbiol.* **61,** 273–277.
5. Reynen, M., Reipen, I., Sahm, H., and Sprenger, G. A. (1990) Construction of expression vectors for the gram-negative bacterium *Zymomonas mobilis*. *Mol. Gen. Genet.* **223,** 335–341.
6. Miller, W. G. and Lindow, S. E. (1997) An improved GFP cloning cassette designed for prokaryotic transcriptional fusions. *Gene* **191,** 149–153.
7. Frillingos, S., Linden, A., Niehaus, F., Vargas, C., Nieto, J. J., Ventosa, A., et al. (2000) Cloning and expression of α-amylase from the hyperthermophilic archeon *Pyrococcus woesei* in the moderately halophilic bacterium *Halomonas elongata*. *J. Appl. Microbiol.* **88,** 495–503.
8. Ventosa, A., Nieto, J. J., and Oren, A. (1998) Biology of moderately halophilic aerobic bacteria. *Microbiol. Mol. Biol. Rev.* **62,** 504–544.

9. Vargas, C., Tegos, G., Drainas, C., Ventosa, A., and Nieto, J. J. (1999) Analysis of the replication region of the cryptic plasmid pHE1 from the moderate halophile *Halomonas elongata*. *Mol. Gen. Genet.* **261,** 851–856.

10. Vargas, C., Tegos, G., Vartholomatos, G., Drainas, C., Ventosa, A., and Nieto, J. J. (1999) Genetic organization of the mobilization region of the plasmid pHE1 from *Halomonas elongata*. *Syst. Appl. Microbiol.* **22,** 520–529.

11. Wolber, P. K. (1992) Bacterial ice nucleation. *Adv. Microb. Physiol.* **31,** 203–237.

12. Margaritis, A. and Singh-Bassi, A. (1991) Principles and biotechnological applications of bacterial ice nucleation. *Crit. Rev. Biotechnol.* **11,** 277–295.

13. Panopoulos, N. (1995) Ice nucleation genes as reporters, in *Biological Ice Nucleation and its Applications* (Lee, R. E. Jr., Warren, G. J. and Gusta, L. V., eds.) APS Press, St. Paul, MN, pp. 271–281.

14. Arvanitis, N., Vargas, C., Tegos, G., Perysinakis, A., Nieto, J. J., Ventosa, A., et al. (1995) Development of a gene reporter system in moderately halophilic bacteria by employing the ice nucleation gene of *Pseudomonas syringae*. *Appl. Environ. Microbiol.* **61,** 3821–3825.

15. Chalfie, M., Tu, Y., Euskirchen, G., Ward, W. W., and Prasher, D. C. (1994) Green fluorescence protein as a marker for gene expression. *Science* **263,** 802–805.

16. Cormark, B. P., Valdivia, R. H., and Falkow, S. (1996) FACS-optimized mutants of the green fluorescent protein (GFP). *Gene* **173,** 33–38.

17. Douka, E., Christoyianni, A., Koukkou, A.-I., Afendra, A.-S., and Drainas, C. (2001) Use of a green fluorescent protein gene as a reporter in *Zymomonas mobilis* and *Halomonas elongata*. *FEMS Microbiol. Lett.* **201,** 221–227.

18. Jørgensen, S., Vorgias, C. E., and Antranikian, G. (1997) Cloning, sequencing, characterization, and expression of an extracellular α-amylase from the hyperthermophilic archaeon *Pyrococcus furiosus* in *Escherichia coli* and *Bacillus subtilis*. *J. Biol. Chem.* **268,** 16335–16342.

19. Dong, G., Vieille, C., Savchenko, A., and Zeikus, J. G. (1997) Cloning, sequencing, and expression of the gene encoding extracellular α-amylase from *Pyrococcus furiosus* and biochemical characterization of the recombinant enzyme. *Appl. Environm. Microbiol.* **63,** 3569–3576.

20. Janecek, S. (1994) Sequence similarities and evolutionary relationships of microbial, plant and animal alpha-amylases. *Eur. J. Biochem.* **224,** 519–524.

21. Janecek, S. (1998) Sequence of archeal *Methanococcus jannaschii* alpha-amylase contains features of families 13 and 57 of glycosyl hydrolases: a trace of their common ancestor? *Folia Microbiol. (Praha)* **43,** 123–128.

22. Vali, G. (1971) Quantitative evaluation of experimental results on the heterologous freezing nucleation of supercooled liquids. *J. Atmos. Sci.* **28,** 402–409.

23. Tegos, G., Vargas, C., Perysinakis, A., Koukkou, A. I., Christogianni, A., Nieto J. J., et al. (2000) Release of cell-free ice nuclei from *Halomonas elongata* expressing the ice nucleation gene *inaZ* of *Pseudomonas syringae*. *J. Appl. Microbiol.* **89,** 785–792.

24. de Palencia, P. F., Nieto, C., Acebo, P., Espinosa, M., and Lopez, P. (2000) Expression of green fluorescent protein in *Lactococcus lactis*. *FEMS Microbiol. Lett.* **183,** 229–234.

25. Lowry, O. H., Rosebrough, N. J., Farr, A. L., and Randall, R. J. (1951) Protein measurement with the Folin phenol reagent. *J. Biol. Chem.* **193,** 265–275.

26. Lissemore, J. L., Jankowski, J. T., Thomas, C. B., Mascotti, D. P., and deHaseth, P. L. (2000) Green fluorescent protein as a quantitative reporter of relative promoter activity in *E. coli*. *BioTechniques* **28,** 82–89.

27. Laderman, K., Davis, B. R., Krutzsch, H. C., Lewis, M. S., Griko, Y. V., Privalov, P. L., et al. (1993) The purification and characterization of an extremely thermostable α-amylase from the hyperthermophilic archaeobacterium *Pyrococcus furiosus. J. Biol. Chem.* **268,** 24394–24401.

28. Koch, R., Spreinat, A., Lemke, K., and Antranikian, G. (1991) Purification and properties of a hyperthermoactive α-amylase from the archaeobacterium *Pyrococcus woesei. Arch. Microbiol.* **155,** 572–578.

29. Coronado, M. J., Vargas, C., Kunte, H. J., Galinski, E., Ventosa, A., and Nieto, J. J. (1995) Influence of salt concentration on the susceptibility of moderately halophilic bacteria to antimicrobials and its potential use for genetic transfer studies. *Curr. Microbiol.* **31,** 365–371.

15

Recombinant Protein Production in Antarctic Gram-Negative Bacteria

Angela Duilio, Maria Luisa Tutino, and Gennaro Marino

Summary

This review reports some results from our laboratory on the setting up of a psychrophilic expression system for the homologous/heterologous protein production in cold-adapted bacteria by using natural plasmids as cloning vectors.

By screening some Antarctic bacteria for the presence of extrachromosomal elements, we identified three new plasmids, pMtBL from *Pseudoalteromonas haloplanktis* TAC125, and pTAUp and pTADw, from *Psychrobacter sp.* TA144.

The latter autoreplicating elements were isolated, cloned, and fully sequenced and their molecular characterisation was carried out; however, we focused our attention on the small multicopy plasmid, pMtBL, from the Gram-negative *P. haloplanktis* TAC125 strain. This episome turned out to be an interesting extrachromosomal element, since it displays unique molecular features as its transcriptional inactivity. Being cryptic, the inheritance of pMtBL totally relied on the efficiency of its replication function. This function was bound to a region of about 850 bp, identified by an in vivo assay based on the possibility to efficiently mobilize plasmidic DNA from a mesophilic donor (*Escherichia coli*) to psychrophilic recipient by intergeneric conjugation. This information was instrumental in the construction of a shuttle vector, able to replicate either in *E. coli* or in several cold-adapted hosts (*clone Q*).

Since the conversion of a cloning system into an expression vector requires the insertion of transcription and translation regulative sequences, the corresponding signals from the aspartate aminotransferase gene isolated from *P. haloplanktis* TAC125 were inserted, generating the pFF vector.

To investigate the possibility of obtaining recombinant proteins in this cold-adapted host, we used the psychrophilic α-amylase from the Antarctic bacterium *P. haloplanktis* TAB23 (previously known as *Alteromonas haloplanktis* A23) as a model enzyme to be produced. Our results demonstrate that the cold-adapted enzyme was not only produced but also efficiently secreted by

From: *Methods in Molecular Biology, vol. 267:*
Recombinant Gene Expression: Reviews and Protocols, Second Edition
Edited by: P. Balbás and A. Lorence © Humana Press Inc., Totowa, NJ

the recombinant *Ph*TAC125 cells. The described expression system represents the first example of heterologous protein production based on a true cold-adapted replicon.

Key Words: Cold-adapted bacteria; plasmid; shuttle vector; gene expression; *Pseudoalteromonas haloplanktis*; conjugation.

1. Introduction

The development of recombinant DNA techniques opened a new era in protein production for both basic research and industrial applications. Recombinant protein production in transformed microorganisms is by far the simplest genetic approach to guarantee an unlimited supply of rare, high-value proteins that are not accessible using conventional protein isolation techniques.

The Gram-negative bacterium *Escherichia coli* is usually the host of choice for the recombinant protein production; there are many reasons that justify this large success. *E. coli* physiology and metabolism have been extensively characterized, so that its ability to grow rapidly and at high cell density could be achieved easily even using simple and quite inexpensive broths. Furthermore, there are many genetic tools readily available for its manipulation. In fact, an increasingly large number of cloning vectors, strong and/or regulated transcriptional promoters, and mutant host strains have been selected in the past years, with the final result of a large availability of efficient and versatile gene expression systems. As a result, the synthesis of heterologous polypeptides in *E. coli* has become a matter of routine.

However, to efficiently produce active biomolecules, a high rate of protein synthesis is necessary, but it is by no means sufficient; sometimes some heterologous proteins are extremely toxic for *E. coli* and in such cases, the induction of a heterologous gene is shortly followed by cessation of protein synthesis. If a newly synthesized polypeptide is recognized by the proteolytic machinery of the cell, its net or steady-state accumulation will be very low because of degradation. Proteolysis of heterologous proteins in bacteria is conceptually like a primitive immune system, and as such deserves far more experimental work; the mechanisms by which *E. coli* recognizes "self" are completely unknown and no generic methods exist for blocking all host proteases.

One of the main problems often occurring during the heterologous protein production in bacteria is undoubtedly the incorrect folding of the nascent polypeptides, resulting in their aggregation and accumulation as insoluble inclusion bodies. The study of inclusion-body formation highlighted the role played by hydrophobic interaction as mainly responsible of driving partially folded protein intermediates to stick and subtracting them from the productive folding pathway (*1,2*).

The misfolding and aggregation of polypeptide chains into insoluble bodies is a serious problem in both biotechnology and basic research; several experimental approaches have been developed to prevent protein precipitation (*3*) or to recover the protein products from the inclusion bodies in an active form (*4*) (*see* Chapter 3). Furthermore, many attempts have been made to produce recombinant proteins in a soluble form by coexpression of "chaperones," which assist folding by preventing the aggregation. Despite numerous experimental efforts, positive results are scarce although in certain cases, a positive effect of coexpression of molecular chaperones or other folding enhancers such as thiol-disulfide-isomerases or peptidyl-prolyl-isomerases has been reported (*5–7*). More recently, the influence of various fusion protein partners for the expression of recombinant proteins in soluble form in *E. coli* has been studied. The choice of protein partners is based on their favorable cytoplasmic solubility characteristics as predicted by a statistical solubility model. Examples of fusion partners that have been touted as solubilizing agents include thioredoxin, gluthatione S-transferase, maltose-binding protein, Protein A, ubiquitin, and DsbA. Although widely recognized and potentially of great importance, this solubilizing effect remains poorly understood. In particular it is not known whether the solubility of many different polypeptides can be improved by fusing them to a highly soluble partner or whether this approach is only effective in a small fraction of cases (*8*).

To minimize the inclusion bodies' formation, many experimental approaches have been explored with some success (*3*), one of which consists in lowering the expression temperature to the physiological limit allowed for the growth of mesophilic hosts (15–18°C for *E. coli*). In fact, lowering the temperature has a pleiotropic consequence on the folding processes, minimizing the so-called hydrophobic effect. Although in some cases this approach has been reported to enhance the yield of recombinant protein production in soluble and active form (*5*), the exploitation of an industrial process implying a suboptimal growth of the expression host can be considered quite uneconomic.

Starting from the above considerations, the use of naturally "cold-adapted" bacteria as hosts for protein production at low temperature (even at around 0°C) might represent a rational alternative to expression in mesophiles. However, since very little information is still available on genetic elements and manipulation procedures for these hosts, this work started with the analysis of the distribution of extrachromosomal elements in a collection of Antarctic bacteria isolated in the surrounding of the French Antarctic station Dumont d'Urville in Terre Adélie (66°40' S, 140°01' E), kindly provided by C. Gerday (University of Liege, Belgium). Several isolates were subjected to a genomic analysis three new plasmids were identified: pMtBL from *Pseudoalteromonas haloplanktis* TAC125, and pTAUp and pTADw, from *Psychrobacter sp.* TA144 (*9*).

These autoreplicating elements were isolated, cloned and fully sequenced and their molecular characterisation was carried out (*10–12*). It turned out that pMtBL plasmid, harbored by the TAC125 strain, is a medium-copy-number element (about 50 copies/cell), whereas pTA plasmids from TA144 are less represented. Due to the usefulness coming from a higher copy number, pMtBL has been the first "cold" genetic element characterized to a good extent; however, the preliminary characterization of the pTA plasmids system opened the way to investigate the "cold adaptation" of a different DNA replication mechanism, the rolling circle (*11,12*).

Before starting the molecular characterization of pMtBL plasmid, some aspects of TAC125 biology were characterised. It is a Gram-negative pink-pigmented bacterium, classified as a *P. haloplanktis* species and isolated from Antarctic seawater. The dependence of TAC125 growth from incubation temperature was analyzed. Although this isolate comes from a permanently cold environment, characterized by an average temperature of −1.8°C, its growth was checked in a wide temperature range, 4–25°C, and the lowest observed doubling time was at 20°C (**9**). The above results suggest that although this Antarctic isolate was originally selected during a screening for true psychrophiles, it can be properly defined a psychrotolerant strain, according to Morita and Russell definitions (**13,14**), since it thrives either at very low temperature or in quite "mesophilic" conditions ($T_{opt} > 15°C$).

As for the molecular characterization of the pMtBL plasmid, its complete nucleotide sequence was determined and directly compared to the entries stored in the GeneBank/EMBL database. However, no significant results were obtained, as none of the sequences stored in the database appeared to be related to the TAC125 plasmid. Furthermore, although the pMtBL sequence (4081 bp) contains several putative or "likely" open reading frames, none of them possesses a homolog in the database, a result in agreement with the observation that the plasmid appears to be transcriptional silent in the tested conditions. These considerations lead to the fascinating hypothesis that the above psychrophilic plasmid represents a sort of genetic relic, i.e., the remains of a larger element possibly mobilized in *Ph*TAC125 by horizontal gene transfer and fated to be functionally mute. Being cryptic, the inheritance of pMtBL totally relied on the efficiency of its replication functions. Due to the putative absence of a pMtBL-coded incompatibility factor, we embarked on setting up an in vivo experiment for the determination of the minimal plasmid sequence responsible for its replication (ARS) (**10**).

Between the conventional techniques generally used for the transfer of genetic materials in microorganisms, conjugation appears to be the less influenced by superficial bacterial properties. Therefore, we developed a suitable conjugation protocol to transform these cells with the desired recombinant vectors. For these reasons an *E. coli* conjugative plasmid was constructed by cloning into a commercial cloning vector the conjugational DNA transfer origin (OriT) from the broad host range plasmid pJB3 (kindly provided by Dr. S. Valla, Trondheim, Norway) (**15**), an RK2 derivative plasmid. The resulting vector, pGEM-T, was successfully mobilised between two *E. coli* strains using the S 17–1(λ*pir*) strain as donor (**16**). On this ground, a conjugation protocol for the cold bacteria transformation was defined (*see* **Subheadings 2.** and **3.**) requiring the intergeneric mobilization of the recombinant DNA by an *E. coli* donor strain.

A pMtBL statistic library was prepared in it and mobilized into TAC125 via conjugation. The analysis of the pMtBL fragments carried by the psychrophilic ampicillin-resistant conjugants lead us to identify the plasmidic ARS. This function can be bound to a region of about 850 bp, the pMtBL ARS, that was defined as the smallest portion of the psychrophilic plasmid able to actively promote the duplication of an *E. coli* vector (unable to replicate by itself) in the cold-adapted host *Ph*TAC125. This novel cold-adapted replication element was used as a tool for the construction of a shuttle vector (*clone Q*) able to replicate either in *E. coli* or in cold-adapted hosts (**10**).

Interestingly, no incompatibility but a simple competition was observed between the endogenous wild-type plasmid and the pMtBL-derived shuttle vector (*clone Q*), actually carrying the selection resistance gene (Amp^R). It might be that other functions, possibly related to a specific plasmid partitioning mechanism, are responsible of the stable pMtBL inheritance at a very low copy number in *Ph*TAC125/*clone Q* transconjugants.

The ability of this cloning vector to replicate in several psychrophilic hosts was analyzed and it resulted in quite broad host range systems, since the number of Antarctic bacteria in which they are able to replicate is steadily increasing. No rearrangements or modifications of the shuttle vector were observed after transfer from *E. coli* to psychrophilic cells and back.

Since the conversion of a cloning system into an expression vector requires the insertion of transcription and translation regulative sequences, the corresponding signals from the aspartate aminotransferase gene isolated from *P. haloplanktis* TAC125 were inserted (*17*), generating the pFF vector (Fig. 1), the first gene expression vector for cold-adapted bacteria (*18*).

To investigate the possibility of obtaining recombinant proteins from these cold-adapted hosts, we used the psychrophilic a-amylase from the Antarctic bacterium *P. haloplanktis* A23 as model enzyme to be produced. This enzyme was chosen for several reasons: (1) it is heat-labile; (2) the psychrophilic α-amylase is a secreted protein; and (3) the selected host *P. haloplanktis* TAC125 lacks the endogenous activity, although it belongs to the same species as the source strain of the amylase. Our results demonstrate that the cold-adapted enzyme was not only overproduced (about 20 mg/L at 15°C) but also was efficiently secreted by the recombinant *Ph*TAC125 cells (*18*).

The complete forthcoming elucidation of the secretion mechanism will give us the possibility to set up new expression systems for the production and secretion of any recombinant protein at low temperatures.

In order to enhance the efficiency and versatility of such expression systems, we isolated and molecularly characterized some cold protein folding factors (a GroEL/ES-like system and a thioredoxin) (*19*). Their corresponding genes were isolated and characterized, and the effect of their coproduction on the yield of the cold expression system is on the way to be evaluated.

In conclusion, the described expression system represents the first example of heterologous protein production based on a true cold-adapted replicon. While the α-amylase production from the pMtBL-derived cold expression system was significantly lower (about one order of magnitude) than that obtained in *E. coli* at 18°C, it is easily foreseen that optimization of several parameters, such as the regulation and efficiency of transcription and translation initiation, can considerably improve the efficiency of the system. The inherent replicative stability of the pMtBL-derived replicon in cold-adapted hosts is likely to be of great value in the development of stable expression systems for high-level production of thermal labile proteins at temperatures as low as 4°C.

2. Materials

2.1. Bacterial Strains

Pseudoalteromonas haloplanktis TAC125 and *P. haloplanktis* TAB23 were kindly provided by C. Gerday, University of Liege, Belgium. The strains were isolated from the seawater surrounding the Dumont d'Urville Antarctic station (66°40' S, 40°01' E) during the 1988 summer 1988 campaign of the Expeditions Polaires Françaises; in Terre Adélie.

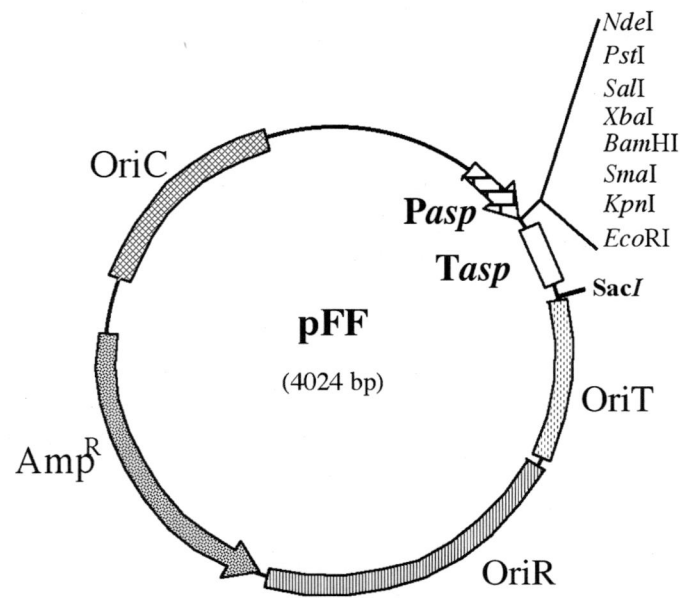

Fig. 1. Map of the pFF vector.

E. coli DH5α [*supE*44, Δ*lac*U169 (φ80 *lacZ*ΔM15) *hsdR*17, *recA*1, *endA*1, *gyrA*96, *thi*-1, *relA*1] was used as host for the gene cloning.

E. coli strain S17-1(λ*pir*) [*thi, pro, hsd* (r⁻ m⁺) *recA*::RP4-2-TCʳ::Mu Kmʳ::Tn7 Tpʳ Smʳ λ*pir*] (**16**) was used as donor in intergeneric conjugation experiments.

2.2. Liquid Media

1. LB medium (**20**) (per liter): To 950 mL of deionized H_2O, add Bacto-Tryptone (10 g), Bacto-yeast extract (5 g), NaCl (10 g) (**20**). Shake until the solutes have dissolved. Adjust the pH to 7.5 with 5 *N* NaOH (10.2 mL). Adjust the volume of the solution to 1L with deionized H_2O. Sterilize by autoclaving for 20 min at 1 atm on liquid cycle.
2. LB medium modified for *E. coli* competent cell preparation: Prepare liquid media according to the recipes given above (item 1). Allow the solution to cool to 40°C or less, and then add a sterile solution of 2 *M* $MgCl_2$ (5 mL) and 0.2 *M* $MgSO_4$ (112 mL).
3. TYP medium (per liter): To 950 mL of deionized H_2O, add Bacto-Tryptone (16 g), Bacto-yeast extract (16 g), Marine mix (10 g). Shake until the solutes have dissolved. Adjust the pH to 7.5 with 5 *N* NaOH (10.2 mL). Adjust the volume of the solution to 1 L with deionized H_2O. Sterilize by autoclaving for 20 min at 1 atm on liquid cycle (**11**).
4. Terrific medium (per liter): To 900 mL of deionized H_2O, add Bacto-tryptone (12 g), Bacto-yeast extract (24 g), glycerol (4 g). Shake until the solutes have dissolved and sterilize by autoclaving for 20 min at 1 atm on liquid cycle. Allow the solution to cool to 60°C or less, and then add 100 mL of a sterile solution of 0.17 *M* KH_2PO_4, 0.72 *M* K_2HPO_4. (This solution is made by dissolving 2.31 g of KH_2PO_4 and 12.54 g of K_2HPO_4 in 90 mL of deionized H_2O.) After the salts have dissolved adjust the volume of the solution to 100 mL with deionized H_2O and sterilize by autoclaving for 20 min at 1 atm on liquid cycle (**20**).

2.3. Media Containing Agar

1. Prepare liquid media according to the recipes given above.
2. Just before autoclaving, add Bacto-agar (15 g/L). Sterilize by autoclaving for 20 min at 1 atm on liquid cycle.
3. When the medium is removed from the autoclave, swirl it gently to distribute the melted agar evenly throughout the solution. The fluid may be superheated and may boil over when swirled.
4. Allow the medium to cool to 50°C before adding thermolabile substances (e.g., antibiotics).
5. To avoid producing air bubbles, mix the medium by swirling. Plates can than be poured directly from the flask; allow about 30–35 mL of medium per 90-mm plate.
6. When the medium has hardened completely, invert the plates and store them at 4°C until needed. The plates should be removed from storage 1–2 h before they are used. If the plates are fresh, they will "sweat" when incubated at 37°C. This allows bacterial colonies to spread across the surface of the plates and increases the chances of cross-contamination. This problem can be avoided by wiping off any condensation from the lids of the plates and then incubating the plates for several hours at 37°C in an inverted position before they are used. Alternatively, the liquid can be removed by shaking the lid with a single quick motion. To minimize the possibility of contamination, hold the open plate in an inverted position while removing the liquid from the lid.

2.4. Storage Medium

1. Bacteria can be stored indefinitely in cultures containing glycerol at low temperature (from −20 to −70°C).
2. To store bacteria, pick a single, well-isolated colony growing on the surface of an agar plate. With a sterile inoculating needle, dip the needle into the liquid broth and grow the bacterial culture up to medium-late exponential phase.
3. To 0.85 mL of bacterial culture, add 0.15 mL of sterile glycerol (sterilize by autoclaving for 20 min at 1 atm on liquid cycle). Vortex the culture to ensure that the glycerol is evenly dispersed.
4. Transfer the culture to a labeled storage tube, freeze the culture in ethanol-dry ice or in liquid nitrogen, and then transfer the tube to −70°C for long term storage.

2.5. CaCl$_2$ Sterile Solution

Prepare a solution of 0.1 M CaCl$_2$ and move on sterilization by filtration through a 0.22-μm filter.

2.6. E. coli Competent Cells

1. Prepare liquid LB media supplemented with 20 mM MgSO$_4$ (*see* **Subheading 2.2.2.**).
2. The *E. coli* cells, strains DH5a and S17-1 (λ*pir*), were made competent to DNA transformation, using the CaCl$_2$ procedure(*see* **Subheading 3.4.1**).
3. Aliquots of the competent bacteria can be stored for several months at −70°C without any significant decrease in efficiency.

2.7. Solutions for Plasmid Miniprep Extraction

1. Solution I: 137 mM sucrose, 1 mM HEPES (pH 8.0), 10% glycerol.
2. Solution II: 0.2 N NaOH (freshly diluted from a 10 N stock) 1% SDS.

3. Solution III: 5 *M* potassium acetate (60 mL), glacial acetic acid (11.5 mL), H_2O (28.5 mL). The resulting solution is 3 *M* with respect to potassium and 5 *M* with respect to acetate.
4. TE 1X: 10 mM Tris-HCl, pH 8.0, 1 m*M* EDTA pH 8.0.
5. TE 1X/ RNAse: 1.5 mL TE 1X, 20 µL RNAse A (10 mg/mL), 5 µL RNAse T1 (1.38 mg/mL).

3. Methods

3.1. Intergeneric Conjugation

3.1.1. Growth of Bacterial Cells

1. To recover the bacteria from the frozen culture, scrape the surface of the culture with a sterile inoculating needle, and then immediately streak the bacteria that adhere to the needle onto the surface of an LB-agar plate containing the appropriate antibiotics, if required.
2. Incubate the plate at appropriate temperature.
3. Transfer one well-isolated colony into 5 mL of suitable broth containing, if transformed, the appropriate antibiotic in a loosely capped 15-mL tube.
4. Incubate the culture under continuous rotary shaking to the adequate temperature up to a medium-late exponential phase.
5. All the cold-adapted strains are Gram-negative and can grow in the above aerobic conditions at 4°C or 15°C in TYP medium or LB medium at pH 7.5, supplemented with 200 µg/mL ampicillin if transformed.
6. *E. coli* cells can be routinely grown in Terrific broth containing 100 µg/mL of ampicillin if transformed.

3.1.2. Preparation of Fresh or Frozen Competent Cells With CaCl₂

The following procedure by Hanahan (*21*) can yield competent cultures of several of *E. coli* strains that can be transformed at frequencies $\geq 5 \times 10^8$ transformed colonies per microgram of supercoiled plasmid DNA.

1. Using a sterile wire, streak *E. coli* strain directly from a frozen stock (stored at −70°C in freezing medium) onto the surface of an LB agar plate. Incubate the plate for 16 h at 37°C.
2. Transfer one well-isolated colony into 1 mL of LB broth containing 20 mM $MgSO_4$. The colonies should be 1–2 mm in diameter. Disperse the bacteria by vortexing at moderate speed, and then dilute the culture in 30–100 mL of LB containing 20 mM $MgSO_4$ in a 1-L flask.
3. Grow the cells for 2.5–3 h at 37°C. For efficient transformation, it is essential that the number of viable cells should not exceed 10^8 cells/mL. To monitor the growth of cultures, determine the OD_{600} every 20–30 min.
4. Aseptically transfer the cells to sterile, disposable, ice-cold 50-mL polypropylene tubes. Cool the cultures to 0°C by storing the tubes on ice for 10 min. All subsequent steps in this procedure should be carried out aseptically.
5. Recover the cells by centrifugation at 4000 rpm for 10 min at 4°C in a Sorvall GS3 rotor (or its equivalent).
6. Decant the media from the cell pellets. Stand the tubes in an inverted position for 1 min to allow the last traces of media to drain away.
7. Resuspend the pellets, by gentle vortexing, in approx 20 mL (per 50-mL tube) of ice-cold transformation buffer (0.1 *M* $CaCl_2$). Store the resuspended cells on ice for 10 min.

8. Recover the cells by centrifugation at 4000 rpm for 10 min at 4°C in a Sorvall GS3 rotor (or its equivalent).
9. Decant the buffer from the cell pellets. Stand the tubes in an inverted position for 1 min to allow the last traces of buffer to drain away.
10. Resuspend the pellets by gentle vortexing in 2 mL (per 50-mL tube) of ice-cold 0.1 M CaCl$_2$. Store the resuspended cells on ice.
11. At this point, the cells can be dispensed into aliquots that can be frozen at −70°C after addition of sterile glycerol (by autoclaving for 20 min at 1 atm on liquid cycle) at a final concentration of 15%. The cells maintain competency under these conditions, although the transformation efficiency may drop slightly during prolonged storage.
12. The cells may be stored at 4°C in CaCl$_2$ solution for 24–28 h. The efficiency of transformation increases four- to sixfold during the first 12–24 h of storage and then decreases to the original level (*22*).

3.1.3. E. coli *Cell Transformation*

1. Using a chilled, sterile pipett tip, transfer 200 µL of each suspension of competent cells to a sterile microfuge tube. Add DNA (no more than 50 ng in a volume of 10 µL or less) to each tube. Mix the contents of the tubes by swirling gently. Store the tubes on ice for at least 30 min. Be sure to include the following controls in your experiment: (a) Competent bacteria that receive a known amount of the standard preparation of supercoiled plasmid DNA; (b) Competent bacteria that receive no plasmid DNA at all.
2. Transfer the tubes to a rack placed in a circulating water bath that has preheated to 42°C. Leave the tubes in the rack for exactly 90 s. Do not shake the tubes.
3. Rapidly transfer the tubes to an ice bath. Allow the cells to chill for 1–2 min.
4. Add 800 µL of LB broth to each tube. Incubate the cultures for 45 min in a water bath set at 37°C to allow the bacteria to recover and to express the antibiotic resistance marker by the plasmid.
5. Transfer the appropriate volume (up to 200 µL per 90-mm plate) of transformed competent cells onto agar LB medium containing 20 mM MgSO$_4$ and the appropriate antibiotic. Add additional broth if small volumes of culture (<10 µL) are transferred. Using a sterile bent glass rod, gently spread the transformed cells over the surface of the agar plate. If more than 200 µL of transformed competent cells are to be plated on a single 90-mm plate, the cells should be concentrated by centrifugation and gently resuspended in an appropriate volume of LB.
6. Leave the plates at room temperature until the liquid has been absorbed.
7. Invert the plates and incubate at 37°C. Colonies should appear in 12–16 h (*see* **Note 1**).

3.1.4. *Intergeneric Mating*

1. Transform the competent *E. coli* S17-1(λpir) cells (donor strain) with the suitable plasmid.
2. Transfer one well-isolated colony into 3 mL of LB broth containing 100 µg/mL of ampicillin in a loosely capped 15-mL tube. Grow the cell in aerobic conditions at 37°C for 16–18 h.
3. Grow one well-isolated colony of the recipient cold-adapted strain, *Ph*TAC125, in 3 mL of TYP medium at pH 7.5 at 15°C for 24–36 h.
4. Mix 100 µL of logarithmic cultures of the donor and the recipient strains into a sterile microfuge tube.
5. Spot, as two drop, onto a TYP plate, the mixture of the cell (donor and recipient).
6. Leave the plate at room temperature until the liquid has been absorbed.
7. Invert the plate and incubate at 15°C for 16–18 h.

3.1.5. Selection of Transformed Cold-Adapted Bacteria

1. Recover the cell of a drop into 100 μL of TYP medium.
2. Psychrophilic transconjugants were selected by plating serial dilutions at 4°C on TYP plates containing 100 μg/mL ampicillin. The mesophilic donors are unable to grow as a colony at 4°C.

3.1.6. Small-Scale Preparations of Plasmid DNA From Transformed Cold-Adapted Bacteria

3.1.6.1. HARVESTING BACTERIA

1. Transfer a single bacterial colony into 2 mL of suitable medium containing the appropriate antibiotic in a loosely capped 15-mL tube. Incubate the culture under continuous rotary shaking to the adequate temperature up to a medium-late exponential phase.
2. Pour 1.5 mL of the culture into a microfuge tube. Centrifuge at 12,000g for 30 s at 4°C in a microfuge. Store the remainder of the culture at 4°C.
3. Remove the medium by aspiration, leaving the bacterial pellet as dry as possible.

3.1.6.2. BACTERIAL LYSIS

1. Resuspend the bacterial pellet (obtained in **step 3** above) in 100 μL of ice-cold Solution I (*see* **Subheading 2.7.**) by vigorous vortexing.
2. Add 200 μL of freshly prepared Solution II. Close the tube tightly, and mix the contents by inverting the tube rapidly several times. Make sure that the entire surface of the tube comes in contact with Solution II. Do not vortex. Store the tube on ice.
3. Add 150 μL of ice-cold Solution III. Close the tube and vortex it gently in an inverted position for 10 s to disperse Solution III through the viscous bacterial lysate. Store the tube on ice for 3–5 min.
4. Centrifuge at 12,000 rpm for 5 min at 4°C in a microfuge. Transfer the supernatant to a fresh tube.
5. Add an equal volume of phenol:chloroform. Mix by vortexing. After centrifuging at 12,000g for 2 min at 4°C in a microfuge, transfer the supernatant to a fresh tube.
6. Precipitate the double-stranded DNA with 2 vol of ethanol at room temperature. Mix by vortexing. Allow the mixture to stand for 2 min at room temperature.
7. Centrifuge at 12,000g for 5 min at 4°C in a microfuge.
8. Remove the supernatant by gentle aspiration. Remove any drops of fluid adhering to the walls of the tube.
9. Rinse the pellet of double-stranded DNA with 1 mL of 70% ethanol at 4°C. Remove the supernatant as described in **step 8** and allow the pellet of nucleic acid to dry in the air for 10 min.
10. Dissolve the nucleic acids in 50 μL of TE (pH 8.0) containing DNAase-free pancreatic RNAase (20 μg/mL). Vortex briefly. Store the DNA at −20°C.

3.2. Recombinant α-Amylase Protein Production

1. The gene coding for the psychrophilic α-amylase isolated from *P. haloplanktis* TAB23 was cloned into the pFF vector (**Fig. 1**), under the transcriptional control of the constitutive *aspC* gene promoter (*18*).

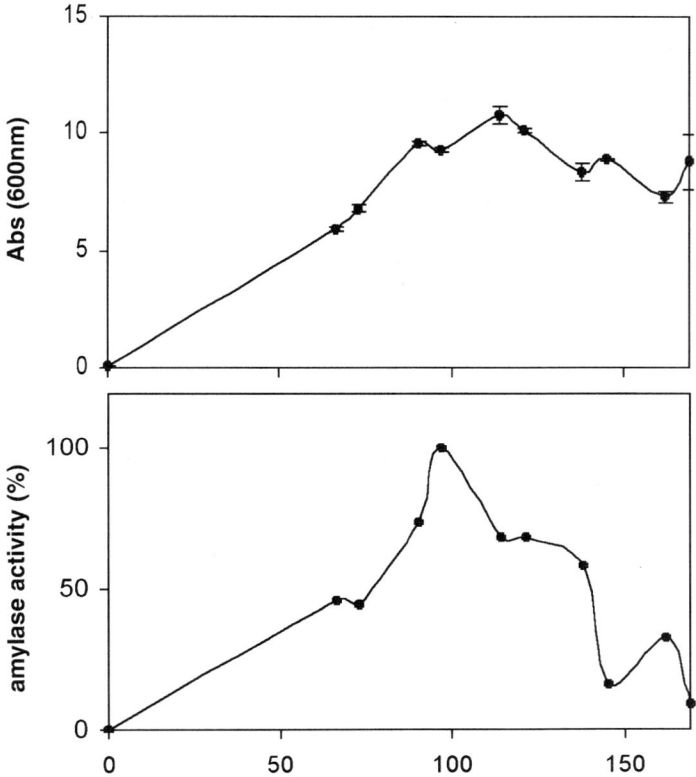

Fig. 2. Production of recombinant α-amylase in *Ph*TAC125 cells at 15°C. Kinetics of bacterial growth (upper panel) and recombinant α-amylase activity (lower panel) in the cell-free supernatant of *Ph*TAC125 cells transformed with pFFamy are presented. Enzyme activity is expressed as percentage of the maximal activity recorded in the cell-free supernatants. The curves were obtained as average results of three independent experiments at 15°C.

2. The resulting vector, pFFamy, was mobilized into *Ph*TAC125 cells by intergeneric conjugation (*see* **Subheading 3.1.**) and psychrophilic transconjugants were grown in liquid medium (as described in **Subheading 3.1.1.**) at 15°C, conditions in which the highest yields of recombinant product are obtained.
3. The growth was ended when the culture reached the late exponential phase, corresponding to a value of optical density of 10–11 O.D.$_{600nm}$, and the production and secretion of recombinant α-amylase was calculated determining the amount of amylase activity in the culture supernatant (**Fig. 2**) using the Boehringer-Roche kit AMYL following the manufacturer instruction.

4. Notes

1. Three factors appear to be critical for the achievement of consistently high frequencies of transformation (defined as the ratio of the number of transformed bacteria per μg of DNA used):

- The purity of the reagents used in the transformation buffers is very critical; in fact, it is important to use reagents of highest quality.
- The cellular viability is decisive; the highest frequencies of transformation are obtained with cultures that have been grown directly from a master stock stored in freezing medium at − 70°C.
- The cleanliness of the glassware and plasticware is crucial because the presence of trace amounts of detergent or other chemicals greatly reduces the efficiency of bacterial transformation. It is best to set set aside a batch of glassware that is used for no other purpose than to prepare competent bacteria.

References

1. Speed, M. A., Wang, D. I., and King, J. (1996) Specific aggregation of partially folded polypeptide chains: the molecular basis of inclusion body composition. *Nat. Biotechnol.* **14**, 1283–1287.
2. Georgiou, G., Valax, P., Ostermeier, M., and Horowitz, P. M. (1994) Folding and aggregation of TEM beta-lactamase: analogies with the formation of inclusion bodies in *Escherichia coli*. *Protein Sci.* **3**, 1953–1960.
3. Baneyx, F. (1999) Recombinant protein expression in *Escherichia coli. Curr. Opin. Biotechnol.* **10**, 411–421.
4. Rudolph, R. and Lilie, H. (1996) *In vitro* folding of inclusion body proteins. *FASEB J.* **10**, 49–56.
5. Georgiou, G. and Valax, P. (1996) Expression of correctly folded proteins in *Escherichia coli. Curr. Opin. Biotechnol.* **7**, 190–197.
6. Jakob, U., Gaestel, M., Engel, K., and Buchner, J. (1993) Small heat shock proteins are molecular chaperones. *J. Biol. Chem.* **268**, 1517–1520.
7. Schwarz, E., Lilie, H., and Rudolph, R. (1996) The effect of molecular chaperones on in vivo and in vitro folding processes. *Biol. Chem.* **377**, 411–416.
8. Schaffner, J., Winter, J., Rudolph, R., and Schwarz, E. (2001) Cosecretion of chaperones and low-molecular-size medium additives increases the yield of recombinant disulfide-bridged proteins. *Appl. Environ. Microbiol.* **67**, 3994–4000.
9. Tutino, M. L., Duilio, A., Fontanella, B., Moretti, M. A., Sannia, G., and Marino, G. (1999) Plasmids from Antarctic bacteria, in *Cold-Adapted Organisms: Ecology, Physiology, Enzymology and Molecular Biology* (Margesin, R. and Schinner, F. eds.), Springer Verlag, Berlin Heidelberg, pp. 335–348.
10. Tutino, M. L., Duilio, A., Parrilli, E., Remaut, E., Sannia, G., and Marino, G. (2001) A novel replication element from an Antarctic plasmid as a tool for the expression of proteins at low temperature. *Extremophiles* **5**, 257–264.
11. Tutino, M. L., Duilio, A., Moretti, M. A., Sannia, G., and Marino, G. (2000) A rolling-circle plasmid from *Psychrobacter sp.* TA144: evidence for a novel Rep subfamily. *Biochem. Biophys. Res. Commun.* **274**, 488–495.
12. Duilio, A., Tutino, M. L., Matafora, V., Sannia, G., and Marino, G. (2001) Molecular characterisation of a recombinant replication protein (Rep) from the Antarctic bacterium *Psychrobacter sp.* TA144. *FEMS Microbiol. Lett.* **198**, 49–55.
13. Morita, R. Y. (1975) Psychrophilic bacteria. *Bacteriol. Rev.* **39**, 144–167.
14. Russell, N. J. (1992) Physiology and molecular biology of psychrophilic micro-organisms, in *Molecular Biology and Biotechnology of extremophiles* (Herbet, R. A. and Sharp, R. J., eds.), Chapman & Hall, New York, NY, pp. 203–221.

15. Blatny, J. M., Brautaset, T., Winther-Larsen, H. C., Haugan, K., and Valla, S. (1997) Construction and use of a versatile set of broad-host-range cloning and expression vectors based on the RK2 replicon. *Appl. Environ. Microbiol.* **63**, 370–379.

16. Tascon, R. I., Rodriguez-Ferri, E. F., Gutierrez-Martin, C. B., Rodriguez-Barbosa, I., Berche, P., and Vazquez-Boland, J. A. (1993) Transposon mutagenesis in Actinobacillus pleuropneumoniae with a Tn10 derivative. *J. Bacteriol.* **175**, 5717–5722.

17. Birolo, L., Tutino, M. L., Fontanella, B., Gerday, C., Mainolfi, K., Pascarella, S., et al. (2000) Aspartate aminotransferase from the Antarctic bacterium *Pseudoalteromonas haloplanktis* TAC125. Cloning, expression, properties, and molecular modelling. *Eur. J. Biochem.* **267**, 2790–2802.

18. Tutino, M. L., Parrilli, E., Giaquinto, L., Duilio, A., Sannia, G., Feller, G., et al. (2002) Secretion of α-amylase from *Pseudoalteromonas haloplanktis* TAB23: two different pathways in different hosts. *J. Bacteriol.* **184**, 5814–5817.

19. Tosco, A., Birolo, L., Madonna, S., Lolli, G., Sannia, G., and Marino, G. (2003) GroEL from the psychrophilic bacterium *Pseudoalteromonas haloplanktis* TAC125: molecular characterization and gene cloning. *Extremophiles* **7**, 17–28.

20. Sambrook, J., Fritsch, E. F., and Maniatis, T. (1989) *Molecular Cloning: A Laboratory Manual*, 2nd ed. Cold Spring Harbor Laboratory Press, Cold Spring Harbor, NY.

21. Hanahan, D. (1983) Studies on transformation of *Escherichia coli* with plasmids. *J. Mol. Biol.* **166**, 557–580.

22. Dagert, M., and Ehrlich, S. D. (1979) Prolonged incubation in calcium chloride improves the competence of *Escherichia coli* cells. *Gene* **6**, 23–28.

III

FUNGI

16

Recombinant Protein Production in Yeasts

Danilo Porro and Diethard Mattanovich

Summary

Recombinant DNA (rDNA) technologies (genetic, protein, and metabolic engineering) allow the production of a wide range of peptides, proteins, and biochemicals from naturally nonproducing cells. This technology, now approx 25 yr old, is becoming one of the most important technologies developed in the 20th century.

Pharmaceutical products and industrial enzymes were the first biotech products on the world market made by means of rDNA. Despite important advances regarding rDNA applications in mammalian cells, yeasts still represent attractive hosts for the production of heterologous proteins. In this review we summarize the advantages and limitations of the main and promising yeast hosts.

Key Words: Yeasts; heterologous proteins; expression; biotechnology.

1. Host Species Commonly Used

High-level production of proteins from engineered organisms provides an alternative to the extraction of the proteins from natural sources. Natural sources of proteins are often limited; furthermore, the concentration of the desired product is generally low, so extraction is regularly very cost-intensive or even impossible. Besides, extraction might bear the danger of toxic or infectious contamination depending on the natural origin of the protein. With the advent of molecular cloning in the mid-1970s, it became possible to produce foreign proteins in new hosts. Recombinant DNA (rDNA) technologies (genetic, protein, and metabolic engineering) allow the production of a wide range of peptides and proteins from naturally non-producing cells. The first biotech products on the world market made by means of rDNA were pharmaceutical products (for example, insulin, interferons, erythropoietin, and vaccine against hepatitis B) and industrial

From: *Methods in Molecular Biology, vol. 267:*
Recombinant Gene Expression: Reviews and Protocols, Second Edition
Edited by: P. Balbás and A. Lorence © Humana Press Inc., Totowa, NJ

enzymes (for example, these used for the treatments of food, feed, detergents, paper pulp, and health care). World sales of the top 20 recombinant pharmaceutical products in 2000 was about 13 billion Euros, while the worldwide market for the industrial enzymes was about 2 billion and it is projected to reach about 8 billion Euros in 2008.

Microorganisms as well as cultured cells from higher organisms (such as mammals, insects, or plants) today represent the most frequently used hosts for the production of heterologous as well as homologous proteins. Microorganisms (prokaryotic as well as eukaryotic) are advantageous hosts for the production of proteins because of high growth rates and, commonly, ease of genetic manipulation. But, in particular, bacterial hosts lack the ability for correct protein processing; in many cases, heterologous proteins build up inclusion bodies inside the bacterial cells, whereupon the proteins are, or could be, lost because their enzymatic activity/3D structure can often not be reconstituted. Because of their incorrect structure, any use of such proteins for the treatment of humans is also excluded, unless they can be correctly refolded in vitro. In this respect, the dominance of *Escherichia coli* as host for the production of heterologous proteins is clearly a reflection of the quantity and quality of the information available about its genetic, molecular biology, biochemical, physiological, and fermentation technologies.

Among the microbial eukaryotic host systems, yeasts can combine the advantages of unicellular organisms (i.e., ease of genetic manipulation and growth) with the capability of a protein processing typical for eukaryotic organisms (i.e., protein folding, assembly, and posttranslational modifications), together with the absence of endotoxins as well as oncogenic or viral DNA. Starting in the early 1980s, the majority of recombinant proteins produced in yeasts have been expressed using *Saccharomyces cerevisiae* (*1*). As for *E. coli*, this first choice was determined by the familiarity of molecular biologists with this yeast together with the accumulated knowledge about its genetics, biochemistry, physiology, and fermentation technologies. Furthermore, *S. cerevisiae* is recognized by the US Food and Drug Administration (FDA) as an organism generally regarded as safe (GRAS). However, this yeast is sometimes not an optimal host for the large-scale production of foreign proteins, especially due to its characteristics regarding technical fermentation needs that require highly sophisticated equipment. In addition, the proteins produced by *S. cerevisiae* are often hyperglycosylated and retention of the products within the periplasmic space is frequently observed (*2,3*). Furthermore, due to the partial retention of the protein in *S. cerevisiae*, a fraction of the protein is commonly degraded. These respective degradation products are generally very difficult to remove from the desired product. Disadvantages such as these have promoted, since the mid-1980s, a search for alternative hosts, trying to exploit the great biodiversity existing among the yeasts, and starting the development of expression systems using the so-called "nonconventional" yeasts. The most established or prominent examples are *Hansenula polymorpha* (*4*), *Pichia pastoris* (*5*), *Kluyveromyces lactis* (*6,7*), *Yarrowia lipolytica* (*8*), *Pichia methanolica* (*9*), *Pichia stipitis* (*10–12*), *Zygosaccharomyces rouxii* (*13*) and *Z. bailii* (*14,15*), *Candida boidinii* (*16*), and *Schwanniomyces (Debaryomyces) occidentalis* (*17*), among others (**Table 1**).

The choice of the yeast host is currently among the most important parameters that can determine the success of the whole process. In this respect, different reviews have

Table 1
Nonmethylotrophic and Methylotrophic Yeast
Species Used for Recombinant Protein Production

Nonmethylotrophic	Methylotrophic
S. cerevisiae	*H. polymorpha*
K. lactis	*P. pastoris*
Y. lipolitica	*P. methanolica*
Z. rouxii	*C. boidinii*
Z. bailii	
S. occidentalis	
P. stipitis	

For references, *see* text.

been published in the last years (*7,18–22*). Usually, yeast hosts are divided into two main categories, (1) conventional and nonconventional or (2) Crabtree-positive and Crabtree-negative yeast hosts. In this respect, *S. cerevisiae* is the only conventional yeast and one of the few Crabtree-positive yeasts (i.e., producing ethanol under aerobic conditions), together with *Z. rouxii* and *Z. bailii*. With respect to fermentation, *Y. lipolytica* represents an exception as the only nonfermenting yeast among the host species mentioned above. Nevertheless, based on the recent developments of heterologous protein production systems, yeast hosts should be grouped in the following two categories: nonmethylotrophic and methylotrophic hosts. This review will analyze the advantages and disadvantages of different yeasts based on this grouping.

1.1. Nonmethylotrophic Yeasts

Many of these wild-type yeasts, lacking both endotoxins and lytic viruses, are known for their established applications for the production of ethanol, beverages, flavors, enzymes, vitamins, organic acids, and single-cell proteins (biomasses). Generally speaking, the main advantage of the nonmethylotrophic hosts is related to the familiarity of molecular biologists, biochemists, microbial physiologists, and fermentation technologists with these yeasts. This is particularly true for the yeast *S. cerevisiae* and, and to a lesser extent, for *K. lactis* and *Y. lipolytica*. Indeed, for *S. cerevisiae*, the goal of having the entire genome fully sequenced has been reached. Further, *K. lactis*, like the already mentioned *S. cerevisiae,* is a GRAS organism. For many of these yeasts, the tremendous advances in yeast molecular genetics have yielded a wide range of vectors, selection markers, promoters, terminators, and secretion signals for the expression and production of heterologous genes. As a result, many problems encountered in the expression of heterologous genes can be understood and solved. The success of this approach is indicated by the high expression levels that have been obtained in shake-flask cultures. Advanced fermentation techniques (continuous and fed-batch processes) and technologies (computer-controlled fermentations) are also available. Another interesting feature is the wide range of carbon and energy sources that can be used during the industrial processes using these yeasts, such as glucose, lactose, maltose, starch, alkanes, and fatty acids. However, at present, most processes for the production of heterologous proteins by these yeasts are still in the development phase. In fact, large-scale processes impose many restrictions and often, these expression systems cannot be implemented in large-scale environments.

The amount of genetic, biochemical, and physiological information and technologies available for the genus *Zygosaccharomyces* is very poor. However, these yeasts seem very promising. Eleven species, which appear to be evolutionary quite close to *S. cerevisiae* and not so far from *K. lactis*, have been classified so far (**23–25**). An exceptional resistance to several stresses renders some of the *Zygosaccharomyces* species potentially interesting for industrial purposes. For example, *Z. rouxii* is known to be salt-tolerant (osmophilic) and *Z. bailii* is known to tolerate high sugar concentrations (osmotolerant) and acidic environments, as well as relatively high temperatures of growth (**26,27**). Despite the fact that these yeasts are classified as Crabtree-positive yeasts, ethanol production under aerobic conditions is quite low when compared to that observed from *S. cerevisiae* cells; indeed, these yeasts can grow to high cell densities even without complex fermentation strategies. Development of these yeasts as hosts for the production of heterologous proteins is quite promising because all these properties could allow for the development of easier and cheaper production strategies.

1.2. Methylotrophic Yeasts

Methylotrophic yeasts constitute a group of yeast species that essentially share a common pathway to metabolize one-carbon compounds, such as methanol, as carbon and energy sources. Initially this group of yeasts was employed in the development of processes for the production of feed protein (single-cell protein) (**28**), but soon they attracted interest as production systems for recombinant proteins (**5**) for two major features: first, they are able to grow to high cell densities even in unsophisticated fermentation processes, and second, their high demand for methanol-oxidizing enzymes (such as alcohol oxidase) endowed them with very strong, and strictly regulated, promoters.

Two species are mainly employed for heterologous protein expression: *Pichia pastoris* (syn. *Komagataella pastoris*) (**29**) and *H. polymorpha* (also known as *Pichia angusta*) (**19**). Additionally, *P. methanolica* and *C. boidinii* appear in the literature as expression systems. In the following, *P. pastoris* and *H. polymorpha* will be discussed in more detail. Although the two yeasts share a common pathway to metabolize methanol, they differ significantly in the specific genetics. *P. pastoris* produces two different alcohol oxidases (AOX1 and AOX2), whereas *H. polymorpha* expresses only one methanol oxidase (MOX).

The first generation of expression systems relied on the *AOX*1 promoter (the stronger one of the two) or the *MOX* promoter, respectively. With respect to methanol utilization, there are three possible phenotypes for *P. pastoris* compared to two for *H. polymorpha*, namely Mut$^+$ (methanol utilization wild-type; both *AOX*1 and *AOX*2 intact), Mut^s (methanol utilization slow; *AOX*1 interrupted, *AOX*2 intact), and Mut$^-$ (methanol utilization deleted, both *AOX*1 and *AOX*2 interrupted). As methanol is employed as inducer of these expression systems, the methanol utilization phenotype plays an important role with respect to the design of the production process. The technical consequences of these phenotypes will be discussed below.

Eventually, novel promoters were isolated and employed for gene expression, namely the formaldehyde dehydrogenase (*FLD*) promoter (30) and the glycerol aldehyde phosphate dehydrogenase (*GAP*) promoter (*31*) of *P. pastoris* and the formate dehydrogenase (*FMD*) promoter of *H. polymorpha* (*32*). These promoters are either regulated, as the *FLD* promoter (nitrogen- and methanol-regulated), or constitutive, as the *GAP* promoter. The type of regulation again plays a major role for the layout of the fermentation process.

Among the many methylotrophic yeast species, some others have been employed as expression systems, mainly *P. methanolica* (*9*) and *C. boidinii* (*16*). The accumulation of comparative data will prove whether these systems show advantages over the established expression systems.

A major drawback of all methylotrophic yeasts as compared to *S. cerevisiae* is the lack of fundamental knowledge of their genetics and molecular biology. The genetic regulation is by far not fully understood, even of the promoters employed for overexpression, and genomic sequences are not yet available. Therefore the engineering of improved expression strains is severely handicapped.

2. Transformation and Vector Systems

S. cerevisiae, *K. lactis*, and *Y. lipolytica* are heterothallic, while *P. pastoris*, *P. methanolica*, *H. polymorpha*, and the *Zygosaccharomyces* species are homothallic. For the *K. lactis* and *Y. lipolytica* heterothallic strains, the genetic manipulations described for *S. cerevisiae* can be carried on with good efficiency, while for the homothallic strains, complementation and tetrad analyses pose different problems.

The expression of a foreign protein in yeasts consists of, first, cloning of a foreign protein-coding DNA sequence within an expression cassette containing a yeast promoter and transcriptional termination sequences; second, transformation and stable maintenance of this DNA fusion in the host. Cloning of a heterologous gene into the cited yeast hosts can be carried out by means of three different approaches: following spheroplast preparation (*33*), by the lithium acetate method (*34*), and by electroporation (*35*). Lithium acetate and electroporation are the methods of choice today. Efficiency of transformation is clearly strain-dependent and detailed studies should be carried out when a high efficiency of transformation is required.

Transformants can be selected by complementation of (1) auxotrophic markers (for example, *URA*3, *LEU*2, *TRP*1, *HIS*3, *HIS*4, *ADE*1 being the most commonly used), by (2) dominant markers (genes conferring resistance to the presence of antibiotic such as G418, phleomycin, hygromycin, zeocin, or others in the culture medium) or by (3) the so-called autoselection systems. For an industrial process, the expression system should be genetically stable without selection pressure. It is strongly advised against the use of antibiotics during the production process. Auxotrophic marker systems would require chemically defined minimal media (which would not be unusual for a large-scale process), but also show the tendency of cross-feeding at high cell densities between marker-containing and auxotrophic cells that have lost the marker.

For *S. cerevisiae*, the *LEU2* gene and the G418 resistance gene are the two most popular markers used, whereas for *P. pastoris*, *HIS4* and zeocin resistance are mainly employed. Autoselection systems are based on the expression of a vital gene/activity in host strains lacking such a gene/activity, like *URA3* in a Δ*fur1* (uracil phosphoribosyl transferase), or *FBA1* (fructose biphosphate aldolase) in a Δ*FBA1*, backgrounds. These systems ensure that plasmid selection is maintained irrespective of culture conditions. While dominant markers can be used with any yeast strain, auxotrophic markers and autoselection systems deserve the availability of the respective deletion strains. The availability of such deletion strains depends much on the abundance of molecular biology research with this species.

The heterologous gene can be introduced into the yeast host cells by means of an integrative plasmid as well as by autonomous or episomal circular plasmids. In the first case, the heterologous gene fate is the integration into the chromosomal DNA by means of recombination events. Directed integration requires homology of the DNA introduced with a chromosomal locus. On the other side, heterologous recombination may occur at random, therefore not directed, positions. In other cases, the heterologous gene will be replicated due to the replication of a circular plasmid. For the autonomous plasmids, replication is governed by the autonomous replicating sequences (ARS), while the episomal plasmids are based on endogenous circular yeast plasmids like the 2 μ plasmid of the yeast *S. cerevisiae*.

Ideally, an expression vector should be stably maintained in the host cells without the need of any selection pressure and (assuming a direct correlation between gene dosage and expression level) with a high copy number per cell. Cloning of centromeric sequences (CEN) in autonomous plasmids yields stable centromeric plasmids, but those are replicated in one or two copies per cell. ARS-based plasmids without centromeric sequences are generally very unstable, and are easily lost by growing the host cells without selective pressure. Episomal plasmids are generally maintained with a high copy number per cell (from 1–10 up to 100 copies per cell). Unfortunately, endogenous plasmids have been isolated and characterized for only a few yeast hosts: *S. cerevisiae*, *K. lactis*, *Y. lipolytica*, *Z. rouxii*, *and Z. bailii*.

Finally, the copy number per cell of the integrative plasmids is related to the strategy of cloning. Multiple integrations can be obtained by targeting the cloned gene to the ribosomal DNA cluster, constructing concatamers of the expression cassette or simply by chance. Nevertheless, it is very important to emphasize that a direct correlation between the gene dosage and the level of expression is not generally observed. Cloning and transformation of the host cells is only the first step required for the production of a heterologous protein. It has been shown that lower level of expression (transcription and translation) could give higher yields in the production of the heterologous proteins.

The vectors used are often hybrids between yeast-derived and bacterial sequences. The bacterial fraction of these vectors bears an origin of replication for the chosen bacteria (essentially, *E. coli*) and selection markers (e.g., antibiotic resistance). Of course, the hybrid fraction is introduced for an easier manipulation of the vector itself.

With regard to the heterologous gene and its codon bias index, it should be noted that codon usage does not appear to be an essential parameter for high production levels even if, in some cases, higher productions have been obtained.

In conclusion, no general rules are known to illustrate the transformation efficiency, vector stability, and copy number, although factors such as vector composition, host strain, transformation method, and selective pressure might influence them.

3. Promoters

One of the important keys for the production of heterologous proteins is related to transcription efficiency of the heterologous gene(s). For efficient transcription of foreign genes a large variety of heterologous and homologous, as well as constitutive and inducible, yeast promoters is available. Generally, homologous promoters originating from the yeast species used as hosts are preferred, as heterologous promoters often do not yield good efficiency of expression.

On one side, inducible promoters can be useful to drive foreign protein expression when, prior to induction, the opportunity to maintain yeast cultures in an "expression off" mode minimizes selection for nonexpressing mutant cells during the cell growth phase. Such a selection can occur as a result of the added metabolic burden placed on cells expressing high levels of a foreign protein or the toxic effect of a foreign protein on the cells. On the other side, constitutive promoters are preferred since their use does not imply the development of complex fermentation strategies and/or the use of specific inducers that could be expensive or could interfere with the isolation of the final product.

In the methylotrophic yeast, because of the lower degree of current studies, only few promoters have been used for the production of heterologous proteins. **Table 2** shows some of the most used promoters.

4. Cytoplasmatic vs Secreted Expression

For the production of heterologous proteins, these are, in almost all cases, either targeted to the cytoplasm or secreted, best into the culture supernatant. While it is not possible to design the optimum expression strategy a priori, some guidelines can be given.

Cytoplasmic expression often leads to very high expression levels, as the potential limitations of the secretion pathways are not involved. Very high expression levels have been reported, e.g., for rubber tree hydroxynitrile lyase, 22 g/L (*36*), or tetanus toxin fragment C, 12 g/L (*37*). However, several limitations are associated with the expression in the cytoplasm. Indeed, cell lysis requires additional process steps during the downstream processing of the product. The cells must be disrupted usually by high-pressure homogenization (and yeast cell walls are known to be very robust), and then the lysate needs to be clarified, but the product is still only a fraction of the total soluble cellular constituents, so that additional purification steps must be employed. In many cases protein folding and processing—mainly disulfide bond formation—pose a severe limitation for cytoplasmic expression. Some authors have described the deposition of recombinant proteins as insoluble aggregates (inclusion bodies) in the cytoplasm of yeasts (*38–41*). These inclusion bodies may be processed as those obtained in *E. coli* (*41*).

Alternatively, proteins can be tagged to be secreted essentially by adding an appropriate secretion signal sequence. The signal sequence used most often (even with non-*Saccharomyces* yeasts) is the *S. cerevisiae* α mating factor (α-MF) signal. High secretion levels have been described for *S. cerevisiae, P. pastoris, H. polymorpha, K. lactis,* and

Table 2
Most Important Promoters Used for Gene Expression in Yeasts

Type	Species	Constitutive	Galactose-induced	Lactose-induced	Ethanol-induced	Starch-induced	Xylose-induced	Methanol-induced
Nonmethyl-otrophic	S. cerevisiae	GAPDH, PGK, TP1, ENO, α-MP	UAS=_GAL 1–10, UAS=_GAL 7	ADH 1				
	K. lactis	PGK		LAC 4				
	S. occidentalis	GAM 1			ADH 4	AMY 1, GAM 1		
	Y. lipolytica	TEF, RPS 7						
	Z. rouxii	GAPDH						
	Z. bailii	TPI						
	P. stipidis						XYL 1	
Methyl-otrophic	P. pastoris	GAP						AOX 1, FLD 1
	H.polymorpha							MOX
	C. boidinii							AOD 1
	P. methanolica							AUG 1

For references see ref. **21**.

Gene nomenclature: α-MP, α-mating factor; ADH1, ADH4, alcohol dehydrogenase; AMY1, α-amylase; AOX1, AUG1, AOD1, and MOX, alcohol oxidase in species shown; ENO, enolase; FLD1, formaldehyde dehydrogenase; UAS=_GAL1-10 and UAS=_GAL7, upstream activating sequence of the GAL1-10 and GAL7 promoters; GAM1, glucoamylase; GAPDH, GAP, glyceraldehyde-3-phosphate dehydrogenase; LAC4, β-galactosidase; PGK, phosphoglycerate kinase; RPS7, ribosomal protein S7; TEF, translation elongation factor-1a; TPI; triose phosphate isomerase.

Z. bailii by using the α-MF signal. Several authors have observed optimum secretion when the entire processing sequence of prepro α-MF was employed. After signal peptide removal, the prepro sequence is cleaved by the Kex2 proteinase in the Golgi, and subsequently two Glu-Ala dipeptides are removed by the Ste13 dipeptidase. However, the dipeptidase cleavage may not be quantitative, resulting in a nonhomogenous product, which is problematic mainly for pharmaceutical applications. Additionally, other yeast signal sequences have been employed (e.g., *S. cerevisiae* invertase, or *P. pastoris* acid phosphatase—the latter often with lower yields), or the heterologous secretion leader of the gene to express was utilized, which functioned efficiently in several cases even with human genes (e.g., human serum albumin) (*42*).

Folding and translocation are the subsequent steps to occur, and appear to be product-dependent to a high degree. Several authors describe that significant amounts of product may be retained in the cell even when targeted to secretion (*36,43,44*). Unfolded proteins in the ER induce the "unfolded protein response" (UPR) (*45*) and may be either rescued by chaperones (*46*) or targeted to proteolytic degradation (*47*). According to Kauffman et al. (*43*), the UPR is necessary for the cell to decrease the level of intracellular product and to resume normal growth. However, a clear relation between UPR induction and secreted product yield was not reported.

After being correctly folded and released from the ER and the Golgi, the product may still be retained within the periplasmic space of the host cell. This effect was observed mainly with *S. cerevisiae*, but this may be due to the larger amount of data available for this yeast species. The distribution of a product between the culture supernatant and the periplasm seems to be product-dependent. Modification of the *S. cerevisiae* cell wall by deletion of the *GAS*1 gene strongly improved the release of IGF1 to the supernatant (*48*).

The glycosylation patterns of yeasts (*S. cerevisiae*, as well as other yeasts, such as *P. pastoris* and *H. polymorpha*) have been reviewed before (*49*). Generally, the glucan chains of yeasts are less complex than those of higher eukaryotes, but tend to be larger in size (high-mannose-type overglycosylation). *P. pastoris* and *H. polymorpha* have been reported to show less overglycosylation than *S. cerevisiae* (*19*), but Muller et al. (*8*), for instance, reported high overglycosylation of several fungal proteins in *H. polymorpha* in contrast to *S. cerevisiae*, *K. lactis*, *S. pombe*, and *Y. lipolytica*.

As a general rule, most proteins that are secreted in their homologous environment will be more easily produced in an active form if secreted from the recombinant host. Problems can be encountered with translocation and/or folding. Proteins containing many disulfide bonds are often (but not always) expressed at a lower yield in yeasts. If secretion and/or folding appear to be the bottleneck in production, an increase of transcription by use of a stronger promoter or increase of the gene dosage will not improve the yield but may even lead to a reduction of productivity and a severe stress on the host cells. A general technological problem encountered with secretion systems is the harvesting of the supernatants from high-cell-density fermentations (up to 500 g/L wet biomass). The possible technical solutions for centrifugation of dense cultures at large scale are limited (e.g., decanters) and involve high investment costs. As an alternative, capturing the product out of the total culture broth by STREAMLINE chromatography has been suggested (*50*).

Since the first use of yeasts as hosts for the production of heterologous proteins, several research teams have addressed the problems of endogenous protease production and developed protease-deficient strains. Such strains are also available in different yeast collections. However, also in this case, no general rules can be written in relation to the efficiency of such strains.

5. Physiological/Metabolic Basis for Process Design

An efficient production of a heterologous protein in yeast first requires:

1. Finding out the genetic determinants ensuring an adequate efficiency of transcription, translation, and localization of the heterologous gene/protein;
2. Finding out the physiological determinants that maximize the potential of the genetic determinants;
3. Developing clean, fast, and reproducible fermentation processes;
4. Developing a fast and cheap downstream process.

High expression levels have been obtained in shake-flask cultures. However, at present, most processes for the production of heterologous proteins by yeasts are still in the development phase. In this respect, the hearth of the biotechnological process, based on yeast hosts, is the stirred tank bioreactor. Any promising result obtained by flask cultures that cannot be reproduced in the bioreactor is—from an industrial point of view—meaningless. The ideal fermentation process should be as short and cheap as possible. These fermentation processes generally require medium- to large-size reactor vessels (several liters to some 10 m^3), a high amount of air/oxygen, and, in turn, a high amount of electricity. Yeast metabolism also generates a large amount of heat and thus requires cooling. Yeasts also release high levels of CO_2 into the environment and generally produce low aqueous titers, necessitating high cell densities and large volumes, and therefore elaborate recovery procedures. The main costs of the overall process are related to the substrate, the cost of the energy required, ecological costs (costs of disposal of wastes) and, mainly, to the purification activities. **Figure 1** shows the correlations between the costs of selling of different biotech products and their concentration at the end of the fermentation process, while **Table 3** reports the average influence of the purification costs (DSP: downstream processing) on the global costs.

Generally, batch or fed-batch are the techniques of choice to obtain high production of recombinant biomass, and hence high production of recombinant proteins.

6. Process Development for Recombinant Protein Production With Yeasts

As said before, the most common fermentation strategy for recombinant microorganisms is fed-batch, with the aim of obtaining high biomass concentrations. The design of an optimized fed-batch protocol depends to a great extent on the physiological parameters of the expressing strain. The major points to consider are: (1) Is the host strain Crabtree-positive or negative on the desired substrate? and (2) Is the expression of the product constitutive or regulated, and if regulated, by which means?

If the host strain is Crabtree-positive, the biomass yield (gram of dry biomass/gram of substrate consumed) in the batch phase will be comparatively low (0.1–0.2 g/g), and

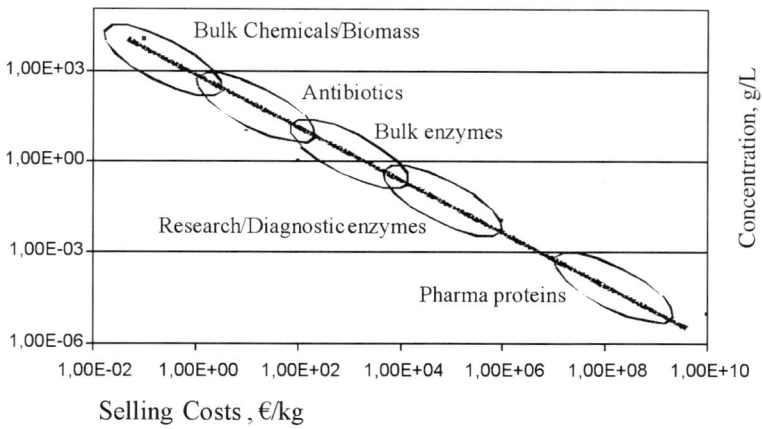

Fig. 1. Correlation between product concentration and selling price for fermentation bio-products.

the byproducts produced (mainly ethanol) should be metabolized to a large extent before starting the fed-batch, which results in an extended time required for the batch phase. The feed rate then needs to be limited so that further aerobic fermentation is prevented. Usually this results in growth rates below 0.1/h. The optimum feed rate must not exceed a critical limit of specific glucose uptake rate, and is best optimized in continuous chemostat cultures (*51*). The second important concern for optimising the feed rate is of course product formation, as it is for Crabtree-negative yeasts.

The main group of Crabtree-negative yeasts employed for recombinant product formation is the methylotrophic yeasts. As described before, the first generation of expression systems developed for methylotrophic yeasts employed alcohol oxidase promoters, which are induced by methanol and repressed by glucose and most other carbon sources. Therefore mainly methanol is employed as a carbon and energy source and as an inducer as well. Quite frequently, feedback loop control strategies employing a methanol sensor to maintain a certain methanol concentration (up to 10 g/L) are described. Several groups postulated that high product yields can be achieved only when such a methanol concentration is maintained (*52,53*). However, this approach will not allow for control of the growth rate at a desired low level as in a carbon-limited culture. Therefore methanol-limited fed-batch cultures are employed, and have been described to yield higher product titers as compared to the above-described feedback-loop-controlled cultures (*29*). A special case is the use of strains with deleted alcohol oxidases. As described above, most methylotrophic yeasts will either have a Mut[+] or a Mut[-] phenotype except for *P. pastoris*, which carries two alcohol oxidases in the wild type, *AOX*1 (the major fraction of alcohol oxidase), and *AOX*2, which is expressed at much lower levels. Deletion of *AOX*1 leads to the Mut[s] phenotype, which allows for a slow growth on methanol, so that these strains can be cultivated on a controlled methanol concentration at a low growth rate. Although some authors describe good yields of recombinant protein in such systems (*36*), other groups found equal or lower

Table 3
Percentage of Purification Costs on the Total Costs of Fermentation Bioproducts

Bioproducts	% DSP
Single-cell proteins/biomass	1–5
Organic acids	10–50
Extracellular enzymes	10–15
Antibiotics	20–50
Recombinant proteins	90–95

yields in Mut^s strains compared to methanol limited cultures of Mut$^+$ strains (*37*; Hohenblum and Mattanovich, unpublished). Alternatively, Mut^s strains and Mut$^-$ strains may be cultivated using mixed feed protocols employing a controlled methanol concentration for induction and a limited feed of glycerol as the carbon source (*54*).

Generally, product formation can be positively growth-related (product formation rate increases with specific growth rate), growth-indifferent, or negatively growth-related (product formation rate decreases with increasing specific growth rate). There are still debates going on as to which relation is observed most commonly for recombinant proteins. Also in general, a negative relation can be observed at high growth rates, so that fast production processes are often disadvantageous concerning product yield. At low growth rates (up to 0.1 or 0.2/h) positive correlations are observed most often, so that at first sight, optimization should mean increasing the growth rate. However, the most important parameter for judging an industrial process is not the productivity of the cells, but the product yield per fermenter volume and time (defining the costs for using the fermentation plant), and the final product titer as a major parameter for the efforts of the initial purification steps.

The yield per fermenter volume and time (or space-time yield, STY) can easily be calculated by dividing the total product achieved by the actual volume and the time consumed, and plotted against the fermentation time, so that an optimum can be deduced, which is usually reached at an earlier time point than the highest product titer is achieved. The progression of the STY will depend on the characteristic development of the growth rate and the product formation rate during the course of a fed-batch fermentation. Typically, the growth rate either is kept constant by employing an exponential feed profile, or it is decreasing when a linear feed profile is used. By modeling fed-batch processes based on either constant or decreasing growth rates (μ) and either constant or decreasing product formation rates (q_p), the consequences of different strategies can be anticipated. In **Fig. 2**, such models are presented based on actual data obtained in lab scale fermentations. At constant μ and constant q_p (**Fig. 2A**), the STY increases steadily up to much higher values than in the other models. While constant q_p at decreasing μ (**Fig. 2C**) leads to an asymptotic approach of STY to a maximum, decreasing q_p results in an optimum STY that decreases at extension of time (**Fig. 2B, C**). It can be assumed from the data of several groups (*44,55*) that q_p will positively correlate with μ in this range of data, so that either model 2A or 1D will practically

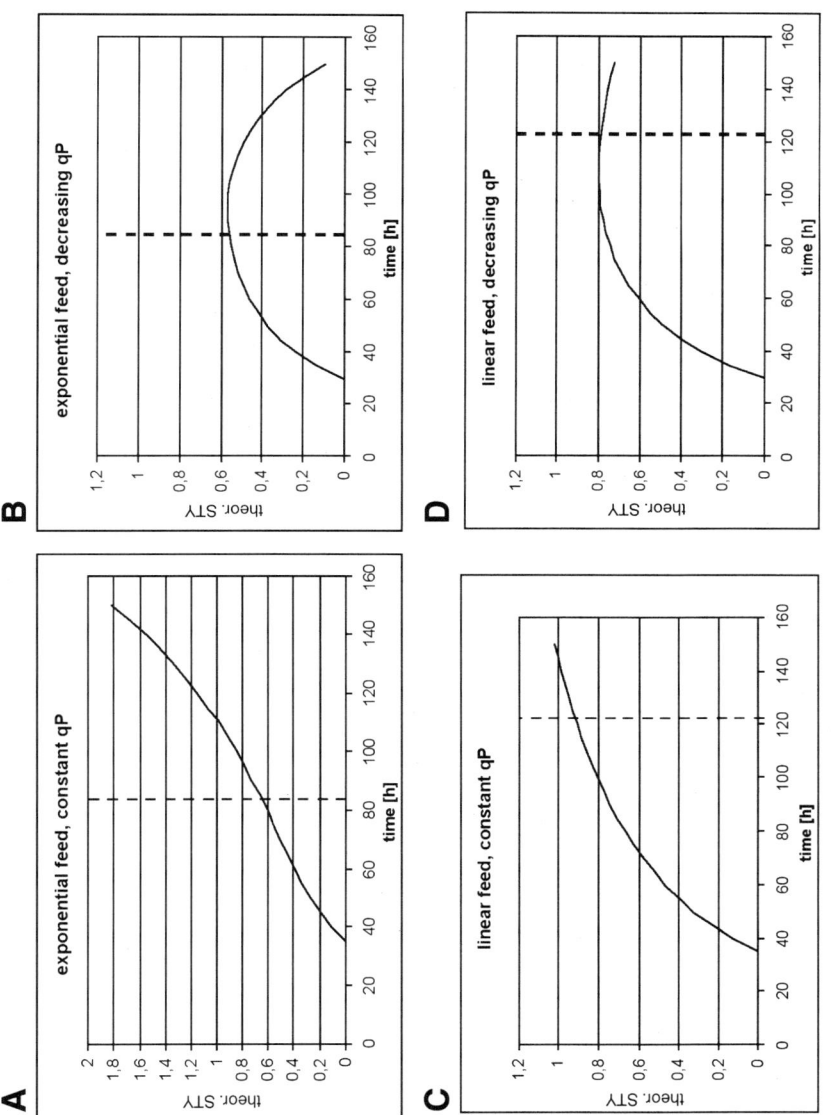

Fig. 2. Theoretical development of space-time yields (STY) of model fed-batch processes for a secreted recombinant protein based on the following assumptions: (A) exponential feed, constant specific product formation rate. (B) exponential feed, steadily decreasing specific product formation rate. (C) linear feed, constant specific product formation rate. (D) linear feed, steadily decreasing specific product formation rate. The dotted line marks the accumulation of 100 g/L yeast dry mass in the model process.

apply; this would speak strongly for model 2A at first sight, as much higher STY is plotted there. However, when considering the maximum of cell density that can be reasonably achieved, as indicated by the vertical line (100 g/L yeast dry mass), one can clearly see that a higher STY is achieved in model 1D with linear feed rate (decreasing μ) and decreasing q_p.

Most protein production processes employing yeasts are based on fed-batch protocols. Continuous processes are not often applied. The reasons for this are divergent: genetic instabilities may lead to a loss of productivity, the risk of a contamination may be higher, and, especially, the biopharmaceutical industry fears an unclear regulatory situation concerning batch definition and process lifetime. However, many recombinant expression systems are extremely stable, e.g., when the heterologous gene is integrated into the host genome or when plasmids including stabilising sequences are employed (e.g., *57*). From a theoretical point of view, significantly higher STY can be achieved in continuous culture; therefore this approach should be considered as an attractive alternative to fed-batch. Several authors have achieved promising results by applying continuous cultivation to different yeast species like *S. cerevisiae* (*56,57*) or *P. pastoris* (*55*). However, it should be noted that product instability at the culture conditions may increase in a continuous process, and deleterious effects of the product itself, or its formation, on the host cells may lead to increased problems in continuous culture compared to fed-batch. Therefore it will be necessary to select the optimum process type for every new product.

7. Conclusions and Outlook

In summary, one more key to the success of genetic engineering for enhancing product formation is to individuate all the possible limiting steps involved in product biosynthesis. The production of heterologous proteins mainly combines genetic engineering and microbial physiology with the objective of increasing the rate of production of a desired product. Such approach is often hampered by the lack of knowledge of the production pathway and its dynamic profile in the producing cells. Therefore, for successful production of heterologous proteins, detailed physiological studies are required devoted to the identification of the different physiological determinants that could maximize the potential of the genetic determinants.

Direct approaches aiming at the extension of substrate range utilization (e.g., starch, lactose, melibiose, xylose), improvement of productivity and yield (e.g., by avoiding the formation of byproduct), improvement of process performance (e.g., modulation of the flocculation process), improvement of cellular properties (e.g., alleviation of glucose repression, alleviation of the Crabtree effect, modulation of the secretion process), or improving product quality (e.g., by modulating the glycosylation patterns) are also possible by metabolic engineering applications. Metabolic engineering involves a direct approach to the application of the rDNA technology for strain improvements defined as follows: "The directed improvement of product formation or cellular properties through the modification of specific biochemical reaction(s) or the introduction of new one(s) with the use of recombinant DNA technology" (*58*).

References

1. Hitzeman, R. A., Hagie, F. E., Levine, H. L., Goeddel, D. V., Ammerer, G., and Hall, B. D. (1981) Expression of a human gene for interferon in yeast. *Nature* **293**, 717–722.
2. Reiser, J., Glumoff, V., Kalin, M., and Ochsner, U. (1990). Transfer and expression of heterologous genes in yeasts other than *Saccharomyces cerevisiae*. *Adv. Biochem. Eng. Biotechnol.* **43**, 75–102.
3. Romanos, M. A., Scorer, C. A. and Clare, J. J. (1992) Foreign gene expression in yeast: a review. *Yeast* **8**, 423–488.
4. Sudbery, P. E., Gleeson, M. A., Veale, R. A., Ledeboer, A. M., and Zoetmulder, M. C. (1988) *Hansenula polymorpha* as a novel yeast system for the expression of heterologous genes. *Biochem. Soc. Trans.* **16**, 1081–1083.
5. Thill, G., Davis, G., Stillman, C., Tschopp, J. F., Craig, W. S., Velicelebi, G., et al. (1987). The methylotrophic yeast *Pichia pastoris* as a host for heterologous protein production, in *Microbial Growth on C1 Compounds*. van Verseveld, H. W., and Duine, J. A., eds. Dordrecht, The Netherlands, pp. 289–296.
6. Blondeau, K., Boze, H., Jung, G., Moulin, G., and Galzy, P. (1994) Physiological approach to heterologous human serum albumin production by *Kluyveromyces lactis* in chemostat culture. *Yeast* **10**, 1297–1303.
7. Gellissen, G. and Hollenberg, C. P. (1997) Application of yeasts in gene expression studies: a comparison of *Saccharomyces cerevisiae*, *Hansenula polymorpha* and *Kluyveromyces lactis*—a review. *Gene* **190**, 87–97.
8. Muller, S., Sandal, T., Kamp-Hansen, P., and Dalboge, H. (1998) Comparison of expression systems in the yeasts *Saccharomyces cerevisiae*, *Hansenula polymorpha*, *Klyveromyces lactis*, *Schizosaccharomyces pombe* and *Yarrowia lipolytica*. Cloning of two novel promoters from *Yarrowia lipolytica*. *Yeast* **14**, 1267–1283.
9. Raymond, C. K., Bukowski, T., Holderman, S. D., Ching, A. F. T., Vanaja, E., and Stamm, M. R. (1998) Development of the methylotrophic yeast, *Pichia methanolica*, for the expression of the 65-kilodalton isoform of human glutamate decarboxylase. *Yeast* **14**, 11–23.
10. Den Haan, R. and Van Zyl, W. H. (2001) Differential expression of the *Trichoderma reesei* beta-xylanase II (xyn2) gene in the xylose-fermenting yeast *Pichia stipitis*. *Appl. Microbiol. Biotechnol.* **57**, 521–527.
11. Hohenblum, H., Naschberger, S., Weik, R., Katinger, H., and Mattanovich, D. (2001) Production of recombinant human trypsinogen in *Escherichia coli* and *Pichia pastoris*. A comparison of expression systems, in Merten, O.-W., Mattanovich, D., Lang, C., Larsson, G., Neubauer, P., Porro, D., et al. eds., Recombinant Protein Production with Prokaryotic and Eukaryotic Cells. A Comparative View on Host Physiology. Kluwer Acad. Publ., Dordrecht, The Netherlands, pp. 339–346.
12. Romanos, M. (1995) Advances in the use of *Pichia pastoris* for high-level gene expression. *Curr. Opin. Biotechnol.* **6**, 527–533.
13. Ogawa, Y., Tatsumi, H., Murakami, S., Ishida, Y., Murakami, K., Masaki A., et al. (1990) Secretion of *Aspergillus oryzae* alkaline protease in an osmophilic yeast, *Zygosaccharomyces rouxii*. *Agric. Biol. Chem.* **54**, 2521–2529.
14. Brambilla, L., Ranzi, B. M., Vai, M., Alberghina, L., and Porro, D. (2000) Production of heterologous proteins from *Zygosaccharomyces bailii*. PCT US Patent. Issued/Filed: 14/1/2000. Serial number No. PCT/RP00/00268.

15. Branduardi, P., Valli, M., Alberghina, L., and Porro, D. (2002) Process for expression and secretion of proteins by the non-conventional yeast *Zygosaccharomyces bailii.* Germany National Patent, Issued/Filed: 10 November 2002. Serial number No. 102 52 245.6.

16. Sakai, Y., Rogi, T., Takeuchi, R., Kato, N., and Tani, Y. (1995) Expression of *Saccharomyces* adenylate kinase gene in *Candida boidinii* under the regulation of its alcohol oxidase promoter. *Appl. Microbiol. Biotechnol.* **42**, 860–864.

17. Buckholz, R. G. and Gleeson, M. A. (1991) Yeast systems for the commercial production of heterologous proteins. *Biotechnology* **9**, 1067–1072.

18. Sudbery, P. E. (1996) The expression of recombinant proteins in yeasts. *Curr. Opin. Biotechnol.* **7**, 517–524.

19. Gellissen, G. (2000) Heterologous protein production in methylotrophic yeasts. *Appl. Microbiol. Biotechnol.* **54**,741–750.

20. Dominguez, A., Ferminan, E., Sanchez, M., Gonzalez, F. J., Perez-Campo, F. M., Garcia, S., et al. (1998) Non-conventional yeasts as hosts for heterologous protein production. *Int. Microbiol.* **1**, 131–142.

21. Cereghino, J. L. and Cregg, J. M. (2000) Heterologous protein expression in the methylotrophic yeast *Pichia pastoris. FEMS Microbiol. Rev.* **24**, 45–66.

22. Giga-Hama, Y. and Kumagai, H. (1999). Expression system for foreign genes using the fission yeast *Schizosaccharomyces pombe. Biotechnol. Appl. Biochem.* (1999) **30**, 235–244.

23. James, S. A., Collins, M. D., and Roberts, I. N. (1994) Genetic interrelationship among species of the genus *Zygosaccharomyces* as revealed by small-subunit rRNA gene sequences. *Yeast* **10**, 871–881.

24. Steels, H., Bond, C. J., Collins, M. D., Roberts, I. N., Stratford, M., and James, S. A. (1999) *Zygosaccharomyces lentus* sp. nov., a new member of the yeast genus *Zygosaccharomyces* Barker. *Int. J. Syst. Bacteriol.* **49**, 319–327.

25. Kurtzman, C. P., Robnett, C. J., and Basehoar-Powers, E. (2001) *Zygosaccharomyces kombuchaensis*, a new ascosporogenous yeast from "Kombucha tea." *FEMS Yeast Res.* **1**, 133–138.

26. Makdesi, A. K. and Beuchat, L. R. (1996) Evaluation of media for enumerating heat-stressed, benzoate-resistant *Zygosaccharomyces bailii. Int. J. Food Microbiol.* **33**, 169–181.

27. Sousa, M. J., Miranda, L., Corte-Real, M., and Leao, C. (1996) Transport of acetic acid in *Zygosaccharomyces bailii*: effects of ethanol and their implications on the resistance of the yeast to acidic environments. *Appl. Environ. Microbiol.* **62**, 3152–3157.

28. Wegener, G. H. and Harder, W. (1987) Methylotrophic yeasts—1986. *Antonie van Leeuwenhoek* **53**, 29–36.

29. Cregg, J. M., Cereghino, J. L., Shi, J., and Higgins, D. R. (2000) Recombinant protein expression in *Pichia pastoris. Mol. Biotechnol.* **16**, 23–52.

30. Shen, S., Sulter, G., Jeffries, T. W. and Cregg, J. M. (1998) A strong nitrogen source-regulated promoter for controlled expression of foreign genes in the yeast *Pichia pastoris. Gene* **216**, 93–102.

31. Waterham, H. R., Digan, M. E., Koutz, P. J., Lair, S. V. and Cregg, J. M. (1997) Isolation of the *Pichia pastoris* glyceraldehyde-3-phosphate dehydrogenase gene and regulation and use of its promoter. *Gene* **186**, 37–44.

32. Hollenberg, C. P. and Gellissen, G. (1997) Production of recombinant proteins by methylotrophic yeasts. *Curr. Opin. Biotechnol.* **8**, 554–560.

33. Burgers, P. M. and Percival, K. J. (1987) Transformation of yeast spheroplasts without cell fusion. *Anal. Biochem.* **163**, 391–397.

34. Agatep, R., Kirkpatrick, R. D., Parchaliuk, D. L., Woods, R. A., and Gietz, R. D. (1998) Transformation of *Saccharomyces cerevisiae* by the lithium acetate/single-stranded carrier DNA/polyethylene glycol (LiAc/ss-DNA/PEG) protocol. *Technical Tips Online* (http://tto. trends.com).
35. Sanchez, M., Iglesias, F. J., Santamaria, C., and Dominguez, A. (1993) Transformation of *Kluyveromyces lactis* by electroporation. *Appl. Environm. Microbiol.* **59**, 2087–2092.
36. Hasslacher, M., Schall, M., Hayn, M., Bona, R., Rumbold, K., Luckl, J., et al. (1997) High-level intracellular expression of hydroxynitrile lyase from the tropical rubber tree *Hevea brasiliensis* in microbial hosts. *Protein Expr. Purif.* **11**, 61–71.
37. Clare, J. J., Rayment, F. B., Ballantine, S. P., Sreekrishna, K., and Romanos, M. A. (1991) High-level expression of tetanus toxin fragment C in *Pichia pastoris* strains containing multiple tandem integrations of the gene. *Bio/Technology* **9**, 455–460.
38. Richardson, P. T., Roberts, L. M., Gould, J. H., and Lord, J. M. (1988) The expression of functional ricin B-chain in *Saccharomyces cerevisiae*. *Biochim Biophys. Acta* **950**, 385–394.
39. Binder, M., Schanz, M., and Hartig, A. (1991) Vector-mediated overexpression of catalase A in the yeast *Saccharomyces cerevisiae* induces inclusion body formation. *Eur. J. Cell Biol.* **54**, 305–312.
40. Choi, S. Y., Lee, S. Y. and Bock, R. M. (1993) High level expression in *Saccharomyces cerevisiae* of an artificial gene encoding a repeated tripeptide aspartyl-phenylyalanyl-lysine. *J. Biotechnol.* **30**, 211–223.
41. Weik, R., Francky, A., Striedner, G., Raspor, P., Bayer, K., and Mattanovich, D. (1998) Recombinant expression of alliin lyase from garlic (*Allium sativum*) in bacteria and yeasts. *Planta Med.* **64**, 387–388.
42. Barr, K. A., Hopkins, S. A., and Sreekrishna, K. (1992) Protocol for efficient secretion of HSA developed from *Pichia pastoris*. *Pharm. Eng.* **12**, 48–51.
43. Kauffman, K. J., Pridgen, E. M., Doyle, F. J. 3rd, Dhurjati, P. S., and Robinson, A. S. (2002) Decreased protein expression and intermittent recoveries in BiP levels result from cellular stress during heterologous protein expression in *Saccharomyces cerevisiae*. *Biotechnol. Prog.* **18**, 942–950.
44. Hohenblum, H., Borth, N., and Mattanovich, D. (2003) Assessing viability and cell-associated product of recombinant protein producing *Pichia pastoris* with flow cytometry. *J. Biotechnol.* **102**, 281–290.
45. Patil, C. and Walter, P. (2001) Intracellular signaling from the endoplasmic reticulum to the nucleus: the unfolded protein response in yeast and mammals. *Curr. Opin. Cell Biol.* **13**, 349–355.
46. Welihinda, A. A., Tirasophon, W., and Kaufman, R. J. (1999) The cellular response to protein misfolding in the endoplasmic reticulum. *Gene Expr.* **7**, 293–300.
47. Casagrande, R., Stern, P., Diehn, M., Shamu, C., Osario, M., Zuniga, M., et al. (2000) Degradation of proteins from the ER of *S. cerevisiae* requires an intact unfolded protein response pathway. *Mol. Cell* **5**, 729–735.
48. Vai, M., Brambilla, L., Orlandi, I., Rota, N., Ranzi, B. M., Alberghina, L., et al. (2000). Improved secretion of native human insulin-like growth factor 1 from gas1 mutant *Saccharomyces cerevisiae* cells. *Appl. Environ. Microbiol.* **66**, 5477–5479.
49. Gemmill, T. R. and Trimble, R. B. (1999) Overview of N- and O-linked oligosaccharide structures found in various yeast species. *Biochim. Biophys. Acta* **1426**, 227–237.
50. Sumi, A., Okuyama, K., Kobayashi, K., Ohtani, W., Ohmura T., and Yokoyama, K. (1999)

Purification of recombinant human serum albumin. Efficient purification using STREAM-LINE. *Bioseparation* **8**, 195–200.

51. Weik, R., Striedner, G., Francky, A., Raspor, P., Bayer, K., and Mattanovich, D. (1999) Induction of oxidofermentative ethanol formation in recombinant cells of *Saccharomyces cerevisiae* yeasts. *Food Technol. Biotechnol.* **37**, 191–194.

52. Hong, F., Meinander, N. Q., and Jonsson, L. J. (2002) Fermentation strategies for improved heterologous expression of lactase in *Pichia pastoris. Biotechnol. Bioeng.* **79**, 438–449.

53. Zhang, W., Smith, L. A., Plantz, B. A., Schlegel, V. L. and Meagher, M. M. (2002) Design of methanol feed control in *Pichia pastoris* fermentations based upon a growth model. *Biotechnol. Prog.* **18**, 1392–1399.

54. Loewen, M. C., Liu, X., Davies, P. L., and Daugulis, A. J. (1997) Biosynthetic production of type II fish antifreeze protein: fermentation by *Pichia pastoris. Appl Microbiol. Biotechnol.* **48**, 480–486.

55. Curvers, S., Brixius, P., Klauser, T., Thommes, J., Weuster-Botz, D., Takors, R., et al. (2001) Human chymotrypsinogen B production with *Pichia pastoris* by integrated development of fermentation and downstream processing. Part 1. Fermentation. *Biotechnol. Prog.* **17**, 495–502.

56. Porro, D., Martegani, E., Ranzi, B. M. and Alberghina, L. (1991) Heterologous gene expression in continuous cultures of budding yeast. *Appl. Microbiol. Biotechnol.* **34**, 632–636.

57. Goodey, A. R. (1993) The production of heterologous plasma proteins. *Trends Biotechnol.* **11**, 430–433.

58. Stephanopoulos, G., Aristodou. A., and Nielsen, J. (1998) *Metabolic Engineering.* Academic Press, Inc., San Diego, CA.

17

Controlled Expression of Homologous Genes by Genomic Promoter Replacement in the Yeast *Saccharomyces cerevisiae*

Kevin J. Verstrepen and Johan M. Thevelein

Summary

Exchange of the promoter of a gene in the genome for another promoter whose expression can be controlled easily can overcome problems associated with the expression of the same gene from a promoter on a plasmid. Some genes are difficult or impossible to clone in plasmid-based vectors and often a stable expression and maintenance of the gene during cell proliferation is desirable. We present a method by which any genomic promoter can be replaced by a promoter of choice to achieve controlled (or constitutive and strong) expression of the gene concerned. The new promoter and a marker gene of choice are amplified by PCR using primers with a tail homologous to the regions adjacent to the site of integration in the genome and primers with a restriction site allowing ligation of the promoter and marker PCR products. After ligation of these PCR products, the ligated construct is transformed into yeast cells and allowed to exchange for the original promoter by homologous recombination. The transformants are selected based on the presence of the marker gene and proper exchange of the original promoter for the new promoter is checked by means of PCR amplification using primers in the new promoter and in the gene under its control.

Key Words: *Saccharomyces cerevisiae*, chromosomal promoter replacement.

1. Introduction

Controlled expression of both homologous and heterologous genes in the yeast *Saccharomyces cerevisiae* is most commonly achieved by using plasmid constructs. These expression vectors typically contain a multiple cloning site situated between a specific promoter and terminator element. The open reading frame (ORF) of interest is then cloned into this multiple cloning site so that its expression is controlled by the plasmid

From: *Methods in Molecular Biology, vol. 267:*
Recombinant Gene Expression: Reviews and Protocols, Second Edition
Edited by: P. Balbás and A. Lorence © Humana Press Inc., Totowa, NJ

promoter element. In most cases, these expression plasmids are so-called shuttle vectors, allowing cloning steps in *Escherichia coli* before introducing the construct in yeast. While the series of available yeast expression vectors is extensive, with a range of different promoters, multiple cloning sites and plasmid types (YRps, YEps, and YIps), there are some possible disadvantages to this system. Firstly, an appropriate plasmid construct, containing the required promoter sequence, multiple cloning site, and selection markers, may not always be available. Constructing such a plasmid may be difficult and time-consuming. Furthermore, not all ORFs are clonable into (expression) plasmids. Several reports of such so-called unclonable sequences exist (*1–3*).

Another disadvantage of the use of plasmids for gene expression is the introduction of extra DNA sequences other than the necessary selection marker, promoter, and ORF in the yeast genome. A common example is the presence of an antibiotic resistance marker gene for selection in *E. coli*. While this may not be a problem for standard laboratory experiments, it can be an issue for the industrial use of the transformed yeast strain, as a safety clearance is needed according to the respective laws and guidelines. This is not the case when the transformed yeast strain contains only DNA sequences found in the host *S. cerevisiae* and does not imply any danger for humans, animals, and plants. In this case, the recombinant strain can be regarded as a so-called self-cloning strain rather than a genetically modified strain. The use of self-cloning microorganisms is not restricted by laws for the contained industrial use of genetically modified organisms (GMOs) (e.g., European Council Directives 90/219/EEC and 91/448/EEC). But even in other cases, where self-cloning organisms are not exempted from the guidelines, the use of self-cloning may help to comply with GMO guidelines. Therefore, self-cloning strategies may offer a possibility to obtain genetically modified organisms that are more easily accepted by both legal authorities and the general public. The PCR-based method for the controlled expression of genes in *S. cerevisiae* may therefore offer a good alternative for the traditional expression vectors. This fast method does not require specific plasmids and there are no intermediary cloning steps in *E. coli*. Furthermore, no special equipment other than standard PCR apparatus, DNA gel electrophoresis, and commercially available enzyme kits is needed. The method allows controlled expression of any *S. cerevisiae* (homologous) gene by replacing its promoter sequence with a suitable (homologous or heterologous) promoter. The method has for example been successfully used for the controlled expression of the unclonable *FLO1* gene in brewer's yeast (*4*).

2. Materials

All solutions, recipients, and equipment should be clean, sterile, and detergent-free. Water should be autoclaved, distilled, and deionized (conductivity <10 μS).

PCR and restriction enzymes and their respective buffer solutions are stored at −30°C and always kept on ice when used. All other solutions are stored at room temperature unless stated otherwise.

2.1. General

1. Tris buffer: 10 mM Tris-HCl in H_2O, pH 7.5.

2.2. PCR Reactions

1. PCR polymerase enzyme with proofreading activity and corresponding PCR reaction buffer (commercially available).
2. Primers: 100 μ*M* stock solution, 20 μ*M* working solution in ultra-pure H_2O.
3. 25 m*M* $MgCl_2$ solution (in H2O).
4. dNTP solution (commercially available from PCR enzyme producer).
5. Template DNA; 0.01–1 μg/μL (plasmid or genomic DNA in 10 m*M* Tris buffer, pH 7.5).

2.3. Restriction, Dephosphorylation, and Ligation Reactions

1. Appropriate enzymes and buffers (commercially available; shrimp alkaline phosphatase is recommended).

2.4 Gel Electrophoresis

1. Loading buffer (10X): 0.25% (w/v) bromophenol blue, 0.25% (w/v) xylene cyanol FF, 40% (w/v) sucrose; store at 4°C.
2. TAE running buffer (50X stock solution, dilute 50 times in H_2O before use): 24.2% (w/v) Tris-HCl pH 7.5, 5.71 % (w/v) acetic acid, 3.72 % (w/v) EDTA·$2H_2O$ (adjust to pH 8.0 with HCl) or TBE running buffer (better but more expensive) (10X stock solution, dilute 20 times in H_2O before use): 0.9 *M* Tris pH 7.5, 0.9 *M* boric acid, 25 mM Na_2EDTA (adjust to pH 8.2 with HCl).
3. DNA size/concentration marker (commercially available, store at −30°C or 4°C, refer to manual for use).

3. Methods

An outline of the procedure is given in **Fig. 1**.

1. Design primers for the selection marker gene (*see* **Note 1**). The forward primer should contain an appropriate restriction site to link the marker PCR product to the promoter PCR product (*see* **Note 2**). The reverse primer should have a 5' tail homologous to part of the sequence in the upstream region of the genomic promoter that is to be replaced (*see* **Note 3**).
2. Design primers for the new promoter sequence that will control the expression of the homologous gene. It is advisable to choose the primers in such a way that the PCR product is of a different length than the marker gene, as this will facilitate step 8. The forward primer should contain an appropriate restriction site to link the PCR product to the marker PCR product (*see* **Note 2**). The reverse primer should have a 5' tail homologous to the most upstream sequence of the genomic ORF that is to be controlled (e.g., from position + 1 to position + 80; *see* **Note 4**).
3. Perform the two PCR reactions for the marker gene and the promoter with the respective primers (*see* **Notes 5–7**). Check a sample of each PCR product (3 μL) by gel electrophoresis and estimate the concentration of PCR product by comparison against a commercially available DNA size/quantity marker.
4. Perform restriction reactions on each of the two PCR products with the appropriate restriction enzymes, corresponding to the sites that were introduced in the primers (*see* **Note 8**).
5. Optional step, necessary only if the restriction sites that were introduced in the two PCR products are the same or when complementary restriction enzymes were used that do not allow optimal cutting in the same buffer: Perform a dephosphorylation reaction on the shorter of the

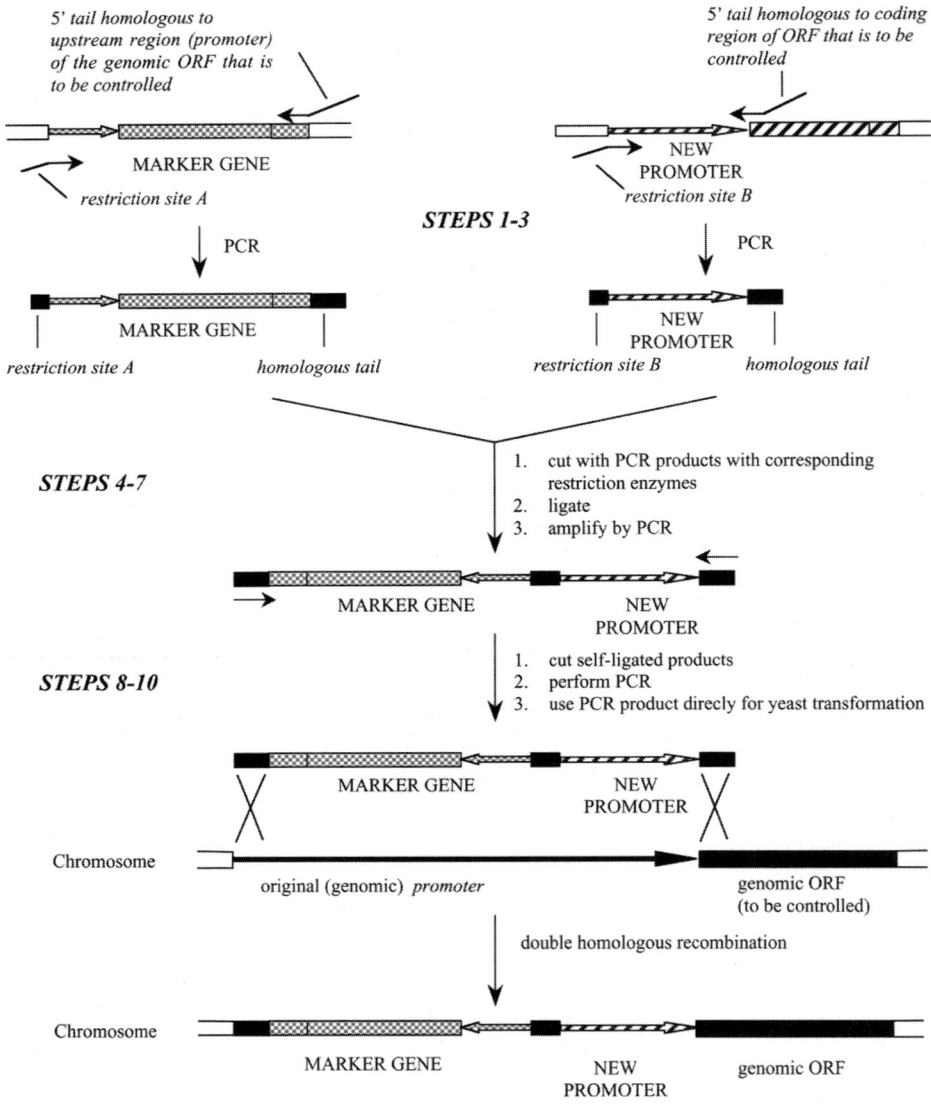

Fig. 1. Strategy overview.

two PCR reaction products (in most cases, this will be the promoter sequence). This step will avoid self-ligation of these shorter products. If shrimp alkaline phosphatase is used (recommended), inactivate the dephosphorylation enzyme by heating the mixture for 15 min at 65°C.

6. Purify the cut PCR products by gel electrophoresis and subsequently isolate the desired DNA band (*see* **Note 9**). Estimate the concentration of the two purified PCR products by performing gel electrophoresis on a sample and compare with known amount of DNA of comparable size (e.g., commercially available size/concentration markers; *see* **Note 8**).

7. Perform a ligation reaction with equimolar amounts of the two PCR products (refer to ligase product manual for details).

8. Optional step, to be performed only when two complementary restriction enzymes were used in **step 4**: Perform a restriction reaction on the ligation product with both complementary restriction enzymes. In most cases, it is sufficient to just add appropriate volumes of restriction buffer and enzyme(s) directly to the ligation mixture. If both enzymes can not be used in a single optimal buffer, perform **step 5** as indicated and use the restriction enzyme that cuts the longest (usually the marker gene) of the two PCR products generated in **step 3**. This step facilitates the PCR reaction in **step 9**.

9. Perform PCR reactions with 0.5–5 μL of the ligation product as a template. Primers are the two reverse (long) primers. Optimize elongation times for the ligation product (product length = marker length + promoter length; *see* **Notes 5–7** for tips to optimize the reaction). Check a sample of each PCR product (3 μL) by gel electrophoresis and estimate size and concentration by comparison against a commercially available DNA size/quantity marker (*see* **Note 8**). If it is unsure that the PCR product is the marker::promoter fusion, perform restriction reactions.

10. Use 1–10 μg of PCR product directly for transformation of *S. cerevisiae* and select for the presence of the marker gene. Different marker genes and selection procedures were previously described by Mount et al. (*5*). All current transformation procedures for *S. cerevisiae* are applicable, the most common and very effective one being the lithium-acetate method (*6*). Transformants should be further evaluated by PCR for correct exchange of the original genomic promoter by the new marker-promoter construct. Use the forward primer for the promoter sequence designed in **step 2** and a reverse primer that anneals in the ORF of which the expression is to be controlled. Always perform a negative control with template genomic DNA extracted from wild-type cells. If the PCR reaction of the negative control results in a product of the same size as the expected size for positive transformants, the PCR reaction should be made more stringent (e.g., by elevating annealing temperatures, by shortening elongation times and/or by adding less $MgCl_2$ (*see* **Notes 5–7**). An additional restriction reaction may also allow to distinguish the correct PCR product from nonspecific products.

4. Notes

1. Make sure the PCR product encodes a complete functional marker gene, including the necessary promoter and terminator sequences. These marker genes can be easily obtained by PCR amplification, using yeast plasmids that contain the desired marker genes as a template. A thoughtful choice of the marker gene can facilitate further selection. Auxotrophic markers such as *LEU2*, *HIS3*, *URA3*, and *LYS2* are commonly used for transformation of most laboratory strains. All these auxotrophic markers are homologous genes and thus suited for self-cloning purposes. However, dominant marker genes, such as CUP^R, KAN^R, and $G418^R$, offer the advantage of being applicable for all strains, including polyploid and aneuploid industrial *S. cerevisiae* strains. An interesting marker is the dominant *SMR1-410* marker gene that allows selection on sulfometuron methyl (SMM)-containing media. *SMR1-410* is an allele of the *S. cerevisiae ILV2* gene, so that the marker is usable for self-cloning strategies and all *S. cerevisiae* strains, including industrial polyploid, aneuploid, and prototrophic strains (*5,7*).

2. If the two PCR products (promoter and marker) are of the same size, it is recommended to choose complementary restriction sites. This allows ligation of the two PCR products while unwanted ligation products (promoter-promoter and marker-marker fusions) can be removed by cutting the ligation mixture with both restriction enzymes prior to the subsequent third PCR reaction (**step 9**). Even if the two PCR products are not of the same length, it is

advisable to choose complementary restriction sites, as this greatly facilitates the subsequent PCR reaction. If the use of two complementary restriction enzymes is impossible, an extra dephosphorylation step (*see* **Subheading 3.5.**) is necessary. To facilitate the restriction reactions, it is very important to add 4–8 nucleotides (standard is 6 nucleotides) at the 5' end of the primer, so that the restriction site is flanked by at least 4 nucleotides at each side. Check restriction enzyme specifications or contact supplier for more details.

3. The homologous primer tails are chosen in such a way that the genomic promoter sequence is completely replaced by the PCR construct. This means that the tail of the downstream primer has to be complementary to the most upstream *coding* sequence of the ORF that is to be controlled (e.g., from position +1 to +80 of the ORF).

4. While some *S. cerevisiae* promoters can be as long as 3 kb, in most cases it is sufficient to replace about 1–2 kb of the promoter sequence (e.g., upstream primer tail homologous to the sequence between −1420 and −1500 of the ORF). To increase the efficiency of double homologous recombination, the homologous tail of the primers should be as long as possible, preferably at least 70–100 nucleotides.

5. It is advisable to use commercially available DNA polymerases with proofreading activity. For long products (>4–5 kb), a special long-template enzyme (commercially available) may be useful. For optimal efficiency of all PCR reactions, follow the protocol provided with the PCR enzyme kit (*see* **Notes 6 and 7**).

6. Dissolve primers to a concentration of 100 picomol/μL in ultra-pure water. From this stock solution, a 20 picomol/μL working solution is made by diluting the stock five times with ultra-pure water. This working solution can be stored at −30°C for months, but it is advisable to prepare a new working solution from the stock after 20 cycles of thawing-refreezing or after 12 months of storage.

7. Store all PCR solutions at −30°C and keep on ice when thawed. The template DNA should be of good quality; e.g., yeast genomic DNA isolated through phenol-chloroform extraction, as described by Piper (*8*).

   To obtain maximal efficiency, it is extremely important to work with an optimal PCR cycle program (see instructions with PCR enzyme). The next cycle program can serve as a useful example:

   a. 1 min 30 s at 94°C.
   b. 18 s at 94°C.
   c. 30 s at annealing temperature (usually 50–60°C).
   d. X s at elongation temperature (usually 68–72°C; X depends on the desired product length. Check PCR enzyme specifications).
   e. Repeat **steps 2–4** 10 times.
   f. 18 s at 94°C.
   g. 30 s at annealing temperature (usually 50–60°C); we usually increase the annealing temperature of **step 3** by 2–4°C.
   h. X s + 5 s/cycle at elongation temperature (usually 68–72°C). For long PCR products, it is extremely important to increase the elongation time with 5 s per cycle.
   i. Repeat **steps 6–8** 15 to 20 times.
   j. 10 min at elongation temperature.
   k. Store at 2°C or freeze at −30°C.

To optimize the PCR efficiency, it may be necessary to try different annealing temperatures and elongation times and to vary the concentration of $MgCl_2$.

8. Usually 0.5–1 μg (5–10 μL) of DNA is cut in a reaction volume of 20 μL. Add 0.5–1 μL (5–10 U) of restriction enzyme and incubate for 4 h at appropriate temperature. For difficult restriction reactions, it is advisable to add another 0.5 μL of enzyme after 4 h of incubation, after which the restriction reaction is continued for another 4 h.
9. Prepare an agarose (electrophoresis grade) gel in fresh 1X TAE or 0.5X TBE (better but more expensive) running buffer. The appropriate agarose concentration depends on the size of the expected band:

Agarose concentration (% w/v)	Efficient separation range (kbp)
0.3	5–60
0.5	1–25
0.7	0.8–10
0.9	0.5–7
1.2	0.4–6
1.5	0.2–4
2.0	0.1–3

Boil the agarose solution until all agarose has dissolved; it is safer to let the solution boil for at least 30 s, even if the solution already appears to be completely clear. Cool the agarose solution to approx 60–70°C and then pour it into the gel tray (refer to manual of gel trays and electrophoresis system for more directions).

References

1. Teunissen, A., Van Den Berg, J., and Steensma, H. Y. (1993). Physical localization of the flocculation gene on chromosome I of *Saccharomyces cerevisiae*. *Yeast* **9**, 1–10.
2. Razin, S. V., Ioudinkova, E. S., Trifonov E. N., and Scherrer, K. (2001). Non-clonability correlates with genomic instability: a case study of a unique DNA region. *J. Mol. Biol.* **307**, 481–486.
3. Voet, M., Defoor, E., Verhasselt, P., Riles, L., Robben, J., and Volckaert, G. (1997). The sequence of a nearly unclonable 22 center dot 8 kb segment on the left arm of chromosome VII from *Saccharomyces cerevisiae* reveals *ARO2, RPL9A, TIP1, MRF1* genes and six new open reading frames. *Yeast* **13**, 177–182.
4. Verstrepen, K. J., Derdelinckx, G., Delvaux, F. R., Michiels, C., Winderickx, J., Thevelein, J., Bauer, F. F., and Pretorius, I. S. (2001). Late fermentation expression of *FLO1* in *Saccharomyces cerevisiae*. *J. Am. Soc. Brew. Chem.* **59**, 69–76.
5. Mount, C. R., Jordan, B. E., and Hadfield, C. Transformation of lithium-treated yeast cells and the selection of auxotrophic and dominant markers. In *Yeast Protocols* (Evans, I. H., ed.), Humana Press, Totowa, NJ, pp. 103–107.
6. Gietz, R. D. and Schiestl, R. H. (1995). Transforming yeast with DNA. *Methods Mol. Cell. Biol.* 5, 255–269.
7. Casey, G. P., Xiao, W., and Rank, G. H. (1988). A convenient dominant selection marker for gene transfer in industrial strains of *Saccharomyces cerevsisiae: SMR1* encoded resistance to the herbicide sulfometuron methyl. *J. Inst. Brew.* **94**, 93–97.
8. Piper, P. (1996). Isolation of yeast DNA. In *Yeast Protocols* (Evans, I. H., ed.), Humana Press, Totowa, NJ, pp. 103–107.

18

High-Throughput Expression in Microplate Format in *Saccharomyces cerevisiae*

Caterina Holz and Christine Lang

Summary

We have developed a high-throughput technology that allows parallel expression, purification, and analysis of large numbers of cloned cDNAs in the yeast *Saccharomyces cerevisiae*. The technology is based on a vector for intracellular protein expression under control of the inducible *CUP*1 promoter, where the gene products are fused to specific peptide sequences. These N-terminal and C-terminal epitope tags allow the immunological identification and purification of the gene products independent of the protein produced. By introducing the method of recombinational cloning we avoid time-consuming re-cloning steps and enable the easy switching between different expression vectors and host systems.

Key Words: *Saccharomyces cerevisiae;* microplate format; *CUP*1 promoter; epitope tagging.

1. Introduction

The translation of genomic sequence information into functional information is an essential step toward understanding biological processes. Due to the large number of genes, which have been catalogued by worldwide genome sequencing projects *(1,2)*, high-throughput technologies that allow parallel expression, purification, and analysis of large numbers of clones representing different proteins had to be developed. Flexible methods that enable rapid DNA cloning are needed to allow easy switching between different vector and host expression systems; the purification protocol has to work independently of the protein produced. This requirement is generally addressed by employing N- and/or C-terminal expression tags for specific high-affinity protein binding to suitable resins *(3)*. A micro-scale process for high-throughput (HT) expression of cloned cDNAs in *Saccharomyces cerevisiae* was developed by miniaturizing the procedures for

From: *Methods in Molecular Biology, vol. 267:*
Recombinant Gene Expression: Reviews and Protocols, Second Edition
Edited by: P. Balbás and A. Lorence © Humana Press Inc., Totowa, NJ

cloning, expression, and purification, enabling all steps to be carried out in 96-well microtiter plates *(4)*. Here we describe the detailed protocol of the developed HT technology based on the epitope expression vector pYEXTHS-BN allowing the copper-inducible expression of cDNA inserts.

2. Materials

1. pYEXbx expression system (BD Biosciences, Clontech, Palo Alto, CA).
2. Human cDNAs (German Resource Center, Berlin, Germany).
3. *E. coli* strain Xl-1 blue (*recA1, end A1, gyrA96, thi-1, hsdR17, supE44, relA1, lac* [F'*proAB lacl*^{q} Δ*ZM15* Tn*10* (Tet^r)].
4. Oligonucleotide primers.
5. ProofStart™ DNA Polymerase (Qiagen, Hilden, Germany).
6. Restriction enzymes and T4 DNA ligase (New England Biolabs, Beverly, MA).
7. Alkaline phosphatase, from shrimp (Roche Molecular Biochemicals, Germany).
8. 96 TubePlate for PCR (Biozym Diagnostik GmbH, Germany).
9. Agarose gel equipment.
10. Montage PCR96 Purification Kit (Millipore, Bedford, MA).
11. Suspension culture plate, 96-well, flat-bottom with lid (Greiner Bio-one, Frickenhausen, Germany).
12. *Saccharomyces cerevisiae* strain AH22ura3pep4 (MATa, *ura3*Δ, *leu*2-23,112, *his*4-519, *can*1, *pep*4Δ).
13. Lithium acetate, sorbitol.
14. PEG 3350.
15. Deoxyribonucleic acid, single-stranded herring sperm (Fluka, Sigma-Aldrich Corp., St. Louis, MO).
16. DMSO (Dimethyl sulfoxide).
17. YNBD w/o Ura medium: 2% dextrose, 0.67% yeast nitrogen base (Difco), 40 mg/L leucine, 40 mg/L histidine.
18. WMVIII medium according to Lang and Looman *(5)*: basic medium (per 1 L): 0.25 g $NH_4H_2PO_4$, 2.8 g NH_4Cl, 0.25 g $MgCl_2 \cdot 6H_2O$, 0.1 g $CaCl_2 \cdot 2H_2O$, 2.0 g KH_2PO_4, 0.55 g $MgSO_4 \cdot 7H_2O$, 0.075 g Myo-Inositol, 10.0 g sodium L-glutamate monohydrate, 50 g sucrose (autoclave). Trace element stock solution: 0.4 m*M* $CuSO_4$, 1.8 mM $FeSO_4$, 0.5 m*M* $MnCl_2$, 0.4 mM Na_2MoO_4, 6mM $ZnSO_4$, 10 m*M* EDTA (filter-sterilize); add 4 mL to 1 L medium. Vitamin stock solution (per 100 mL): 62.5 mg biotin, 1.25 g Ca-panthotenate, 250 mg nicotinic acid, 626 mg pyridoxine, 250 mg thiamine (filter-sterilize and store at 4°C), add 4 mL to 1 L medium.
19. 96 deep-well plates (Whatman, Maidstone, Kent, UK).
20. $CuSO_4$.
21. PMSF (phenylmethylsulfonyl fluoride).
22. PBS: 8.4 m*M* Na_2HPO_4, 1.6 m*M* KH_2PO_4, 150 m*M* NaCl.
23. Lysis buffer: 100 m*M* (NaH_2PO_4), pH 8.0, 300 m*M* NaCl, 1% TritonX-100.
24. Acid-washed glass beads (0.5 mm).
25. MultiScreen-HV Clear Plates (Millipore).
26. 4X SDS PAGE sample buffer: 0.2 *M* Tris-HCl, pH 6.8, 8% (w/v) SDS, 40% (v/v) glycerol, 5% (v/v) 2-mercaptoethanol and 0.4% (w/v) bromphenol blue.
27. 50% Ni-NTA agarose (Qiagen).

28. Wash buffer: 50 mM NaH$_2$PO$_4$, pH 8.0, 300 mM NaCl, 5 mM imidazole.
29. Elution buffer: 50 mM NaH$_2$PO$_4$, pH 8.0, 300 mM NaCl, 250 mM imidazole.
30. SDS-PAGE equipment.
31. Equipment for Western blotting, 3 MM Whatman filter paper, polyvenylidene (PVDF) membrane.
32. BSA (bovine serum albumin).
33. PBST: 8.4 mM Na$_2$HPO$_4$, 1.6 mM KH$_2$PO$_4$, 150 mM NaCl, and 0.1% (v/v) Tween-20.
34. Avidin (Sigma-Aldrich Corp., St. Louis, MO).
35. StrepTactin HRP conjugate (IBA, Göttingen, Germany).
36. Anti-Penta His antibody (Qiagen).
37. Nonfat dried milk.
38. Rabbit-anti-mouse IgG/HRP conjugate (DAKO, Glostrup, Denmark).
39. Western Lightning Chemiluminescence Reagent Plus (Perkin-Elmer Life Science, Boston, MA).

3. Methods

The methods described here are divided into the following sections: (1) cloning of the cDNAs in *E. coli*, (2) subcloning of the cDNAs by recombinational cloning in *S. cerevisiae*, (3) verification of yeast transformants, (4) protein expression and purification, and (5) characterization of the protein.

3.1. Expression Plasmid

The design of the yeast expression plasmid and the steps for the cloning of the cDNAs are described in **Subheadings 3.1.1.** and **3.1.2.**

3.1.1. pYEXTHS-BN Expression Vector

The epitope tag expression vector pYEXTHS-BN (**Fig. 1**) is based on the episomal plasmid pYEXbx (Clontech). The transcription of the cDNAs inserted in the cloning site is controlled by the Cu^2+-regulated *CUP1* promoter from the yeast metallothionein gene. The promoter is rapidly induced by copper sulfate (0.01 to 1 mM, depending on the copper resistance of the host strain) *(6)*. The pYEXTHS-BN was generated from pYEXbx by a series of modifications *(4)*. The oligohistidine domain (His$_6$) including a translation initiation sequence was inserted N-terminally of the cloning site and the StrepII epitope *(7)* sequence was introduced at the C-terminus. The multiple cloning site was changed into a unique *Bam*HI/*Not*I cloning site to avoid additional amino acids from a translated polylinker left on the produced fusion protein. The system is designed for intracellular expression and yields the target protein as a dual-tagged gene product that can be purified by both immobilized-metal affinity chromatography (IMAC) *(8)* and StrepTactin sepharose (see **Note 1**).

3.1.2. Cloning of cDNAs

DNA manipulations were performed according to standard recombinant DNA methods *(9)*. Individual cDNA clones obtained from the German Resource Center, Berlin (RZPD) as bacterial stab cultures were inoculated into fresh LB medium containing the

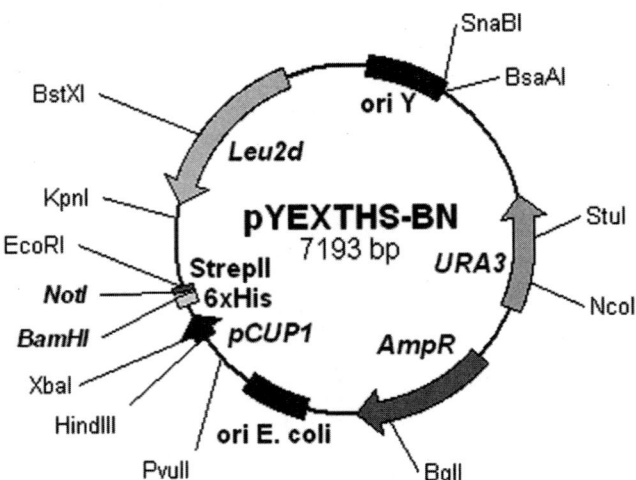

Fig. 1. Schematic drawing of the pYEXTHS-BN expression vector generated from the episo-mal plasmid pYEXbx. The modified cloning site contains unique *Bam*HI/*Not*I restriction sites. The His6 tag including a translation initiation sequence was inserted N-terminally of the cloning site and the StrepII tag was introduced at the C-terminus.

appropriate antibiotic grown overnight at 37°C. The target cDNAs were amplified in a first PCR with gene-specific primers, which contain an additional *Bam*HI and *Not*I restriction site overhang at the 5' ends of the sense and the antisense primers, respec-tively. The purified PCR products were cut with *Bam*HI/*Not*I and directionally ligated into the entry vector for recombinational cloning, which is the *E. coli* expression vector pQStrep2 in our case (GenBank accession AY028642) *(10)*. This vector, which carries the inserted cDNAs integrated in a His$_6$/StrepII expression cassette that is identical to that of the target vector pYEXTHS-BN, served as template for the second PCR. In this amplification step recombination sites required for the homologous recombination pro-cedure consisting of 60 nucleotides are attached to the target cDNAs (**Fig. 2**).

3.2. Recombinational Cloning and Transformation of S. cerevisiae

The next steps in this process involve the second PCR applying the two recombination primers 5'CupPHishomrec (5'-AATCATCACATAAAATATTCAGCGAATTGGATCT AAAATG<u>TCGCATCACCATCACCATCACGG</u>-3') and 3'StrepCupThomrec (5'-CTCA TGACCTTCATTTTGGAAGTTAATGAATTCA<u>TTTTTCGAACTGCGGGTGGCTC-CAAGCG</u>-3') *(see* **Note 2**) and the co-transformation of the PCR products and the linearized expression plasmid pYEXTHS-BN in the expression host *S. cerevisiae* AH22*ura3pep4 (see* **Note 3**) using a modified lithium acetate transformation protocol *(11)*.

1. Transfer 50-μL reaction aliquots of PCR-premix consisting of 5'CupPHishomrec and 3'StrepCupThomrec primer, dNTPs, polymerase reaction buffer, and ProofStart™ DNA polymerase or any other proof-reading polymerase into each well of a 96-well PCR plate. Follow manufacturers' instructions for units of enzyme and concentrations of other com-

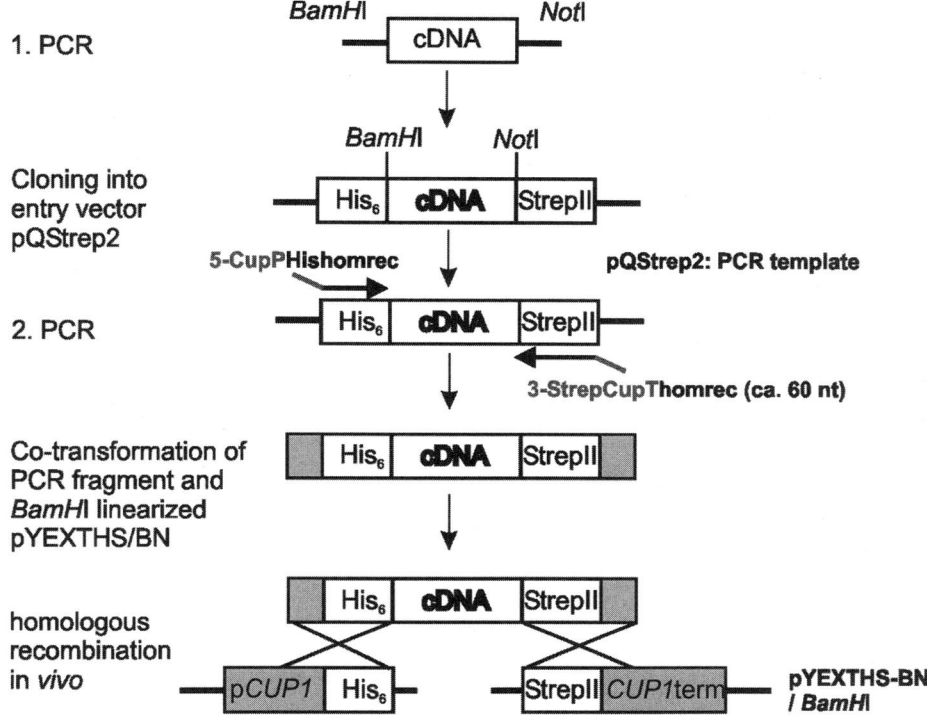

Fig. 2. Schematic diagram showing the homologous recombination procedure for cloning of heterologous cDNAs into the yeast expression plasmid pYEXTHS-BN. In a first PCR the target cDNAs are amplified with a *Bam*HI and *Not*I restriction site overhang. The PCR products are cut by *Bam*HI/*Not*I and are directionally ligated into the entry vector pQStrep2 (GenBank accession AY028642) that is used as a template for the second PCR. The complete expression cassette containing the target cDNA is amplified with the two *homrec* primers. The grey segment of the forward primer (5'CupPHishomrec) represents the region that is identical to a part of the *CUP1* promoter region of pYEXTHS-BN and the gray segment of the reverse primer (3'StrepCupThomrec) is homologous to the terminator sequence of the expression cassette in pYEXTHS-BN. The black segments in these primers are homologous to the tag sequences His$_6$ and StrepII present in the expression cassette of pQStrep2. The resulting PCR products are co-transformed with the linearized empty vector pYEXTHS-BN into the yeast cell and the expression cassette is integrated by homologous recombination in vivo.

pounds added to the reaction mix. Add 1 µL bacterial culture carrying the entry vector (pQEStrep2 with cDNA insert) to each well or alternatively inoculate the PCR reactions by using a 96-pin replicator.

2. Perform the PCR reaction: 94°C for 4 min; 5 cycles of 94°C for 45 s, 60°C for 45 s, 72°C for 2 min; 30 cycles of 94°C for 45 s, 72°C for 45 s, 72°C for 2 min and followed by 72°C for 5 min.

3. Digest the empty expression plasmid pYEXTHS-BN with *Bam*HI and dephosphorylate the cut DNA using standard molecular biology methods *(9)*. Store the controlled linearized plasmid at −20°C.

4. Inoculate 20 mL YPD medium in a 100-mL shake flask with a single colony of *S. cerevisiae* AH22*ura3pep4* as preculture, and grow overnight with shaking (200 rpm) at 28°C.

5. Inoculate 200 mL YEPD medium in a 1-L shake flask with 6–10 mL from the preculture to adjust the culture to an initial OD_{600} of 0.3. Grow with shaking (200 rpm) at 28°C for 4–6 h to reach an OD_{600} of 0.8–1.

6. Harvest cells by centrifugation (10 min, 2500g, 4°C).

7. Wash cell pellet in 50 mL cold distilled water. Repeat the spin as in step 6, discard the supernatant, and resuspend the cells in 6 mL 0.1 M lithium acetate/1 M sorbitol.

8. Incubate cell suspension for 30 min with shaking at 28°C.

9. Take an aliquot of 4 mL of these competent cells and add 23 mL PEG 3350 (60% w/v)/0.1 M lithium acetate and 500 µL herring sperm DNA (10 mg/mL) to prepare the transformation pre-mix.

10. Transfer aliquots of 275 µL of the pre-mix into each well of a 96-well suspension culture plate and add 5 µL (approx 200 ng) of each PCR reaction mix (**step 2**) and 200 ng of the linearized pYEXTHS-BN DNA (**step 3**). For re-ligation control, add the linearized vector DNA only to one sample. Incubate plates for 30 min at 28°C.

11. Add 30 µL DMSO to each well and perform heat shock for 7 min at 42°C.

12. Centrifuge the plates (10 min, 1280g, 4°C) and resuspend cell pellets in 200 µL distilled water.

13. Plate cells on YNB medium containing 2% glucose, leucine, and histidine and incubate 4 d at 28°C to select URA$^+$ transformants (*see* **Note 4**).

3.3. Verification of Correct Yeast Transformants by Colony PCR

Select 3 independent URA$^+$ colonies of each transformation approach and check for correct integration of the insert into the expression vector by colony PCR. Primers used are universal and hybridize at the *CUP1* promoter and terminator region of the expression cassette of pYEXTHS-BN (pYEXbx5' CACATAAAATATTCAGCGAATTGG and pYEXbx3' GCATTGGCACTCATGACC TTC). Transfer a small number of cells by toothpick into 20 µL PCR reaction mix and subsequently transfer the colonies onto YNBD plates containing only histidine.

Analyze aliquots of the PCR products by agarose gel electrophoresis (**Fig. 3**).

3.4. Protein Expression and Purification

The next steps in this process involve the induction of protein expression by copper sulfate, the extraction of the proteins from yeast cells, and the purification of the target proteins.

3.4.1. Micro-Scale Protein Expression

1. Transfer selected yeast transformants into 96-deep-well microplate filled with 1 mL WMVIII medium with 40 mg/L histidine per well and incubate 2 d at 28°C with shaking at 180 rpm.

2. Inoculate 1.4 mL WM VIII medium with 40 mg/L histidine in a 96-deep-well microplate with 20 µL of the pre-culture and incubate 24 h at 28°C with shaking.

3. Induce protein expression of the cultures by adding 1mM $CuSO_4$ and shake plate for another 24–48 h until the average OD_{600} of the cultures reaches 5–7.

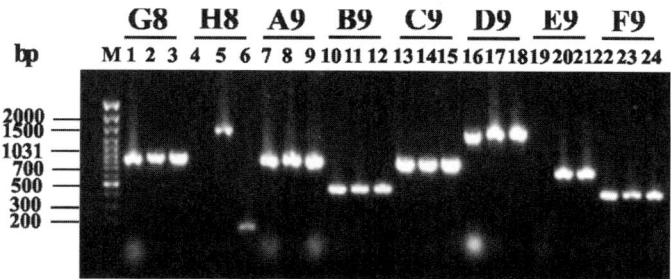

Fig. 3. Verification of correct *S. cerevisiae* transformants by colony PCR. Three independent transformants of each cDNA insert were checked for correct integration of the insert into the yeast expression vector. The PCR products from 24 yeast clones that are labeled with numbers resulting from eight different cDNA inserts (G8, H8, A9, B9, C9, D9, E9, and F9) are shown as an example. In rare cases, the product of the empty expression vector is observed (clones 6 and 16).

4. Harvest cells by centrifugation (10 min, 1280*g*, 4°C) and resuspend cell pellets in 1 mL PBS.
5. Repeat the centrifugation; discard the supernatant. The washed cell pellets can either be subjected to cell disruption directly or stored at −70°C for at least a month.

3.4.2. Cell Disruption

1. Resuspend the cells in 300 μL ice-cold lysis buffer and add PMSF (phenylmethylsulfonyl fluoride) to a final concentration of 1 m*M*.
2. Add one volume of sterile, acid-washed glass beads and disrupt cells by shaking 15 min at 500 rpm.

3.4.3. Protein Extraction

1. Remove beads and cellular debris by centrifugation (20 min, 1280*g*, 4°C).
2. Transfer the cell extract into a 96-well MultiScreen-HV Clear Plate (Millipore, 0.45 μm Durapore PVDF) and centrifuge as in step 1. Collect the filtrate in a 96-well plate.
3. Take 30 μL sample per well and add 10 μL 4 X SDS-PAGE sample buffer, heat the sample at 95°C for 5 min, and store at −20°C for SDS PAGE and Western blot analysis (*see* **Note 5).**

3.4.4. Purification

1. Apply 200 μL of each lysate filtrate to a fresh MultiScreen plate and add 35 μL of a 50% Ni-NTA resin to the protein extract and mix gently by shaking for 60 min at 4°C.
2. Spin plate for 2 min at 150*g* and collect the flow-through in a collection microtiter plate.
3. Wash resin three times with 150 μL wash buffer (50 m*M* NaPO$_4$, pH 8.0, 300 m*M* NaCl, 5 m*M* imidazole) per well. Centrifuge as in **step 2** and discard the filtrate.
4. Elute proteins twice with 75 μL of elution buffer. Collect elution fractions and analyze by SDS-PAGE and Western blotting.

3.5. Protein Characterization by Western Blotting and Immunodetection

1. Take 15 μL of the eluates and add 5 μL 4X SDS-PAGE sample buffer, heat the sample at 95°C for 5 min, and separate the proteins by 12.5% SDS-PAGE (*12*).

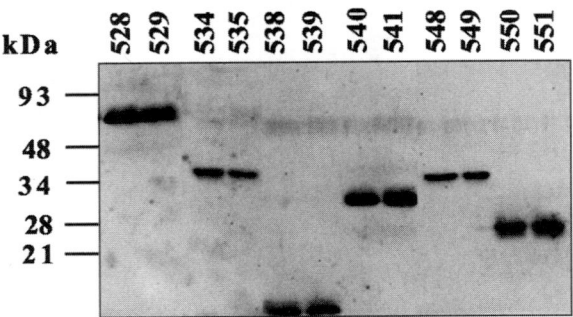

Fig. 4. Expression of different human cDNAs in *S. cerevisiae* in microtiter format. Cleared cell lysates were prepared from 1.4-mL cultures of 12 different cDNA expressing strains. The clones are labeled with the last three digits of a 10-digit bar code. Each two clones side by side contain the same cDNA insert. The dual-tagged proteins were purified via Ni-NTA, separated by SDS-PAGE, and detected with StrepTactin-HRP conjugate.

2. Transfer proteins in the polyacrylamide gel to polyvenylidene difluoride (PVDF) membrane by semi-dry electroblotting according to manufacturer's instructions (*see* **Note 6**).

3. Block membrane for at least 1 h with 2% BSA in PBST (8.4 mM Na$_2$HPO$_4$, 1.6 mM KH$_2$PO$_4$, 150 mM NaCl, and 0.1% [v/v] Tween 20). If using StrepTactin HRP conjugate to detect the C-terminally fused StrepII tag, pretreat the membrane in PBST containing 2 μg/mL avidin for 30 min to block biotinylated proteins.

4. Incubate in primary antibody (penta-His antibody) diluted to 1:1000/1:2000 or StrepTactin HRP conjugate diluted to 1:4000 in blocking solution at room temperature for 1 h (*see* **Note 7**).

5. Wash membrane three times for 10 min each time in PBST at room temperature. When using StrepTactin HRP conjugate, proceed directly to **step 8**.

6. Incubate with secondary antibody solution for 1 h at room temperature. Either alkaline phosphatase (AP) or horseradish peroxidase (HRP) conjugated anti-mouse IgG may be used (*see* **Note 8**). Dilute according to manufacturer's recommendations. Use 5% nonfat dried milk in PBST for incubation with the secondary antibody when using the chemiluminescent detection method.

7. Wash filter four times for 10 min each time in PBST at room temperature.

8. Perform detection reaction with AP or HRP chemiluminescence reagent (*see* **Note 9**), and expose to X-ray film or detect by a luminescent imaging system according to the manufacturer's recommendations (for results *see* **Fig. 4**).

4. Notes

1. The introduction of the N-terminal His$_6$-tag and the C-terminal StrepII-tag enables the immunological identification of full-length gene products and allows the purification of non-truncated proteins by a two-step affinity chromatography.

2. The forward primer 5'CupPHishomrec is a chimeric primer composed of 40 nucleotides homologous to the *CUP*-1 promoter region of the pYEXTHS-BN (including the start codon) plus 23 nucleotides annealing to the His tag sequence of the pStrep2 expression cassette

(underlined sequence). The reverse primer 3'StrepCupThomrec is composed of 34 nucleotides derived from the terminator region of the pYEXTHS-BN plus 28 nucleotides annealing to the StrepII tag sequence of the pQStrep2 (underlined sequence).

3. Additional *S. cerevisiae* strains suitable for copper-inducible protein expression are DY150 (MAT a, *ura3-52, leu2-3, 112, trp1-1, ade2-1, his3-11, can1-100*) and DBY747 (MAT a, *his3-11, leu2-3, 112, trp1-289, ura3-52, CUP1*ᴿ). We suggest using protease-deficient yeast strains to avoid degradation of the target protein during the cultivation or the cell harvest and recovery procedures and thus improve the quality and the yield of heterologous protein. Otherwise the addition of protease inhibitor cocktails is recommended.

4. Transformants should first be selected on medium lacking only uracil. Afterwards, cells are cultivated in selective medium lacking both uracil and leucine for protein expression to achieve a higher copy number of the expression vector pYEXTHS-BN. Copy numbers of 100–400 per cell are obtained by using the partially defective selectable maker *leu2-d*, which has an intact coding sequence but a truncated promoter (*13*). Therefore, a high copy number of the *leu2-d* gene is required to synthesize sufficient gene product for the transformants to grow in the absence of leucine.

5. The recombinant target gene product cannot be detected directly from the total yeast protein after gel electrophoresis and staining due to the low proportion of recombinant protein. This requires purifying the recombinant protein from crude extract or detecting the target protein by immunoblotting.

6. We used the blotting apparatus from H. Hölzel GmbH for semi-dry electroblotting. The current density required is determined by the size of the gel: $1 mA/cm^2$ is recommended (1 h transfer).

7. Do not use dilution buffer containing milk powder for anti-His antibody. This will reduce sensitivity. Do not use milk powder for StrepTactin HRP conjugate either, as it contains large quantities of cross-reacting biotin.

8. Secondary anti-mouse IgG antibodies conjugated either with alkaline phosphatase or horseradish peroxidase are available from many suppliers. We obtained good results with rabbit-anti-mouse IgG/HRP-conjugate from DAKO (Denmark). It is important that the secondary antibodies used recognize mouse IgG1 and that they are used at the highest dilution recommended to avoid nonspecific signals. Detection with anti-His HRP and Ni-NTA conjugates available from Qiagen does not require the use of secondary antibodies.

9. Chromogenic or chemiluminescent substrates for detection are applicable. In contrast to chromogenic substrates, chemiluminescent substrates allow multiple exposures so that the result obtained can be optimized and are more sensitive. In our experience the Western Lightning Chemiluminescence Reagent Plus (Perkin-Elmer) yields good results. It is essential to use suitable negative and positive controls for the blotting and detection procedure. As a negative control, a cleared cell lysate from a yeast clone, which carries the empty expression vector without insert, is usually applied. As a positive control, a purified His6- and StrepII-tagged protein, such as GFP (green fluorescence protein), that is easy to express in large amounts and to detect may be used.

Acknowledgments

The authors thank Natalia Bolotina for excellent technical assistance and Roslin Bensmann for help with the English. This work was supported by the Bundesministerium für Bildung und Forschung through the Leitprojekt Proteinstrukturfabrik (to C. L.).

References

1. Lander, E. S., Linton, L. M., Birren, B., Nusbaum, C., Zody, M. C., Baldwin, J., et al. (2001) Initial sequencing and analysis of the human genome. *Nature* **409**, 860–921.
2. Venter, J. C., Adams, M. D., Myers, E. W., Li, P. W., Mural, R. J., Sutton, G. G., et al. (2001) The sequence of the human genome. *Science* **291**, 1304–1351.
3. Nilsson, J., Stahl, S., Lundeberg, J., Uhlen, M., and Nygren, P. A. (1997) Affinity fusion strategies for detection, purification, and immobilization of recombinant proteins. *Protein Expr. Purif.* **11**, 1–16.
4. Holz, C., Hesse, O., Bolotina, N., Stahl, U., and Lang, C. (2002) A micro-scale process for high-throughput expression of cDNAs in the yeast *Saccharomyces cerevisiae*. *Protein Expr. Purif.* **25**, 372–378.
5. Lang, C. and Looman, A. C. (1995) Efficient expression and secretion of *Aspergillus niger* RH5344 polygalacturonase in *Saccharomyces cerevisiae*. *Appl. Microbiol. Biotechnol.* **44**, 147–156.
6. Etcheverry, T. (1990) Induced expression using yeast copper metallothionein promoter. *Methods Enzymol.* **185**, 319–329.
7. Voss, S. and Skerra, A. (1997) Mutagenesis of a flexible loop in streptavidin leads to higher affinity for the Strep-tag II peptide and improved performance in recombinant protein purification. *Protein Eng.* **10**, 975–982.
8. Porath, J., Carlsson, J., Olsson, I., and Belfrage, G. (1975) Metal chelate affinity chromatography, a new approach to protein fractionation. *Nature* **258**, 598–599.
9. Sambrook, J., Fritsch, E. F., and Maniatis, T., (1989) *Molecular Cloning, A Laboratory Manual*, 2nd ed. Cold Spring Harbor Laboratory Press, Cold Spring Harbor, NY.
10. Holz, C., Prinz, B., Bolotina, N., Sievert, V., Büssow, K., Simon, B., et al. (2003) Establishing the yeast *Saccharomyces cerevisiae* as a system for expression of human proteins on a proteome-scale. *J. Struct. Funct. Genomics* **4,** 97–108.
11. Gietz, D., St Jean, A., Woods, R. A., and Schiestl, R. H. (1992) Improved method for high efficiency transformation of intact yeast cells. *Nucleic Acids Res.* **20**, 1425.
12. Laemmli, U. K. (1970) Cleavage of structural proteins during the assembly of the head of bacteriophage T4. *Nature* **227**, 680–685.
13. Erhart, E. and Hollenberg, C. P. (1983) The presence of a defective LEU2 gene on 2 mu DNA recombinant plasmids of *Saccharomyces cerevisiae* is responsible for curing and high copy number. *J. Bacteriol.* **156**, 625–635.

19

High-Throughput Expression
in Microplate Format in *Pichia pastoris*

Mewes Böttner and Christine Lang

Summary

The methylotrophic yeast *Pichia pastoris* has become a powerful host for the heterologous expression of proteins. To serve the increasing demand for clones expressing different cDNAs, we developed a cultivation and induction protocol amenable to automation to increase the number of clones screened for protein expression. Therefore cDNAs are cloned for intracellular expression. The resulting fusion proteins carry affinity tags (6*HIS and StrepII, respectively) at the N- and C-terminus for the immunological detection and chromatographic purification of full-length proteins. Expression is controlled by the tightly regulated and highly inducible alcohol oxidase 1 (AOX1) promoter. The screening procedure is based on a culture volume of 2 mL in a 24-well format. Lysis of the cells occurs via chemical lysis without mechanical disruption. Using the optimized feeding and induction protocol, we are now able to screen for and identify expression clones that produce heterologous protein with a yield of 2 mg per L culture volume or higher.

Key Words: *Pichia pastoris;* posttranslational modifications; expression screening; microplates; AOX1 promoter.

1. Introduction

The methylotrophic yeast *Pichia pastoris* has become a powerful system for the heterologous expression of proteins during recent years. High yields of recombinant protein have been reported in many cases. Molecular genetic manipulations are performed using techniques similar to those well established for *Saccharomyces cerevisiae*. Both *S. cerevisiae* and *P. pastoris* are able to introduce eukaryotic posttranslational modifications in proteins (reviewed in *1,2*). The chain length and the distribution of *N*-linked oligosaccharides are, however, different between these yeasts. As the chain length is

From: *Methods in Molecular Biology, vol. 267:*
Recombinant Gene Expression: Reviews and Protocols, Second Edition
Edited by: P. Balbás and A. Lorence © Humana Press Inc., Totowa, NJ

significantly shorter in *P. pastoris*, it is an interesting alternative for the extracellular expression of human proteins *(3)*.

Using defined minimal media, *P. pastoris* can easily be grown to high cell densities. Routinely, the alcohol oxidase 1 (AOX 1) promoter is employed for the expression of recombinant proteins. This promoter is tightly regulated and highly inducible. It is induced by methanol, which also serves as the main carbon source during the expression phase *(4)*. To enable parallel production and testing of expression clones carrying the same or different cDNAs, we established a screening system that allows analysis of the expression of multiple different proteins in parallel in a microplate based small-scale assay.

2. Materials

1. *E. coli* strain JM 109 (Stratagene, La Jolla, CA).
2. Expression vector pPICHS *(5)*(**Fig. 1**).
3. cDNAs of interest.
4. Oligonucleotide primers.
5. *Taq* polymerase, restriction enzymes, alkaline phosphatase (Boehringer, Germany), T4 DNA ligase.
6. Proofreading *Taq* polymerase (Qiagen, Hilden, Germany).
7. Sterile toothpicks.
8. Thermocycler for polymerase chain reaction.
9. Agarose gel equipment.
10. 96-well PCR purification kit (Millipore, Billerica, MA).
11. LB plates containing 100 μg/ml ampicillin.
12. 96-well Plasmid Preparation Kit (Millipore).
13. *Pichia pastoris* strain GS115 (Invitrogen, Carlsbad, CA).
14. Electroporation device.
15. YNB media plates (Becton, Dickinson, Franklin Lakes, NJ).
16. Trace element stock solution: 0.4 mM $CuSO_4$, 1.8 mM $FeSO_4$, 0.5 mM $MnCl_2$, 0.4 mM Na_2MoO_4, 6 mM $ZnSO_4$, 10 mM EDTA; filter-sterilize.
17. Vitamin stock solution (per 100 mL): biotin 62.5 mg, Ca-panthotenate 1.25 g, nicotinic acid 250 mg, pyridoxine 626 mg, thiamine 250 mg; filter-sterilize and store at 4°C.
18. WM9 medium: 10 g/L sodium glutamate, 75 mg/L inositol, 25 mg/L Mg_2Cl_2, 10 mg/L $CaCl_2$, 55 mg/L $MgSO_4$; after autoclaving add 40 mL/L 0.5 M phosphate buffer, pH 6.8, 1 mL/L trace element stock solution and 4 mL/L vitamin stock solution; add carbon source stock solution as outlined under methods.
19. Methanol.
20. 24-well plates (Whatman, Kent, UK).
21. Microplate centrifuge.
22. SDS-PAGE (sodium dodecyl sulfate-polyacrylamide gel electrophoresis) equipment.
23. Western blotting equipment.
24. PVDF (polyvinylidene fluoride) membrane (Millipore).
25. BSA (bovine serum albumin).
26. AntiPenta His antibody (Qiagen, Germany).
27. Secondary antibody (swine anti-rabbit, peroxidase conjugated; DAKO, Denmark [*see* **Note 1**].
28. Streptactin-horseradish peroxidase conjugate (IBA, Germany).

29. Chemiluminescence substrate for peroxidase.
30. Chemiluminescence imaging system or X-ray film.
31. Western blot stripping buffer: 62.5 mM Tris-HCl, pH 6.7, 2% SDS, add 14 μL β-mercaptoethanol per 20 mL of solution prior to use.
32. Glass beads 0.25–0.5 mm (Roth, Karlsruhe, Germany).
33. Ni-NTA agarose matrix (Qiagen).
34. StrepII-sepharose matrix (IBA, Goettinger, Germany).
35. Empty polypropylene columns (Qiagen).
36. PBS buffer: 8.4 mM NaH$_2$PO$_4$, 1.6 mM KH$_2$PO$_4$, 150 mM NaCl.
37. PMSF (phenylmethylsulfonyl fluoride).
38. Imidazole.
39. Wash buffer IMAC: 50 mM Na$_2$HPO$_4$, 300 mM NaCl, pH 8.0, 20 mM imidazole.
40. Elution buffer IMAC; same as wash buffer, but with 250 mM imidazole.
41. Desthiobiotine.
42. Wash buffer for StrepII sepharose matrix: 100 mM Tris-HCl, pH 8.0, 1 mM EDTA.
43. Elution buffer for StrepII sepharose matrix: same as wash buffer, but with 5 mM desthiobiotin.

3. Methods

3.1. Expression Plasmid

The expression vector pPICHS (**Fig. 1**) *(5)* is based on the plasmid pPIC3.5 (Invitrogen). The expression of the inserted cDNA is controlled by the AOX 1 promoter. Transcription is induced by methanol *(4)* and repressed by most other carbon sources *(6)*. Due to the fact that different proteins are to be detected and purified using a universal protocol, the proteins of interest are expressed as fusion proteins carrying affinity tags at the N- and the C-terminus. We use an N-terminal six histidine (6*His) tag and a C-terminal StrepII tag *(7)*. cDNAs are amplified by PCR using gene specific primers with overhangs generating *Bam*HI/*Not*I restriction sites at the 5' and 3' ends, respectively (*see* **Notes 2** and **3**). Digested PCR products are inserted in-frame into the *Bam*HI/*Not*I restriction sites of the multiple cloning site.

Plasmids used for *P. pastoris* transformation do not carry a host-specific origin of replication, which would result in a stable propagation of the plasmids in the host cell. Therefore the plasmid has to be integrated into the host chromosome. Integration is targeted into the nonfunctional *his4* locus of the host strain used by linearizing the plasmid within the *HIS4* gene. This can be done by either *Sal*I or *Stu*I restriction (*see* **Note 4**). Integration restores the *HIS4* locus whereby selection of transformants is based on this restored histidine prototrophy.

3.2. Cloning of cDNAs

3.2.1. PCR Amplification of cDNA

The primers are designed to ensure in-frame cloning of the cDNA of interest into the expression cassette. Therefore, sequences providing the restriction sites for cloning (*Bam*HI and *Not*I respectively) are fused to gene specific sequences (*see* **Notes 2** and **3**). cDNAs are directly amplified from the respective cDNA library clones by colony PCR.

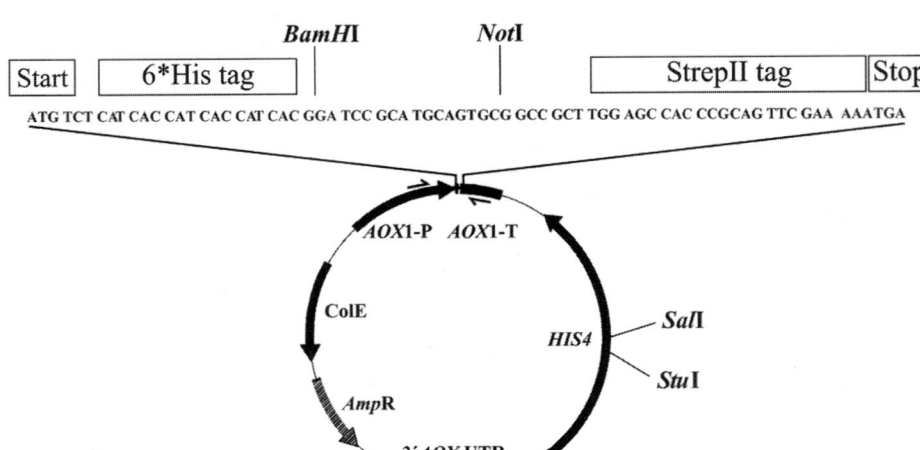

Fig. 1. Schematic drawing of pPICHS expression plasmid and the nucleotide sequence of the multiple cloning site including the tag coding sequences. The black arrows indicate the hybridization sites of the 5'-AOX1-f and 3'-AOX1-r primers, respectively.

Using a sterile toothpick, a few cells of these *E. coli* clones are transferred into 50 µL of PCR reaction mix. The amplification is done using proofreading polymerase to ensure minimal mutation of the sequence during the PCR reaction.

The success of the PCR and the size of the product are checked by gel electrophoresis using a 1.5% agarose gel.

3.2.2. Cloning

DNA manipulations are performed by standard recombinant DNA methods *(8)*. The vector is prepared by sequential cutting with *Bam*HI and *Not*I, respectively. The ends are subsequently dephosphorylated using alkaline phosphatase to reduce self-ligation due to incomplete restriction.

Prior to digestion with *Bam*HI and *Not*I, the PCR products are purified using a Millipore 96-well PCR purification kit according to the manufacturer's instructions.

Ligation reactions are set up using about 200 ng plasmid and about 150 ng PCR product, adding T4 ligase according to the manufacturer's instructions. Ligation takes place at room temperature for two hours and the reaction is transformed into *E. coli* JM109 cells using standard methods. Transformants are selected on LB plates containing 100 µg/mL ampicillin after overnight growth at 37°C.

Transformants are checked for the presence of a cloned insert of the expected size by colony PCR. Primers used are universal and hybridise at the promoter and the terminator region of the expression cassette (5'-AOX1-f: 5'-GACTGGTTCCAATTGACAAGC-3' and 3'-AOX1-r: 5'-CAAATGGCATTCTGACATCC-3', **Fig. 1**), respectively. Plasmid DNA of *E. coli* clones carrying an insert of the expected size is isolated using a 96-well Plasmid Preparation Kit (Millipore).

3.3. Screening of P. pastoris for Protein Expression

3.3.1. Transformation of P. pastoris

1. Linearize 3–5 μg of the isolated plasmid DNA by *Sal*I or *Stu*I (*see* **Note 4**). The restricted DNA is purified and desalted using a Millipore 96-well PCR purification kit.
2. Electrocompetent (*see* **Note 5**) *P. pastoris* cells are prepared as follows: Inoculate 5 mL YPD medium with cells from a glycerol stock and incubate for 24 h at 28°C at 180 rpm.
3. Inoculate 4 × 500 mL of YPD medium with 10–100 μL from the preculture each (*see* **Note 6**) and grow them to an OD $_{600}$ of 1.2.
4. Harvest cells by centrifugation at 3345*g* for 10 min at 4°C.
5. Wash cells twice in 50 mL ice-cold sterile water and pool them.
6. Wash pooled cells in 30 mL ice-cold 1 *M* sorbitol.
7. Resuspend pellet in 1.5 mL ice-cold 1 *M* sorbitol; competent cells can be stored in aliquots at −70°C up to several months.
8. Mix plasmid DNA with 40 μL of competent cells and keep the mix on ice for 5 min.
9. Transfer cells to an ice-cold 0.2-cm electroporation cuvette.
10. Transformation is performed using a Bio-Rad GenePulser II. Settings are 1500 V, 50 μF, and 200 Ω. Immediately after pulsing, resuspend cells in 1 mL 4°C cold 1 *M* sorbitol.
11. Plate cells on YNB plates containing 2% glucose and incubate for 3–4 h at 28°C.

3.3.2. Verification of Transformants by Colony PCR

Our experience has shown that not all histidine prototrophic colonies carry the expression cassette. It is therefore necessary to verify the integration and stable occurrence of this cassette. To avoid false positive clones, 5 His[+] clones per transformation are cultivated in 96-well plates in 100 μL YNB at 28°C without shaking for another 2 d at 28°C. Two μL of these cultures are subjected to PCR as described for the *E. coli* colony PCR (*see* **Subheading 3.2.2.**).

Yeast clones are stored by adding sterile glycerol to a culture to a final concentration of 20%. After mixing, the culture is frozen at −70°C.

3.3.3. Cultivation in 2-mL Format

Cultivation is done in 2-mL culture volume in a 24-well format using Whatman 24-deep-well plates.

1. Inoculate 2 mL of WM9 containing 2% glucose as carbon source with 10 μL from a glycerol stock.
2. Incubate plate at 28°C for 3 h at 180 rpm.
3. Centrifuge plate at 1280*g* for 10 min and discard the supernatant.
4. Resuspend cells in 2 mL WM9 medium without carbon source.
5. Add 1% (v/v) methanol and 0.1% (w/v) glucose.
6. Incubate plate at 28°C for 24 h at 180 rpm.

3.3.4. Harvesting the Cells

1. Transfer 750 μL of the culture to a reaction tube and harvest cells by centrifuging at 10,000*g* for 5 min at 4°C.
2. Wash pellet once in 500 μL PBS buffer. The pellet can be stored at −70°C for a few days.

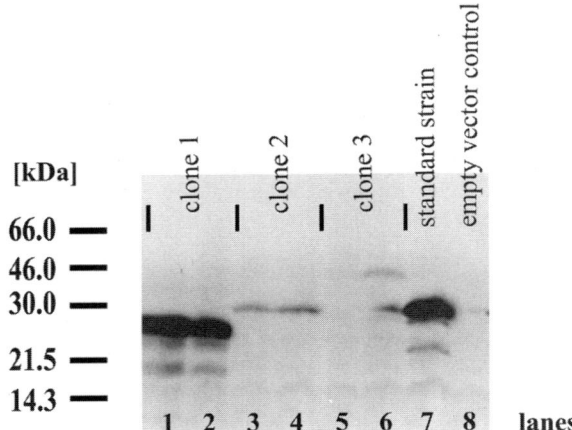

Fig. 2. Example for a Western blot detection using AntiHis antibody. Extracts of two transformants each for three different cDNAs are shown. Extract applied to lane 7 is from a standard strain expressing human med7 protein as a control, lane 8 is a negative control (strain transformed with the empty vector).

3. Resuspend pellet in 30 µL of SDS-PAGE sample buffer according to Laemmli *(9)*.
4. Incubate cells in a boiling water bath for 5 min.
5. Centrifuge at 10,000*g* for 5 min at 4°C.
6. Apply 10 µL of the supernatant to an SDS-PAGE.
7. Transfer proteins to PVDF membrane by Western blotting (*see* **Note 7**).

3.3.5. Protein Detection

Perform all washing and hybridization steps in an amount of solution sufficient to cover the membrane well. Perform washing steps for 10 min each with moderate shaking. *See* **Fig. 2** for an example of protein detection using PentaHis antibody.

1. Block membrane in PBS + 0.1% Tween 20 + 2% BSA either overnight at 4°C or at room temperature for 1 h.
2. Incubate membrane with PentaHis antibody diluted to 1:2000 in PBS + 2% BSA for 1 h at room temperature with moderate shaking.
3. Wash three times with PBS + 0.05% Tween-20.
4. Incubate with secondary antibody (horseradish-peroxidase conjugated anti-mouse; *see* **step 2**).
5. Wash three times with PBS + 0.05% Tween-20.
6. Incubate with peroxidase substrate according to manufacturer's instructions (*see* **Note 1**).
7. Visualize results either by exposure to X-ray film or by imaging with luminescence detection system.

    We routinely test for the presence of the C-terminal StrepII tag to ensure that the proteins synthesised are full length. The membrane can be stripped to remove the anti-His antibodies and re-probed as outlined below. To do so, the membrane must not dry after the first hybridization procedure. The membrane can be stored for several days prior to stripping in PBS at 4°C.
8. Incubate membrane in stripping buffer for 30 min at 50°C.

9. Wash membrane twice with PBS + 0.1% Tween-20.
10. Block membrane in PBS + 0.1% Tween-20 + 2% BSA either overnight at 4°C or at room temperature for 1 h.
11. Incubate for 30 min at room temperature in PBS + 0.05% Tween-20 + 2 μg/mL avidin with moderate shaking.
12. Add StrepTactin-peroxidase conjugate to the solution to obtain a final dilution of 1:4000.
13. Wash three times with PBS + 0.05% Tween-20.
14. Wash three times with PBS.
15. *See* **steps 6** and **7** for detection.

3.4. Purification of Recombinant Protein

The procedures outlined below typically lead to a protein yield of several micrograms, an amount sufficient for most analytical purposes, e.g., MALDI-MS.

3.4.1. Growth of Cells and Induction

Cultivation on a 50-mL scale.

1. Inoculate 50 mL WM9 medium containing 2% glucose in baffled shake flasks (*see* **Note 8**) with 10 μL of a glycerol stock and incubate for 3 d at 28°C at 180 rpm. The OD_{600} should be between 35 and 40.
2. Harvest cells by centrifugation at 3345g for 10 min.
3. Resuspend cells in 50 mL WM9 medium without carbon source.
4. Add methanol to a final concentration of 1% (v/v) and glucose to a final concentration of 0.1% (w/v) twice a day.
5. Cells are usually harvested after 2 d. To increase protein yield, the induction phase can be extended up to 5 d.
6. Harvesting is done by centrifugation at 3345 g for 5 min at 4°C in 10 mL aliquots.
7. Cells are washed once in 5 mL PBS and can be stored at −70°C up to several days.

3.4.2. Cell Lysis

Lysis of cells for the preparation of native proteins is done by disruption with glass beads.

1. Resuspend cells in 10 mL PBS + 0.05% Tween-20 + 1 mM PMSF.
2. Add one volume (10 mL) of glass beads.
3. Disrupt cells by 10 cycles of full speed vortexing for 1 min/1 min cooling on ice.
4. Remove beads and cellular debris by centrifugation at 10,000g for 15 min at 4°C.

3.4.3. IMAC Purification

To monitor the purification procedure, samples should be taken at each step including the cleared lysate. Routinely, we apply 10 μL of each fraction per lane of an SDS-PAGE. **Figure 3** shows a typical example for the distribution of the protein across the different fractions.

1. Add 400 μL Ni-NTA sepharose matrix—an amount of matrix that we have found to work well—to the cleared lysate.
2. Incubate for 1 h at 4°C with moderate shaking.

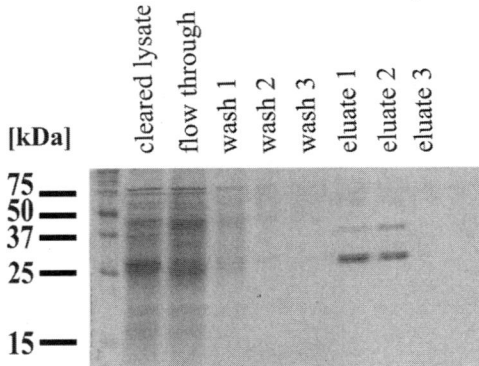

Fig. 3. Coomassie-stained gel as an example for IMAC purification. Ten microliters of the indicated fractions were applied per lane.

3. Pour mixture into an unpacked column and let the column empty by gravity flow.
4. Wash column three times with two bed volumes of wash buffer.
5. Elute protein with five fractions of elution buffer each of half a bed volume.

3.4.4. StrepII Purification

This can be performed initially (for an example, *see* **Fig. 4**) or as a second step after the IMAC purification to further enhance the purity of the protein. For further analysis of the purification procedure, samples should be taken from each step including the cleared lysate (*see* **Subheading 3.4.3.**). If the StrepTactin affinity chromatography is carried out as the initial purification step, prepare the lysate as described in **Subheading 3.4.2.**

1. Pour 400 μL StrepTactin sepharose matrix—an amount of matrix that we have found to work well—into an unpacked column.
2. Equilibrate the column with 3 bed volumes of wash buffer.
3. Apply the cleared lysate—or alternatively, the IMAC elution fractions containing your protein—to the column and let the column empty by gravity flow.
4. Wash five times with half a bed volume of buffer W.
5. Elute protein with five fractions of elution buffer, 100 μL each.

4. Notes

1. Different techniques can be used to detect a Western blot. In our experience the chemiluminescence detection method works well. This requires special equipment, such as an imaging system or the development of X-ray films. Alternatively, colorimetric detection can be performed. In this case, you have to choose a secondary antibody linked to an enzyme capable of converting your substrate of choice.
2. The *Bam*HI restriction site in pPICHS allows ligation with an *Bgl*II overhang. If the target cDNA contains a *Bam*HI restriction site, a *Bgl*II site is attached via the primer and used for cloning. The occurrence of *Not*I sites is very rare.

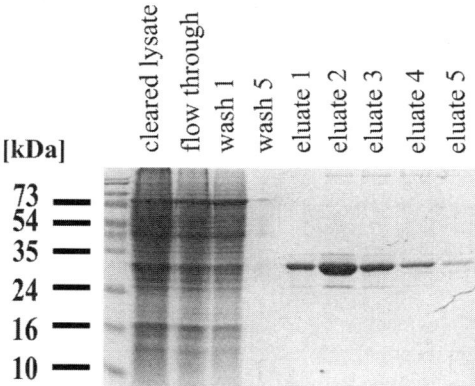

Fig. 4. Coomassie-stained gel as an example for StrepII affinity chromatography. Ten microliters of the indicated fractions were applied per lane.

3. The gene-specific sequences within the primers are designed to have a length resulting in an annealing temperature of at least 50°C (counting every G or C as contributing 4°C and every A or T as contributing 2°C). 5' to the restriction sites bases have to be added to ensure effective digestion of the PCR products. In case of *Bam*HI or *Bgl*II we add two bases; in case of *Not*I we add three bases.

4. Routinely, we use *Stu*I for linearization. Plasmids carrying cDNAs containing a *Stu*I site are linearised using *Sal*I. Incomplete digestion reduces the number of yeast transformants. We routinely check the restriction reaction on a 1% agarose gel.

5. As an alternative to using electrocompetent *P. pastoris* cells, one can prepare competent cells for transformation by chemical methods or spheroplasting. These protocols are described in *(10)*. We have found electrotransformation to be a reliable and reproducible method.

6. It is important to harvest the cells in the appropriate growth phase (early logarithmic phase). The actual doubling time in YPD medium depends on the manufacturer and the batch of the ingredients you use. To have at least one culture at the recommended OD after overnight incubation, inoculate the main culture with different amounts of the preculture. It is usually a good choice to vary the amount of the inoculum between 10 μL and 100 μL.

7. There are different methods for protein transfer while Western blotting. We use a semi-dry electrotransfer device (Hölzel, Wörth, Germany).

8. The methanol metabolism consumes a high amount of oxygen. Therefore, a sufficient aeration of the cultures is necessary.

Acknowledgments

The authors thank Christina Steffens for the cloning of the cDNAs and the screening of the yeast clones. We also thank Roslin Bensmann for help with the English. This work was supported by the Bundesministerium für Bildung und Forschung through the Leitprojekt "Proteinstrukturfabrik" (to C.L.).

References

1. Cereghino, J. L. and Cregg, J. M. (2000) Heterologous protein expression in the methylotrophic yeast *Pichia pastoris*. *FEMS Microbiol Rev.* **24**, 45–66.
2. Faber, K. N., Harder, W., Ab, G., and Veenhuis, M. (1995) Review: methylotrophic yeasts as factories for the production of foreign proteins. *Yeast* **11**, 1331–1344.
3. Grinna, L. S. and Tschopp, J. F. (1989) Size distribution and general structural features of N-linked oligosaccharides from the methylotrophic yeast *Pichia pastoris*. *Yeast* **5**, 107–115.
4. Ellis, S. B., Brust, P. F., Koutz, P. J., Waters, A. F., Harpold, M. M., and Gingeras, T. R. (1985) Isolation of alcohol oxidase and two other methanol regulatable genes from the yeast *Pichia pastoris*. *Mol. Cell. Biol.* **5**, 1111–1121.
5. Boettner, M., Prinz, B., Holz, C., Stahl, U., and Lang, C. (2002) High-throughput screening for expression of heterologous proteins in the yeast *Pichia pastoris*. *J Biotechnol.* **99**, 51–62.
6. Tschopp, J. F., Brust, P. F., Cregg, J. M., Stillman, C. A., and Gingeras, T. R. (1987) Expression of the *lacZ* gene from two methanol-regulated promoters in *Pichia pastoris*. *Nucleic Acids Res.* **15**, 3859–3876.
7. Voss, S. and Skerra, A. (1997) Mutagenesis of a flexible loop in streptavidin leads to higher affinity for the Strep-tag II peptide and improved performance in recombinant protein purification. *Protein Eng.* **10**, 975–982.
8. Sambrook, J., Fritsch, E., and Maniatis, T. (1989) *Molecular Cloning: A Laboratory Manual*, 2nd ed. Cold Spring Harbor Laboratory Press, Cold Spring Harbor, NY.
9. Laemmli, U. K. (1970) Cleavage of structural proteins during the assembly of the head of bacteriophage T4. *Nature* **227**, 680–685.
10. Invitrogen. (1997) *A Manual of Methods for Expression of Recombinant Proteins in Pichia pastoris*. Invitrogen Corporation, San Diego, CA.

20

Multiple Gene Expression by Chromosomal Integration and CRE-*lox*P-Mediated Marker Recycling in *Saccharomyces cerevisiae*

Björn Johansson and Bärbel Hahn-Hägerdal

Summary

Multiple gene expression can be introduced in a yeast strain with using only two markers by means of the two new vectors described, the expression vector pB3 PGK and the CRE recombinase vector pCRE3. The pB3 PGK has a zeocin-selectable marker flanked by *lox*P sequences and an expression cassette consisting of the strong *PGK1* promoter and the *GCY1* terminator. The gene of interest (*YFG1*) is cloned between the promoter and terminator of pB3 PGK. The pB3 PGK-*YFG1* is integrated into the genome by a single restriction cut within the *YFG1* gene and integrated in the *YFG1* locus. The strain is further transformed with the pCRE3 vector. The CRE recombinase expressed from this vector removes the zeocin marker and makes it possible to use the pB3 PGK vector over again in the same strain after curing of the pCRE3 vector. The 2μ-based pCRE3 carries the aureobasidin A, zeocin and *URA3* markers. pCRE3 is easily cured by growth in nonselective medium without active counterselection. The screening for loss of the chromosomal zeocin marker, as well as curing of the pCRE3 vector, is done in one step, by scoring zeocin sensitivity. This can be done because the zeocin marker is present in both the pB3 PGK and pCRE3. The *S. cerevisiae* pentose phosphate pathway genes *RKI1*, *RPE1*, *TAL1*, and *TKL1* were cloned in pB3 PGK and integrated in the locus of the respective gene, resulting in simultaneous overexpression of the genes in the xylose-fermenting *S. cerevisiae* strain TMB3001.

Key Words: Chromosomal insertion; CRE recombinase; *lox* sites; zeocin; aureobacidin A; *URA3*.

From: *Methods in Molecular Biology, vol. 267:*
Recombinant Gene Expression: Reviews and Protocols, Second Edition
Edited by: P. Balbás and A. Lorence © Humana Press Inc., Totowa, NJ

1. Introduction

Relatively few single-gene deletions in *Saccharomyces cerevisiae* have any obvious phenotype; even fewer are harmful to the cell. This has been attributed to redundancy in the genome *(1–3)*—i.e., more than one gene is capable of carrying out any given task in the cell. The hexose transporter gene family of *S. cerevisiae* is a good example of this. Although the 20 hexose transporter genes are expressed at different conditions, it took 20 consecutive gene disruptions to obtain a yeast strain incapable of hexose transport *(4)*. The same dominant marker was recycled 20 times, since the number of genetic modifications exceeds the number of available markers.

Besides functional analysis of a redundant genome, metabolic engineering requires a system where many rounds of genetic engineering are possible *(5)*. Metabolic engineering is defined as the directed change of the metabolism using genetic engineering *(5)*. The theory of metabolic control analysis (MCA) predicts that all enzymes along a metabolic pathway share the control of the metabolite flow through that pathway *(6,7)*. There are indeed few examples where overproduction of a single enzyme results in improvement of the desired trait. The overproduction of single glycolytic enzymes *(8)* did not result in higher glycolytic flux. Over-production of multiple enzymes is usually necessary to attain significant improvements *(9–12)*. This has also been predicted by theory *(13,14)*.

We used the pB3 PGK/pCRE3 system to overexpress four genes in the pentose phosphate pathway of a xylose-fermenting *S. cerevisiae* strain *(15)*. This chapter describes how to use the expression vector pB3 PGK and the CRE recombinase expression vector to change the expression level or regulatory characteristics of any number of genes in *S. cerevisiae*.

1.1. The pB3 PGK Vector

The pB3 PGK *(15)* vector contains the strong *S. cerevisiae* *PGK1* promoter *(16)*, a multicloning site and the *S. cerevisiae* *GCY1* terminator *(17)*, the *bla* gene (ampicillin resistance) for *E. coli* maintenance, and the zeocin resistance gene (Invitrogen) for selection in both *E. coli* and *S. cerevisiae*. The zeocin gene is flanked by two *lox*P sites. The strategy to change the expression of any gene is a follows: A yeast gene (*YFG1*) is cloned between the promoter and terminator of the pB3 PGK, resulting in pB3 PGK *YFG1* (**Fig. 1A**). The vector is linearized using a unique restriction site within the coding sequence of *YFG1*. The linear vector is used to transform yeast, directing the integration to the locus of *YFG1* (**Fig. 1B**). The vector can also be cut in the ribosomal DNA sequence (rDNA) (**Fig. 1A**) to facilitate integration in the repetitive rDNA sequences to obtain multiple copies *(18)*. The *PGK1* promoter is excisable with *Sac*I and *Xba*I and can be exchanged for any of the following promoters, offering a wide range of promoters with diverse characteristics shown in **Table 1**.

1.2. The pCRE3 Vector

To further modify the strain with the integrated pB3 PGK vector, the zeocin resistance marker must be lost. This is accomplished by transformation with the vector

Table 1
PGK Promoters Excisable With *SacI* and *XbaI*

Promoter	Characteristics	Reference
TDH3	Strong	*19*
TEF2	Strong, constitutive	*19*
ADH1	Intermediate	*19*
CYC1	Weak	*19*
CUP1	Copper-induced	*20*
CTR1	Copper-repressed	*20*
CTR3	Copper-repressed	*20*
GAL1	Galactose-induced	*21*

pCRE3, that carries the CRE recombinase under the control of the *GAL1* promoter *(15)*. CRE catalyzes the specific recombination between genetic elements called the *lox*P sites. The *lox*P site consists of two 13-bp inverted binding sites situated 6 bp apart. The pCRE3 is 2μ-based and contains the URA3, AUR1-C, and zeocin-resistance genes. The URA3 or AUR1-C gene is used for selection of the vector. The zeocin marker facilitates simultaneous scoring of marker and pCRE3 loss by zeocin sensitivity.

2. Materials

1. *S. cerevisiae*/*E. coli* shuttle vector pB3PGK *(15)* (*see* **Note 1**).
2. Zeocin (Invitrogen, cat. nos. R250-01 or R250-05; CAYLA, prod. no. ZEOCL0001).
3. Aureobasidin (TaKaRa, prod. no. TAK 9000).
4. PEG 3350 MW 3350 (50% w/v) (Sigma cat. no. P3640).
5. 1.0 *M* lithium acetate.
6. Salmon sperm DNA 2.0 mg/mL (Sigma cat. no. D1626)
7. YPD liquid medium: 20 g/L peptone, 10 g/L yeast extract, 20 g/L glucose, pH 5.5.
8. YPD plates: same as YPD liquid medium, adding 20 g/L agar. The pH should be 7.5 when selecting for zeocin resistance.
9. One airtight plastic box (approx 2L volume, such as an ice-cream box; *see* **Note 2**).
10. General molecular biology reagents and equipment, such as restriction enzymes, PCR cleanup kit, agarose gel, and PCR equipment.

3. Methods

3.1. Test Host Strain Sensitivity to Zeocin

Make YPD (pH 7.5) plates with different zeocin concentrations (25, 50, 100, and 150 mg/L). Note that the pH of the YPD medium for selective plates should be set to 7.5, as the zeocin is less active at lower pH. We routinely use 50 mg/L for the laboratory strain CEN.PK. Industrial and wild-type strains may require higher levels. The pB3 PGK vector is also available with a marker other than zeocin if zeocin is inefficient (*see* **Note 3**). Resuspend a loopful of yeast cells in 1 mL sterile water, spread a loop of the suspension on each plate, and check growth after 2 to 3 d at 30°C.

3.2. Amplification of the Gene of Interest by PCR and Cloning in pB3 PGK

Amplify the gene of interest (*YFG1*) with a proofreading DNA polymerase with primers that incorporate a restriction site in the primer. We usually incorporate *Bam*HI if possible, since this site is compatible with the *Bgl*II in pB3 PGK1. See **Fig. 1A** for alternative enzymes. Purify the PCR product with a PCR clean-up kit. It is not necessary to perform agarose gel purification, given that the PCR was reasonably specific. Digest the restriction sites that were incorporated by the primers and do a second PCR clean-up. Alternatively, a phenol-chloroform extraction can be used to remove the restriction enzymes. The PCR product is then ligated to the linearized pB3 PGK1 vector. The reverse primer used for cloning the gene can be used together with the BJ5756 primer to score correct *E. coli* clones (*see* **Note 4** for the sequence of the BJ5756 primer).

3.3. Transform S. cerevisiae *Host Strain With pB3 PGK YFG1*

Digest approx 0.3–1 µg pB3 PGK vector DNA with the appropriate unique restriction enzyme. The restriction cut should be made with a unique enzyme cutting within the coding region of *YFG*1 if possible (*see* **Note 5**). The digested DNA can be used in yeast transformation without further purification or enzyme inactivation. The frequency of transformation and integration is high, so there is no need for very high efficiency transformation. We use the "Quick and Easy TRAFO Protocol" as described by Dr. Daniel Gietz (*22*) with minor modifications (*see* **Note 6** for the web link to the original protocol). It is our experience that the purity of the water used for the preparation of the reagents is critical; therefore, use the purest available water.

1. Collect 25 µL yeast inoculum for EACH transformation from a fresh plate (less than 5 d) with a sterile loop.
2. Wash cells with 1mL water in an Eppendorf tube and centrifuge.
3. Resuspend the cell pellet in 1 mL of 100 m*M* lithium acetate and incubate for 5 min at 30°C and divide the cell suspension to the number of planned transformations.
4. Spin cells at top speed in a microfuge for 5 s and remove the supernatant with a pipet.
5. Loosen the pellet by vortexing briefly. This vortex step helps resuspend the cells after the heat shock (see below).
6. Add the following components into the tube on top of the cells:

    a. 240 µL of PEG (50% w/v);
    b. 36 µL 1.0 M. LiAc;
    c. 50 µL SS-DNA (2.0 mg/mL);
    d. 5.0 µL of vector DNA (100 ng to 5 µg);
    e. 29 µL of sterile water.

7. Vortex for at least 1 min to resuspend the cell pellet in the transformation mix and incubate at 42°C for 20 min.
8. Pellet the cells at top speed in a microcentrifuge for 10 s and remove the supernatant with a pipet.
9. Wash the cells with 1 mL water by slowly pipetting up and down and pellet the cells at top speed in a microcentrifuge for 10 s.

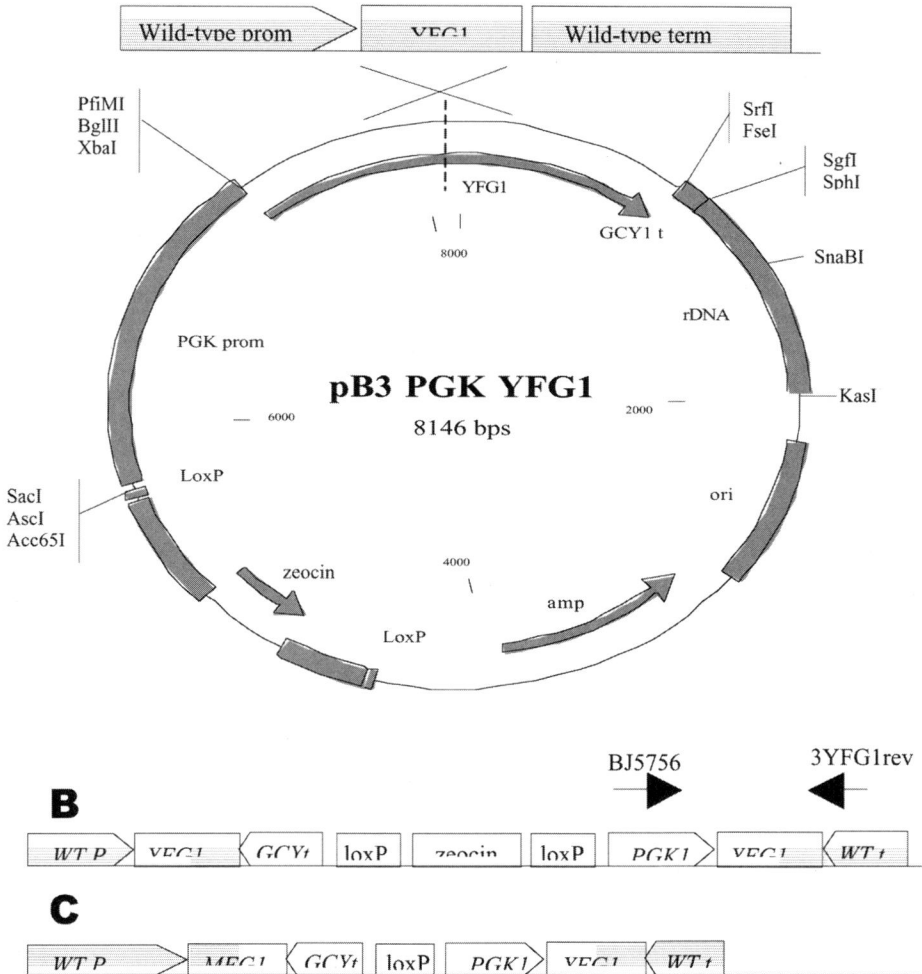

Fig. 1. Schematic view of expression of a *S. cerevisiae* gene using the pB3PGK vector. (**A**) pB3PGK YFG1 is digested with a unique restriction enzyme cutting inside the coding sequence of *YFG1*. The linear fragment is integrated at the *YFG1* locus by a single cross-over event. (**B**) The correct integration is verified by PCR using the BJ5756 primer and the reverse primer for *YFG1*. (**C**) The zeocin marker is lost by looping out the DNA between the *lox*P sites. Annotations: PGK prom = *PGK1* promoter; *YFG1* = Your Favorite Gene; GCYterm = *GCY1* terminator; rDNA = ribosomal RNA; Ori = *E. coli* origin of replication; Amp = Ampicillin resistance gene; TEF-prom = *TEF1* promoter; zeocin = zeocin resistance gene; LoxP = *Lox*P recombination site.

10. Add 1 mL nonselective YPD medium and resuspend carefully.
11. Transfer cells to a 15-mL or 50-mL Falcon tube or another large volume tube with a tight fit-ting cap. A large screw-cap tube will hold the cells safely. Do not use Eppendorf tubes, since the lid may pop open from the pressure.

12. Incubate the tube shaking at 30°C for 3 h to overnight.
13. Transfer cells to a 1.5-mL Eppendorf tube and pellet the cells at top speed in a microcentrifuge for 10 s. Remove supernatant and wash cells with 1 mL water.
14. Resuspend the cells in 200 µL water and incubate on ice for 1 h prior to spreading on selective plates. This step may lower the background, if high background is a problem.
15. Spread 20 µL and 100 µL on selective plates. The remainder may be refrigerated for several weeks. It is good to save an aliquot of transformants if the number of transformants on the plates is too high or too low.

3.4. Analytical Colony PCR

1. The analytical PCR step can also be performed after the marker has been looped out (*see* **Note 7**). The PCR master mix should be prepared and stored on ice so that the master mix can be immediately added to the cells.

   PCR mix (50 µL):

   a. 1 µ*M* reverse primer for cloning of *YFG1*
   b. 1 µ*M* BJ5756
   c. 5U *Taq* polymerase
   d. 100 µ*M* dNTPs, 2 m*M* MgCl$_2$ and 1X PCR buffer

2. The background on zeocin is typically much lower than that for geneticin. Most colonies that appear are true transformants. Pick a number of big, well-grown colonies with a sterile pipet tip and streak a small amount of yeast cells on the side of a PCR tube; streak the remainder on a new selective plate.
3. Incubate the PCR tube containing yeast cells in a microwave oven at full power for 1–2 min. Be careful not to take too many cells, as this will inhibit the PCR reaction. It is important to keep the reactions on ice after the addition of the master mix; otherwise, the proteases from the lysed yeast may degrade the DNA polymerase. These are essentially the same methods as recommended by the EUROFAN project.
4. Spot 10 µL of the PCR product on an agarose gel with the appropriate size marker (*see* **Note 6** for web link to the original protocol).

3.5. Transformation and Curing of pCRE3

The same procedure for transformation as described above can be used to transform the correct integrants with the CRE expression vector pCRE3. The pCRE3 contains the selectable markers URA3, zeocin, and AUR1-C. The selection of the vector must be done using the AUR1-C or the URA3 markers, since the strains are already zeocin-resistant. When aureobasidin-resistant or uracil prototrophic clones have been obtained, the pCRE3 can be cured immediately. Note that it is not necessary to grow the cells in a medium containing galactose, since the background activity of the CRE recombinase seems to be efficient enough to obtain high frequency of looped-out clones.

1. Inoculate 100 mL YPD with one colony (1 mm) from the pCRE3 transformed cells.
2. Incubate 1–2 d or until stationary phase.
3. Dilute 1 µL of cell culture in 1 mL sterile water and plate 10, 50, and 100 µL of the dilution on three YPD plates. This procedure should yield 100–500 cells on the plate with the least cells.

4. Replica plate the transformants with sterile velvet onto two fresh plates after two days, one YPD medium with zeocin and the other with YPD only. It is a good practice to replicate the YPD plate after the YPD/zeocin, since the number of cells replicated on the second plate is usually lower. This practice ensures that colonies are not mistakenly scored as zeocin-sensitive because an insufficient number of cells was replicated onto the YPD/zeocin plate. Ten to 20% of the colonies are zeocin-sensitive, following this procedure. Since the zeocin marker is present on both the chromosomal construct and the pCRE3 vector, zeocin sensitivity marks loss of both elements.

3.6. Simultaneous Overexpression of the S. cerevisiae Genes RKI1, RPE1, TAL1, and TKL1

We used the pB3 PGK/pCRE3 system to create strains overexpressing the *S. cerevisiae* nonoxidative pentose phosphate pathway genes *RKI1*, *RPE1*, *TAL1*, and *TKL1* *(12)*. The genes were individually cloned in pB3 PGK and integrated in the locus of the respective gene, resulting in overexpression of the genes. The pCRE3 vector was used to create strains simultaneously overexpressing the *RKI1*, *RPE1*, *TAL1*, and *TKL1* by successive integrations and removal of the *lox*P-zeocin-*lox*P cassette. **Figure 2** shows an agarose gel loaded with the PCR products from the analytical colony PCR procedure, confirming the correct integrations. The *lox*P-zeocin-*lox*P cassette has also been used successfully to create xylose-fermenting yeasts from industrial strains lacking auxotrophic markers *(23,24)*.

3.7. Simultaneous Expression and Silencing of Wild-Type Gene Copy

Underproduction of a protein, i.e., production at lower level than that of the wild type, requires the wild-type gene to be removed or silenced. The pB3 PGK vector can be used to simultaneously silence the copy of the gene that is controlled by the wild-type promoter. This is achieved by cloning the 5' part of the gene (about 80%) in the pB3 PGK vector. Integration with this construct disrupts the gene expressed under the original promoter, while the gene controlled by the promoter in the pB3 PGK vector remains intact. This strategy was successfully used to create strains based on TMB3001, with underexpressed or conditionally expressed *ZWF1* gene *(25)*.

4. Notes

1. To obtain the vectors described in this chapter, contact Bärbel Hahn-Hägerdal, Department of Applied Microbiology, Lund University, P.O. Box 124, 221 00 Lund, Sweden; e-mail: barbel.hahn-hagerdal@tmb.lth.se.
2. Zeocin and aureobasidin are expensive. Using less medium in the selective plates is a way of lowering costs. Petri dishes with only 5 mL medium can be used if some measure is taken to prevent overdrying of the plates. A way of doing this is to incubate the plates in an airtight container with moist air. Put the plates upside down in an empty ice cream box at some distance from the bottom. Pour 10 mL boiling water into the box and tighten the lid. In this way the plates can be incubated for many days without drying.
3. The pB3 PGK vector is also available with the AUR1-C marker instead of the zeocin marker (pB1 PGK). This vector is useful if a particular yeast strain is not sensitive to zeocin. The

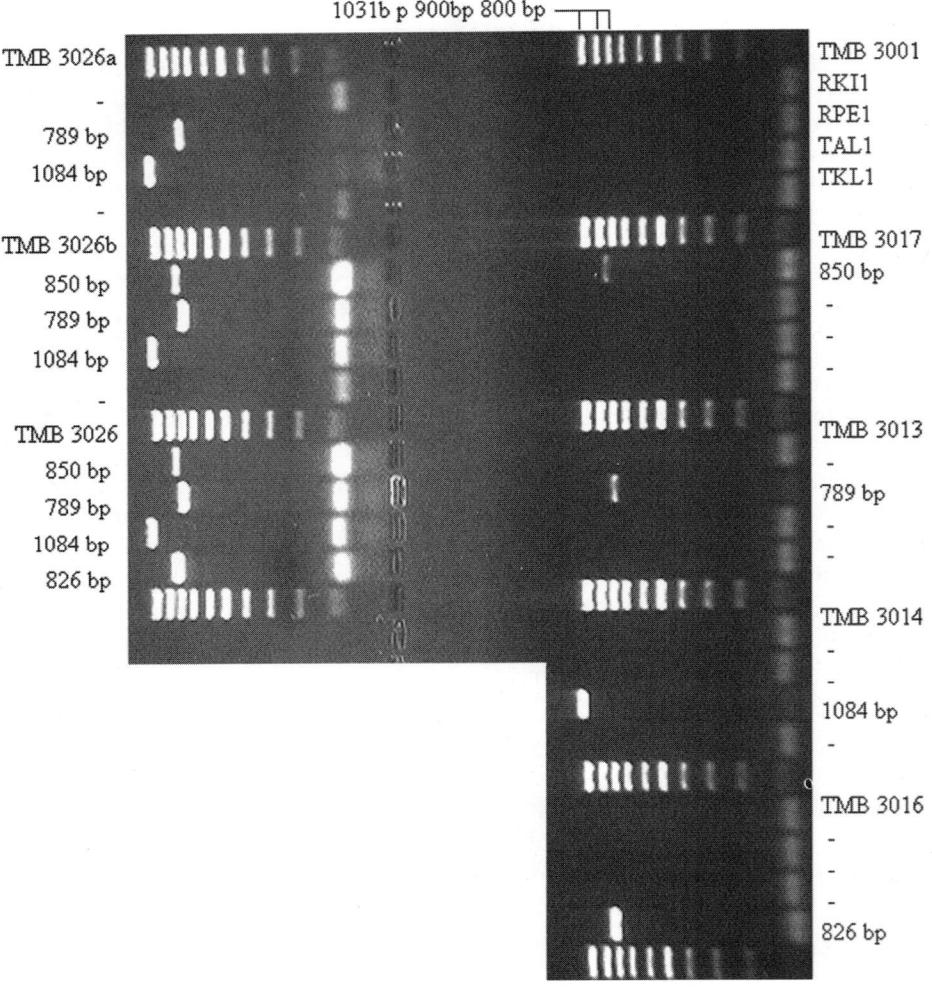

Fig. 2. Colony PCR using primer BJ5756 (**Note 4**) and the reverse primer for each gene. The *S. cerevisiae* genes *RKI1, RPE1, TAL1,* and *TKL1* were inserted one by one (TMB 3017, 3013, 3014 and 3016). The TMB3026a carries the overexpressed *TAL1* and *TKL1*. TMB3026 carries *TAL1, TKL1,* and *RPE1* and TMB3026 carries *RKI1, RPE1, TAL1,* and *TKL1*.

AUR1-C is a homologous marker, so a higher frequency of erroneous integration can be expected than with the heterologous zeocin marker.

4. The sequence of the PGK promoter-specific primer BJ5756 is 5'-CAT CAA GGA AGT AAT TAT CTA CT-3' Tm = 51°C.

5. If a heterologous gene is to be expressed, the gene itself cannot be used as a target of integration (since it does not exist in the yeast genome). If the yeast already contains an integrated vector containing the bacterial ampicillin resistance gene, this can be used as integration target. The pB3 PGK may be linearized in the Amp gene with *Nde*I. Alterna-

tively, the integration can be targeted to the rDNA sequences with pB3 PGK linearized with *Sna*BI.

6. Web links to protocols and suppliers. Transformation protocols: The Gietz Lab Yeast Transformation Home Page, www.umanitoba.ca/faculties/medicine/biochem/gietz/Trafo.html; The EUROFAN colony PCR protocols, mips.gsf.de/proj/eurofan/eurofan=_1/ b0/home=_requi sites /guideline/sixpack.html. Suppliers: CAYLA, www.cayla.com; Invitrogen, www.invitrogen.com; TaKaRa, bio.takara.co.jp.

7. The frequency of correct integrants is very high with the pB3 PGK vector as is the frequency of marker loss with the pCRE3 vector. Therefore it may not be necessary to perform the analytical PCR after the integration, but to wait until after picking zeocin-negative cells after transforming and curing of the pCRE3 vector. The advantage of doing this is that the positive identification of correct integration is done in the end of the process. The probability of having incorrect contaminating clones is much reduced.

Acknowledgments

This work was financially supported by The Nordic Energy Research Program and The Swedish National Energy Administration and the postdoctoral scholarship (SFRH/BPD/7145/2001) from Fundacao Para a Ciência e a Tecnologia, Portugal.

References

1. Oliver, S. G. (1996) From DNA sequence to biological function. *Nature* **379,** 597–600.
2. Mewes, H. W., Albermann, K., Bahr, M., Frishman, D., Gleissner, A., Hani, J., et al. (1997) Overview of the yeast genome. *Nature* **387,** 7–65.
3. Wolfe, K. H., and Shields, D. C. (1997) Molecular evidence for an ancient duplication of the entire yeast genome. *Nature* **387,** 708–713.
4. Wieczorke, R., Krampe, S., Weierstall, T., Freidel, K., Hollenberg, C. P., and Boles, E. (1999) Concurrent knock-out of at least 20 transporter genes is required to block uptake of hexoses in *Saccharomyces cerevisiae. FEBS Lett.* **464,** 123–128.
5. Bailey, J. E. (1991) Toward a science of metabolic engineering. *Science* **252,** 1668–1675.
6. Kacser, H., and Burns, J. A. (1973) The control of flux. *Symp. Soc. Exp. Biol.* **27,** 65–104.
7. Heinrich, R., and Rapoport, T. A. (1974) A linear steady-state treatment of enzymatic chains: General properties, control and effector-strength. *Eur. J. Biochem.* **42,** 89–95.
8. Schaaff, I., Heinisch, J., and Zimmermann, F. K. (1989) Overproduction of glycolytic enzymes in yeast. *Yeast* **5,** 285–290.
9. Niederberger, P., Prasad, R., Miozzari, G., and Kacser, H. (1992) A strategy for increasing an *in-vivo* flux by genetic manipulations. The tryptophan system of yeast. *Biochem. J.* **287,** 473–479.
10. Hauf, J., Zimmermann, F. K., and Müller, S. (2000) Simultaneous genomic overexpression of seven glycolytic enzymes in the yeast *Saccharomyces cerevisiae. Enzyme Microb. Technol.* **26,** 688–698.
11. Smits, H. P., Hauf, J., Müller, S., Hobley, T. J., Zimmermann, F. K., Hahn-Hägerdal, B., et al. (2000) Simultaneous overexpression of enzymes of the lower part of glycolysis can enhance the fermentative capacity of *Saccharomyces cerevisiae. Yeast* **16,** 1325–1334.
12. Johansson, B., and Hahn-Hägerdal, B. (2002) The non-oxidative pentose phosphate pathway controls the fermentation rate of xylulose but not of xylose in *Saccharomyces cerevisiae* TMB3001. *FEMS Yeast Research* **2,** 277–282.

13. Kacser, H., and Acerenza, L. (1993) A universal method for achieving increases in metabolite production. *Eur. J. Biochem.* **216,** 361–367.
14. Fell, D. A., and Thomas, S. (1995) Physiological control of metabolic flux: the requirement for multi-site modulation. *Biochem. J.* **311,** 35–39.
15. Johansson, B., and Hahn-Hägerdal, B. (2002) Overproduction of pentose phosphate pathway enzymes using a new CRE-*loxP* expression vector for repeated genomic integration in *Saccharomyces cerevisiae. Yeast* **19,** 225–231.
16. Mellor, J., Dobson, M. J., Roberts, N. A., Tuite, M. F., Emtage, J. S., White, S., et al. (1983) Efficient synthesis of enzymatically active calf chymosin in *Saccharomyces cerevisiae. Gene* **24,** 1–14.
17. Hermann, H., Hacker, U., Bandlow, W., and Magdolen, V. (1992) pYLZ vectors: *Saccharomyces cerevisiae/Escherichia coli* shuttle plasmids to analyze yeast promoters. *Gene* **119,** 137–141.
18. Nieto, A., Prieto, J. A., and Sanz, P. (1999) Stable high-copy-number integration of *Aspergillus oryzae* alpha-AMYLASE cDNA in an industrial baker's yeast strain. *Biotechnol. Prog.* **15,** 459–466.
19. Mumberg, D., Müller, R., and Funk, M. (1995) Yeast vectors for the controlled expression of heterologous proteins in different genetic backgrounds. *Gene* **156,** 119–122.
20. Labbé, S., and Thiele, D. J. (1999) Copper ion inducible and repressible promoter systems in yeast. *Methods Enzymol.* **306,** 145–153.
21. Güldener, U., Heck, S., Fielder, T., Beinhauer, J., and Hegemann, J. H. (1996) A new efficient gene disruption cassette for repeated use in budding yeast. *Nucleic Acids Res.* **24,** 2519–2524.
22. Gietz, R. D. and Woods, R. A. (1994) High efficiency transformation in yeast, in *Molecular Genetics of Yeast: Practical Approaches* (Johnston, J. A., ed.), Oxford University Press, Oxford, UK, pp. 121–134.
23. Johansson, B. (2001) Metabolic engineering of the pentose phosphate pathway of xylose fermenting *Saccharomyces cerevisiae*, Lund University, Lund, Sweden.
24. Zaldivar, J., Borges, A., Johansson, B., Smits, H. P., Villas-Boas, S. G., Nielsen, J., et al. (2002) Fermentation performance and intracellular metabolite patterns in laboratory and industrial xylose-fermenting *Saccharomyces cerevisiae. Appl. Microbiol. Biotechnol.* **59,** 436–442.
25. Jeppsson, M., Johansson, B., Hahn-Hägerdal, B., and Gorwa-Grauslund, M. F. (2002) Reduced oxidative pentose phosphate pathway flux in recombinant xylose-utilizing *Saccharomyces cerevisiae* strains improves the ethanol yield from xylose. *Appl. Environ. Microbiol.* **68,** 1604–1609.

21

Three Decades of Fungal Transformation

Key Concepts and Applications

Vianey Olmedo-Monfil, Carlos Cortés-Penagos, and Alfredo Herrera-Estrella

Summary

Filamentous fungi include a large, heterogeneous group of heterotrophic organisms with profound influence in human activities. In spite of their economic and scientific relevance, little information is available at the molecular level about their biology. The development of genetic transformation protocols has contributed greatly to the molecular dissection of fungal behavior. The use of this approach in combination with large-scale genome sequencing projects has now provided the basis for gaining insight into the function of fungal genes. This chapter reviews the technology of the transformation of filamentous fungi. The protocols for gene transference by protoplasting/PEG, LiAc, electroporation, biolistics and *A. tumefaciens* are described, and possible mechanisms for transformation are discussed. A brief description of previously reported selection systems is also included. The application of transforming protocols concerning the study of gene function and manipulation are discussed in relation to some fungal species.

Key Words: Gene transference; fungal transformation; selection markers; gene inactivation.

1. Introduction

The potential of molecular genetics has been extensively exploited in the study and manipulation of microorganisms. This has led to the development of new technologies applied to both eukaryotic and prokaryotic species. In this sense, genetic transformation has played a significant role not only in our understanding of life, but also in the enhancement of particular traits. Among microorganisms, the budding yeast *Saccharomyces cerevisiae* has contributed enormously to the improvement of genetic tools

From: *Methods in Molecular Biology, vol. 267:*
Recombinant Gene Expression: Reviews and Protocols, Second Edition
Edited by: P. Balbás and A. Lorence © Humana Press Inc., Totowa, NJ

including the transference of genetic material. Since the first report of yeast transformation in 1978 (*1*), several transformation systems for many other fungi have been reported.

Filamentous fungi, as a group, exhibit unique characteristics due to their cellular organization and their lifestyle. The relative small genome size and the low frequency of redundant DNA have made these organisms very attractive for molecular genetic research. From the Ascomycota, *Neurospora crassa* and *Aspergillus nidulans* have been especially valuable as model organisms, mainly because it is possible to combine molecular approaches with classical genetics. Nowadays, numerous fungal species have a deep impact in human activities. Clearly, the future will see an expansion of the number of fungal species and their products that are commercially exploited. Consequently, molecular genetic technology will play a major role in the adjustment and improvement of products and processes. The application of molecular genetics to fungi has been quite slow because of the need to develop protoplasting techniques, transformation systems, and appropriate vectors. These are still the most important restrictions for some fungal species. Genetic transformation was first achieved in *N. crassa* in 1973 (*2*); this report is now recognized as the earliest for fungi. Traditionally, fungal transformation has relied largely on variations of the same basic protocol. The extent to which this technology can be applied is in direct proportion to the amount of fundamental knowledge of the species concerned.

The information contained in this chapter is intended to be a complementary resource for fungal genomics experts and nonexperts alike, as it puts together recent advances in fungal transformation systems. It includes examination of traditional transformation protocols highlighting the application for relevant fungal species.

2. Transformation Methods

All DNA transference experiments have two separate components: a method for transferring the genetic material to the recipient cells and a selection method for isolating recipient cells that have received the DNA. Most transference methods result in a considerable number of cells receiving the desired material; consequently, very powerful selection techniques are required and frequently selection is the limiting component during genetic transformation. The majority of fungal species are susceptible to transformation. At present, several transformation methods have been described. Most of them have been modified in order to make them easier and more precise for each fungal strain. It is worth noting that the main obstacle in any transformation system is the fungal cell wall.

2.1. Protoplasts/PEG

Generally, a transformation procedure starts with the generation of osmotically sensitive cells (OSCs) or protoplasts. Yeast cells, germinating conidia, or young mycelia are treated with hydrolytic enzyme mixtures in order to remove the fungal cell wall. Accordingly, the enzyme cocktails are a complex combination of β1,3-gluconases and chitinases as predominant activities. In the market there are some commercial products that have demonstrated to be highly effective in protoplast generation. Helicase and Glusulase are the commercial names of the enzymatic mixture obtained from nail gut; they have been used in protoplast generation of *Cryptoccocus neoformans* and *S. cerevisiae* (*3*) and in the filamentous fungi *N. crassa* (*4*). Zymolase T100 contains concentrated enzymes produced by *Artrobacter luteus*. This product was found to be useful in *S. cerevisiae* generating protoplasts (*5*). Of all commercially available products, Novozyme 234 has been the most popular when protoplasts are required either from yeast or filamentous fungi (*6–8*). Novozyme is composed of enzymes secreted by *Trichoderma atroviridae*. Currently, this enzymatic mixture is commercially available with a different name, Glucanex.

The protoplast-obtaining process could be tedious, mainly because cell-wall-free fungal cells are osmotically sensitive and therefore must be handled with care. Protoplasts should be protected at all times by solutions providing osmotic support. Therefore, solutions containing chlorides, magnesium sulfate, mannitol, sorbitol, or sucrose are normally used at concentrations ranging from 0.8 to 1.2 M (*9,10*). Once protoplasts are properly formed and protected, they are incubated in a solution mix that includes polyethylene glycol, calcium chloride, and the transforming DNA. Calcium ions could participate in the formation of pores or channels in the cell membrane, which allow internalization of exogenous DNA. On the other hand, PEG promotes cell agglomeration to make DNA uptake easier. The relevance of PEG during transformation has been proven to be crucial (*11*). After calcium-PEG-DNA treatment, protoplasts are transferred to selective media that allow cell wall regeneration and the proper isolation of transgenic lines.

2.2. Lithium Acetate

In theory, it is possible to generate protoplasts from any fungal species but in particular cases this protocol does not work. If it is difficult to obtain protoplasts or if their regeneration is not possible, transformation of intact fungal cells offers an alternative method for those species. It is possible to induce cell permeability by using high concentrations of lithium acetate in combination with PEG. This procedure was initially developed for *S. cerevisiae* transformation (*12*) and has been successfully applied in *N. crassa* (*13*), *Coprinus lagopus* (*14*), and *Ustilago violacea* (*15*). However, the use of this technique has been so far limited to a few species.

2.3. Electroporation

Application of an electric pulse offers an additional alternative to induce transient cell permeability and, consequently, DNA uptake. When cells are exposed to an electric field, polarization of structural components of the cell membrane occurs. If the applied voltage overcomes a threshold level, pores are formed in specific areas, allowing diffusion of molecules (*16*). Electroporation methodology has been applied in both intact cells and protoplasts of different species, from bacteria to mammalian cells, including filamentous fungi (*17–20*). During DNA integration mediated by *in vivo* action of restriction enzymes (REMI), electroporation permits uptake of plasmids together with the selected restriction enzyme (*21*).

2.4. Biolistics

Biolistics involves DNA transference by using microparticles of colloidal gold or tungsten as a vehicle. The DNA-covered particles are shot to the sample at high speed through a special device. Bombardment methodology was originally developed for plant tissues (22,23). Subsequently, it was adapted for the transformation of a wide range of cells, such as bacteria (24), yeast (25), fungal conidia (26), and animal cells (27). Furthermore, successful chloroplast and mitochondria transformation has been reported by using biolistic protocols (28–30). Depending on the cell type, various parameters must be considered before using the biolistic-transformation procedure. Transforming-DNA quantity and microparticle speed are the most important. The required pressure to penetrate without causing irreversible cellular damage is different if tissues, cell cultures, or conidia are bombarded. The required equipment is the limiting factor for the application of biolistic-based transformation protocols. However, it is possible to construct gun devices at low cost if suitable technical knowledge is available.

2.5. Agrobacterium tumefaciens *Mediated Transformation*

Protocols have recently been applied for filamentous fungi that take advantage of the infective strategy of *A. tumefaciens*. This Gram-negative, soil-living bacterium is capable of producing tumors in plants. Tumors are developed as a consequence of the expression of genes transferred from bacteria into the infected plant. This event is an interesting transformation mechanism that occurs in nature. Bacteria transfers T-DNA contained in the Ti (tumor-inducing) plasmid into the genome of plant cells by a conjugation-like process. DNA transfer depends on the expression of genes contained in the Ti plasmid, *vir*-genes, involved in virulence of *A. tumefaciens*. Some of these genes encode proteins, essential for the formation of the T-complex, the actual form of the transferred DNA. The *vir* genes are involved in the formation of the conjugation-like tube or sensing the host. Signaling involves phenolic compounds produced by wounded plants, such as acetosyringone, which induce *vir* gene expression. The T-DNA is flanked by two short direct repeats (25 bp), from which the right border is essential for the process. In nature, the T-DNA contains genes leading to the production of opines, compounds that provide the main nitrogen and carbon sources for *Agrobacterium* (31). Thus, T-DNA integration into the plant genome represents an ecological advantage to the bacteria. At the beginning, application of this technology was restricted to dicotyledonous species (32–35), natural hosts of the bacteria. Subsequently, protocols were adapted to monocotyledonous plants (36,37) and fungi, including yeast (38–41). The vehicle for transforming DNA consists of modified versions of the Ti plasmid, where the desired DNA replaces the complete region contained between the T-DNA borders. One of the most important characteristics of this methodology is that it allows random integration of a single copy of the T-DNA. This represents a useful tool in the generation of banks of unique mutants. Moreover, by using PCR technology, it is possible to identify the nucleotide sequence affected by integration of exogenous DNA in the transformed organism (42).

3. Selection Markers

Independent of the transformation methodology used, transgenic line selection is one of the most important limitations. In the first event of transformation in the early 1970s, the fungal model selected contained a mutant genetic background. They were auxotrophic for different compounds such as inositol, uracil, and leucin, so screening for transgenic lines was based on the reversion toward wild-type phenotype. There are many well-characterized auxotrophic strains, but they are limited to a few species (*S. cerevisiae, N. crassa*, and *A. nidulans*). In all cases, selection markers consist of those genes encoding products that allow transition from mutant to prototrophic phenotype. Thus, the use of selection procedures based on complementation of auxotrophic strains seems to be difficult for other fungal species. Particular auxotrophic mutants could be isolated by positive selection using chemical compounds. Only mutant strains affected in *pyrG* (*A. nidulans*), *pyr4* (*N. crassa*) (**43**), or *ura3* (*S. cerevisiae*) (**44**) genes and consequently, lacking orotidine 5'-monophosphate carboxylase activity are capable to growth in culture media containing a toxic analog of the fluoroorotic acid. In *A. nidulans*, the use of chlorate allows selection of strains affected in nitrate reductase activity, encoded by the *niaD* gene. The isolated mutants are incapable of using nitrate as a sole nitrogen source (**45**). Selective pressure exerted by selenate or fluoroacetate present in culture media is useful in the selection of mutants without ATP sulfurylase (**46**) or acetyl CoA synthetase (**47**) activities, respectively. Finally, in *A. nidulans*, by using fluoracetamide, a toxic analog of acetamide, strains affected in the *amdS* gene may be obtained that are incapable of metabolizing acetamide as a sole nitrogen or carbon source (**48**). Most fungal species cannot metabolize acetamide. This characteristic makes the *amdS* gene a good heterologous selection marker (**49**). Nowadays, many other genes are available as dominant selection markers. They confer resistance to different chemical drugs. **Table 1** shows some of the genes commonly used as heterologous selection markers in filamentous fungi.

The main advantage in the use of selection markers that confer drug resistance is that it is not necessary to have prior knowledge of the genotype of the fungus subjected to transformation. On the other hand, it demands the measurement of the dosage of the drug required to completely inhibit the growth of the wild-type population. In some cases, fungi show natural drug resistance and consequently, high drug concentration is required. However, this kind of selection markers represents the best choice when auxotrophic strains are not available.

4. Fate of DNA

To obtain stable transgenic lines it is not only necessary that DNA gets into the nuclei, but the DNA also must remain unchanged in the genome. This means that the DNA introduced into the transformed cells has to be inherited to the daughter cells after mitotic and/or meiotic divisions. In order to achieve this condition, there are two alternatives: (1) exogenous DNA should replicate in an autonomous way, so vectors containing an origin of replication are required; or (2) stable integration of the exogenous DNA into the chromosomes should occur.

Table 1
Dominant Selectable Markers Commonly Used for Transformation in Filamentous Fungi

Marker gene	Source	Resistance	Recipient organism
Hph	E. coli	Hygromicin B	A. nidulans (50) Magnaporthe grisea (51) Arthrobotrys oligospora (52) Trichoderma spp (8)
NphII	Bacterial Tn5	Kanamycin, G418	Phanerochaete chrysosporium (53) Yeast (54)
β-tubulin; tub2	N. crassa U. maydis Trichoderma viride	Benomyl	N. crassa (55) U. maydis (56) T. viride (57)
OliCR	A. niger	Oligomycin	A. niger (58) P. chrysogenum (59)
ble; sh-ble	Bacterial Tn5 S. hygroscopicus	Bleomycin, phleomycin	N. crassa; A. nidulans (60) P.chrysogenum (61) U. maydis (56)
SulII	Enterobacteria	Sulfonamide	Penicillium crysogenum (62)
Bar	S. hygroscopicus	Glufosinate ammonium (Bialaphos)	Metharhizium anisopliae (63) Pleurotus ostreatus (64) N. crassa (65)
CbxR	U. maydis	Carboxin	U. maydis (66)

4.1. Autonomously Replicating Vectors

These vectors are commonly derived from mitochondrial plasmids or contain autonomously replicating sequences (ARSs). Vectors with ARSs promote a high transformation frequency when used in S. cerevisiae; for this reason they represent powerful tools in the isolation and cloning of yeast genes (67). However, ARSs arising from yeast do not seem to work in filamentous fungi. Many efforts have been made in order to generate truly autonomously replicating vectors in fungi. In some cases, it has been attempted to introduce replicons from fungal plasmids into transformation vectors, but only a few have been reported successful (68–70).

Another approach for vector construction is the use of ARSs or sequences that confer a similar function as reported by Balance and Tuner in 1985 (71). They made use of three fungal models in order to isolate an A. nidulans origin of replication. Hybrid vectors containing genomic Aspergillus DNA fragments and the N. crassa pyr4 gene, which is poorly expressed in *Saccharomyces*, were constructed. The yeast was transformed with the generated plasmids and unstable transgenic strains that lost the plasmid under nonselective conditions were considered as strains containing autonomously replicating plasmids. The rescued vectors were used to transform *A. nidulans* and, only one, namely ARS1, increased the transformation frequency 50- to 100-fold as compared with the original plasmid lacking this sequence. Although the isolated sequence behaved like an autonomously replicating sequence in *S. cerevisiae*, it was not functional in *Aspergillus*. Thus, sequences may behave like ARS in yeast but are nonfunctional in the organism from which they were isolated. Nevertheless investigations are still aiming at isolating true ARSs in filamentous fungi.

Tsukuda et al. (*72*) reported a 389-bp sequence arising from *Ustilago maydis* similar to yeast ARSs previously reported. The use of this sequence in *U. maydis* generated high plasmid copy number and mitotic instability in the transgenic lines. Van Heeswijck (*73*) and Roncero et al. (*74*) demonstrated the presence of possible ARSs sequences in the *Mucor circinelloides* genome, by transforming a *leu⁻* mutant. They obtained a high number of mitotically unstable transgenic strains and it was possible to reisolate the transforming DNA, which was present as an extrachromosomal element. Gems et al. (*75*) reported the AMA-1 sequence from *A. nidulans*, which could be an origin of replication. Transformation using AMA-1-containing plasmids generated high frequencies of unstable transgenic lines. This sequence has been used in *Penicillium crysogenum* (*76*), *Gaeumannomyces graminis* (*77*), and other *Aspergilli*, such as *A.niger* and *A. orizae* (*75*). In spite of all efforts attempting to obtain autonomously replicating systems, neither of the reports has been conclusive. In part, this may be due to the limited occurrence of ARSs among filamentous fungi. An alternative explanation may be that the plasmids used with ARSs did not contain telomeric and centromeric sequences of adequate size, so they could not behave as true chromosomes (*67*). Furthermore, centromeric sequences that provide stability to the replicating vectors in filamentous fungi have not been isolated. In relation to telomeric sequences, it has recently been described that some fungi are capable of adding telomeric repeats to exogenous DNA. This phenomenon seems to be widespread, because it has been recorded in unrelated species, such as the pathogenic yeast *Cryptoccocus neoformans* and filamentous fungi such as *Fusarium oxysporum* (*78*) and *Pestalotiopsis microspora* (*79*). Moreover, telomeric sequences from humans (*80*) or *F. oxysporum* (*78*) have been included in plasmids used for transformation of heterologous systems, showing autonomous replication. Even though at the moment there are not many reliable autonomously replicating vectors for fungal transformation, their development seems to be quite close.

4.2. Integration of DNA Into Chromosomes

Depending on the purpose of obtaining a particular transgenic line, it may be more adequate to obtain strains with the transformed DNA integrated into their genome, avoiding the need of maintaining selective pressure or strains in which DNA may be lost upon removal of the pressure. However, in most cases stable integration of transforming DNA into chromosomes is desired (i.e., the generation of banks of mutants). Consequently, the majority of plasmids used in transformation of filamentous fungi do not contain an origin of replication. These vectors integrate into the genome at either homologous or heterologous sequences.

Integrative events are separated into three main classes (*1*). Class I includes integration processes that occurred into a region of homology within the genome; this is known as homologous recombination. The Class II group's integration events have opposite characteristics to those already mentioned; it includes those where there is not apparent homology between the transforming DNA and its integration site into the recipient genome. This integration type is known as heterologous recombination. The mechanism through which heterologous recombination takes place is not well understood. Apparently, short sequences with fortuitous identity are enough to allow DNA integration (*81*). Class III contains those integration events where transforming and endogenous DNA interact and only DNA with high homology is integrated, replacing part of the recipient genome. Thus, it is not possible to detect plasmid sequences in the transformed cells. This process is known as double homologous recombination. It is a useful tool that allows direct integration toward particular sequences generating gene conversion. In some cases, the integration event gives mitotic stability to the transforming DNA but it is still meiotically unstable, just as has been observed in *N. crassa* (*82*) and *A. nidulans* (*83*). In the case of fungi lacking a sexual phase, this instability does not represent any problem.

Integration of multiple plasmid copies is a characteristic commonly observed in transgenic lines. Tandem integrations could be generated by two mechanisms: free plasmids form circular oligomers by homologous recombination and then all copies are integrated into the genome. Alternatively, single plasmid increases the homology with copies of the plasmid making a second integration event easier (*10*).

In order to increase the frequency of transformation, some protocols recommend the use of lineal plasmids. Nevertheless, the results obtained by using them are contrasting, depending on the fungal species and the method of transformation. In the case of *S. cerevisiae* and *U. maydis*, lineal DNA gives higher efficiencies of transformation than circular plasmid (*84,85*). In *N. crassa* the results are similar whether lineal or circular DNA is used (*86*). It is worth noting that in some fungal species transformation with linearized versions of the plasmid increases the frequency of homologous integration and consequently gene conversion, which in some cases is desirable (*87,88*).

5. Applications of Transformation

5.1. Gene Inactivation

Classical fungal genetic approaches for gene identification rely on inactivation of a gene leading to a recognizable phenotype. This approach continues to be extremely useful, yielding mutations that result in obvious phenotypes reflecting the role of the corresponding gene. The process of identifying a gene's function subsequent to establishing its DNA sequence is known as reverse genetics, for which a number of methods have been explored. These methods include homologous recombination, insertional mutagenesis, and gene silencing.

High-throughput sequencing projects are providing large quantities of genomic and

EST sequences for a considerable number of fungal species. This newly compiled database confronts the scientific community with the problem of understanding the function of thousands of previously unknown genes. Comparative analysis using the gene-bank database can designate putative functions by association to earlier identified genes, but only the characterization of mutants can offer definitive answers. In most fungal systems, knockout mutations are now routinely obtained by promoting the homologous recombination of null gene constructs with the genomic wild-type sequence by using the appropriate transformation system. Gene knockouts, or null mutations, are important because they provide a straight practice to determining the function of a gene product. Gene inactivation by homologous recombination has been used in all major groups of fungi, including different zygomycetes, ascomycetes, and basidiomycetes, in addition to several deutermycetes. However, these processes cannot be applied as such for high-throughput mutagenesis.

Insertional mutagenesis is an alternative method for large-scale reverse genetics and is based on the insertion of foreign DNA into the gene of interest by using mobile elements of various origins. Individual clones carrying a particular insertional-DNA mutation can then be easily identified by a polymerase chain reaction (PCR) method (*89*). *A. tumefaciens* T-DNA and transposable elements appear to be the most suitable methods for generating insertional mutations in filamentous fungi (*90,91*). Heterologous transposition of retrotransposons *impala* and MAGGY from *F. oxysporum* and *M. grisea*, respectively, have been reported in different fungal hosts (*92–94*). Methods for insertional mutagenesis were extended to include restriction-enzyme-mediated integration (REMI). This technique, originally described in *S. cerevisiae* (*21*), involves the addition to the transformation reaction of an active restriction enzyme that efficiently targets the transforming DNA to corresponding restriction site. In fungi, REMI can be used in combination with both protoplast- (*95*) and electroporation-mediated transformation (*96*). The REMI technique has been extensively used among plant pathogens, that including *C. heterostrophus* (*95*), *U maydis* (*97*), *M. grisea* (*98*), and *C. graminicola* (*99*).

Techniques established on posttranscriptional gene silencing (PTGS) appear promising for the study of fungal gene function. PTGS based on the expression of an antisense RNA molecule has been successfully used in *C. albicans* (*100*), *C. neoformans* (*101*), and *Aspergillus* species (*102*). Similarly, when a small double-stranded RNA corresponding to a sense sequence of an endogenous mRNA is introduced into a cell, the cognate mRNA is degraded and the gene silenced. This type of PTGS, called RNA interference (RNAi), was first discovered in *C. elegans* (*103*) and has been recently used in the specific inhibition of genes involved in capsule synthesis in *C. neoformans* (*104*). Despite these promising data, gene inactivation through such methods has not been extensively tested in fungi, so the interpretation is sometimes complex.

5.2. Gene Regulation

The use of gene fusions with a reporter gene in combination with the proper transformation system allows gene activity to be monitored. This procedure was first used in bacteria 24 yr ago (*105*). In this way genes could be recognized based on their pattern of expression over time or under a selected condition. Several reporter genes have been successfully applied in fungi (*106–110*). The green fluorescent protein (GFP) from the jellyfish *Aequorea victoria* has proven to be a powerful tool in fungal genetic transformation studies, although modifications to the original sequence were required to obtain GFP versions that fluoresce proficiently in fungal cells (*111*). GFP-expressing transformants of *U. maydis*, *C. albicans*, and *T. atroviride* have been evaluated *in situ* during interaction with their respective hosts (*111–114*).

5.3. Gene Improvement

Genetic transformation seems to be an essential step by which useful traits are engineered into fungi. Species belonging to the genus *Trichoderma* (*115*), *Aspergillus* (*116*) and *Penicillium* (*117*) have been studied extensively in order to understand and manipulate enzyme production and secondary metabolism. Thus, recombinant DNA technologies could be successfully applied to most commercially useful proteins. Yeast and some filamentous fungi are already well established in fermentation procedures and are able to secrete glycosylated and modified proteins to provide the biologically active product (*118,119*).

6. Conclusions

Several powerful methods for studying fungal biology have been developed in recent years, many of which have been used in combination with transformation systems. Technical improvements to basic transformation protocols, in particular selection markers, have allowed the manipulation of a significant number of fungal species. Each of the transformation systems discussed in this chapter has its own advantages and disadvantages, but most of them are potentially suitable for any given species. No general rule could be applied for the transformation efficiency, although factors such as transformation method, vector, host strain, and selection system might influence it.

In the near future, the sequences of numerous genes will be available; consequently the assignment of their biological function is one of the most challenging goals in fungal biology. It is likely that genetic transformation will continue playing a major role in the next phase of fungal genomics—determining gene function.

References

1. Hinnen, A., Hicks, J. B., and Fink, G. R. (1978) Transformation of yeast. *Proc. Natl. Acad. Sci. USA.* **75**, 1929–1923.
2. Mishra, N. C. sand Tatum, E. L. (1973) Non-Mendelian inheritance of DNA-induced inositol independence in *Neurospora. Proc. Natl. Acad. Sci. USA.* **70**, 3875–3879.
3. Peterson, E. M., Hawley, R. J., and Calderone, R. A. (1976) An ultrastructural analysis of protoplast-spheroplast induction in *Cryptoccocus neoformans. Can. J. Microbiol.* **10**, 1518–1521.
4. Case, M. E., Schweizer, M., Kushner, S. R., and Giles, N. H. (1979) Efficient transformation of *Neurospora crassa* by utilizing hybrid plasmid DNA. *Proc. Natl. Acad. Sci. USA.* **76**, 5259–5263.
5. Evans, C. T. and Conrad, D. (1987) An improved method for protoplast formation and its application in the fusion of *Rhodotorula rubra* with *Saccharomyces cerevisiae. Arch. Microbiol.* **1**, 77–82.

6. Bailey, A. M., Mena, G. L., and Herrera-Estrella, L. (1991) Genetic transformation of the plant pathogens *Phytophtora capsici* and *Phytophtora parasitica*. *Nucleic Acids Res.* **15**, 4273–4278.

7. Rhodes, J. C. and Kwon-Chung, K. J. (1985) Production and regeneration of protoplast from *Cryptococcus*. *Sabouraudia* **1**, 77–80.

8. Herrera-Estrella, A., Goldman, G. H. and Van Montagu, M. (1990) High-efficiency transformation system for the biocontrol agents, *Trichoderma spp. Mol. Microbiol.* **4**, 839–843.

9. May, G. S. (1992) Fungal technology, in *Applied Molecular Genetics of Filamentous Fungi* (Kinghorn, J. R. and Turner, G., ed.), Blacie Academic and Professional, Glasgow, pp. 1–27.

10. Finchman, J. R. S. (1989) Transformation in fungi. *Microbiol Rev.* **53**, 148–170.

11. Timberlake, W. E. and Marshall, M. A. (1989) Genetic engineering of filamentous fungi. *Science.* **244**, 1313–1317.

12. Ito, H., Fukuda, Y. Murata, K., and Kimura, A. (1983) Transformation of intact yeast cells treated with alkali cations. *J. Bacteriol.* **153**, 163–168.

13. Dhawale, S. S., Paietta, J. V., and Marzluf, G. A. (1984) A new, rapid and efficient transformation procedure for *Neurospora. Curr. Genet.* **8**, 77–79.

14. Binninger, D. M., Skrzynia, C., Pukkila, P. J., and Casselton, L. A. (1987) DNA-mediated transformation of the basidiomycete *Coprinus cinereus. EMBO J.* **6**, 835–840.

15. Bej, A. K. and Perlin, M. G. (1989) A high efficiency transformation system for the basidiomycete *Ustilago violacea* employing hygromycin resistance and lithium-acetate treatment. *Gene* **80**, 171–176.

16. Shigekawa, K. and Dower, W. J. (1988) Electroporation of eukaryotes and prokaryotes: a general approach to the introduction of macromolecules into cells. *Biotechniques* **6**, 742–751.

17. Ward, M. Kodama, K. H. and Wilson, L. J. (1989) Transformation of *Aspergillus awamori* and *A. niger* by electroporation. *Exp. Mycol.* **13**, 289–293.

18. Goldman, G. H., Van Montagu, M., and Herrera-Estrella, A. (1990) Transformation of *Trichoderma harzianum* by high-voltage electric pulse. *Curr. Genet.* **17**, 169–174.

19. Chakraborty, B. N., Patterson, N. A., and Kapoor, M. (1991) A electroporation-based system for high efficiency transformation of germinated conidia of filamentous fungi. *Can. J. Microbiol.* **37**, 858–863.

20. Edman, J. C. and Kwon-Chung, K. J. (1990) Isolation of the *URA5* gene from *Cryptococcus neoformans* and its use as a selective marker for transformation. *Mol. Cell. Biol.* **10**, 4538–4544.

21. Schiestl, R. H. and Petes, T. D. (1991) Integration of DNA fragments by illegitimate recombination in *Saccharomyces cerevisiae. Proc. Natl. Acad. Sci. USA.* **88**, 7585–7589.

22. Klein, T. M., Wolf, E. D., Wu, R., and Sanford, J. C. (1987) High-velocity microprojectiles for delivering nucleic acids into living cells. *Nature* (London) **327**, 70–73.

23. Sanford, J. C. (1990) Biolistic plant transformation. *Physiol. Plant* **79**, 206–209.

24. Smith, F. D., Harpending, P. R., and Sanford, J. C. (1991) Biolistic transformation of prokaryotes: factors that affect biolistic transformation of very small cells. *J. Gen. Microbiol.* **138**, 239–248.

25. Armaleo, D., Ye, G. N., Klein, T. M., Shark, J. C., and Johnston, S. A. (1990) Biolistic nuclear transformation of *Saccharomyces cerevisiae* and other fungi. *Curr. Genet.* **17**, 97–103.

26. Rocha-Ramírez, V., Omero, C., Chet, I., Horwitz, B., and Herrera-Estrella, A. (2002) A *Trichoderma atroviride* G protein a subunit gene, *tga1*, involved in mycoparasitic coiling and conidiation. *Eukaryotic Cell* **1**, 594–605.

27. Williams, R. S., Johnston, S. A., Riedy, M., De Vit, M. J., McElligott, S. G., and Sanford, J. C. (1991) Introduction of foreign genes into tissues of living mice by DNA-coated microprojectiles. *Proc. Natl. Acad. Sci., USA.* **88**, 2726–2730.

28. Fox, T. D., Sanford, J. C., and McMullin, T. W. (1988) Plasmids can stably transform yeast mitochondria lacking endogenous mtDNA. *Proc. Natl. Acad. Sci. USA* **85**, 7288–7292.

29. Johnston, S. A., Anziano, P. Q., Shark, K., Sanford, J. C., and Butow, R. A. (1988) Mitochondrial transformation in yeast by bombardment with microprojectiles. *Science* **240**, 1538–1541.

30. Daniell, H., Vivekananda, J., Nielsen, B. L., Ye, G. N., and Sanford, J. C. (1990) Transient foreign gene expression in chloroplast of cultured tobacco cells after biolistic delivery of chloroplast vectors. *Proc. Natl. Acad. Sci. USA* **87**, 88–92.

31. Gold, J., Good, L., Herrera-Estrella, A., Diener, T. O., and Martinez-Soriano, J. P. (1999) Plant pathogen interactions. In *Molecular Biotechnology for Plant Food Production* (Paredes-Lopez, O., ed.), Technomic Publishing Co., London, pp. 131–186.

32. Herrera-Estrella, L. (1983) Transfer and expression of foreign genes in plants. PhD thesis. Laboratory of Genetics. Ghent University, Belgium.

33. Bent A. F. and Clough, S. J. (1988) *Agrobacterium* germ-line transformation: Transformation of *Arabidopsis* without tissue culture. In *Plant Molecular Biology Manual* (Gelvin, S. B. and Schilperoort, R. A., eds.), Kluwer Academic Publishers, Dordrecht, The Netherlands, section B7, pp. 1–14.

34. Desiderio, A., Aracri, B., Leckie, F., Mattei, B., Salvi, G., Tigelaar, H., et al. (1997) Polygalacturonase-inhibiting proteins (PGIPs) with different specificities are expressed in *Phaseolus vulgaris. Mol. Plant Microbe Interact.* **10**, 852–860.

35. Franco-Lara, L. F., McGeachy, K. D., Commandeur, U., Martin, R. R., Mayo, M. A., and Barker, H. (1999) Transformation of tobacco and potato with cDNA encoding the full-length genome of potato leafroll virus: evidence for a novel virus distribution and host effects on virus multiplication. *J. Gen. Virol.* **80**, 2813–2822.

36. Grimsley, N., Hohn, T., Davies, J. W., and Hohn, B. (1987) *Agrobacterium* mediated delivery of infectious maize streak virus into maize plants. *Nature* **325**, 177–179.

37. Hiei, Y., Ohta, S., Komari, T., and Kumashiro, T. (1994) Efficient transformation of rice (*Oryza sativa* L.) mediated by *Agrobacterium* and sequence analysis of the boundaries of the T-DNA. *Plant J.* **6**, 271–282.

38. De Groot, M. J. A., Bundock, P., Hooykaas, P. J. J., and Beijerbergen, A. G. M. (1998) *Agrobacterium tumefaciens*-mediated transformation of filamentous fungi. *Nature Biotech.* **16**, 839–842.

39. Gouka, R. J., Gerk, C., Hooykaas, P. J. J., Bundok, P., Musters, W., Verrips, C. T., et al. (1999) Transformation of *Aspergillus awamori* by *Agrobacterium tumefaciens*-mediated homologous recombination. *Nature Biotech.* **17**, 598–601.

40. Mullins, E. D., Chen, X., Romaine, P., Raina, R., Geiser, D. M., and Kang, S. (2001) *Agrobacterium*-mediated transformation of *Fusarium oxysporum*: An efficient tool for insertional mutagenesis and gene transfer. *Techniques* **91**, 173–180.

41. Bundock, P., den Dulk-Ras, A., Beijerbergen, A. and Hooykaas, P. J. J. (1995) Transkingdom T-DNA transfer from *Agrobacterium tumefaciens* to *Saccharomyces cerevisiae. EMBO J.* **14**, 3206–3214.

42. Liu, Y. G., Mitsukawa, N., Oosumi, T., and Whittier, R. F. (1995) Efficient isolation and mapping of *Arabidopsis thaliana* T-DNA transfer junctions by thermal asymmetric interlaced PCR. *Plant J.* **8**, 457–463.

43. D'Enfert, C. (1996) Selection of multiple disruption events in *Aspergillus fumigatus* using the orotidine-5'-decarboxylase gene, *pyrG*, as a unique transformation marker. *Curr. Genet.* **30**, 76–82.

44. Alani, E., Cao, L., and Kleckner, N. (1987) A method for gene disruption that allows repeated use of *URA3* selection in the construction of multiply disrupted yeast strains. *Genetics.* **116**, 541–545.

45. Daboussi, M. J., Djeballi, A., Gerlinger, C., Blaiseau, P. L., Bouvier, I., Cassan, M., et al. (1989) Transformation of seven species of filamentous fungi using the nitrate reductase gene of *Aspergillus nidulans. Curr. Genet.* **15**, 453–456.

46. Bouxton, F. P., Gwynne, D. I., and Davies, R. W. (1989) Cloning of a new bidirectionally selectable marker for *Aspergillus* strains. *Gene* **84**, 329–334.

47. Gouka, R. J., Van Hartingsveldt, W., Bovenberg, R. A. L., Van Zegil, C. M. J., Van den Hondel, C. A. M. J. J., and Van Gorcom, R. F. M. (1993) Development of a new transformant selection system for *Penicillium crysogenum*: isolation and characterization of the *P. crysogenum* acetyl-coenzime A synthetase gene (*facA*) and its use as a homologous selection marker. *Appl. Microbiol. Biotechnol.* **38**, 514–519.

48. Hynes, M. J., and Pateman, J. A. (1970) The genetic analysis of regulation of amidase synthesis in *Aspergillus nidulans.* II. Mutants resistant to fluoroacetamide. *Mol. Gen. Genet.* **108**, 107–116

49. Debets, A. J., Swart, K., Holub, E. F., Goosen, T., and Bos, C. J. (1990) Genetic analysis of *amdS* transformants of *Aspergillus niger* and their use in chromosome mapping. *Mol. Gen. Genet.* **222**, 284–290.

50. Punt, P. J., Oliver, R. P., Digemanse, M. A., Pouwels, P. H., and van den Hondel, C. A. (1987) Transformation of *Aspergillus* based on the hygromycin B resistance marker from *Escherichia coli. Gene* **56**, 117–124.

51. Leung, H., Lethtinen, U., Karjelainen, R., Skinner, D., Tooley, P., Leong, S., et al. (1990) Transformation of the rice blast fungus *Magnaporthe grisea* to hygromycin B resistance. *Curr. Genet.* **17**, 409–411.

52. Tunlid, A., Ahman, J., and Oliver, R. P. (1999) Transformation of the nematode-trapping fungus *Arthrobotrys oligospora. FEMS Microbiol. Lett.* **173**, 111–116.

53. Randall, T. and Reddy, C. A. (1991) An improved transformation vector for the lignin-degrading white rot basidiomycete *Phanerochaete chrysosporium. Gene* **103**,125–130.

54. De Antoni, A. and Gallwitz, D. (2000) A novel multi-purpose cassette for repeated integrative epitope tagging of genes in *Saccharomyces cerevisiae. Gene* **246**,179–185.

55. Orbach, M. S., Porro, E. B., and Yanafsky, C. (1986) Cloning and characterization of the gene of β-tubulin from a benomyl-resistant mutant of *Neurospora crassa* and its use as a dominant selectable marker. *Mol. Cell. Biol.* **6**, 2452–2461.

56. Gold, S. E., Bakkeren, G., Davies, J. E., and Kronstand, J. W. (1994) Three selectable markers for transformation of *Ustilago maydis. Gene* **142**, 225–230.

57. Goldman, G. H., Temmerman, W., Jacobs, D., Contreras, R., Van Montagu, M., and Herrera-Estrella, A. (1993) A nucleotide substitution in one of the β-*tubulin* genes of *Trichoderma viride* confers resistance to the antibiotic drug methyl benzimidazole-2-yl-carbamate. *Mol. Gen. Genet.* **240**, 73–80.

58. Ward, M., Wilson, L. J., Carmona, C. L., and Turner, G. (1988) The *oliC3* gene of *Aspergillus niger*: isolation, sequence and use as a selectable marker for transformation. *Curr. Genet.* **14**, 37–42.

59. Bull, J. H., Smith, D. J., and Turner, G. (1988) Transformation of *Penicillium chrysogenum* with a dominant selectable marker. *Curr. Genet.* **13**, 377–382.

60. Austin, B., Hall, R. M., and Tyler, B. M. (1990) Optimized vectors and selection for transformation of *Neurospora crassa* and *Aspergillus nidulans* to bleomycin and phleomycin resistance. *Gene* **93**,157–162.

61. Kolar, M., Punt, P. J., van den Hondel, C. A., and Schwab, H. (1988) Transformation of *Penicillium chrysogenum* using dominant selection markers and expression of an *Escherichia coli lacZ* fusion gene. *Gene* **62**,127–134.

62. Carramolino, L., Lozano, M., Perez-Aranda, A., Rubio, V., and Sánchez, F. (1989) Transformation of *Penicillium chrysogenum* to sulfonamide resistance. *Gene* **77**, 31–38.

63. Inglis, P. W. (2000) Biolistic co-transformation of *Metarhizium anisopliae* var. *acridum* strain CG423 with green fluorescent protein and resistance to glucosinate ammonium. *FEMS Microbiol. Lett.* **191**, 249–254.

64. Yanai, K., Yonekura, K., Usami, H., Hirayama, M., Kajiwara, S., Yamazaki, T., et al. (1996) The integrative transformation of *Pleurotus ostreatus* using bialaphos resistance as a dominant selectable marker. *Biosci. Biotechnol. Biochem.* **60**, 472–475.

65. Avalos, J., Geever, R. F., and Case, M. E. (1989) Bialaphos resistance as a dominant selectable marker in *Neurospora crassa*. *Curr. Genet.* **16**, 369–372.

66. Keon, J. P., White, G. A., and Hargreaves, J. A. (1991) Isolation, characterization and sequence of a gene conferring resistance to the systemic fungicide carboxin from the maize smut pathogen, *Ustilago maydis*. *Curr. Genet* **19**, 475–481.

67. Goldman, G. H., Van Montagu, M., and Herrera-Estrella, A. (1995) Filamentous fungi, in *Transformation of Plants and Soil Microorganisms* (Lynch, J., ed.), Cambridge University Press, New York, NY, pp. 34–49.

68. Stahl, U., Tudzynski, P., Kück, U., and Esser, K. (1982) Replication and expression of a bacterial-mitochondrial hybrid plasmid in the fungus *Podospora anserina*. *Proc. Natl. Acad. Sci. USA* **79**, 3641–3645.

69. Stohl, L. L. and Lambowitz, A. M. (1983) Construction of a shuttle vector for the filamentous fungus *Neurospora crassa*. *Proc. Natl. Acad. Sci. USA* **80**, 1058–1062.

70. Esser K., Kück, U., Stahl, U., and Tudzynski, P. (1983) Cloning vectors of mitochondrial origin for eukaryotes: a new concept in genetic engineering. *Curr. Genet.* **13**, 327–330.

71. Balance, D. J. and Turner, G. (1985) Development of a high-frequency transforming vector for *Aspergillus nidulans*. *Gene* **36**, 321–331.

72. Tusukuda, T., Carleton, S., Fotheringham, S., and Holloman, W. K. (1988) Isolation and characterization of an autonomously replicating sequence from *Ustilago maydis*. *Mol. Cell. Biol.* **8**, 3703–3709.

73. Van Heeswijck, R. (1986) Autonomous replication of plasmids in *Mucor* transformants. *Carlsber Res. Commun.* **51**, 433–443.

74. Roncero, M. I. G., Jepsen, L. P., Stroman, P., and Van Heeswijck, R. (1989) Characterization of a *leuA* gene and an ARS element from *Mucor circinelloides*. *Gene* **84**, 335–343.

75. Gems, D. H., Johnstone, I. L., and Clutterbuck, A. J. (1991) An autonomously replicating plasmid transforms *Aspergillus nidulans* at high frequency. *Gene* **98**, 61–67.

76. Fierro, F., Kosalková, K., Gutiérrez, S., and Martín, J. F. (1996) Autonomously replicating plasmids carrying the *AMA1* region in *Penicillium chrysogenum*. *Curr. Genet.* **29**, 482–489.

77. Bowyer, P., Osbourn, A. E., and Daniels, M. J. (1994) An "instant gene bank" method for heterologous gene cloning: complementation of two *Aspergillus nidulans* mutants with *Gaeumennomyces graminis* DNA. *Mol. Gen. Genet.* **242**, 448–454.

78. Powell, W. A. and Kistler, H. C. (1990) In vivo rearrangement of foreign DNA by *Fusarium oxysporum* produces lineal self-replicating plasmids. *J. Bacteriol.* **172**, 3163–3166.

79. Long, D. M., Smidansky, E. D., Archer, A. J., and Strobel, G. A. (1998) In vivo addition of telomeric repeats to foreign DNA generates extrachromosomal DNAs in the taxol-producing fungus *Pestalitiopsis microspora*. *Fungal Genet. Biol.* **24**, 335–344.

80. Aleksenko, A. Y. and Ivanova, L. (1998) In vivo linearization and autonomous replication of plasmids containing human telomeric DNA in *Aspergillus nidulans*. *Mol. Gen. Genet.* **260**, 159–164.

81. Ward, M. (1991) *Aspergillus nidulans* and other filamentous fungi as genetic systems, in *Modern Microbial Genetics* (Streips, U. N. and Yasbin, R. E., eds.), Wiley-Liss, New York, NY, pp. 455–496.

82. Selker, E. U., Cambareri, E. B., Jensen, B. C., and Haak, K. R. (1987) Rearrangement of duplicated DNA in specialized cells of *Neurospora*. *Cell* **51**, 741–752.

83. Tilburn, J., Scazzocchio, C., Taylor, G. G., Zabicky-Zissman, J. H., Lockington, R. A., and Davies, R. W. (1983) Transformation by integration in *Aspergillus nidulans*. *Gene* **26**, 205–221.

84. Suzuki, K., Imai, Y., Yamashita, I., and Fukui, S. (1983) In vivo ligation of linear DNA molecules to circular forms in the yeast *Saccharomyces cerevisiae*. *J. Bacteriol.* **155**, 747–754.

85. Wang, J., Holden, D. W., and Leong, S. A. (1988) Gene transfer system for the phytopathogenic fungus *Ustilago maydis*. *Proc. Natl. Acad. Sci. USA* **85**, 865–869.

86. Huiet, L. and Case, M. (1985) Molecular biology of the *qa* gene cluster in *Neurospora crassa*, in *Gene Manipulations in Fungi* (Bennett, J. W. and Lasure, L. L., eds.) Academic Press, Orlando, FL., pp. 229–244.

87. Boylan, M. T., Mirabito, P. M., Wilett, C. E., Zimmerman, C. R., and Timberlake, W. E. (1987) Isolation and physical characterization of three essential conidiation genes from *Aspergillus nidulans*. *Mol. Cell. Biol.* **7**, 3113–3118.

88. Aramayo, R., Adams, T. H., and Timberlake, W. E. (1989) A large cluster of highly expressed genes is dispensable for growth and development in *Aspergillus nidulans*. *Genetics* **122**, 65–71.

89. Ballinger, D. G. and Benzer, S. (1989) Targeted gene mutations in *Drosophila*. *Proc. Natl. Acad. Sci. USA* **86**, 9402–9406.

90. Li Nestri Nicosia, M. G., Brocard-Masson, C., Demais, S., Hua Van, A., Daboussi, M. J., and Scazzochio, C. (2001) Heterologous transposition in *Aspergillus nidulans*. *Mol. Microbiol.* **39**, 1330–1344.

91. De Groot, M. J., Bundock, P., Hooykaas, P. J., and Beijersbergen, A. G. (1998) *Agrobacterium tumefaciens*-mediated transformation of filamentous fungi. *Nat. Biotechnol.* **16**, 839–842.

92. Hua-Van, A., Pamphile, J. A., Langin, T., and Daboussi, M. J. (2001) Transposition of autonomous and engineered impala transposons in *Fusarium oxysporum* and a related species. *Mol. Gen. Genet.* **264**, 724–731.

93. Villalba, F., Lebrun, M. H., Hua-Van, A., Daboussi, M. J., and Grosjean-Cournoyer, M. C. (2001) Transposon impala, a novel tool for gene tagging in the rice blast fungus *Magnaporthe grisea*. *Mol. Plant Microbe Interact.* **14**, 308–315.

94. Nakayashiki, H., Ikeda, K., Tosa, Y., and Mayama, S. (1999) Transposition of the retrotransposon MAGGY in heterologous species of filamentous fungi. *Genetics* **153**, 693–703.

95. Lu, S., Lyngholm, L., Yan, G., Bronson, C., Yoder, O. C., and Turgeon, B. G. (1994) Tagged mutation at the *Tox1* locus of *Cochliobolus heterostrophus* by restriction enzyme-mediated integration. *Proc. Natl. Acad. Sci. USA* **91**, 12,649–12,653.

96. Kuspa, A. and Loomis, W. F. (1992) Tagging developmental genes in *Dictyostelium* by restriction enzyme-mediated integration of plasmid DNA. *Proc. Natl. Acad. Sci. USA* **89**, 8803–8807.

97. Bölker, M., Böhnert, H. U., Heinz Braun, K., Görl, J., and Kahmann, R. (1995) Tagging pathogenicity genes in *Ustilago maydis* by restriction enzyme-mediated integration (REMI).

Mol. Gen. Genet. **248**, 547–552.

98. Sweigard, J. A., Carroll, A. M., Farral, L., Chumley, F. G., and Valent, B. (1998) *Magnaporthe grisea* pathogenicity genes obtained through insertional mutagenesis. *Mol. Plant-Microbe Interact.* **11**, 404–412.

99. Thon, M. R., Nuckles, E. M., and Vaillancourt, L. J. (2000) Restriction enzyme-mediated integration used to produce pathogenicity mutants of *Colletotrichum graminicola. Mol. Plant-Microbe Interact.* **13**, 1356–1365.

100. DeBacker, M. D., Nelissen, B., Logghe, M., Viaene, J., Loonen, I., Vandoninck, S., et al. (2001) An antisense-based functional genomics approach for identification of genes critical for growth of *Candida albicans. Nat. Biotechnol.* **19**, 235–241.

101. Gorlach, J. M., McDade, H. C., Perfect, J. R., and Cox, G. M. (2002) Antisense repression in *Cryptoccocus neoformans* as a laboratory tool and potential antifungal strategy. *Microbiology* **148**, 213–219.

102. Bautista, L. F., Aleksenko, A., Hentzer, M., Santerre-Henriksen, A., and Nielsen, J. (2000) Antisense silencing of the *creA* gene in *Aspergillus nidulans. Appl. Environ. Microbiol.* **66**, 4579–4581.

103. Fire, A., Xu, S., Montgomery, M. K., Kostas S. A., Driver, S. E., and Mello, C. C. (1998) Potent and specific genetic interference by double-stranded RNA in *Caenorhabditis elegans. Nature* **404**, 804–808.

104. Liu, H., Cottrell, T. R., Pierini, L. M., Goldman, W. E., and Doering, T. L. (2002) RNA interference in the pathogenic fungus *Cryptoccocus neoformans. Genetics* **160**, 463–470.

105. Casadaban, M. J. and Cohen, S. N. (1979) Lactose genes fused to exogenous promoters in one step using a Mu-lac bacteriophage: in vivo probe for transcriptional control sequences. *Proc. Natl. Acad. Sci. USA* **76**, 4530–4533.

106. Ilmen, M., Onnela, M. L., Klemsdal, S., Keranen, S., and Penttila, M. (1996) Functional analysis of the cellobiohydrolase I promoter of the filamentous fungus *Trichoderma reesei. Mol. Gen. Genet.*. **253**, 303–314.

107. Hynes, M. J., Draht, O. W., and Davis, M. A. (2002) Regulation of the *acuF* gene, encoding phosphoenolpyruvate carboxyikinase in the filamentous fungus *Aspergillus nidulans. J. Bacteriol.* **184**, 183–190.

108. Roberts, I. N., Oliver, R. P., Punt, J. P., and van den Hondel, C. A. (1989) Expression of the *Escherichia coli* beta-glucuronidase gene in industrial and phytopathogenic filamentous fungi. *Curr. Genet.* **15**, 177–180.

109. Snoeijers, S. S., Vossen, P., Goosen, T., van den Broek, H. W. and De Witt, P. J. (1999) Transcription of the avirulence gene *avr9* of the fungal tomato pathogen *Cladosporium fulvum* is regulated by a GATA-type transcription factor in *Aspergillus nidulans. Mol. Gen. Genet.* **261**, 653–659.

110. Mach, R. L., Peterbauer, C. K., Payer, K., Jaksits, S., Woo, S., Zeilinger, S., Kullning, C. M., Lorito, M., and Kubicek, C. (1999) Expression of two major chitinase genes of *Trichoderma atroviride* (*T. harzianum* P1) is triggered by different regulatory signals. *Appl. Environ. Microbiol.* **65**, 1858–1863.

111. Fernández-Ábalos, J. M., Fox, H., Pitt, C., Wells, B., and Doonan J. H. (1998) Plant-adapted green fluorescent protein is a versatile vital reporter for gene expression, protein localization and mitosis in the filamentous fungus, *Aspergillus nidulans. Mol. Microbiol.* **27**, 121–130.

112. Spellig, T., Bottin, A., and Kahmann, R. (1996) Green fluorescent protein (GFP) as a new vital marker in the phytopathogenic fungus *Ustilago maydis. Mol. Gen. Genet.* **252**, 503–509.

113. Cormack, B. P., Bertram, G., Egerton, M., Gow, N. A. R., Falkow, S., and Brown A. J. P. (1997) Yeast-enhanced green fluorescent protein (yEGFP) a reporter of gene expression in

Candida albicans. Microbiology. **143**, 303–311.

114. Olmedo-Monfil, V., Mendoza-Mendoza, A., Gómez, I., Cortés, C., and Herrera-Estrella, A. (2002) Multiple environmental signals determine the transcriptional activation of the mycoparasitism related gene *prb1* in *Trichoderma atroviride. Mol. Genet. Genomics.* **267**, 703–712.

115. Harkki, A., Uusitalo, J., Bailey, M., Pentilla, M., and Knowles, J. K. C. (1989) A novel fungal expression system: secretion of active calf chymosin from the filamentous fungus *Trichoderma reesei. BioTechnol.* **7**, 596–603.

116. Penalva, M. A., Rowlands, R. T., and Turner, G. (1998) The optimization of penicillin biosynthesis in fungi. *Trends Biotechnol.* **11**, 483–489.

117. MacCabe, A. P., Orejas, M., Tamayo, E. N., Villanueva, A., and Ramon, D. J. (2002) Improving extracellular production of food-use enzymes from *Aspergillus nidulans. Biotechnol.* **96**, 43–54.

118. Holz, C., Hesse, O., Bolotina, N., Stahl, U., and Lang C. (2002) A micro-scale process for high-throughput expression of cDNAs in the yeast *Saccharomyces cerevisiae. Protein Expr. Purif.* **25**, 372–378.

119. Punt, P. J., van Biezen, N., Conesa, A., Albers, A., Magnus, J., and van den Hondel, C. (2002) Filamentous fungi as cell factories for heterologous protein production. *Trends Biotechnol.* **20**, 200–206.

22

Three Decades of Fungal Transformation

Novel Technologies

Sergio Casas-Flores, Teresa Rosales-Saavedra, and Alfredo Herrera-Estrella

Summary

Fungi are lower eukaryotes that play important roles in many human activities, including biotechnological processes, phytopathology, and biomedical research. In addition, they are excellent models for molecular and genetic studies. An important key in the advancement of genetics and molecular biology of a given organism is the development of genetic transformation systems. This technology makes possible the analysis and manipulation of the genome of the organism of interest. Thirty years from the first report of transformation of a fungus, transformation of many other fungi has been achieved. However, the development of gene tagging systems generally applicable to a wide range of filamentous fungi has remained elusive. A widely used gene tagging strategy for filamentous fungi is restriction enzyme mediated integration (REMI). In recent years numerous reports have been published describing the effective application of REMI. However, REMI shows certain disadvantages for some fungi. Recently a very promising alternative strategy has been reported based on the use of the soil bacterium *Agrobacterium tumefaciens*. Using this system a well-defined DNA segment (T-DNA) is transferred, which integrates by illegitimate recombination and is 100–1000 times more efficient than conventional methods. The T-DNA can be used as an efficient tool to generate recombinant strains where DNA is integrated as a single copy, allowing the generation of collections of gene-tagged mutants of the fungus of interest.

Key Words: *Agrobacterium*, ATMT, gene transfer, tagging, mutagenesis, filamentous fungi, insertional mutagenesis.

From: *Methods in Molecular Biology, vol. 267:*
Recombinant Gene Expression: Reviews and Protocols, Second Edition
Edited by: P. Balbás and A. Lorence © Humana Press Inc., Totowa, NJ

1. Introduction

Fungi constitute a morphologically and physiologically diverse group of eukaryotes. Among the attributes that distinguish the fungi are a hyphal or yeast-like form; a rigid frequently chitinous cell wall; apical growth; adsorptive heterotrophic metabolism, reproduction by spores; and small genome size.

As a group, the fungi include organisms with diverse ecological roles. Fungi are also of obvious importance in biotechnological processes due to the fact that they produce a large variety of useful secondary metabolites. Many of them are important pathogens of humans, animals, and plants and some others are used as biological control agents of phytopathogens and insects. Finally, some of them represent model systems for basic research. In order to obtain knowledge about fungi to either combat or exploit their biological properties efficiently, we must understand their physiology and the molecular events underlying these processes.

Recombinant DNA technology has provided us with techniques to isolate genes and analyze their function for a large group of organisms; these techniques were first employed successfully during the 1970s in *Escherichia coli* and then in eukaryotes with the baker's yeast *Saccharomyces cerevisiae*. Transformation of fungi was first described, three decades ago, for the inositol gene in *Neuropora crassa* by Mishra and Tatum (*1*), utilizing total DNA. The outcome of their experiments constituted a milestone for the establishment of gene transfer technologies, a fundamental process for the development of molecular genetics in filamentous fungi. This process is usually carried out nowadays by the use of a vector containing a marker gene that allows the selection of successfully transformed cells.

Another important process in genetic analysis is cloning genes of interest. Classic genetic analysis using mutagens such as chemicals or ultraviolet (UV) light has yielded a wealth of knowledge on fungal physiology and development. These mutagens mainly generate base pair deletions or substitutions, which can result in the loss or alteration of gene function and their relative lack of specificity allows saturation of a genome with mutations. Subsequent genetic analyses of mutant strains can, however, be time-consuming because they usually involve isolation of the mutated gene by complementation using a genomic DNA library from the wild-type strain (*2*). Other techniques that have been used to clone genes include reverse genetics, heterologous probes, differential hybridization, immunodetection, and functional complementation in *E. coli* or *S. cerevisiae* (*3–6*). An attractive approach for cloning and studying genes of interest is insertional mutagenesis, which has many useful features; basically, because the mutation of interest is physically marked, it is easy to isolate and analyze the DNA region flanking the site of the insertion. For many fungi, transformation with heterologous plasmid DNA carrying a selectable marker has been used as an insertional approach; however, for some fungi, plasmid DNA integrates in multiple copies, either at different sites in the genome or in tandem in a single site (*3,7–9*). Additionally, it has been shown that these methods can result in a high proportion of significant genomic rearrangements, hampering a high-throughput search for genes. Alternatively, mobile elements such as transposons have been exploited to create insertion mutations with great success in many bacterial (*10*) and a few eukaryotic systems (*11–13*). Unfortunately, transposons are not uniformly found in fungi, especially in laboratory strains, and those found do not show the relevant characteristics, such as stability and random integration required to prove useful for genetic analysis (*14,15*). In filamentous fungi, transformation with nonhomologous DNA results in illegitimate integration of DNA into the fungal genome (ectopic integration). This characteristic opens the possibility of using the transforming DNA as an insertional mutagen to disrupt and tag genes.

An advantage of this technique over chemical or UV mutagenesis is that the mutated gene is tagged with the transforming DNA and can be used to clone the wild-type gene by rescue of the plasmid or by the polymerase chain reaction (PCR). In the majority of fungi, ectopic integration of transforming DNA is much more frequent than homologous, targeted integration (*3*). Alternatives to the traditional DNA-mediated transformation for the production of insertional mutants in fungi can be found in the use of restriction enzyme mediated integration (REMI) and *Agrobacterium tumefaciens* mediated transformation (ATMT).

REMI, a method for generating nonhomologous integration of transforming DNA into the chromosome that facilitates single-copy integration of the transforming DNA, has been used for insertional mutagenesis in several eukaryotes, including fungi (*16*). This process requires the addition to the transformation reaction of a restriction enzyme recognizing a single site in the transforming plasmid. REMI frequently increases transformation efficiency compared to conventional transformation (*17–19*). How REMI works is unclear, but this integration event can be interpreted as follows; First, during transformation, the restriction enzyme enters the nucleus along with linearized transforming DNA whose ends have been generated by the added enzyme. Second, the restriction enzyme digests the host's DNA at its specific recognition sites. Third, the complementary ends of the chromosomal DNA and the transforming fragment are ligated in vivo to generate a non homologous integration event (*17*).

In recent years numerous papers have been published describing the effective application of REMI to study gene function (*18,20–30*). However, REMI has great disadvantages for certain fungi. Reports indicate that one third of the mutants generated by REMI were found untagged with the selectable marker gene (*18,22,25,31*). It has been proposed that digestion of the chromosome by a restriction enzyme followed by imperfect DNA repair during transformation is responsible for these untagged mutations (*32*). The generation of such untagged mutations can make cloning of the gene causing the mutant phenotype substantially tedious and time-consuming. For fungi with a sexual stage, genetic crosses can be employed to identify untagged mutants, in which the mutant phenotype fails to co-segregate with the selectable marker on the inserted plasmid. However for fungi lacking a sexual stage the cloning of a tagged gene and subsequent functional complementation by transformation is necessary to verify whether a given mutant was tagged.

Agrobacterium tumefaciens is a Gram-negative soil bacterium able to transfer part of its tumor-inducing (Ti) plasmid, the transferred DNA or T-DNA, to a plant during tumorigenesis (*33*). T-DNA transfer is dependent on vir gene induction, a group of genes located on the Ti plasmid. Phenolic compounds secreted by wounded plants such as acetosyringone induce these genes. *A. tumefaciens* has been successfully used to transfer genes in to a wide variety of plants and has been used as an insertional tool in *Arabidopsis thaliana* (*33–35*). Recently, several fungi have been transformed using the *A. tumefaciens* mediated transformation system (*36–51*). One of the main advantages that *Agrobacterium* offers compared with conventional techniques is versatility in the selection of the material to be transformed. *Agrobacterium* has been successfully used to transform protoplasts, hyphae, spores, and mycelial tissue (*41*).

Agrobacterium transfers and integrates the DNA located between the right and left borders of the T-DNA by illegitimate recombination and is 100–1000 times more efficient than conventional methods for fungal transformation (*41*). The T-DNA can be used as an efficient tool to generate recombinant strains where DNA is integrated as a single copy, allowing the generation of a collection of gene-tagged mutants of the fungus of interest. ATMT can also be used as an efficient tool to generate recombinant fungal strains with multiple copies of a gene integrated at a predetermined site in the fungal genome. When DNA located between the T-DNA borders shares homology with the fungal genome, integration can also occur by homologous recombination; on this basis, it is possible to obtain recombinant strains with industrial and economical importance. Gouka and coworkers (*45*) obtained a recombinant *Aspergillus awamori* strain by introducing multiple copies in tandem of a *Fusarium solani pisi pyrG* gene in the *A. awamori* genome, using ATMT with successful results. This transformation system will potentially stimulate market acceptance of derived products by preventing the introduction of undesirable foreign DNA into the fungus.

Recovery of the sequences flanking the site of insertion in gene tagging is a requisite for gene isolation and molecular genetic analysis of putative tagged transformants. The thermal asymmetric interlaced PCR (TAIL-PCR) method has been used successfully to recover the flanking sequences of T-DNA insertional mutants in plants and fungi (*50,52–54*). The principle of this approach resides in the use of nested sequence-specific primers together with a shorter arbitrary degenerated primer so that the relative amplification efficiencies of specific and nonspecific products can be thermally controlled. One low-stringency PCR cycle is carried out to create annealing site(s) adapted for the arbitrary primer within the known target sequence bordering the known segment. This sequence is then preferentially and geometrically amplified over nontarget ones by interspersion of high-stringency PCR cycles with reduced-stringency PCR cycles.

2. Materials

2.1. Biological Material

The appropriate fungal strain.

A suitable *Agrobacterium tumefaciens* strain (the most commonly used strains are LBA126, LBA100 or AGL-1).

2.2. Plasmids

An adequate binary vector for *A. tumefaciens* (based preferentially on pCAMBIA vectors with a useful selective marker in between the T-DNA borders, e.g., pCAMBIA 1300 vectors series) (*see* **Note 1**).

2.3. Reagents, Solutions, and Culture Media

1. *Agrobacterium* minimal medium: 20X phosphate buffer (100 mL), [KH_2PO_4 (6.0 g), NaH_2PO_4 (2.0 g)]; 20X salt solution (100 mL), [NH_4Cl (2.0 g), $MgSO_4 \cdot 7H_2O$ (0.6 g), KCl (0.3 g), $CaCl_2$ (0.02 g) and $FeSO_4 \cdot 7H_2O$ (0.005 g)]; adjust to pH 7.0 with HCl; medium is sterilized by autoclaving. Prepare the medium in a flow bench from the stock solutions at 1X in autoclaved sucrose/water 5%; add chloroform as a preservative.

2. Induction medium (IM): Prepare the next stock solutions: 1 *M* MES adjusted to pH 5.6 with 5 *M* KOH, sterilized by filtration; 1 *M* glucose, sterilized by autoclaving; 50% glycerol in water, autoclaved, and acetosyringone 0.5 *M* in DMSO (store at −20°C). For 20 mL IM: In a flow bench add to 15 mL sterilized water 1 mL of 20X minimal medium as in **Subheading 2.3.1.**, 0.8 mL 1 *M* MES, 0.2 mL 1 *M* glucose, 0.2 mL 50% glycerol, and 8 µL 0.5 *M* acetosyringone; adjust volume to 20 mL with sterilized water. We advise to prepare the IM and co-cultivation media the same day that the experiment is going to be carried out, due to the short lifetime of acetosyringone.

3. Co-cultivation medium: Same as IM supplemented with 1.5 g/L agar. Sterilize agar with half of the total water required in order to avoid the degradation by high temperature of some compounds in the IM, allow to cool down to 50°C, and add the other solution in a flow bench.

4. Selection medium: Usually rich medium for the fungus of interest with the appropriate antibiotic for selection of transformants, and cefotaxime 200 µ*M* to kill *Agrobacterium* (*see* **Note 2**).

5. *Agrobacterium* YEB medium: 0.5% beef extract, 0.1% yeast extract, 0.5% bactopeptone, 0.5% sucrose, 0.8 g $MgSO_4$. For solid medium add 1.5% agar; to maintain the strain add the appropriate antibiotic (medium is sterilized by autoclaving).

6. 0.45-µ*M* pore size, 0.45-mm diameter nitrocellulose filters or sterilized cellophane disks.

7. Cefotaxime stock prepared in water at 200 m*M*.

8. Sterilized Miracloth filters.

9. 10X PCR buffer: 100 m*M* Tris-HCl, pH 8.3, 50 m*M* KCl, and 0.001% gelatin.

10. 50 m*M* $MgCl_2$.

11. 10X dNTPs (2 mM each).

12. The three designed nested specific primers and arbitrary primers.

13. *Taq* polymerase.

3. Methods
3.1. Transformation of Fungi

1. Inoculate cultures of the selected fungus in a flask or Petri dish in the appropriate medium and incubate at 28°C until mature spores are produced.

2. Five days before the spores reach the mature stage, streak out the *Agrobacterium* strain containing the appropriate binary vector on fresh YEB plates containing the appropriate antibiotic at 28°C.

3. Two days before the spores reach the mature stage inoculate a single colony from the fresh plate to 5 mL of *Agrobacterium* minimal medium containing antibiotics and grow it for 1–2 at 28°C with agitation (200 rpm). Dilute the *Agrobacterium* culture to OD 0.15 at 600 nm in liquid induction media (IM) and grow for 6 at 28°C with agitation with the appropriate antibiotics. Do not forget to include controls without acetosyringone.

4. While *Agrobacterium* grows collect the spores of the fungus by adding 5 mL of induction medium to the flask or Petri dish to wash off conidia and pass the suspension through a sterilized

piece of Miracloth to eliminate mycelium and cellular debris. Count the number of spores and determine concentration using a hemocytometer. Prepare dilution from 1×10^{-4} per mL to 1×10^{-7} per mL in induction medium.

5. Mix 100 μL of the *Agrobacterium* culture with 100 μL of the spore suspension. Plate out 200 μL onto nitrocellulose filters or cellophane disks on solid induction media. Experimental plates should contain 200 μ*M* of acetosyringone; for the control, plate out the mix onto medium without acetosyringone (you would not expect to obtain colonies under this condition, or only a few) (*see* **Note 3**; if you are using fast-growing fungi, *see* **Note 4**).

6. Incubate at the appropriate temperature (usually 28°C) for approx 2–3 d.

7. After the incubation time on induction media transfer the membranes or cellophane to the rich medium for growth of the fungus containing the antibiotic for selection of transformants and cefotaxime 200 μ*M* to kill *Agrobacterium*.

8. Leave the plates at the appropriate temperature (the optimal for the fungus) for two days and subculture to a fresh selective medium and leave it until colonies begin to appear on the Petri dish.

9. Count the transformants on the Petri dish and calculate the frequency of transformation by multiplying the number of transformants by the dilution used.

10. Pick transformation candidates and transfer to fresh rich medium containing antibiotic.

11. Extract genomic DNA of the transformants by the routine method in your lab in order to analyze the DNA by Southern blotting to determine the copy number of inserted T-DNA or to rescue the DNA flanking T-DNA sequences by TAIL-PCR (*see* **Subheading 3.2.**).

3.2. Rescue of Genomic DNA Flanking T-DNA Sequences by Thermal Asymmetric Interlaced (TAIL)-PCR Technique

TAIL-PCR uses three nested specific primers in successive reactions together with a shorter arbitrary degenerated primer so that the relative amplification efficiencies of specific and nonspecific products can be thermally controlled (*see* **Note 5**).

1. Set up the primary reaction (20 μL volume):
   1 μL DNA (20–100 ng); 2 μL 10X PCR buffer; 0.8 μL MgCl$_2$ (50 m*M*); 0.5 μL 10X dNTPs (2 m*M* each); 2 μL First specific primer (2 pmole/μL); 3 μL Degenerated primer (20 pmole/μL); 0.2 μL *Taq* polymerase (5 U/μL); 10.5 μL H$_2$O.

2. Use the following cycling conditions for primary PCR:
   2 min 95°C; 5 cycles (30 s 94°C, 1 min 62°C, 2.5 min 72°C); 1 cycle (30 s 94°C, 3 min 25°C, 3 min ramp to 72°C, 2.0 min 72°C); 15 supercycles (10 s 94°C, 1 min 68°C, 2.5 min 72°C, 10 s 94°C, 1 min 68°C, 2.5 min 72°C, 10 s 94°C, 1 min 44°C, 2.5 min 72°C); 5 min 72°C; and ∞ at 4°C.

3. Dilute the primary PCR product 1:50 in distilled sterile water.

4. Set up the secondary PCR reaction (20 μL volume):
   1 μL DNA (1:50 dilution 1 product); 2 μL 10X PCR buffer; 0.8 μL MgCl$_2$ (50 m*M*); 2 μL 10X dNTPs (2 m*M* each); 2 μL Second specific primer (2 pmole/μL); 3 μL Degenerated primer (20 pmole/μL); 9.05 μL H$_2$O; 0.15 μL *Taq* polymerase (5U/μL).

5. Use the following cycling conditions for secondary PCR:
   15 supercycles (10 s 94°C, 1 min 64°C, 2.5 min 72°C, 10 s 94°C, 1 min 64°C, 2.5 min 72°C, 10 s 94°C, 1 min 44°C, 2.5 min 72°C); 5 min 72°C; ∞ at 4°C.

6. Dilute the secondary PCR product 1:50 in distilled sterile water.

7. Set up the tertiary PCR reaction (20 μL volume):
   1 μL DNA (1:50 dilution 2 product); 2 μL 10X PCR buffer; 0.8 μL MgCl$_2$ (50 m*M*);

2 µL 10X dNTPs (2 m*M* each); 2 µL Third specific primer (2 pmole/µL); 3 µL Degenerated primer (20 pmole/µL); 9.05 µL H_2O; 0.15 µL *Taq* polymerase (5U/µL).

8. Use the following cycling conditions for tertiary PCR:
   25–30 cycles (15 s 94°C, 1 min 44°C, 2.5 min 72°C); 5 min 72°C; ∞ at 4°C.
9. Load 15 to 20 µL of the secondary and tertiary products on a 2% agarose gel.
10. Products can be sequenced directly if amplified with low dNTP concentration or after column purification.

4. Notes

1. The pCAMBIA 1300 vector series can be used for routine cloning of gene cassettes of interest comprising a promoter, coding sequence of the gene, and a terminator (the pUC18 multiple cloning site is proximal to the right border). The new constructs can then be used for transformation of fungi via *Agrobacterium*. The vectors have been tested successfully for *Agrobacterium*-mediated transformation of fungi (*50*).

2. Selective media vary according to the fungus to be transformed. In this section we include media formulations for three representative fungi.

   Neurospora crassa. Minimal medium (N), for one liter in 750 mL distilled water, dissolve successively with stirring at room temperature: Na$_3$ citrate·5½ H_2O (150 g); KH$_2$PO$_4$ anhydrous (250 g); NH$_4$NO$_3$ anhydrous (100 g); MgSO$_4$·H$_2O$ (10 g); CaCl$_2$·2 H_2O (5 g). Add with stirring 5 mL trace element, 2.5 mL biotin solution. Add 2 mL Chloroform as a preservative. For the working solution, this medium is diluted 50-fold with distilled water. Medium is supplemented with a suitable carbon source such as sucrose (20 g/L), and the supplemented medium is sterilized by autoclaving. The trace element solution (containing citric acid as a solubilizing agent) is made up as follows: in 95 mL distilled water, dissolve successively with stirring at room temperature [Citric acid, 1 H_2O (5.0 g); ZnSO$_4$· 7 H_2O (5 g); Fe (NH$_4$)$_2$(SO$_4$)$_2$·6 H_2O (1 g); CuSO$_4$·5 H_2O (0.25 g); MnSO$_4$·1H$_2O$ (0.05 g); H$_3$BO$_3$ anhydrous (0.05 g); Na$_2$MoO$_4$·2 H_2O (0.05 g)]. Add 1 mL chloroform as a preservative. The biotin solution is prepared by dissolving 5.0 mg biotin (Merck) in 50 mL distilled water. The solution obtained is dispensed into test tubes and stored at −20°C. To prepare "complete medium" supplement Medium N with a carbon source, 0.5% yeast extract, and 0.5% N-Z-Case (Sheffield).

   Fusarium oxysporum and *Aspergillus* minimal medium (*56*): for 1 L 20X stock solution: NaNO$_3$ (120 g); KCl (10.4 g); MgSO$_4$·7H$_2O$ (10.4 g), and KH$_2$PO$_4$ (30.4 g); 100 mL Hutner's trace elements: ZnSO$_4$·7H$_2O$ (2.2 g); H$_3$BO$_3$ (1.1 g); MnCl$_2$·4H$_2O$ (0.5 g); FeSO$_4$·7H$_2O$ (0.5 g); CoCl$_2$·6H$_2O$ (0.16 g); CuSO$_4$·5H$_2O$ (0.16 g); (NH$_4$)$_6$Mo$_7$O$_{24}$·4H$_2O$ (0.11 g), and Na$_2$EDTA (5 g). To prepare trace elements, heat to boiling; cool to 60°C; add KOH, adjusting pH to 6.5–6.8. The solution goes to deep purple after standing for several days. If in titrating the pH exceeds 7.0, discard; salts will precipitate, thus it is best to start over. For one liter working solution add 50 mL stock salt solution, 2 mL Hutner's trace elements, 10 g glucose.

   For solid medium add 15.0 g agar. Complete media (glucose 2.5%, yeast extract 0.5%, and agar 1.5%).

3. The transformation frequency by ATMT can be improved by increasing the *Agrobacterium tumefaciens* cell number. Pre-growth of *Agrobacterium* cells in the presence of acetosyringone will also increase the number of transformants generated. Substantial bacterial growth during the co-cultivation period is also critical for good transformation. If the fungus germinates and covers the membrane, which appears to suppress bacterial growth, you

would not get a high number of transformants. Co-cultivation time could be another important point to analyze for improving the transformation frequency; long co-cultivation periods are not recommended for fast-growing fungi, because fungal growth can suppress bacterial growth. Additionally, long co-cultivation periods could result in the production of transformants with more than one T-DNA insertion. You should try different co-cultivation times in order to get the best results.

4. Many filamentous fungi can generate problems due to their growth rate and they can overgrow the bacterial cells, yielding low transformation frequencies (*see* Note 1). Furthermore, transformants may grow to confluence, making it difficult to select individual transformants. These points can be solved by including compounds that restrict colony growth in the culture medium. Triton X-100 and Na-deoxycolate have been used to reduce the growth rate of many fungi and sorbose 0.1% is widely used for *Neurospora crassa*.

5. The thermal cycles proposed in this protocol are for the Perkin Elmer 9600 model of thermal cyclers; adjust the settings for the thermal cycler in your lab to avoid failure in the TAIL-PCR reaction. To achieve adequate thermal asymmetry, the specific and degenerate primers should be designed to have melting temperatures (Tms) of 57–62°C and 44–46°C, respectively. The Tms should be calculated according to the following formula; $69.3 + 0.41(\%GC) - 650/L$ (*54*), where L is primer length. The three nested specific primers should be designed over the right and left borders of the T-DNA going from the 5' to 3' end in both cases. For the degenerated primers we advise the use of the primers described by Liu et al. 1995, AD1 (5'-NTCGA (G/C) T(A/T)T(G/C)G(A/T)GTT-3'; AD2(5'- NGTCGA(G/C)(A/T)GANA(A/T)GAA-3') and AD3 (5'-(A/T)GTGNAG(A/T)ANCANAGA). If you have a high background of nonspecific products, it is important to increase the annealing temperature in order to obtain specific products with lower backgrounds of nonspecific amplifications. For sequencing of TAIL-PCR products you can use the specific primer designed of the right and left borders of the T-DNA.

References

1. Mishra, N. C. and Tatum, E. L. (1973) Non-Mendelian inheritance of DNA-mediated inositol independence in *Neurospora. Proc. Natl. Acad. Sci. USA*. **70**, 3875–3879.
2. Brown, J. S. and Holden, D. W. (1998) Insertional mutagenesis of pathogenic fungi. *Curr. Opin. Microbiol.* **1**, 390–394.
3. Finchman, J. R. S. (1989) Transformation of fungi. *Microbiol. Rev.* **53**, 148–170.
4. Balance, D. J. (1991) Transformation systems for filamentous fungi and an overview of fungal gene structure. In Molecular Industrial Mycology Systems and Applications for Filamentous Fungi (Leong, S. A., Berka, R. M., eds.) Marcel Dekker, New York, p. 1.
5. Kinghorn, J. R. and Unkles, S. E. (1999) Molecular genetics and expression of foreing proteins in the genus *Aspergillus* (Smith, J. E., ed.) Plenum Press, London, p. 65.
6. Hogan, L. H. and Klein, B. S. (1997) Transforming DNA integrates at multiple sites in the dimorphic fungal pathogen *Blastomyces dermatitidis. Gene* **186**, 219–226.
7. Turner, G. (1990) Strategies for cloning genes from filamentous fungi. In *Applied Molecular Genetics of Fungi* (Peberdy, J. F., Caten, C. E., Ogden, J. E., Bennett, J. W., eds.) Cambridge University Press, Cambridge, p. 29.
8. Wang, J., Holden, W. and Leong, S. A. (1988) Gene transfer system for the phytopathogenic fungus *Ustilago maydis. Proc. Natl. Acad. Sci. USA*. **85**, 865–869.
9. Worsham, P. L. and Goldman, W. E. (1990) Development of a genetic transformation system for *Histoplasma capsulatum*, complementation of uracil auxotrophy. *Mol. Gen. Genet.* **221**, 358–362.

10. Varma, A., Edman, J. C., and Kwong-chong, K. J. (1992) Molecular and genetic analysis of *URA* transformants of *Cryptococcus neoformans*. *Infect. Immun.* **60**, 1101–1108.
11. Hensel, M. and Holden, D. W. (1996) Molecular genetic approaches for the study of virulence in both pathogenic bacteria and fungi. *Microbiology* **142**, 1049–1058.
12. Kemken, F. and Kuck, U. (1996) Restless, an active Ac-like transposon from the fungus *Tolypocladium inflatum*, structure, expression, and alternative RNA splicing. *Mol. Cell Biol.* **16**, 6563–6572.
13. Langing, T., Capy, P., and Daboussi, M. J. (1995) The transposable element impala, a fungal member of the Tc1-mariner superfamily. *Mol. Gen. Genet.* **246**, 19–28.
14. Daboussi, M.J., Djaballi, A., and Gerlinger, C. (1989) Transformation of seven species of filamentous fungi using nitrate reductase gene of *Aspergillus nidulans*. *Curr. Genetics* **15**, 453–456.
15. Kinsey, J. A. and Helber, J. (1989) Isolation of transposable element from *Neurospora crassa*. *Proc. Natl. Acad. Sci. USA* **86**, 1929–1933.
16. Devine, S. E. and Boeke, J. D. (1996) Integration of the yeast retrotransposon Ty1 is targeted to regions upstream of genes transcribed by RNA polymerase III. *Genes Dev.* **10**, 620–633.
17. Riggle, P. J. and Kumamoto, C. A. (1998) Genetic analysis in fungi using restriction-enzyme-mediated integration. *Curr. Opin. Microbiol.* **1**, 395–399.
18. Lu, S., Lyngholm, L., Yang, G., Bronson, C., Yoder, O. C., and Turgen, B. G. (1994) Tagged mutations at the *Tox1* locus of *Cochliobolus heterostrophus* by restriction enzyme-mediated integration. *Proc. Natl. Acad. Sci. USA* **91**, 12649–12653.
19. Granado, J. D., Kertesz-Chaloupkova, K., Aebi, M., and Kues, U. (1997) Restriction enzyme-mediated DNA integration in *Coprinus cinereus*. *Mol. Gen. Genet.* **256**, 28–36.
20. Itoh, Y. and Scott, B. (1997) Effect of de-phosphorylation of linearised pAN7-1 and addition of restriction enzyme on plasmid integration in *Penicillium paxilli*. *Curr. Genet.* **32**, 147–151.
21. Akamatsu, H., Itoh, Y., Kodama, M. and Kohmoto, K. (1997). AAL-toxin deficient mutants of *Alternaria alternata* tomato pathotype by restriction enzyme mediated integration. *Phytopathology* **87**, 967–972.
22. Yun, S. H., Turgeon, B. G., and Yoder, O. C. (1998) REMI-induced mutants of *Mycosphaerella zeae-maydis* lacking the PM-toxin are deficient in pathogenesis to corn. *Physiol. Mol. Plant Pathol.* **52**, 53–66.
23. Sweigard, J. A., Carrol, A. M., Farrall, L., Chumley, F. G., and Valent, B. (1998) *Magnaporte grisea* pathogenicity genes obtained through insertional mutagenesis. *Mol. Plant-Microbe Interact.* **11**, 404–412.
24. Balhadere. P. V., Foster, A. J., and Talbot, N. J. (1999) Identification of pathogenicity mutants of the rice blast fungus *Magnaporte grisea* by insertional mutagenesis. *Mol. Plant-Microbe Interact.* **12**, 129–142.
25. Redman, R. S., Ranson, J. C., and Rodriguez, R. J. (1999) Conversion of the pathogenic fungus *Colletotrichum magna* to a nonpathogenic endophytic mutualist by gene disruption. *Mol. Plant-Microbe Interact.* **12**, 969–975.
26. Bölker, M., Böhnert, H. U., Braun, K. H., Görl, J. and Kahmann, R. (1995) Tagging pathogenicity genes in *Ustilago maydis* by restriction enzyme-mediated integration (REMI). *Mol. Gen. Genet.* **248**, 247–552.
27. Linnemannstöns, P., Vob, T., Hedden., P., Gaskin, P., and Tudzynski, B. (1999) Deletions in the gibberellin biosynthesis gene cluster of *Gibberella fujikuroi* by restriction enzyme mediated integration and conventional transformation-mediated mutagenesis. *Appl. Env. Microbiol.* **65**, 2558–2564.

28. Epstein, L., Lusnak, K., and Kaur, S. (1988) Transformation-mediated developmental mutants of *Glomerella graminicola (Colletotrichum graminicola). Fun. Gen. Biol.* **23**, 189–203.

29. Thon, M. R., Nuckles, E. M., and Vaillancourt, L. J. (2000) Restriction enzyme mediated integration used to produce pathogenicity mutants of *Colletotrichum graminicola. Mol. Plant-Microbe Interact.* **13**, 1356–1365.

30. Namiki, F., Matsunaga, M., Okuda, M., Inoue, I., Nishi, K., Fujita, Y., et al. (2001) Mutation of an arginine biosynthesis gene causing reduced pathogenicity in *Fusarium oxysporum* f. sp. Melonis. *Mol. Plant-Microbe Interact.* **14**, 580–584.

31. Kahmann, R. and Basse, C. (1999) REMI (restriction enzyme mediated integration) and its impact on the isolation of pathogenecity genes in fungi attacking plants. *Eur. J. Plant Pathol.*. **105**, 221–229.

32. Sweigard, J. A., Carrol, A. M., Farral, L., Chumley, F. G., and Valent, V. (1998) *Magnaporte grisea* pathogenicity genes obtained through insertional mutagenesis. *Mol. Plant-Microbe Interact.* **11**, 404–412.

33. Feldman, K. A. (1991) T-DNA insertion mutagenesis in *Arabidopsis*, mutational spectrum. *Plant J.* **1**, 71–82.

34. Koncz, C., Németh, K., Rédei, G. P., and Schell, J. (1992) T-DNA insertional mutagenesis in *Arabidopsis. Plant Mol. Biol.* **20**, 963–976.

35. Krysan, P. J., Young, J. C., and Sussman, M. R. (1999) T-DNA as an insertional mutagen in *Arabidopsis. Plant Cell* **11**, 2283–2290.

36. Bundock, P., den Dulk-Ras, A, Beijersbergen, A., and Hooykaas, P. J. (1995) Trans-kingdom T-DNA transfer from *Agrobacterium tumefaciens* to *Saccharomyces cerevisiae. EMBO J.* **13**, 206–14.

37. Bundock, P. and Hooykaas, P. J (1996) Integration of *Agrobacterium tumefaciens* T-DNA in the *Saccharomyces cerevisiae* genome by illegitimate recombination. *Proc. Natl. Acad. Sci. USA* **26**,15272–15275.

38. Risseeuw, E., Franke-van Dijk, M.E. and Hooykaas, P. J. (1996) Integration of an insertion-type transferred DNA vector from *Agrobacterium tumefaciens* into the *Saccharomyces cerevisiae* genome by gap repair. *Mol. Cell Biol.* **10**, 5924–5932.

39. Sawasaki, Y., Inomata, K., and Yoshida, K. (1996) Trans-kingdom conjugation between *Agrobacterium tumefaciens* and *Saccharomyces cerevisiae*, a bacterium and a yeast. *Plant Cell Physiol.* **1**, 103–106.

40. Piers, K.L., Heath, J. D., Liang, X., Stephens, K. M., and Nester, E. W. (1996) *Agrobacterium tumefaciens*-mediated transformation of yeast. *Proc. Natl. Acad. Sci. USA* **4**, 1613–1618.

41. de Groot, M. J., Bundock, P., Hooykaas, P. J., and Beijersbergen, A. G. (1998) *Agrobacterium tumefaciens*-mediated transformation of filamentous fungi. *Nat. Biotechnol.* **9**, 839–842.

42. Dunn-Coleman, N. and Wang, H. (1998) *Agrobacterium* T-DNA, a silver bullet for filamentous fungi? *Nat. Biotechnol.* **9**, 817–818.

43. Bundock, P., Mroczek, K., Winkler, A. A., Steensma, H. Y., and Hooykaas, P. J. (1999) T-DNA from *Agrobacterium tumefaciens* as an efficient tool for gene targeting in *Kluyveromyces lactis. Mol. Gen. Genet.* **1**, 115–121.

44. Bundock, P., Mroczek, K., Winkler, A. A, Steensma, H. Y., and Hooykaas, P. J. (1999) T-DNA from *Agrobacterium tumefaciens* as an efficient tool for gene targeting in *Kluyveromyces lactis. Mol. Gen. Genet.* **1**, 115–121.

45. Gouka, R. J., Gerk, C., Hooykaas, P. J., Bundock, P., Musters, W., Verrips, C. T., and de Groot, M. J. (1999) Transformation of *Aspergillus awamori* by *Agrobacterium tumefaciens*-mediated homologous recombination. *Nat. Biotechnol.* **6**, 598–601.

46. Chen, X., Stone, M., Schlagnhaufer, C., and Romanie, P. (2000). A fruiting body tissue method for efficient *Agrobacterium*-mediated transformation of *Agaricus bisporus*. *Appl. Env. Microbiol.* **66**, 4510–4513.

47. Abuodeh, R. O., Orbach, M. J., Mandel, M. A., Das, A., and Galgiani, J. N. (2000) Genetic transformation of *Coccidioides immitis* facilitated by *Agrobacterium tumefaciens*. *J. Infect. Dis.* **181**, 2106–2110.

48. Mikosch, T. S., Lavrijssen, B., Sonnenberg, A. S., and van Griensven, L. J. (2001) Transformation of the cultivated mushroom *Agaricus bisporus* (Lange) using T-DNA from *Agrobacterium tumefaciens*. *Curr. Genet.* **1**, 35–39.

49. Zwiers, L. H., and De Waard M. A.(2001) Efficient *Agrobacterium tumefaciens*-mediated gene disruption in the phytopathogen *Mycosphaerella graminicola*. *Curr. Genet.* **6**, 388–393.

50. Mullins, E. D., Chen, X., Romaine, P., Raina, R., Geiser, D. M., and Kang, S. (2001) *Agrobacterium*-mediated transformation of *Fusarium oxysporum*, an efficient tool for insertional mutagenesis and gene transfer. *Phytopath.* **91**, 173–180.

51. Bundock, P., van Attikum, H., den Dulk-Ras, A., and Hooykaas, P. J. (2002) Insertional mutagenesis in yeasts using T-DNA from *Agrobacterium tumefaciens*. *Yeast* **19**, 529–536.

52. Liu, Y. G. and Whittier, R. F. (1995) Thermal asymmetric interlaced PCR, automatable amplification and sequencing of inserted end fragments from P1 and YAC clones for chromosome walking. *Genomics* **25**, 674–681.

53. Liu, Y. G., Mitzukawa, N, Oosumi, T., and Whittier, R. F. (1995) Efficient isolation and mapping of *Arabidopsis thaliana* T-DNA insert junctions by thermal asymmetric interlaced PCR. *Plant J.* **8**, 457–463.

54. Mazars, G. R., Moyret, C., Jeanteur, P., and Theillet, C. G. (1991) Direct sequencing by thermal asymmetric PCR. *Nucleic Acids Res.* **19**, 4783.

55. Hooykaas, P. J. J., Roobol, C., and Schilperoort, R. A. (1979) Regulation of the transfer of Ti-plasmid of *Agrobacterium tumefaciens*. *J. Gen. Microbiol.* **110**, 99–109.

56. Bennet, J. W. and Lasure, L. L. (1991) Growth media. in More Gene Manipulations in Fungi. (Bennet, J. W. and Lasure, L. L., eds), Academic Press, San Diego, CA, pp. 441–458.

IV

PLANTS AND PLANT CELLS

23

Gene Transfer and Expression in Plants

Argelia Lorence and Robert Verpoorte

Summary

Until recently, agriculture and plant breeding relied solely on the accumulated experience of generations of farmers and breeders—that is, on sexual transfer of genes between plant species. However, recent developments in plant molecular biology and genomics now give us access to knowledge and understanding of plant genomes and the possibility of modifying them. This chapter presents an updated overview of the two most powerful technologies for transferring genetic material (DNA) into plants: *Agrobacterium*-mediated transformation and microparticle bombardment (biolistics). Some of the topics that are discussed in detail are the main variables controlling the transformation efficiency that can be achieved using each one of these approaches; the advantages and limitations of each methodology; transient versus stable transformation approaches; the potential of some *in planta* transformation systems; alternatives to developing transgenic plants without selection markers; the availability of diverse genetic tools generated as part of the genome sequencing of different plant species; transgene expression, gene silencing, and their association with regulatory elements; and prospects and ways to possibly overcome some transgene expression difficulties, in particular the use of matrix-attachment regions (MARs).

Key Words: Plant transformation; gene expression; *Agrobacterium*; biolistic; transgenic plants; transient expression; stable expression; gene silencing; matrix-attachment regions.

1. Introduction

Three decades after the first success (*1*), the production of genetically transformed plant tissues and plants is a central aspect of experimental plant science and a basic component of agricultural biotechnology. In the future, the proportion of acreage planted with transgenic crops, and the range of transgenic crops, are sure to increase, as

From: *Methods in Molecular Biology, vol. 267:*
Recombinant Gene Expression: Reviews and Protocols, Second Edition
Edited by: P. Balbás and A. Lorence © Humana Press Inc., Totowa, NJ

indicated by the fact that the worldwide area committed to transgenic crops has risen from 1 million to more than 67 million hectares between 1997 and 2003 (*2*).

The powerful combination of genetic engineering and conventional breeding programs permits useful traits encoded by one or more transgenes to be introduced into crops within an economically viable time frame. There is great potential for genetic manipulation of crops to enhance productivity through increasing resistance to diseases, pests, and environmental stress and by quantitatively changing the seed composition. Plant cell "factories" are also being designed for high-volume production of vaccines (*3*), pharmaceutical proteins and other pharmaceuticals, nutraceuticals, and several beneficial chemicals (*4*) (*see* Chapter 24). Moreover, with the establishment and expansion of genomic programs, a much broader range of genes and regulatory sequences with potential for crop improvement is being identified and, in some cases, tailored and/or redesigned for further enhancement of their properties within specific plant species (*see* Chapter 27). This has further intensified the interest in developing efficient plant chromosomal and chloroplast transformation and selection technologies to concurrently test and capture the value of these genes (*5*) (*see* Chapters 25 and 26).

Beyond crop improvement, the ability to engineer transgenic plants is also a powerful and informative means for studying gene function and the regulation of physiological and developmental processes. Transgenic plants are being used as assay systems for the modification of endogenous metabolism or gene inactivation. Advances in tissue culture, combined with improvements in transformation technology, have resulted in increased transformation efficiencies. In recent years, many crops previously classified as recalcitrant because they were stubbornly resistant to the overtures of genetic engineering have now been transformed.

There is no general protocol for gene transfer into plants. Each cell type, tissue, and plant species needs careful characterization to ensure optimal transfer conditions to reach the highest efficiencies and reproducibility in terms of gene expression. In the great majority of experimental biological systems, including plants, we are constantly confronted by two variables when transferring genetic information: (1) the great variability in the expression levels of exogenous DNA (transgenes) and (2) the progressive extinction of transgene expression, a phenomenon known as gene silencing.

An updated overview of the two most powerful technologies for transferring DNA into plants, *Agrobacterium*-mediated transformation and microparticle bombardment (biolistics), is presented in this chapter, which discuses (1) the main variables controlling the transformation efficiency that can be achieved using each one of these approaches; (2) the advantages and limitations of each methodology; (3) the potential of some *in planta* transformation systems; (4) the availability of diverse genetic tools generated as part of the genome sequencing of different plant species; (5) transgene expression, gene silencing, and their association with regulatory elements; and (6) prospects and ways to possibly overcome some transgene expression difficulties.

2. Tissue Culture Prerequisite

Recovery of transgenic plants requires transgenes to be targeted at tissues capable of (1) receiving and integrating the introduced DNA, (2) undergoing selection for the successful transgenic events, and (3) regenerating to produce fertile, phenotypically normal plants under tissue culture conditions. In the case of transgenic crop improvement programs, they must also facilitate the recovery of hundreds of genetically modified plants. Embryogenic and meristematic tissues are the most commonly employed target tissues for the production of transgenic plants.

Embryogenic tissues result from the inherent ability of plant cells to return to the totipotent state when appropriately manipulated in vitro. Such cultures consist of rapidly proliferating tissues in which totipotent cells are located on the surface of small embryogenic units. The cells that have successfully integrated and expressed the selectable marker gene can be identified, isolated, and encouraged to proliferate under the selection pressure. Regeneration of whole plants from transgenic calli lines generated in this manner is achieved by subsequent manipulation of growth regulators within the culture medium. Three drawbacks are associated with these culture systems. Firstly, initiation and maintenance of high-quality embryogenic tissues is a labor-intensive process, requiring significant input from skilled personnel. Secondly, a more serious issue is the genotype-specific nature of this kind of systems. While a few cultivars within a given species may respond well in tissue culture and produce high-quality embryogenic target tissues for transgene insertion, the majority of them remain difficult to manipulate in this manner. A third concern regarding the use of embryogenic cultures is reduction of fertility and increased incidence of phenotypically abnormal plants recovered from long-term callus cultures. Scientists have developed tissue culture protocols, which minimize the time spent by the target tissues in a disorganized state, before and after transformation (*6*).

Shoot meristems were identified as an alternative target tissue for the production of transgenic plants, as they provide the potential to genetically engineer all genotypes from any given species. Shoot meristems consist of undifferentiated cells that divide and develop to produce all the plant's aerial tissues, including germline cells. Under the correct tissue culture and selection procedures, fully transformed plants can be generated from the initial transformation events (*6*).

3. Methods for Gene Transfer

Several methods for transformation are currently available for delivering exogenous DNA to plant cells (*6,7*); the focus here will be in the two most robust approaches: *Agrobacterium*-mediated transformation and microprojectile bombardment.

3.1. Transient vs Stable Expression

The use of plants as expression systems for valuable compounds and recombinant proteins often involves integration of a transgene into the plant genome to obtain stable expression. However, transient expression systems are also useful because they are fast, flexible, and unaffected by chromosomal positional effects and can be used in fully differentiated plant tissues. Transient expression allows researchers to study the effect of mutations, introns, codon use, and efficacy of promoters, terminators, and other regulatory elements, determining how these affect gene expression in various plant tissues. In this kind of systems, the tissues are usually assessed 24 to 48 h after transformation, but continued viability and division of the transgenic cells is not required, the status of the target tissue is not critical, and minimal tissue preparation is necessary prior to gene insertion. In contrast, if the aim is to recover a useful number of stably transformed cell lines and/or transgenic plants, strict axenic culture conditions must be maintained, and the choice of the target tissues and their treatment before and after transformation becomes paramount.

3.2. Agrobacterium tumefaciens-*Mediated Transformation*

Agrobacterium tumefaciens is a plant pathogen that belongs to the family Rhizobiaceae (**Note 1**). It is a ubiquitous soil organism and etiological agent of the plant crown gall disease. The gall results from the transfer, integration, and expression of a discrete set of genes (T-DNA) located on the tumor-inducing (Ti) plasmid (for a historical review *see* **8,9**). The genes within the T-DNA can be replaced by any DNA sequence, making *A. tumefaciens* an ideal vehicle for gene transfer and an essential tool for plant research and transgenic crop production. *Agrobacterium* is the only known organism capable of trans-kingdom DNA transfer, transforming mainly plants but also other eukaryotic species, such as yeast (**10,11**), fungi (**12,13**), and human cells (**14**).

Because *Agrobacterium* represents a major tool for plant molecular breeding, the molecular mechanism by which it genetically transforms the host cells has been intensively studied for the past three decades (**15–19**).

Genetic transformation by *Agrobacterium* results from the transfer of the T-DNA from the Ti plasmid to the plant cell, followed by T-DNA integration into the host cell genome and expression of the introduced genes in the transformed host cells. Several *Agrobacterium* chromosomal virulence (*chv*) genes and a series of Ti plasmid-encoded virulence (*vir*) genes have been identified as participants in the different stages of the *Agrobacterium*-plant interaction process. The biological functions of most of these bacterial virulence proteins have been well characterized, but the roles that host proteins might play in the *Agrobacterium* infection are mostly unknown.

The T-DNA element is a specific DNA fragment located on the Ti plasmid and delimited by two 25-bp direct repeats, termed left and right T-DNA borders (LB and RB, respectively). Following induction of the *Agrobacterium* Vir protein machinery by specific host signals (phenolic compounds), the VirD1 and VirD2 proteins nick both borders at the bottom strand of the T-DNA, resulting in a single-stranded T-DNA molecule (T-strand), which, together with several Vir proteins, is exported into the host cell cytoplasm through a channel formed by *Agrobacterium* VirD4 and VirB proteins. The T-strand with one VirD2 molecule covalently attached to its 5' end and coated with many VirE2 molecules forms a T-DNA transport complex (T-complex). This complex is then imported into the host cell nucleus with the help of VirD2 and VirE2, which might also facilitate, directly or indirectly, the subsequent integration of the T-strand into the host genome. The entire transformation process can be considered in eight distinct steps (**19**): (1) *Agrobacterium* recognition of and attachment to the host cell; (2) sensing of specific plant signals by the *Agrobacterium* two-component (VirA-VirG) signal-transduction system machinery; (3) VirG-mediated signal transduction and *vir* gene activation; (4) generation of the T-strand; (5) generation of the VirB-VirD4 transporter complex, and transport of T-strands and Vir proteins into the host cell cytoplasm; (6) formation of the mature T-complex; (7) T-complex nuclear import facilitated by the AtKAPα, VIP1, and Ran proteins of the host cell; and (8) intranuclear transport of the T-complex to the host chromosome, and T-DNA integration into the host cell genome mediated by VirD2 and/or VirE2 and by host factors. The incoming T-DNA integrates at random positions into the plant genome by a process of nonhomologous recombination.

The wild-type T-DNA carries genes involved in the synthesis of plant growth hormones and the production of opines, tumor-specific compounds formed by the condensation of an amino acid with a keto acid or a sugar. It is the production of growth hormones in the transformed host cells that induces the formation of tumors. These tumors then synthesize opines, a major carbon and nitrogen source for *Agrobacterium*. Agrobacteria are usually classified based on the type of opines specified by the bacterial T-DNA, the most common strains being octopine-or nopaline-specific. Opine import into and the subsequent catabolism within the bacterial cell require specialized enzymes. Since these enzymes are encoded by the Ti plasmid, and practically no other soil microorganism can metabolize opines, a favorable biological niche for *Agrobacterium* is created.

Agrobacterium-mediated transformation systems take advantage of this natural plant transformation mechanism. Removal of all the genes within the T-DNA does not impede the ability of the bacteria to transfer the T-DNA but does prevent the formation of tumors. Ti plasmids and their host *Agrobacterium* strains that are no longer oncogenic are called "disarmed." There are two key advances that have made *Agrobacterium* transformation the method of choice: the development of binary Ti vectors, and a range of disarmed *Agrobacterium* strains. The *vir* gene functions are provided by the disarmed Ti plasmid resident in the *A. tumefaciens* strain, while the T-DNA containing the gene(s) to be transferred is provided in the vector.

Most in vitro gene manipulation techniques use *Escherichia coli*, and consequently binary Ti vectors replicate in both *E. coli* and *Agrobacterium* (**20,21**). Gradual refinements have been implemented since the introduction of the first vectors. These refinements have reduced the size of the vectors and aiding the development of more "user-friendly" formats to facilitate in vitro gene manipulation. There is a polarity of T-DNA transfer from *Agrobacterium*to the plant cell: RB precedes LB. Most of the early binary Ti vectors have their selectable marker genes at the RB (e.g., pBIN19). More recently constructed binary Ti vectors contain their plant selectable marker gene nearest to the LB, to ensure that the gene(s) of interest will be transferred before the marker gene.

Strains of *Agrobacterium* that are useful for Ti-vector-based plant transformation are defined by their chromosomal background and resident Ti plasmid (**Table 1**). The C58 chromosomal background has proven to be popular in plant transformation and now harbors several kinds of wild-type and disarmed Ti plasmids, including strains that are effective in transforming cereals.

Significant modifications to the virulence of *Agrobacterium* strains have expanded the range of plant species that are susceptible to T-DNA transformation by improving the frequency of T-DNA transfer. The main modification enhancing virulence has been

Table 1
Chromosomal Background and Ti Plasmid Harbored
by Disarmed *Agrobacterium tumefaciens* Strains Used in Plant Transformation

Strain	Back-ground	Chromosomal marker gene	Ti plasmid	Ti plasmid marker gene	Opine
LBA4404	TiAch5	*rif*	pAL4404	*spec, strep*	Octopine
GV2260	C58	*rif*	pGV2260 (pTiB6S3Δ T-DNA)	*carb*	Octopine
C58C1	C58	–	Cured	–	Nopaline
GV3100	C58	–	Cured	–	Nopaline
A136	C58	*rif, nal*	Cured	–	Nopaline
GV3101	C58	*rif*	Cured	–	Nopaline
GV3850	C58	*rif*	pGV3850 (pTiC58Δonc. gGenes)	*carb*	Nopaline
GV3101:: pMP90	C58	*rif*	pMP90 (pTiC58Δ T-DNA)	*gent*	Nopaline
GV3101:: pMP90RK	C58	*rif*	pMP90RK (pTiC58Δ T-DNA)	*gent, kan*	Nopaline
EHA101	C58	*rif*	pEHA101 (pTiBo542Δ T-DNA)	*kan*	Nopaline
EHA105	C58	*rif*	pEHA105 (pTiBo542Δ T-DNA)	–	Succi-namopine
AGL-1	C58, RecA	*rif, carb*	pTiBo542Δ T-DNA	–	Succi-namopine

Adapted from **ref. 2**.

Abbreviations: *rif*, rifampicin resistance; *gent*, gentamycin resistance; *nal*, nalidixic acid resistance; *kan*, kanamycin resistance; *carb*, carbenicillin and ampicillin resistance; *spec*, spectinomycin resistance; *strep*, streptomycin resistance;–no marker gene present.

to boost the expression of, or to introduce a change in, the activation state of the *virG* product, which activates transcription of the rest of the *vir* cluster (**22**). Another modification is the enhancement of *virE1* expression. VirE1 and VirG can be limiting when large sections of DNA (>50 kb) need to be transferred using specialized binary Ti vectors called binary bacterial artificial chromosome (BiBAC) plasmids. In both cases, virulence increase is achieved by placing the *vir* genes on a co-resident plasmid that is compatible with the binary Ti vector, thus boosting their expression by increasing their total copy number in *Agrobacterium* (**20**).

Achieving a high expression of the introduced foreign gene in plant cells is still a challenging task. Especially in those projects aiming to develop plants as "bio-factories" to produce novel enzymes, pharmaceuticals, and industrial compounds, it is critical that transgenes produce the corresponding proteins in large amounts to make the venture economically feasible. The level of gene expression is, in part, a function of the promoter to which the coding region of the gene is fused. The use of the most popular promoter in plant molecular biology research, the 35S promoter from the cauliflower mosaic virus (CaMV), usually results in the constitutive production of the foreign protein at rates of less than 1 percent of the total protein. Improvements to the 35S CaMV promoter such as the duplication of specific sequences and the addition of enhancers have boosted expression (*23*), but for some applications, the levels need to be still higher. Recently, a new stronger promoter has been developed. This "super promoter" is a hybrid construct combining a triple repeat of the octopine synthase (*ocs*) activator sequence along with mannopine synthase (*mas*) activator elements fused to the *mas* promoter. In leaf tissue, this chimeric promoter is approx 156-fold and 26-fold stronger than are the CaMV 35S and the "enhanced" double CaMV promoters, respectively (*24*). Monsanto owns the patent on use of the CaMV 35S promoter, while the super-promoter can be obtained from the Biotechnology Research and Development Corporation (BRDC, Peoria, IL). The cost of licensing of proprietary technology for use on a broader scale is a critical aspect that should be analyzed at the beginning of any trans-formation program (**Note 2**).

In current transformation systems, a selectable marker gene is co-delivered with the gene of interest to identify and separate rare transgenic cells from nontransgenic cells. This is because during transformation only a few plant cells accept the integration of foreign DNA, while most of cells remain nontransgenic. Usually, a conditional domi-nant gene, which has no influence on the growth or morphology of the plants, is used as a selectable marker because it remains in the transgenic plant after transformation. **Table 2** presents a list of the most commonly used selectable marker genes and selec-tion agents in plant biotechnology. Most of the selectable marker genes encode proteins that confer resistance to antibiotics or herbicides. Antibiotics are generally applied to the culture medium, while herbicides can be applied either to the medium or directly to the plants. The amount of antibiotic or herbicide to be used needs to be determined experimentally for each plant species and tissue to be transformed. For example, in the case of kanamycin, one of the most-used selective agents, *Arabidopsis* or tobacco plants can be selected using concentrations around 100–200 mg/L, while other more insensitive or tolerant species, such as *Catharanthus roseus*, require concentrations as high as 8 g/L to be able to separate the transformed and nontransformed cells or tissues. Problems associated with the use of selectable marker genes and alternatives are dis-cussed in **Subheading 7** of this chapter. Visual reporter genes—including the *uidA* (GUS) gene (*25*), luciferase (*26,27*), and more recently, green fluorescent protein (*28*) (Chapters 24 and 25)—are powerful components of transient and stable studies, allow-ing qualitative and quantitative assessment of transgene expression levels (**Note 3**). A novel, environmentally friendly selection strategy is presented in Chapter 26.

Typically, *Agrobacterium*-mediated transformation of dicots is performed using ster-

Table 2
Selectable Markers Currently Used in Transgenic Plants

Gene	Gene product	Selection agent(s)	Gene source
nptII/neo	Neomycin phosphotransferase	Kanamycin, neomycin, geneticin	*Escherichia coli* transposon Tn5
bar	Phosphinothricin acetyltransferase	Glufosinate, L-phosphinothricin, bialaphos	*Streptomyces hygroscopicus*
pat	Phosphinothricin acetyltransferase	Glufosinate, L-phosphinothricin, bialaphos	*Streptomyces viridochromogenes*
bla	β-Lactamase	Penicillin, ampicillin	*E. coli*
aadA	Aminoglycoside-3-adenyltransferase	Streptomycin, spectinomycin	*Shigella flexneri*
hpt	Hygromycin phosphotransferase	Hygromycin B	*E. coli*
nptIII	Neomycin phosphotransferase III	Amikacin, kanamycin, neomycin, geneticin (G418), paromomycin	*Streptococcus faecalis* R plasmid
epsps/aro	5-Enoylpyruvate shikimate-3-phosphate	Glyphosate	*Agrobacterium CP4, Zea mays, Petunia hybrida and other plant* species
gox	Glyphosate oxidoreductase	Glyphosate	Achromobacter LBAA
bxn	Bromoxynil nitrilase	Bromoxynil	*Klebsiella pneumoniae*
als	Acetolactate synthase imidazolines, thiazolopyrimidines	Sulfonylureas,	Multiple plant species
cat	Chloramphenicol acetyltransferase	Chloramphenicol	Bacteriophage P1
tdc	Tryptophan decarboxylase	4-Methyltryptophan	*Catharanthus roseus*
uidA/GUS	β-Glucuronidase	Cytokinin glucoronides	*E. coli*
xylA	Xylulose isomerase	D-Xylose	*Thermoanaer obacterium thermosulforogenes*
manA	Phosphomannose isomerase	Mannose-6-phosphate	*E. coli*
BADH	Betaine aldehyde dehydrogenase	Betaine aldehyde	*Spinacia olearacea*

Adapted from **ref. 52**.
Also *see* Chapters 24–27.

ile leaf pieces, cotyledons, stem segments, calli cultures, cell suspension cultures, or germinating seeds. Because of having to deal with two different biological elements, many parameters should be tested to satisfy both partners and guarantee a successful outcome. These variables include the use of feeder cells, alternative *Agrobacterium* strains, infiltration of the bacteria, the duration and temperature of co-cultivation of the tissue and the bacteria, the redox state of the co-cultivation media, and the presence of absence of light during co-cultivation. Pre-wounding the tissue with microprojectiles, glass beads, or a razor blade (*3*) has proven to be beneficial for the transformation process. This phenomenon could be explained by the attraction of *Agrobacterium* to wounded sites and the increased access of bacteria to plant cells. Subjecting plant tissues to brief periods of ultrasound by sonication, which allows *Agrobacterium* entry throughout the tissues, is yet another method of increasing transformation efficiency (*29*).

Some crops appear to react, or be hypersensitive, to *Agrobacterium* inoculation by forming necrotic barriers. In grapes and rice, this reaction has been remedied by the addition of antioxidants, such as polyvinylpyrrolidone and dithiotrietol. An increase in the frequency of T-DNA delivery to targeted cotyledonary node explants of *A. tumefaciens*-infected soybean co-cultivated on 400 mg/L of L-cystein has been reported. Similar results were obtained when immature zygotic embryos of corn were infected with *A. tumefaciens* strain EHA101 (*30*).

In general, the light conditions during co-culture vary considerably in different transformation procedures. In a recent study, light has been identified as an important factor that strongly promotes gene transfer from *Agrobacterium* to cells of *Arabidopsis thaliana* and *Phaseolus acutifolius*. GUS production due to the transient expression of an intron-containing *uidA* gene was used to evaluate T-DNA transfer under different light conditions. In all situations, *uidA* expression correlated highly and positively with the light period used during co-culture: it was inhibited severely by darkness and enhanced more under continuous light than under a 16 light/8 dark photoperiod. The positive effect of light was observed with *Agrobacterium* strains harboring either a nopaline- or octopine- or an agropine/succinamopine-type nononcogenic helper Ti plasmid. These results could be due to a positive effect of light on the cell competence for transformation (*31*). Incubation in darkness seems to improve the morphogenic capacity of callus or explants, essentially by preserving endogenous light-sensitive hormones or by preventing accumulation of phenolic compounds. Therefore, transformation protocols can be improved by incubating co-cultures in the light, but keeping the original dark incubation conditions before and after co-culture in order to protect the productivity of co-cultured plant material.

3.3. Biolistics

Microparticle bombardment or biolistics is a technique by which micron-sized metal particles are coated with DNA and accelerated into target cells at velocities sufficient to penetrate the cell wall without causing lethal damage. In this manner, desired DNA can be transported into the cell's interior, where it becomes detached from the microprojectile and integrated into the nuclear or plastidic genome (*see* Chapter 25). Application of this methodology developed rapidly during the 1990s and currently is used successfully to produce transgenic plants in a wide range of different plant species. Protocols have been adapted to transfer exogenous DNA to bacteria, fungi, algae, insects, and mammals (*7*).

There are at least five different particle delivery devices on the market (*7*). The most widely used is called PDS-1000/He and is commercially available from Bio-Rad (Hercules, CA). Details of the system are available at the Bio-Rad website (http://www.bio-rad.com). As the name suggests, helium gas is used as the propellant. This is continuously fed into a chamber at the top of the device, and held back by a plastic "rupture" disc. A variety of rupture disks are available, designed to break at pressures ranging from 450 to 2200 psi. DNA is precipitated onto gold or tungsten particles (microcarriers) approx 1 μm in diameter and spread evenly on a circular plastic film (the macrocarrier) placed below the rupture disk in the main vacuum chamber. When the rupture disk breaks a shock wave is released; this impacts the macrocarrier, causing it to fly a short distance before striking a wire-mesh stopping screen. The macrocarrier is arrested and the microcarriers launched through the mesh into the plant target tissues situated on a shelf in the chamber below. The gold or tungsten particles impact the tissue in a circular pattern with a diameter of one to several centimeters, depending on the distance at which the tissue is positioned below the stopping screen.

Successful gene transfer using particle bombardment technology requires the development and co-optimization of numerous variables. An outline of the major variables is presented below.

3.3.1. Preparation of the Microcarriers

Genetic material for direct delivery by particle bombardment must be cloned into a suitable vector (small size, high copy number), amplified, and purified to produce sufficient quality and quantity of the desired DNA for replicated experiments. If stably transformed tissues and plants are to be recovered, a selectable marker gene, most often expression cassettes coding for antibiotic or herbicide resistance, will be transferred to the plant genome in addition to the gene of interest. Purified DNA is coated onto washed, disaggregated tungsten or gold particles, mixed with spermidine, and precipitated onto the microcarriers by the addition of $CaCl_2$. The supernatant is removed and replaced with 100% ethanol, and aliquots are spread onto the macrocarrier membranes, where they dry to form a finely dispersed layer. The whole process must be performed under sterile conditions if prolonged tissue culture of the target tissue is intended after transgene insertion. Choice of microparticle type and size is important, as this will determine the mass, and thus depth of penetration, of the accelerated microcarrier. Although tungsten was widely utilized in early experiments and is significantly cheaper than gold, the latter has become more commonly used. Some groups have reported toxic effects of tungsten, while the denser nature and more even surface of the gold particles allow better control of penetration into the plant tissue. Gold particles at sizes varying from 0.6 to 3.0 μm in diameter are commercially available.

3.3.2. Acceleration and Delivery of Microparticles

For successful transgene integration, the complex of microcarrier and DNA must be delivered into the target tissue without causing excessive injury or stress to the plant cells. The degree of penetration required will depend on the thickness of the cell wall, the type of tissue being transformed, and the depth of the target cell layers. Variations in the helium pressures, level of vacuum generated, size of the particles, and position of the target tissues below the stopping screen within the chamber will determine the momentum and thus penetrating power with which the microprojectiles strike the tissue. All these parameters are under the experimenter's control, and must be optimized for a given target tissue.

3.3.3. Treatment of Target Tissue

Treatment of the target tissues prior to and after particle bombardment can have a significant effect on the frequency of successful transformation events, and most especially, the number of recoverable transgenic cell lines and plants. The exact tissue culture conditions required to generate high quality embryogenic tissues vary between species and even between genotypes within a species. Efforts to develop optimum totipotent cells through classical plant tissue culture techniques therefore remain a central component of most plant genetic transformation programs. Factors such as the age of the target tissue and the time since the last subculture must be considered, as it is well known that actively dividing cells are the most effective targets for transgene insertion. In the case of apical meristems, the treatment and physiological status and age of the mother plants prior to excision of the explants must be taken into consideration. Use of an osmotic pretreatment or partial drying of the target cells prior to bombardment is a commonly used technique to increase the frequency of successful transformation. Tissues are exposed to an osmotic agent such as mannitol or sorbitol for several hours prior to transgene insertion to cause partial plasmolysis of the cell. This is thought to prevent or reduce cell death due to extrusion of the protoplasm through the cell wall and wound sites created by incoming particles (7).

3.4. Advantages of Microparticle Bombardment Technology

3.4.1. Co-Transformation With Multiple Transgenes

Most agronomic characteristics are polygenic in nature. Expression of one transgene in a plant can confer beneficial characteristics, but the ability to integrate and co-express multiple transgenes is highly desirable. In this way, complex metabolic pathways can be manipulated, defense mechanisms more effectively optimized, novel compounds synthesized, and multiple beneficial traits built into crop plants. Two genes can be fused within the same plasmid and bombarded into the target tissues. Alternatively, co-transformation is carried out in which the two genes are cloned into separate vectors and then mixed together, prior to coating onto the microcarriers. Varying the ratio in which the two plasmids are mixed influences the number of transgenic plants that can be recovered, with frequencies as high as 80% co-integration achievable. There are reports of simultaneous insertion of transgenes via biolistics into soybean, rice, wheat, and periwinkle (*C. roseus*, *32*), among other species.

3.4.2. Integration of Large DNA Inserts

The ability to insert large DNA fragments into the plant genome is an attractive proposition, most especially for use with map-based cloning of agriculturally important genes. Integration of yeast artificial chromosomes (YACs) into the plant genome by particle bombardment has been successful with inserts of up to 150 kb achieved (*33*). Although clearly requiring further development, integration of large DNA fragments promises to be an important tool in future plant research and crop biotechnology.

3.4.3. Plastid Transformation

Plant cells contain plastid organelles that possess a circular double-stranded DNA genome between 120 and 160 kb in size. Approximately 120 genes are encoded on the prokaryotic-like genome. Genetic transformation of the chloroplast genome occurs via homologous recombination with the incoming transgene and can occur at high copy numbers. After subsequent selection pressure for the transgenic events, and with each plant cell containing as many as 50 plastids, high expression levels for a desired protein are possible. Up to 45% of the total soluble cellular protein can be of transgenic origin from plastid transformation, making this technology attractive for the accumulation of pharmaceutical and industrial products within plant tissues. In addition, because chloroplasts are not transmitted through the pollen in many plant species, the spread of transgenes from transgenic crops of this type would not pose the same degree of concern as it would for plants in which the nuclear genome has been modified. Particle bombardment remains the most effective way to genetically engineer plastids. Recent advances in removing the antibiotic marker (*34*) (*see* **Subheading 7.**) from transgenic chloroplasts and adaptation to tomato and potato indicate that this technology holds significant potential for applications where the desired product can be compartmentalized and stored within the plastids (*35*) (Chapter 25).

Direct gene transfer systems are prone to integrating multiple copies of the desired transgene, in addition to superfluous DNA, sequences associated with the plasmid vector. This occurs across all species studied, with between 1 and 20, and occasionally more, copies of the desired transgene being integrated. In addition, several different insertion sites can be generated in the plant genome by the bombardment and subsequent integration processes. Multicopy and superfluous DNA insertions are disadvantageous for a number of reasons and are recognized drawbacks associated with the use of particle bombardment technology. "Clean gene" technology, by which nonessential plasmid DNA is removed by enzymatic digestion prior to insertion by biolistics, has been developed to address this problem (*36*). In this way, only the desired coding region with its control elements are "shot" into the target cells.

Genetically transformed plants recovered from tissue culture sometimes show phenotypic and genotype variations from the mother plant, in addition to the intended effects of the inserted transgene(s). These variations can be due to tissue culture-induced mutagenesis (somaclonal variation F), insertional mutagenesis, pleiotropic effects of the transgene, or a combination of these phenomena. In all cases, detrimental effects might be imparted to the plant metabolism, potentially leading to changes in the range of characteristics, such as development, response to stress, and nutritional qualities. Strict biosafety regulations of transgenic crops favor the acceptance of genetically modified plants with single copy insertions, as these are easier to fully analyze and are statistically less likely to suffer from the problems outlined above.

Development of "agrolistics" (*see* **Subheading 6.**), redesigning transgenes and vectors to reduce the incidence of recombination, and reduction in the amount of DNA bombarded into each cell are examples of efforts being made to alleviate the problem of multicopy insertions resulting from particle bombardment (*7*).

4. Gene Silencing

In most cases, high levels of constitutive, tissue-specific, or inducible transgene expression are required over the life of a genetically modified plant. Reports of significant variation in the levels of transgene expression between independent genetic transformation events are common in the literature, and are an accepted aspect of transgenic plant production. As a result, between dozens and hundreds of genetically transformed plants must be produced in the initial stages of any transformation program. Plant lines with stable transgene expression and normal phenotype are then selected for further study and development toward the desired end product. Better understanding and control of transgene expression is therefore desirable to increase the efficiency by which useful transgenic plants are produced.

Agrobacterium-mediated transformation is considered to favor insertion of transgenic material into transcriptionally active regions of the plant DNA. Conversely, particle bombardment appears to cause integrations at random locations throughout the genome, even if homologous regions are included in the transferred construct, leading to what is known as "positional effect." In this way, transgenes that are inserted at positions of highly condensed chromatin or that become methylated will not be expressed, or will be expressed at a lower level than those integrated into sites of active transcription.

Suppression of transgene expression through silencing is a relatively common phenomenon in genetically transformed plants, and it is exacerbated by the presence of multiple copies within the plant genome. Both cytosine methylation and co-suppression (also called post-transcriptional gene silencing, or PTGS) are known to operate, resulting in down-regulation of transgene expression, sometimes in an inconsistent and unpredictable manner. Silencing at the transcriptional level is thought to occur primarily by methylation of promoter sequences, thereby interfering with assembly of the transcription factors, and/or by attracting chromatin remodeling proteins to these sites. Co-suppression operates at the RNA level, and involves the production of double-stranded RNA, which acts as a trigger to initiate degradation of a target RNA, thereby resulting in silencing. In plants, PTGS operates as an adaptive immune system target against viruses; as a counterdefensive strategy, many plant viruses have evolved proteins that suppress various steps of the mechanism. It has been recently shown that PTGS is a general plant response that limits the efficiency of *Agrobacterium*-mediated transient expression. It was also demonstrated that transient co-expression of the p19 protein of tomato bushy stunt virus, a suppressor of gene silencing, prevents the onset of PTGS in the transformed tissues and allows high level of transient expression (*37*). Gene silencing and its implications for transgene expression is an area of intense research at this time, and the reader is directed to recent reviews of this subject (*38–40*).

Until recently, gene silencing has been considered as a problem for plant genetic transformation, since it prevented reliable expression of a desired phenotype within transgenic plants. However, with increasing knowledge of the mechanisms underlying the phenomena, and realization that it can be used to down-regulate native genes within the plants, there is no doubt that it will become a powerful tool in future transgene applications, specially to gain a better understanding of the function of plant gene families, as well as to block genes involved in competitive metabolic pathways (*41*). The development of HANIBAL, KANNIBAL, and HELLSGATE (*42*), vectors designed to produce gene-specific silencing, constitute a very powerful tool for this purpose.

5. *In Planta* Transformation

As mentioned above, the generation of genetically homogeneous plants carrying the same transformation event in all cells has typically presented two separate hurdles: the transformation of the plant cells and the regeneration of intact, reproductively competent plants from those transformed cells. It is often necessary to generate and screen a dozen or more independent plant lines transformed with the same construct to find the ones that carry a simple insertion event. Since the late 1980s a number of laboratories have pursued plant transformation methods that avoid tissue culture and regeneration. This type of approach has been particularly successful for two plant model systems: *A. thaliana* (*43*) and *Medicago truncatula* (*44*).

In the early 1990s the most popular method to transform *Arabidopsis* was the "vacuum infiltration" protocol. *Arabidopsis* plants at the early stages of flowering were uprooted and placed en masse into a jar in a suspension of *Agrobacterium*. A vacuum was applied and then released, causing air trapped within the plant to bubble off and be replaced with the *Agrobacterium* suspension. Plants were transplanted back to soil, grown to seed, and in the next generation stably transformed lines could be selected using the antibiotic or herbicide appropriate for the selectable marker gene. Transformation rates often exceeded 1% of the seeds tested. Since then variations of this simple method have been widely adopted by *Arabidopsis* researchers.

Currently the most used transformation protocol for *Arabidopsis* plants is known as "floral dip" method (*see* Chapter 26). In this protocol, *Arabidopsis* plants are grown in pots to a stage when they have just started to flower. The plants are dipped briefly (15–20 min) in a suspension of *Agrobacterium* and a strong surfactant with low toxicity to plants (e.g., Silwet L-77). The plants are then maintained under normal conditions for a few more weeks until maturity and then progeny seeds are harvested. These seeds are after germinated in selective medium (usually Murashige and Skoog (MS) media plus the appropriate antibiotic or herbicide) to identify successfully transformed progeny. Evidence provided by three independent laboratories has shown that developing ovules are the primary target of productive transformation in the *Arabidopsis* floral dip or vacuum infiltration transformation procedures. Transformants from the same plant or even from the same silique (seed pod) are usually independent (*43*).

Trieu and coworkers published two rapid and simple *in planta* transformation methods for *M. truncatula*. A number of research groups have selected *M. truncatula* as a model legume for molecular genetic analyses due to its relatively small genome and rapid life cycle. The first transformation protocol is based on the method developed for *A. thaliana* and involves infiltration of flowering plants with a suspension of *Agrobacterium*. The main difference for *M. truncatula* was the inclusion of a vernalization treatment, which induces the plants to flower early. The second method involves infiltration of young seedlings with *Agrobacterium*. In both cases a proportion of the progeny of the infiltration plants was transformed. The transformation frequency ranged from 4.7% to 76% for the flower infiltration method, and from 2.9% to 27.6% for the seedling infiltration method. It is important to say that both *in planta* transformation methods worked successfully with a range of binary vectors, providing that the T-DNA contains the *bar* gene as selectable marker. A variety of *A. tumefaciens* strains were used, including EHA105, ASE1, and GV3101; the only strain that was unsuccessful was LBA4404, a strain widely used in the *M. truncatula* tissue culture transformation methods. The underlying transformation mechanisms in *M. truncatula* are not known, but the authors suggest that in the flower infiltration and seedling infiltration procedures the targets of transformation are the meristem cells of the axillary buds that later develop into peduncles and inflorescences, and the germ-line cells, respectively (*44*).

6. Other Novel Technologies and Tools
6.1. Agrolistics

Agrolistics is a refinement to microparticle bombardment technology developed to counter the frequency with which broken fragments of the transgene and superfluous DNA are integrated into the plant genome. In this strategy virulence genes from *Agrobacterium* that facilitate release of the T-DNA and contain nuclear targeting sequences are co-bombarded into the target tissue along with the selectable marker and gene of interest. Use of agrolistics has been shown to increase the number of transgenic plants receiving "clean" or precisely defined transgene inserts and to reduce the frequency of degraded transgene integrations (*45*). This technology requires further development but has the potential to address one of the major drawbacks of particle bombardment technology.

6.2. Collections of T-DNA Mutants

The ability to knock out genes or suppress their expression has been a powerful tool for plant biologists. Insertional mutagenesis is one of the approaches most widely used for this purpose. Collections of random T-DNA or transposable element insertion mutants are currently available to the *Arabidopsis* research community through a variety of sources. These collections total more than 200,000 independently derived T-DNA mutants. The flanking DNAs of many of these insertions have been sequenced, and mutants in specific genes can be identified via a BLAST analysis at the public databases. Other mutants can be identified via PCR (*46*) (http://signal.salk.edu/ and www.tmri.org). A similar collection of T-DNA mutants has been developed for rice (*47*). These collections are tremendously useful, but they do not always produce null alleles, and smaller gene targets might not be represented at all. As an alternative, a high-throughput program that combines random chemical mutagenesis with PCR-based screening to identify point mutations in regions of interest has been developed (*48*). The TILLING (target induced local lesion in genomes) program offers its service to the scientific community at http://tilling.fhcrc.org:9366/.

6.3. Activation Tagging

The primary tool for dissecting a genetic pathway is the screen for loss-of-function mutations that disrupt such a pathway. However, a limitation of lost-of-function screens is that they rarely identify genes that act redundantly. The problem of functional redundancy has become particularly apparent during the past few years, as sequencing of eukaryotic genomes has revealed the existence of many duplicated genes that are very similar in both their coding regions and their regulatory regions. A second class of genes whose entire function is difficult to identify with conventional mutagens, which primarily induce loss-of-function mutations, are those that are required during multiple stages of the life cycle and whose loss of function results in early embryonic or in gametophytic lethality. Genes that are not absolutely required for a certain pathway can still be identified through mutant alleles, if such genes are sufficient to activate that pathway. Similarly, genes that are essential for early survival might be identified through mutant alleles if ectopic activation of the pathways they regulate is compatible with survival of the organism. The key in either case is the availability of gain-of-function mutations. A few years ago a directed way to induce such mutations was developed, involving construction of a T-DNA vector with four copies of an enhancer element from the CaMV 35S gene. These enhancers can cause transcriptional activation of nearby genes and, because activated genes will be associated with a T-DNA insertion, this approach has become known as activation tagging. New activation tagging vectors that confer resistance to kanamycin or glyphosate have been used to generate several tens of thousands of transformed *Arabidopsis* plants. Analysis of a subset of mutants has shown that overexpressed genes are almost always found immediately adjacent to the inserted enhancers, at distances ranging from 380 bp to 3.6 kb (*49*). A collection of activated-tagged mutants is available to the research community (Arabidopsis Biological Resource Center, ABRC, http://www.biosci.ohio-state.edu/lplantbio/Facilities/abrc/abrchome.htm).

6.4. Oligonucleotide-Directed Gene Targeting

Self-complementary chimeric DNA/2'-*O*-methyl RNA (chimeras) and modified DNA oligonucleotides direct site-specific base changes in chromosomal and episomal targets in cells. Designed to pair with a homologous sequence within genomic DNA, these molecules introduce single base changes at specific targets. A proposed mechanism by which these molecules act involves the DNA strand of the chimera functioning as a template for gene repair acted on by the host DNA repair machinery, while the RNA strand enhances targeting efficiency by stabilizing complex formation with the target DNA sequence. The utility of chimeras to introduce site-specific base changes has been demonstrated in plants. Chimeras were introduced into tobacco and maize to target endogenous genes and transgenes resulting in selectable phenotypes. Unlike the high efficiency level observed in mammalian systems (1% to 20%), oligonucleotide-directed gene conversion occurs at a low frequency (10^{-4}) in plants. Improvements in vector design and an increase in our knowledge of plant DNA repair mechanisms will expedite the applicability of this technology (*50*).

7. Transgenic Plants Without Selection Marker

Systems that use antibiotics or herbicides to select transgenic plants have potential pitfalls: (1) These selective agents can decrease the ability of transgenic cells to proliferate and differentiate into transgenic plants, (2) the presence of marker genes in transgenic plants precludes the use of the same marker gene for gene stacking through retransformation, and (3) the recent public concerns on the release of antibiotic resistance genes limit their use for commercialization of transgenic crops.

It is difficult to introduce a second gene of interest into a transgenic plant that already contains a resistance gene as a selectable marker. As mentioned previously, a large number of desirable traits and genes are worth incorporating into plants, but only a limited number of selectable marker genes are available for practical use. The problem becomes even more difficult if several genes need to be integrated. It is desirable therefore, to develop marker-free transgenic plants that are environmentally safer and that can be used to express a number of transgenes by repeated transformation. Currently there are two main strategies to achieve this goal: one approach is to excise or segregate marker genes from the host genome after regeneration of transgenic plants (for review *see 51,52*), and the second is based on so-called marker-free transformation (*53*) (*see* **Note 4**).

Among many attempts for marker gene removal from transgenic plants, the site-specific excision of a transgenic DNA sequence containing the marker gene is most commonly used. These transformation systems consist of a site-specific recombination system (*CRE/lox*) or a phage-attachment region (*attP*) to remove the selectable marker gene, and a transposable element system (*Ac*) or a co-transformation system to segregate the gene of interest from the selectable marker gene. Once the selectable marker gene is eliminated from transgenic plants by the removal systems, selective agents cannot be used to identify the marker-free transgenic plants. Therefore, it is necessary to use DNA analysis in order to identify the nonchimeric marker-free transgenic plants.

Removal systems combined with a positive marker (*ipt*), which are called MAT (multi-auto-transformation) vectors (*54*) used in a one-step protocol that needs no selective agent and no sexual crossing to identify the marker-free transgenic plants have been recently published (*55*). The adoption of this kind of approach to generate marker-free transgenic plants in economically important crops is very likely to occur in the near future.

8. Nuclear Matrix Attachment Regions and Plant Gene Expression

DNA sequences necessary for the proper formation, maintenance, or regulation of chromatin structure, including those that define domain boundaries, are considered "chromatin elements." Matrix attachment regions (MARs) are a class of chromatin elements operationally defined by the ability to bind specifically to isolated nuclear matrices. The nuclear matrix is a network of proteinaceous fibers that permeates the nucleus and presumably functions to organize chromatin into a series of topologically isolated loop domains. MARs have been isolated and studies from diverse range of eukaryotes, including mammals, birds, insects, and plants (*see* Chapters 29 and 30). Data from several laboratories indicate that MARs enhance transgene expression to varying extents in different plant systems (from 2- to 60-fold, reviewed in **refs. 56,57**) and are consistent with the hypothesis that MARs can reduce or eliminate some forms of gene silencing. Most of the experiments with MARs in plant systems have focused almost entirely on test constructs involving the CaMV 35S promoter. It has been recently shown that the RB7 MAR, a 1.2-kb fragment isolated from the 3' flanking region of the RB7 root-specific gene of tobacco, caused a significant increase in the activity of highly active promoters (CaMV 35S, nopaline synthase promoter [NOS], and octopine synthase promoter [OCS]) as compared to controls. However, the presence of RB7 MAR did not significantly increase the activity of weak promoters. Importantly, most transgenes flanked by RB7 MAR showed a large reduction in the number of low-expressing transformants relative to control constructs without MARs, suggesting that these elements can reduce the frequency of gene silencing in primary transformants (*58*).

9. Conclusions and Perspectives

Although the basic elements involved in the *A. tumefaciens*-mediated transformation have been known for some time, detailed information on the mechanism of infection and DNA transfer is still lacking. The availability of the *A. tumefaciens* genome sequence (*59,60*) represents a significant advance. By facilitating the identification of genes not previously known to play a role in DNA transfer, the genome sequence will likely increase our ability to exploit *A. tumefaciens* as a vector for the genetic engineering of plants and other organisms (*61*).

Agrobacterium floral transformation procedures have been a tremendous success with *Arabidopsis* and *M. truncatula*. Such successes, along with the information about the targets for *Arabidopsis* transformation most likely will inspire efforts to adapt these methods to the transformation of other plant species. The benefits are clear: transformation without tissue culture can provide a high-throughput method that requires minimal labor, expense, and expertise. Simplified transformation protocols will facilitate positional cloning, insertional mutagenesis, and other transformation-intensive procedures, reducing the effort required to test any given DNA construct within plants (*43*).

Despite the progress obtained regarding transformation protocols, there are many economically important crops, elite cultivars, and tree species that remain highly recalcitrant to *Agrobacterium* infection. Although attempts have been made to improve transformation by altering the bacterium, future successes might come from manipulation of the plant. Genetic and protein interaction approaches, combined with genomic methods have identified more than 100 *Arabidopsis* genes involved in *Agrobacterium*-mediated transformation. These genes encode proteins involved in bacterial attachment, T-DNA transfer, nuclear targeting, and T-DNA integration. It seems likely that some of these genes could serve as targets for genetic manipulation to increase transformation of recalcitrant crops (**62**).

The use of MARs is likely to increase the predictability and stability of transgenic traits. However these studies are still in an early phase. The full potential of MAR technology will not be known until several different MARs have been tested in plant breeding programs spanning multiple generations (**57**).

10. Notes

1. Recently, it has been demonstrated that species of *Agrobacterium tumefaciens* (syn. *Agrobacterium radiobacter*), *Agrobacterium rhizogenes*, *Agrobacterium rubi*, and *Agrobacterium vitis*, together with *Allorhizobium undicula*, form a monophyletic group with all *Rhizobium* species, based on comparative 16S rDNA analyses (**63**). The proposed new name for *A. tumefaciens* is *Rhizobium radiobacter*. It is likely that this new name will be adapted in the scientific literature in the near future.
2. The reader is strongly encouraged to review the collective public intellectual property databases at the Center for Application of Molecular Biology to International Agriculture (CAMBIA) web page (www.cambiaip.org) and at the Public Sector Intellectual Property Resource for Agriculture (PIPRA) web page (www.pipra.org).
3. For the use of some of these selectable markers in other organisms, *see* Chapters 5, 9 and 12.
4. For marker-free chromosomal insertion of genetic material in *E. coli* and *Saccharomyces cerevisiae* (*see* Chapters 9 and 20, respectively).

Acknowledgments

A.L. thanks Consejo Nacional de Ciencia y Tecnología (CONACYT) for support to her laboratory while working at CEIB/UAEM. The authors thank Dr. Fabricio Medina-Bolivar for critical review of the manuscript.

References

1. Herrera-Estrella, L., Depicker, A., Van Montagu, M., and Schell, J. (1982) Expression of chimaeric genes transferred into plant cells using a Ti-plasmid-derived vector. *Biotechnol.* **24**, 377–381.
2. James, C. (2004) *Global status of commercialized transgenic crops*. ISAAA Briefs No. 30, ISAAA, Ithaca, NY.
3. Medina-Bolivar, F., Wright, R., Funk, V., Sentz, D., Barroso, L., et al. (2003) A non-toxic lectin for antigen delivery of plant-based mucosal vaccines. *Vaccine* **21**, 997–1005.
4. Verpoorte, R., van der Heijden, R., and Memelink, J. (2000) Engineering the plant cell factory for secondary metabolite production. *Transgenic Res.* **9**, 323–343.
5. Job, D. (2002) Plant biotechnology in agriculture. *Biochimie* **84**, 1105–1110.
6. Hansen, G. and Wright, M.S. (1999) Recent advances in the transformation of plants. *Trends Plant Sci.* **4**, 226–231.
7. Taylor, N. J., and Fauquet, C. M. (2002) Microparticle bombardment as a tool in plant science and agricultural biotechnology. *DNA Cell Biol.* **21**, 963–977.
8. Chilton, M. D. (2001) *Agrobacterium*. A memoir. *Plant Physiol.* **125**, 9–14.
9. Binns, A. N. (2002) T-DNA of *Agrobacterium tumefaciens*: 25 years and counting. *Trends Plant Sci.* **7**, 231–233.

10. Bundock, P., Den Delk-Ras, A., Beijersbergen, A., and Hooykaas, P. J. J. (1995) Transkingdom T-DNA transfer from *Agrobacterium tumefaciens* to *Saccharomyces cerevisiae*. *EMBO J.* **14**, 3206–3214.

11. Van Attikum, H. and Hooykaas, P. J. J. (2003) Genetic requirements for the targeted integration of *Agrobacterium* T-DNA in *Saccharomyces cerevisiae*. *Nucleic Acid Res.* **31**, 826–832.

12. Gouka, R. J., Gerk, C., Hooykaas, P. J. J., Bundok, P., Musters, W., Verrips, C. T., and de Groot, M. J. (1999) Transformation of *Aspergillus awamori* by *Agrobacterium tumefaciens*-mediated homologous recombination. *Nature Biotech.* **17**, 598–601.

13. Groot, M. J. A., Bundock, P., Hooykaas, P. J. J., and Beijerbergen, A. G. M. (1998) *Agrobacterium tumefaciens*-mediated transformation of filamentous fungi. *Nat. Biotech.* **16**, 839–842.

14. Kunik, T., Tzfira, T., Kapulnik, Y., Gafni, Y., Dingwall, C., and Citovsky, V. (2001) Genetic transformation of HeLa cells by *Agrobacterium*. *Proc. Natl. Acad. Sci. USA.* **98**, 1871–1876.

15. Sheng, J., and Citovsky, V. (1996) *Agrobacterium*-plant cell DNA transport: have virulence proteins, will travel. *Plant Cell* **8**, 1699–1710.

16. Zupan, J., Muth, T. R., Draper, O., and Zambryski, P. (2000) The transfer of DNA from *Agrobacterium tumefaciens* into plants: a feast of fundamental insights. *Plant J.* **23**, 11–28.

17. Zhu, J., Oger, P. M., Schrammeijer, B., Hooykaas, P. J. J., Farrand, S. K., and Winans, S. C. (2000) The bases of crown gall tumorigenesis. *J. Bacteriol.* **182**, 3885–3895.

18. Gelvin, S. B. (2000) *Agrobacterium* and plant genes involved in T-DNA transfer and integration. *Annu. Rev. Plant Physiol. Mol. Biol.* **51**, 223–256.

19. Tzfira, T., and Citovsky, V. (2002) Partners-in-infection: host proteins involved in the transformation of plant cells by *Agrobacterium*. *Trends Cell Biol.* **12**, 121–129.

20. Hellens, R., Mullineaux, P., and Klee, H. (2000) A guide to *Agrobacterium* binary Ti vectors. *Trends Plant Sci.* **5**, 446–451.

21. Karimi, M., Inze, D., and Depicker, A. (2002) GATEWAY vectors for *Agrobacterium*-mediated plant transformation. *Trends Plant Sci.* **7**, 193–195.

22. van der Fitz, L., Deakin, E. A., Hoge, J. H. C., and Memelink, J. (2000) The ternary transformation system: constitutive *vir*G on a compatible plasmid dramatically increases *Agrobacterium*-mediated plant transformation. *Plant Mol. Biol.* **43**, 495–502.

23. Fang, R. X., Nagy, F., Sivasubramaniam, S., and Chua, N. H. (1989) Multiple *cis* regulatory elements for maximal expression of the cauliflower mosaic virus 35S promoter in transgenic plants. *Plant Cell* **1**, 141–150.

24. Ni, M., Cui, D., Einstein, J., Narasimhulu, S., Vergara, C. E., and Gelvin, S. B. (1995) Strength and tissue specificity of chimeric promoters derived from the octopine and mannopine synthase genes. *Plant J.* **7**, 661–676.

25. Jefferson, R. A., Kavanagh, T. A., and Bevan, M. W. (1987) GUS fusions: beta-glucuronidase as a sensitive and versatile gene fusion marker in higher plants. *EMBO J.* **6**, 3901–3907.

26. Millar, A., Short, S., Hiratsuka, K., Chua, N., and Kay, S. (1992) Firefly luciferase as a reporter of regulated gene expression in higher plants. *Plant Mol. Biol. Rep.* **10**, 324–337.

27. Ow, D. W., Wood, K. V., De Luca, M., de Wet, J. R., Helinski, D. R., and Howell, S. H. (1986) Transient and stable expression of the firefly luciferase gene in plant cells and transgenic plants. *Science* **234**, 856–859.

28. Davis, S. and Vierstra, R. (1998) Soluble, highly fluorescent variants of green fluorescent protein (GFP) for use in higher plants. *Plant Mol. Biol.* **36**, 521–528.

29. Santarém, E. R., Trick, H. N., Essig, J. S., and Finer, J. J. (1998) Sonication-assisted *Agrobacterium*-mediated transformation of soybean immature cotyledons: optimization of transient expression. *Plant Cell Rep.* **17**, 752–759.

30. Frame, B. R., Shou, H., Chikwamba, R. K., Zhang, A., Xiang, C., et al. (2002) *Agrobacterium tumefaciens*-mediated transformation of maize embryos using a standard binary vector system. *Plant Physiol.* **129**, 13–22.

31. Zambre, M., Terryn, N., De Clercq, J., De Buck, S., Dillen, W., Van Montagu, M., et al. (2003) Light strongly promotes gene transfer from *Agrobacterium tumefaciens* to plant cells. *Planta* **216**, 580–586.

32. Leech, M. J., May, K., Hallard, D., Verpoorte, R., De Luca, V., and Christou, P. (1998) Expression of two consecutive genes of a secondary metabolic pathway in transgenic tobacco: molecular diversity influences levels of expression and product accumulation. *Plant Mol. Biol.* **38**, 765–774.

33. Mullen, J., Allen, G., Blowers, A., and Earle, E. (1998) Biolistic transfer of large DNA fragments to tobacco cells using YACs retrofitted for plant transformation. *Mol. Breeding* **4**, 449–457.

34. Corneille, S., Lutz, K., Svab, Z., and Maliga, P. (2001) Efficient elimination of selectable marker genes from the plastid genome by the CRE-*lox* site-specific recombination system. *Plant J.* **27**, 171–178.

35. Maliga, P. (2002) Engineering the plastid genome of higher plants. *Curr. Opin. Plant Biol.* **5**, 164–172.

36. Fu, X., Tan Duc, L., Fontana, S., Ba Bong, B., Tinjuangjun, P., Sudhakar, D., et al. (2000) Linear transgene constructs lacking vector backbone sequences generate low-copy-number transgenic plants with simple integration patterns. *Transgenic Res.* **9**, 11–19.

37. Voinnet, O., Rivas, S., Mestre, P., and Baulcombe, D. (2003) An enhanced transient expression system in plants based on suppression of gene silencing by the p19 protein of tomato bushy stunt virus. *Plant J.* **33**, 949–956.

38. Fagard, M., and Vaucheret, H. (2000) (Trans)Gene silencing in plants: how many mechanisms? *Annu. Rev. Plant Physiol. Plant Mol. Biol.* **51**, 167–194.

39. Vaucheret, H., Béclin, C., and Fagard, M. (2001) Post-transcriptional gene silencing in plants. *J. Cell Science* **114**, 3083–3091.

40. Wang, M. B. and Waterhouse P. (2002) Application of gene silencing in plants. *Curr. Opin. Plant Biol.* **5**, 146–150.

41. Verpoorte, R., and Memelink, J. (2002) Engineering secondary metabolite production in plants. *Curr. Opin. Biotechnol.* **13**, 181–187.

42. Wesley, S. V., Helliwell, C. A., Smith, N. A., Wang, M., Rouse, D. T., Liu, Q., et al. (2001) Construct design for efficient, effective and high-throughput gene silencing in plants. *Plant J.* **27**, 581–590.

43. Bent, A. F. (2000) *Arabidopsis in planta* transformation. Uses, mechanisms, and prospects for transformation of other species. *Plant Physiol.* **124**, 1540–1547.

44. Trieu, A. T., Burleigh, S. H., Kardailsky, I. V., Maldonado-Mendoza, I. E., Versaw, W. K., et al. (2000) Transformation of *Medicago truncatula* via infiltration of seedlings of flowering plants with *Agrobacterium*. *Plant J.* **22**, 531–541.

45. Hansen, G. and Chilton, M. D. (1996) "Agrolistic" transformation of plant cells: integration of T-strands generated *in planta*. *Proc. Natl. Acad. Sci. USA* **93**, 14978–14983.

46. Krysan, P. J., Young, J. C., and Sussman, M. R. (1999) T-DNA as an insertional mutagen in *Arabidopsis*. *Plant Cell* **11:** 2283–2290.

47. Jeon, J. S., Lee, S., Jung, K. H., Jun, S. H., Jeong, D. H., Lee, J., et al. (2000) T-DNA insertional mutagenesis for functional genomics in rice. *Plant J.* **22**, 561–570.

48. Till, B. J., Reynolds, S. H., Greene, E. A., Codomo, C. A., Enns, L. C., et al. (2003) Large-scale discovery of induced point mutations with high-throughput TILLING. *Genome Res.*

13, 524–530.

49. Weigel, D., Ahn, J. H., Blázquez, M. A., Borevitz, J. O., Christensen, S. K., Fankhauser, C., et al. (2000) Activation tagging in Arabidopsis. *Plant Physiol.* **122**, 1003–1013.

50. Britt, A. B., and May, G. D. (2003) Re-engineering plant gene targeting. *Trends Plant Sci.* **8**, 90–95.

51. Hohn, B., Levy, A. A., and Puchta, H. (2001) Elimination of selection markers from transgenic plants. *Curr. Opin. Biotechnol.* **12**, 139–143.

52. Hare, P. D., and Chua, N. H. (2002) Excision of selectable marker genes from transgenic plants. *Nature Biotech.* **20**, 575–580.

53. Zuo, J., Niu, Q. W., Ikeda, Y., and Chua, N. H. (2002) Marker-free transformation: increasing transformation frequency by the use of regeneration-promoting genes. *Curr. Opin. Biotechnol.* **13**, 173–180.

54. Ebinuma, H., Sugita, K., Matsunaga, E., and Yamakado, M. (1997) Selection of marker-free transgenic plants using the isopentenyl transferase gene. *Proc. Natl. Acad. Sci.USA* **94**, 2117–2121.

55. Endo, S., Sugita, K., Sakai, M., Tanaka, H., and Ebinuma, H. (2002) Single-step transformation for generating marker-free transgenic rice using the *ipt*-type MAT vector system. *Plant J.* **30**, 115–122.

56. Holmes-Davis, R., and Comai, L. (1998) Nuclear matrix attachment regions and plant gene expression. *Trends Plant Sci.* **3**, 91–97.

57. Allen, G. C., Spiker, S., and Thompson, W. F. (2000) Use of matrix attachment regions (MARs) to minimize transgene silencing. *Plant Mol. Biol.* **43**, 361–376.

58. Mankin, S. L., Allen, G. C., Phelan, T., Spiker, S., and Thompson, W. F. (2003) Elevation of transgene expression level by flanking matrix attachment regions (MAR) is promoter dependent: a study of the interactions of six promoters with the RB7 3' MAR. *Transgenic Res.* **12**, 3–12.

59. Wood, D. W., Setubal, J. C., Kaul, R., Monks, D. E., Kitajima, J. P. et al. (2001) The genome of the natural genetic engineer *Agrobacterium tumefaciens* C58. *Science* **294**, 2317–2323.

60. Goodner, B., Hinkle, G., Gattung, S., Miller, N., Blanchard, M., Qurollo, B., et al. (2001) Genome sequence of the plant pathogen and biotechnology agent *Agrobacterium tumefaciens* C58. *Science* **294**, 2323–2328.

61. Zupan, J., Ward, D., and Zambryski, P. (2002) Inter-kingdom DNA transfer decoded. *Nat. Biotech.* **20**, 129–131.

62. Gelvin, S. B. (2003) Improving plant genetic engineering by manipulating the host. *Trends Biotechnol.* **21**, 95–98.

63. Young, J. M., Kuykendall, L. D., Martínez-Romero, E., Kerr, A., and Sawada, H. (2001) A revision of *Rhizobium* Frank 1889, with an emended description of the genus, and the inclusion of all species of *Agrobacterium* Conn 1942 and *Allorhizobium undicola* de Lajudie *et al.* 1998 as new combinations: *Rhizobium radiobacter, R. rhizogenes, R. rubi, R. undicola* and *R. vitis. Int. J. Syst. Evol. Microbiol.* **51**, 89–103.

24

Production of Recombinant Proteins by Hairy Roots Cultured in Plastic Sleeve Bioreactors

Fabricio Medina-Bolívar and Carole Cramer

Summary

Plant-based expression of recombinant proteins offers significant advantages over transgenic animal-and cell-based systems. Unlike bacteria, plants perform the complex protein-processing steps required to produce eukaryotic proteins in active form. In order to facilitate protein production and purification we used hairy root cultures as a secretion-based in vitro plant system. We utilized the green fluorescent protein (GFP) as our model protein and expressed it for secretion in tobacco hairy root cultures. For large scale production of GFP, we adapted the Life-Reactor™ plastic sleeve bioreactor for growth of hairy roots in cultures containing up to 5 L of medium. Yields higher than 800 µg of GFP per liter of culture were obtained after 21 d of incubation, representing almost 20% of the total secreted protein. The use of the plastic sleeve bioreactor system for expression of proteins in hairy roots allows for continuous or inducible production and recovery, while maintaining absolute containment, of the recombinant product.

Key Words: Hairy roots; GFP; secreted proteins; bioreactor.

1. Introduction

A large number of recombinant proteins have been expressed in plants. However, the extraction and purification of proteins from plant tissue is laborious and expensive. To address this issue, secretion-based systems utilizing hairy root cultures show significant advantages *(1)*. Secretion often makes it possible to increase yields and maintain higher recoveries during purification without denaturization and loss of the protein's biological activity. Generation of genetically stable transgenic hairy roots can be accomplished efficiently by a combination of *Agrobacterium tumefaciens*-and *A. rhizogenes*-mediated transformation. Secretion of the recombinant protein into the medium of hairy root

From: *Methods in Molecular Biology, vol. 267:*
Recombinant Gene Expression: Reviews and Protocols, Second Edition
Edited by: P. Balbás and A. Lorence © Humana Press Inc., Totowa, NJ

cultures is done by genetic fusion of a DNA sequence encoding a signal peptide to the gene of interest. Upon targeting to the ER, the protein will continue the default secretory pathway *(2)* into the apoplast and medium. We have used this system for the secretion of GFP-fusion proteins from transgenic tobacco hairy roots cultured in Erlenmeyer flasks *(3)*. Here we describe the application of plastic sleeve bioreactors (LifeReactor™, Osmotek, Israel) for growth of hairy roots and production of recombinant proteins. This approach provides a nonlaborious, low-cost, scalable alternative for the continuous nondestructive recovery of recombinant proteins from hairy root cultures.

2. Materials
2.1. Construction of Plasmids

1. cDNA: m-GFP5, patatin signal peptide.
2. Promoter: dual enhanced 35S promoter.
3. Cloning vector: pBC (Stratagene, La Jolla, CA).
4. Binary vector: pBIB-Kan.
5. Oligonucleotide primers (Invitrogen, Carlsbad, CA).
6. DNEasy Plant Mini kit (Qiagen, Valencia, CA).
7. LB medium: Luria-Bertani medium (Sigma, St. Louis, MO).
8. Antibiotics: filter-sterilized chloramphenicol, streptomycin, kanamycin.
9. PureTaq™ Ready-to-Go PCR™ beads (Amersham Pharmacia, Piscataway, NJ).
10. PCR RoboCycler (Stratagene).
11. Power supply, electrophoresis chamber.
12. Ethidium bromide prestained 1% agarose gels.
13. TAE: 10X stock, 40 mM Tris-HCl, 1 mM EDTA (pH 8.0), 11.42 mL/L glacial acetic acid.

2.2. Generation of Transgenic Hairy Roots and Establishment of Cultures in Plastic Sleeve Bioreactors

1. Surface-sterilized (20% bleach solution for 5 min) seeds of tobacco cv. Xanthi.
2. mMS medium: Murashige & Skoog basal medium *(4)*, 0.4 g/L MgS04.7H$_2$O, 3% sucrose, 0.4% Phytagel (Sigma, St. Louis, MO). The pH of the medium is adjusted to 5.7 before addition of Phytagel.
3. Regeneration medium: mMS medium supplemented with 0.1 mg/L NAA and 1.0 mg/L BAP, 250 mg/L kanamycin and 500 mg/L carbenicillin. All agents are autoclaved, except for kanamycin and carbenicillin, which are filter-sterilized and added before pouring the medium into plates.
4. B5 medium: Gamborg's B5 basal medium *(5)*, 2% sucrose, pH 5.7; add 0.4% Phytagel for solid medium.
5. *Agrobacterium rhizogenes* strain 15834 (ATCC).
6. *A. tumefaciens* strain LBA4404.
7. YEP medium: 10 g/L Bacto-peptone, 10 g/L yeast extract, 5 g/L NaCl, 1.5% agar-pH 7.0.
8. Antibiotics: filter-sterilized streptomycin, kanamycin, cefotaxime, carbenicillin.
9. NA medium: Bacto nutrient agar medium (Becton Dickinson, Sparks, MD).
10. LifeReactor system: 1.5-L and 5-L plastic sleeve bioreactors (Osmotek, *see* **Note 1**).
11. Erlenmeyer flasks (50 and 250 mL), disposable Petri dishes (100 × 15 mm for hairy root and 100 × 20 mm for regeneration medium, respectively).

12. Plant tissue culture boxes Phytatray II (Sigma).
13. Orbital platform shaker, plant growth incubator, laminar flow hood.

2.3. Protein Concentration, Western Blot, and ELISA

1. Nitrocellulose membrane.
2. Polyacrylamide gel 12% (Novex, San Diego, CA).
3. Mighty Small II electrophoresis apparatus (Hoefer, San Francisco, CA).
4. SDS-PAGE running buffer.
5. Protein sample buffer (6X): 8X Tris-HCl/SDS pH 6.8 (3.5 mL), 30% glycerol, 1 g SDS, 0.93 g dithiothreitol (DTT), 1.2 mg bromophenol blue. Complete to 10 mL with distilled water.
6. Electrophoretic transfer unit: Mini Trans-Blot cell (Bio-Rad, Hercules, CA).
7. Protein extraction buffer: 0.1 M K-phosphate buffer pH 7.0, 5 mM DTT, complete protease inhibitor cocktail. 1 tablet/50 mL buffer (Boehringer-Mannheim, Indianapolis, IN).
8. Centricon YM-10 (Millipore, Bedford, MA).
9. Labscale™ TFF ultrafiltration system (Millipore).
10. Pellicon XL Biomax 10X filter (Millipore).
11. PBS: 10X stock, 137 mM NaCl, 2.7 mM KCl, 10.1 mM Na_2HPO_4, 1.76 mM KH_2PO_4, pH 7.4.
12. PBST: PBS with 0.1% Tween-20.
13. Transfer buffer: 20 mM Tris-HCl, 150 mM glycine, 20% methanol.
14. Antibodies: Anti-GFP antibody (Living Colors™, Clontech, Palo Alto, CA); goat-anti rabbit IgG alkaline phosphatase conjugate (Promega, Madison, WI).
15. Chemiluminescence reagents: CDP-StarTM (Boehringer-Mannheim), Nitroblock Enhancer II (Tropix, Bedford, MA).
16. CDP Star detection buffer: 0.1 M Tris-HCl, 0.1 M NaCl.
17. Film, darkroom, developer.

3. Methods

The methods described here outline (1) the construction of plasmids for *Agrobacterium*-mediated transformation, (2) generation of transgenic tobacco hairy roots and establishment of cultures in plastic sleeve bioreactors, and (3) protein characterization in hairy root cultures.

3.1. Construction of Plasmids

Constructs are made first in high-copy-number small vectors, such as pBS or pBC (Promega, Madison, WI). The latter has the advantage of using chloramphenicol for selection rather than ampicillin (chloramphenicol is more stable). Also, the use of small cloning vectors at this stage facilitates sequencing of junctions (derived from ligations) or any PCR products required for subcloning. All PCR products will have incorporated restriction sites for subcloning of fragments and an additional of 3–4 nucleotide bases at 5' and 3' ends of the DNA sequence to facilitate restriction enzyme digestion. After the final construct is made it can be removed from the plasmid by restriction enzyme digestion and gel extraction, and further ligated into specific sites of a binary vector such as pBIB-Kan (*6*). This vector is a pBIN19 (*7*) derivative and contains the NPTII gene for selection in plants and the left and right borders flanking the T-DNA.

3.1.1. DNA Source

Plasmid (pPATP-35S) containing the dual-enhanced 35S promoter, TEV and patatin signal (Pat), was provided by Dr. E. A. Grabau (Virginia Tech). Construction of this plasmid was previously described *(8)*. Accordingly, the dual-enhanced 35S promoter sequence *(9)* was obtained from plasmid pRTL2 and ligated to the tobacco etch virus translational enhancer, TEV *(10)*. A portion of the sequence from *Nco*I to *Sst*I in pRTL2 was deleted to eliminate the ATG initiation site. The promoter-TEV sequence was ligated to the region of the patatin gene encoding the signal peptide "pat" *(11)*. The signal peptide contains the initiation methionine codon; it is flanked by *Kpn*I at the 5' end and *Xba*I site at the 3' end (*see* **Note 2**). To amplify GFP, we used as template a vector containing m-GFP5-ER *(12)*, provided by Dr. Grabau.

3.1.2. Cloning

DNA manipulations were performed by standard recombinant DNA methods *(13)* and are not described in detail due to space limitations. The promoter-pat sequence was excised with *Hin*dIII/*Xba*I digestion, gel purified, and ligated into pBC vector (Promega, Madison, WI) yielding plasmid R1-1. The sequence for GFP was PCR-amplified from the m-GFP5-ER containing vector with primers GFP5a (5'-GGCC<u>TC TAGA</u>AGTAAAGGAGAAGAACTT-3') and GFP3a (5'-TGC<u>GAGCTC</u>TCATTTGTAT AGTTCATCCAT-3'). The PCR product was then digested with *Xba*I/*Sst*I (sites provided by the primers, sequence underlined) and subcloned into plasmid R1-1 to yield plasmid R8-2. The promoter-pat-GFP DNA sequence was then excised with *Hin*dIII/*Sst*I and ligated into binary vector pBIB-Kan (*see* **Note 3**). The plasmids were then mobilized into *A. tumefaciens* LBA4404 by a modified freeze-thaw method *(14)* (*see* **Note 4**).

3.2. Generation of Transgenic Cultures

Transgenic plants were developed by a simplified direct *A. tumefaciens* inoculation in the petiole of whole leaves. Hairy roots were then obtained by *A. rhizogenes* infection of leaf explants from high-expresser transgenic plants. Alternatively, stable transgenic hairy root lines can be developed by direct transformation with *A. rhizogenes* containing a binary vector with the gene of interest. However, we found that selection of transgenic hairy roots is not efficient; therefore, the establishment of hairy roots from well-characterized transgenic plants is highly advantageous.

3.2.1. Generation of Transgenic Plants

1. Grow *A. tumefaciens,* containing the R8-2 construct in pBIB Kan plasmid (**Fig. 1**), for single colonies on YEP medium with 30 mg/L streptomycin and 100 mg/L kanamycin (*see* **Note 5**).
2. Take a single colony with a scalpel (use blade N. 11 to facilitate wounding) and make a longitudinal cut along the petiole (*see* **Note 6**) of excised leaves of tobacco plantlets grown on mMS medium (**Fig. 2**).
3. Place the *Agrobacterium*-infected leaves in mMS medium. Insert the petiole in the medium (as shown in **Fig. 2**) and incubate for 48 h. At this time a halo in the base of the petiole should be evident.

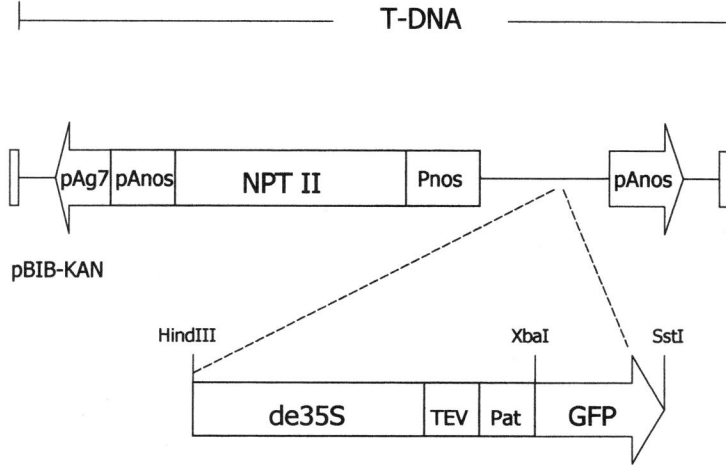

Fig. 1. Plant expression vector pBIB-KAN containing GFP cDNA under the control of the dual-enhanced 35S promoter and the translational enhancer from tobacco etch virus (TEV). The ER targeting signal from potato storage protein patatin (Pat) is fused in frame to allow for protein secretion. The neomycin phosphotransferase II (*nptII*) gene confers resistance to kanamycin and is under the control of the nopaline synthase (Pnos) promoter. The polyadenylation sequences pAnos and pAg7 are derived from the nopaline synthase and gene 7 (octopine T-DNA), respectively.

4. Transfer the leaves to regeneration medium supplemented with 500 mg/L carbenicillin and 250 mg/L kanamycin (*see* **Note 7**). After 7 d, the base of the petiole will start to enlarge, and from days 10–14 the initiation of shoot buds should be obvious. Fully developed shoots will be ready for harvesting from days 18–21 (*see* **Note 8**).
5. Transfer the regenerated shoots to mMS medium supplemented with 500 mg/L carbenicillin and 200 mg/L kanamycin. Rooting will initiate from 2 d up to 2 wk (*see* **Note 9**). The insertion of the T-DNA is random and affects distinctively the levels of transgene expression; therefore it is recommended to screen an average of 30 regenerated rooted shoots in order to identify the highest expressors.

3.2.2. Generation of Hairy Root Cultures

Hairy roots were developed from high-GFP-expressing plants (*see* **Subheading 3.3**).

1. Grow *A. rhizogenes* ATCC 15834 for single colonies on nutrient agar medium (Difco Laboratories, Detroit, MI) at 28°C.
2. Take a single colony with a scalpel and make a longitudinal cut (approx 1 cm) along the midrib of excised leaves placed on B5 medium (**Fig. 2**). Hairy roots will develop at the infection site after 14 d (*see* **Note 10**).
3. Transfer the hairy roots to B5 medium supplemented with 600 mg/L cefotaxime (*see* **Notes 11** and **12**).

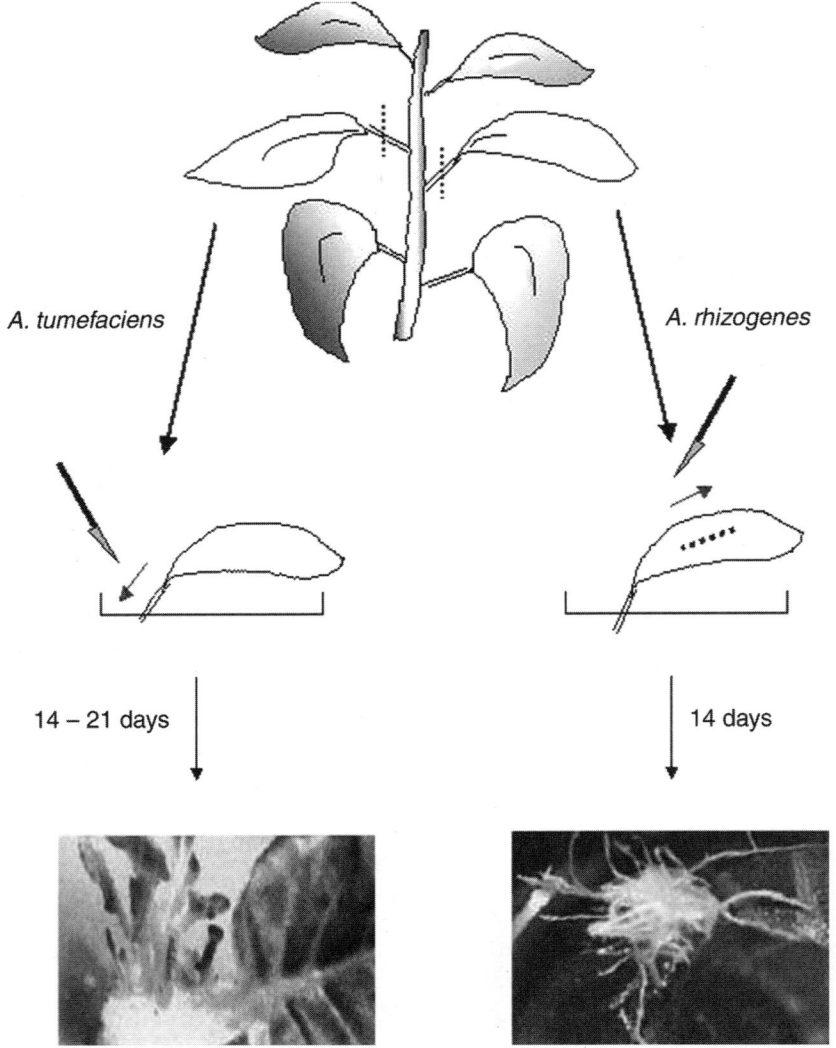

Fig. 2. Generation of transgenic plants and hairy roots. *Left:* Regeneration of putative trans-genic plants from the petiole of whole leaves of tobacco inoculated with *A. tumefaciens* LBA4404 harboring binary vector pBIB-KAN with construct R8-2. *Right:* Development of trans-genic hairy roots from leaves inoculated with *A. rhizogenes* strain 15834.

4. Subculture the root tips to a fresh plate with the same antibiotics after 2 wk, followed by a subculture in antibiotics-free B5 medium (*see* **Note 13**).
5. Transfer approx 20 root tips (approx 1 cm, white tips) into 250-mL flasks with 50 mL B5 medium. Maintain all cultures under continuous light (roots could also be grown under con-tinuous dark) and shaking at 90 rpm in an orbital platform shaker and subculture every 12 d (*see* **Note 14**).

3.2.3. Establishment of Hairy Root Cultures in Plastic Sleeve Bioreactors

We adapted the LifeReactor (Osmotek) system for large-scale cultures of hairy roots. The system uses either 1.5- or 5-L (working volume) presterilized plastic sleeve bioreactors that include an integral sparger, a large port for loading and harvesting plant material. The reactor also harbors an integral drain port with a screen to maintain the culture in the reactor while removing medium (**Fig. 3**), as well as an air exhaust port and a medium inlet port (*see* **Note 15**). This bioreactor has been previously used for propagation of shoots and cells (*15,16*); this is the first demonstration of its application for culture of hairy roots.

Root tips (2 to 5 cm) from six or twelve 14-d cultures (250-mL flask) were used as inoculum for 1.5-L or 5-L bioreactors, respectively. Inoculation of the bioreactors was done as described by the manufacturer, and special care was directed for maintenance of sterile conditions. After inoculation, the roots were dispersed gently inside the plastic sleeve.

Two configurations were tested to set up the bioreactor, including the vertical stand as described by Osmotek's procedures and a flat setup on an orbital platform shaker (**Fig. 3**). Drastic improvement in the growth rate was obtained when the reactor was placed flat on an orbital shaker at constant 40 rpm. This facilitated tremendously the dispersion of oxygen in the culture provided by the sparger. In addition, shaking prevented the formation of root clamps or "balls" observed when the reaction was set up in the vertical stand. The bioreactors were checked daily to ensure proper operation of the sparger. Hairy root line 2 was identified as the fastest-growing line. A yield of approx 250 g (FW) of root material was obtained after 21 d (35 g inoculum) in a 1.5 L-bioreactor. The yields of FW were consistently reproduced in three independent cultures of the same line.

3.3. Characterization of Transgenic Plants and Root Cultures

3.3.1. Screening for Transgene DNA

DNA was extracted from leaves (100 mg) of rooted plantlets using the DNEasy Plant kit (Qiagen, Valencia, CA) following manufacturer's procedures. For PCR, primers GFP5a and GFP5b were used with the following conditions: denaturation 94°C (1 min), annealing 56°C (1 min) and extension 72°C (1 min) for 30 cycles.

3.3.2. Protein Characterization in Transgenic Cultures

3.3.2.1. Protein Extraction From Leaves, Hairy Roots, and Culture Medium

To verify expression of the transgene, proteins were recovered from the secreted fraction of leaves, intracellular fractions of roots, and media of hairy root cultures.

Extraction of secreted leaf proteins:

1. Excise three leaves from PCR-positive plantlets and place them in a Petri dish.
2. Cover the leaves with 15 mL of protein extraction buffer.
3. Place the plate in a vacuum dessicator and vacuum-infiltrate for 5 min.

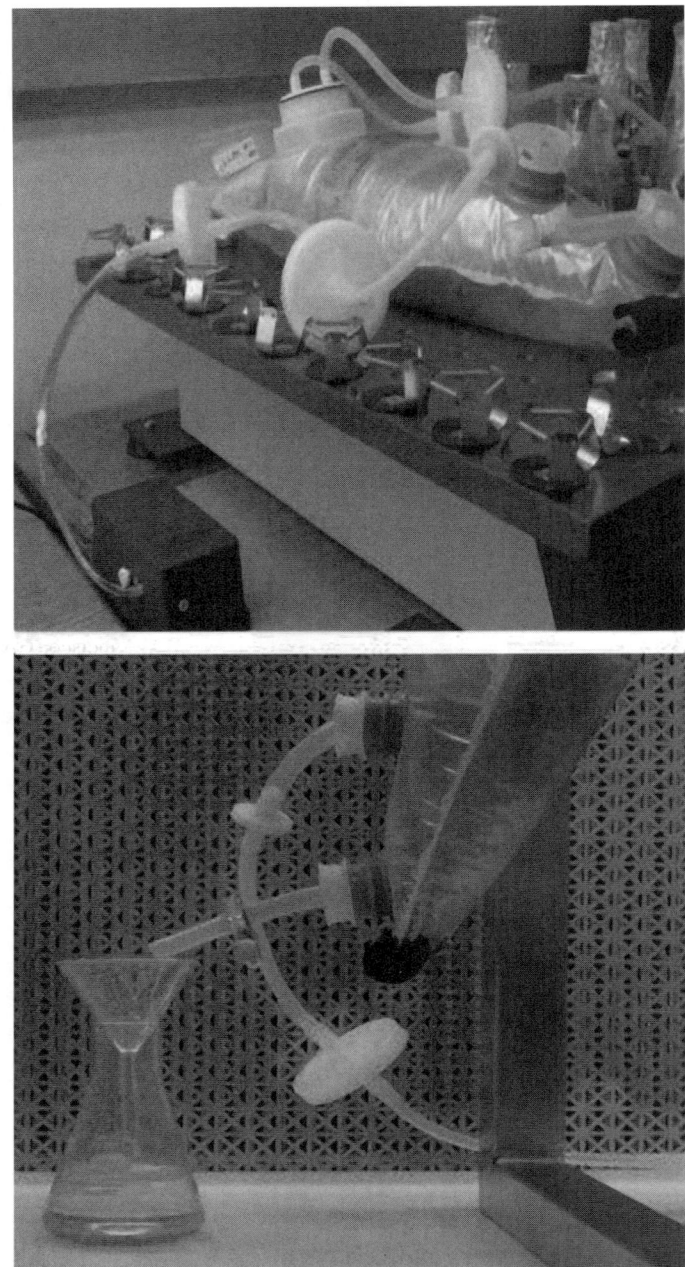

Fig. 3. Hairy roots cultured using the LifeReactor system. *Top:* Hairy roots of tobacco cultured in a 1.5-L plastic sleeve bioreactor. Note that the bioreactor is maintained on an orbital platform shaker. *Bottom:* Removal of medium from the bioreactor.

4. Incubate at room temperature for 24 h.
5. Concentrate the sample through Centricon YM-10 centrifugal filters (Millipore) following manufacturer's procedures.

Extraction of proteins from roots:

1. Grind the roots in a mortar with pestle containing liquid nitrogen.
2. Transfer the powdered tissue to a cold corex tube containing protein extraction buffer (3 mL of buffer per gram of tissue).
3. Disrupt the samples in a water bath sonicator for 30 s.
4. Centrifuge for 10 min at 18,500g and collect the supernatant through Miracloth™.
5. Concentrate the sample with centricon YM 10 as described above.
   To analyze proteins from the culture medium, the liquid is concentrated using Labscale™ TFF ultra-filtration system with a Pellicon XL Biomax 10 filter (Millipore) followed by a second concentration using centricon YM 10 filters as before (*see* **Note 16**). Aliquots of the concentrated protein extracts are run on SDS-PAGE gels and analyzed by Western blot (**Fig. 4**).

3.3.2.2. SDS-PAGE AND WESTERN BLOT

1. Combine the concentrated protein fractions with 6X protein sample buffer and boil for 5 min. Spin down any precipitates.
2. Load the soluble fraction on a 12% SDS-PAGE gel (Novex, San Diego, CA).
3. Run the samples on a Mighty Small II apparatus (Hoefer, San Francisco, CA) following manufacturer's procedures.
4. Transfer the proteins to a nitrocellulose membrane overnight using a Bio-Rad unit at 25 volts and 4°C, following manufacturer's procedures.
5. Block the membrane with 3% BSA in PBST for 1 h.
6. Incubate the membrane in 3% BSA-PBST with Living Colors™ antibody (1/1000, Clontech, Palo Alto, CA) for 1 h, wash three times with PBST.
7. Incubate with goat-anti rabbit IgG alkaline phosphatase conjugate (1/3000, Promega, Madison, WI) in 3% BSA-PBST for 1 h, wash three times with PBST.
8. Detect GFP by chemiluminescence using CDP-Star™ (Boehringer Mannheim, Indianapolis, IN) and Nitroblock enhancer II (Tropix, Bedford, MA) following manufacturer's procedures.

3.3.3. Quantification of GFP

A modified enzyme-linked immunosorbent assay (ELISA) is used for quantification of GFP from the medium of hairy roots cultured in the plastic sleeve bioreactors. The ELISA procedure has been previously described for quantification of GFP fusion proteins (*3*). Accordingly, Immulon 4 plates (Dynatech Laboratories, Chantilly, VA) are coated overnight at 4°C with 3.5 µg/mL Living Colors polyclonal antibody (Clontech Laboratories) in PBS. Blocking is done as described (*17*). Duplicate GFP standards are run on each plate starting with 2 ng in the first well and diluting 1:2 down the plate. All samples and GFP standards are boiled for 5 min, then set on ice before use. The secondary antibody is anti-GFP monoclonal (Clontech Laboratories) used 1:500 dilution in PBST + 1% BSA. A tertiary antibody, goat anti-mouse IgG conjugated to horseradish peroxidase, is used at 1:5000. Detection is done as detailed before (*17*).

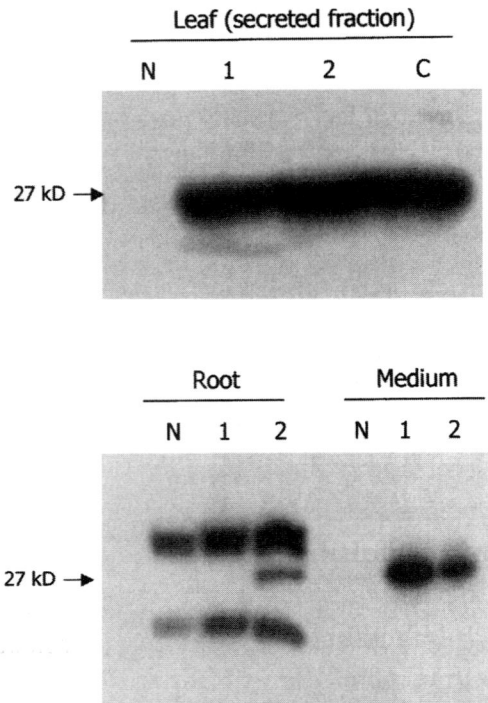

Fig. 4. Detection of recombinant GFP protein by Western blot using anti-GFP antibodies. *Top:* Immnunoblot of secreted proteins from leaves. N; non-transformed plant; C; GFP control protein; 1 and 2; transgenic plants. *Bottom:* Immunoblot analysis from hairy roots and culture medium. N; hairy roots and media from nontransformed plants; 1 and 2; hairy root and media from transgenic lines expressing GFP.

The total amount of recombinant protein produced was measured in hairy root line 2, previously identified as the fastest grower. Media was collected and concentrated, as previously indicated, after 21 d in culture. The total amount of secreted protein was estimated by Bradford's assay and GFP by ELISA, respectively. The amount of GFP was estimated as the average of three independent cultures of line 2 in 1.5 L bioreactors. The average amount of GFP was 810 µg per liter of culture, representing approximately 20% of the total secreted protein. We consistently reproduced the yields of GFP in independent experiments using this root line.

4. Notes

1. We obtained the Osmotek LifeReactor system directly from Osmotek in Israel (http://www.osmotek.com). Since March 2002, Osmotek products can be also obtained from the US distributor Caisson Laboratories (http://www.caissonlabs.com).
2. We have used the patatin signal peptide as a mechanism to deliver proteins to the ER and thereon into the apoplast through the secretory pathway. We have observed that different pro-

teins followed a distinct mechanism of secretion from the apoplast into the culture medium. These conditions seem to be dependent on the size and stability of the protein and developmental stage of the culture. Therefore, conditions need to be optimized for a particular protein.

3. The binary vector pBIB-Kan is a low-copy-number plasmid of approx 12 Kb. We have used the plasmid miniprep kit (Qiagen) with a minimum of 5 mL of bacterial culture to obtain enough plasmid for further cloning steps.

4. We always used freshly made competent cells of *A. tumefaciens*. We have observed reduction in the rate of transformation using frozen stocks of the bacteria. Five-milliliter cultures of *A. tumefaciens* at OD = 0.1 were used to make competent cells as described *(14)*. It is important to observe the formation of whitish precipitates during culture of *A. tumefaciens* LBA4404.

5. Bacterial stocks can be maintained at −80°C in YEP medium containing 70 μL of dimethyl-sulfoxide (DMSO) per milliliter of culture. After addition of DMSO to the culture, mix thoroughly and place immediately in liquid nitrogen.

6. The kind of leaf and length of the petiole are important to ensure a high percentage of regeneration. Fully developed leaf explants should be taken from 3 to 5 wk micropropagated plantlets. We have observed a reduced rate of regeneration when using cotyledons. The optimal size of the petiole is approx 1 cm and the cut should be at a middle position between the axillary meristem and the base of the leaf lamina as described in **Fig. 2**. Shorter petioles will delay the regeneration from 2 to 4 wk and reduce efficiency of regeneration. The cut should be approx 1 cm away from the axillary meristem to avoid carrying over of meristematic cells into the explant.

7. Control experiments should be added to ensure appropriate levels of growth regulators and antibiotics. One control should use leaf explants cultured in regeneration media without antibiotics. The efficiency of regeneration should be 100% within 21 d. To guarantee proper selection, incubate leaf explants (without bacterium inoculation) in regeneration media with antibiotics. No shoots should regenerate under these conditions.

8. We have repeated this procedure several times with different DNA constructs contained in the binary vector pBIB-Kan. In order to completely reproduce the timing of regeneration, particular attention should be considered to the explant source as indicated in **Note 6**. On average, we have observed 95% efficiency in transformation (number of transgenic shoots from total number of rooted shoots) and 100% regeneration efficiency.

9. Some shoots may take more than 2 wk for rooting. If shoots look healthy but no rooting is observed, subculture the apical node into a fresh rooting medium in order to facilitate rooting.

10. We have noticed that after 10 d of incubation most leaf explants will initiate adventitious roots at the base of the petiole. These are not hairy roots. The latter will arise only from the wounded site in the midrib of the lamina and will start after 14–18 d post-infection with *A. rhizogenes*.

11. It is very important to use cefatoxime at 600 mg/L to eliminate *A. rhizogenes*. We have observed that using carbenicillin allows for the bacteria to remain latent with a possibility of regrowth in the late stage of the culture. Also, carbenicillin has a callus-inducing effect on tobacco hairy roots.

12. We normally mark the edges of the roots with a marker (under the plate) and observe for new growth. After 2 d, growth should be evident. Each hairy root should be considered as an indepent clone. Different clones may have different number of T-DNAs inserted and this may affect the growth rate of the roots.

13. Bacteria should not grow in this medium. If bacteria are still growing, transfer again to antibiotic-containing B5 medium. Liquid cultures should not be initiated until the hairy root

line is free of *A. rhizogenes*. Attempts to clean *A. rhizogenes* in the liquid stage are very inefficient. To initiate liquid cultures, transfer 6–10 tips of roots from a particular line into a 50-mL Erlenmeyer flask with 10 mL of medium. Once the roots are actively growing these can be used as mother culture for 250-mL flask cultures.

14. For subculturing hairy roots, transfer the packed roots from the flask to a sterile petri dish containing enough medium to cover the roots. It is important to keep the roots hydrated. Cut the tips (approx 1–2 cm) and transfer to fresh medium.

15. Full description of the LifeReactor™ can be found at the following internet site: http://www.gross.co.il/randd.htm.

16. To collect media from the bioreactor, transfer the whole unit to the clean bench and remove the media through its outlet port. The media is poured through a funnel containing two layers of Miracloth (Calbiochem, San Diego, CA) in order to remove any root debris remaining the culture (**Fig. 3**).

Acknowledgments

This work was supported by NIH TDRU Program Grant P01 AI-44962.

References

1. Sharp, J. and Doran, P. (2001) Characterization of monoclonal antibody fragments produced by plant cells. *Biotech. Bioeng.* **73,** 338–346.
2. Denecke, J., Botterman, J., and Deblaere, R. (1990) Protein secretion in plant cells can occur via a default pathway. *Plant Cell* **2,** 51–59.
3. Medina-Bolivar, F., Wright, R., Funk, V., Sentz, D., Barroso, L., Wilkins, T., et al. (2003) A non-toxic lectin for antigen delivery of plant-based mucosal vaccines. *Vaccine* **21,** 997–1005.
4. Murashige, T. and Skoog, F. (1962) A revised medium for rapid growth and bioassays with tobacco tissue cultures. *Physiol. Plant.* **15,** 473–479.
5. Gamborg, O., Miller, R. A. and Ojima, K. (1968) Nutrient requirements of suspension cultures of soybean root cells. *Exp. Cell. Res.* **50,** 151–158.
6. Becker D. (1990) Binary vectors which allow the exchange of plant selectable markers and reporter genes. *Nucl. Acid Res.* 18, 203–210.
7. Bevan, M. (1984) Binary *Agrobacterium* vectors for plant transformation. *Nucleic Acids Res.* 12, 8711–8721.
8. Li, J., Hegeman, C. E., Hanlon, R. W., Lacy, G. H., Denbow, M. D., and Grabau, E. A. (1997) Secretion of active recombinant phytase from soybean cell-suspension cultures. *Plant Physiol.* 114, 1103–1111.
9. Kay, R., Chan, A., Daly, M., and McPherson, J. (1987) Duplication of CaMV 35S promoter sequences creates a strong enhancer for plant genes. *Science* 236, 1299–1302.
10. Carrington, J. and Freed, D. (1990). Cap-independent enhancement of translation by a plant potyvirus 5' nontranslated region. *J. Virol.* 64, 1590–1597.
11. Bevan, M., Barker, R., Goldsbrough, A., Jarvis, M., Kavanagh, T., and Iturriaga, G. (1986) The structure and transcription start site of a major potato tuber protein gene. *Nucleic Acids Res.* 14, 4625–4638.
12. Haseloff, J., Siemering, K. R., Prasher, D. C., and Hodge, S. (1997) Removal of a cryptic intron and subcellular localization of green fluorescent protein are required to mark transgenic *Arabidopsis* brightly. *Proc. Natl. Acad. Sci. USA* 94, 2122–2127.
13. Sambrook, J., Fritsch, E., and Maniatis, T. (1989) *Molecular Cloning: A Laboratory Manual.*

Cold Spring Harbor Laboratory Press, Cold Spring Harbor, NY.

14. Holsters, M., de Waele, D., Depicker, A., Messens, E., Van Montangu, M., and Schell, J. (1978) Transfection and transformation of *A. tumefaciens*. *Mol. Gen. Genet.* 163, 181–187.

15. Ziv, M. and Shemesh, D. (1996) Propagation and tuberization of potato bud clusters from bioreactor culture. *In Vitro Cell. Dev. Biol.-Plant.* 3, 31–36.

16. Ziv, M., Ronen, G., and Raviv, M. (1998) Proliferation of meristematic clusters in disposable presterilized plastic bioreactors for the large scale micropropagation of plants. *In Vitro Cell. Dev. Biol.-Plant.* 34, 152–158.

17. Vines, R., Perdue, S., Moncrief, J., Sentz, D., Barroso, L., Wright, R., et al. (2001) Fragilysin, the enterotoxin from *Bacteroides fragilis*, enhances the serum antibody response to antigen co-administered by the intranasal route. *Vaccine* 19, 655–660.

25

Engineering the Chloroplast Genome for Hyperexpression of Human Therapeutic Proteins and Vaccine Antigens

Shashi Kumar and Henry Daniell

Summary

The chloroplast genome is ideal for engineering because it offers a number of attractive advantages, including high-level gene expression, the feasibility of expressing multiple genes or pathways in a single transformation event, and transgene containment due to lack of pollen transmission. The chloroplast-based expression system is suitable for hyperexpression of foreign proteins, oral delivery of vaccine antigens and therapeutic proteins, via both leaves and fruits. Through the refinement of expression vectors and use of chaperones, chloroplasts produce up to 47% of foreign protein in the total cellular protein in transgenic tissues. This chapter describes various techniques for creating chloroplast transgenic plants and their biochemical and molecular characterization. Suitable examples for application of chloroplast genetic engineering in human medicine are provided.

Key Words: Chloroplast genetic engineering; plastid transformation; genetically modified crops; edible vaccines; human therapeutic proteins; biopharmaceuticals; medical molecular farming.

1. Introduction

The prokaryotic chloroplast genome has been explored for genetic manipulation, which offers several major advantages over nuclear transformation, including high-level expression of foreign genes (*1,2*), containment of transgenes (*3,4*), expression of multigene operons (*1,5*) in a single transformation event, and the absence of gene silencing or position effect due to site-specific integration of transgenes (*6,7*). Chloroplast transformation has now opened doors to improving agriculture and medicine in an environmentally friendly manner (*2*). This technology has been used successfully for expression of herbicide (*8*), insect (*1,9,10*), and disease resistance (*11*), salt and drought tolerance (*6*),

From: *Methods in Molecular Biology, vol. 267:*
Recombinant Gene Expression: Reviews and Protocols, Second Edition
Edited by: P. Balbás and A. Lorence © Humana Press Inc., Totowa, NJ

phytoremediation of toxic metals (*12*), and production of biopharmaceuticals (*13–16*) and vaccines (*17*) in transgenic plants. Stable integration of foreign genes is achieved through homologous recombination by targeting the chloroplast genome instead of the nuclear genome. Plastid genome is highly polyploid and genetic transformation allows the integration of thousands copies of transgenes in each plant cell, enabling transgenic plants to produce extraordinarily high levels of foreign proteins. Chloroplast transformation utilizes intergenic spacer regions for transgene integration and flanking genes for homologous recombination. Foreign genes regulated by a suitable promoter, along with appropriate untranslated regions (5'UTR and 3'UTR), cloned between two chloroplast DNA flanking sequences, are integrated into the chloroplast genome by homologous recombination, precisely at a predetermined location. Plastid transformation enables expression of foreign genes at high levels and is ideal for the improvement of crops and the development of plants for oral delivery of vaccines or therapeutic proteins (*2,5*).

2. Materials

2.1. Isolation of Genomic DNA From Plants

1. Mortar and pestle.
2. Liquid nitrogen.
3. Fresh dark green leaves.
4. DNeasy Plant Mini Kit (Qiagen).

2.2. PCR Amplification of Chloroplast Flanking Sequence

1. Genomic DNA (50–100 ng/μL).
2. dNTPs.
3. 10X *pfu* buffer.
4. Forward primer (10 μ*M*).
5. Reverse primer (10 μ*M*).
6. Autoclaved distilled H_2O.
7. Turbo *pfu* DNA polymerase.

2.3. Vector Construction

1. Plasmid pUC19 or pBlueScript SK (+/–).
2. Species-specific PCR-amplified chloroplast DNA flanking sequences.
3. A promoter functional in plastids, 5'UTR of chloroplast gene, selectable marker gene, gene of interest, and chloroplast 3'UTR.
4. Restriction enzymes and buffers.
5. T4 DNA polymerase to remove 3' overhangs to form blunt ends and fill-in of 5' overhangs to form blunt ends or Klenow large fragment (fill-in of 5' overhangs to form blunt ends), alkaline phosphatase for dephosphorylation of cohesive ends, DNA ligase to form phosphodiester bonds and appropriate buffers.
6. Water baths or incubators set at different temperatures.

2.4. Preparation for Biolistics

1. Autoclaved Whatman filter paper no. 1 (55 mm in diameter) dried in oven.
2. 100% ethanol.
3. Autoclaved tips in box, autoclaved Kimwipes tissues wrapped in aluminum foil.

4. Sterile gold particles stored at −20°C in 50% glycerol (*see* **Notes 1 and 2**).
5. Sterile rupture discs (1100 psi) and macrocarriers sterilized by dipping in 100% ethanol.
6. Autoclaved steel macrocarrier holders and stopping screens.
7. Freshly prepared 2.5 *M* CaCl$_2$: weigh 1.84 g and dissolve in 5 mL H$_2$O and filter-sterilized with 0.2-μm filter.
8. 0.1 *M* spermidine (highly hygroscopic): dilute 1 *M* spermidine stock to 10X and aliquot 100 μL in 1.5-mL Eppendorf tubes to store at −20°C. Discard each tube after single use.

2.5. Media Preparation for Plant Tissue Culture

1. *For all plant growth media:* Adjust to pH 5.8 with 1 *N* KOH or 1 *N* NaOH and add 4 g/L phytagel (Sigma) before autoclaving at 121°C for 20 min. For preparation of 1 mg/mL stock of BAP, IAA, IBA, NAA, ZR respectively: weigh 10 mg powder and dissolve first in 1 or 2 drops of 1 *N* NaOH and make up the final volume to 10 mL; store all plant growth regulators at 4°C for 1–2 mo.
2. *Tobacco:* Medium for 1000 mL: 4.3 g MS salts (Invitrogen), H$_2$O (molecular biology grade), 100 mg/L myo-inositol, 1 mg/L thiamine-HCl, 3% sucrose for shoot induction and 2% sucrose for root induction, 1 mg/L 6-benzyl aminopurine (BAP; use 1 mL from 1 mg/mL stock), 0.1 mg/L indole-3-acetic acid (use 0.1 mL from 1 mg/mL IAA stock), 1 mg/L indole-3-butyric acid for root induction (use 1 mL from 1 mg/mL IBA stock). Add 500 mg/L spectinomycin in autoclaved medium when it cools to 45°C–50°C (use 5 mL filter sterilized spectinomycin from 100 mg/mL stock).
3. *Potato:* Medium for 1000 mL: 4.3 g MS salts, B5 vitamins (make 100X solution in 100 mL H$_2$O by dissolving 1 g myo-inositol, 10 mg nictonic acid, 10 mg pyridoxine-HCl, 100 mg thiamine-HCl; use 10 mL, store remaining solution at 4°C), 0.5 mg/L zeatin riboside (use 0.5 mL from 1 mg/mL ZR stock), 0.1 mg/l α-napthaleneacetic acid (use 0.1 mL from 1 mg/mL NAA stock), 40 to 500 mg/L spectinomycin.
4. *Tomato:* Medium for 1000 mL: 4.3 g MS salts, B5 vitamins (10 mL from 100X stock), 0.2 mg/l indole-3-acetic acid (use 0.2 mL from 1 mg/mL IAA stock), 3 mg/L of 6-benzylaminopurine (use 3 mL from 1 mg/mL BAP stock). 300 or 500 mg/L spectinomycin.

2.6. Molecular Analysis of Transgenic Plants

2.6.1. PCR Analysis for Gene Integration into Tobacco Chloroplasts

PCR reaction for 50 μL: 1.0 μL genomic DNA (50–100 ng/μL), 1.5 μL dNTPs (stock 10 m*M*), 5.0 μL (10X PCR buffer), 1.5 μL forward primer (to land on the native chloroplast genome; stock 10 μ*M*), 1.5 μL reverse primer (to land on the transgene; stock 10 μ*M*), 39.0 μL autoclaved distilled H$_2$O, and 0.5 μL *Taq* DNA polymerase.

2.6.2. Analysis of Homoplasmy by Southern Blots

1. Depurination solution: 0.25 *N* HCl (use 0.4 mL HCl from 12.1 *N* HCl, Fisher Scientific USA, to make up final volume 500 mL with distilled H$_2$O).
2. Transfer buffer: 0.4 *N* NaOH, 1 *M* NaCl (weigh 16 g NaOH and 58.4 g NaCl and dissolve in distilled H$_2$O to make up the final volume to 1000 mL).
3. 20X SSC: 3 *M* NaCl, 0.3 *M* sodium citrate trisodium salt (weigh 175.3 g NaCl, 88.2 g Na$_3$C$_6$H$_5$O$_7$·2H$_2$O 900 in mL H$_2$O and adjust pH 7.0 using 1 *N* HCl and make up the final volume to 1000 mL with distilled H$_2$O and autoclave).
4. 2X SSC: Add 20 mL of 20X SSC in 180 mL of distilled H$_2$O.

2.6.3 Protein Analysis by Western Blots

1. Acrylamide/Bis: ready made from Fischer (USA), stored at 4°C.
2. 10% SDS: dissolve 10 g SDS in 90 mL deionized water, make up the volume to 100 mL, store at room temperature.
3. Resolving gel buffer: 1.5 M Tris-HCl (add 27.23 g Tris base in 80 mL water, adjust to pH 8.8 with 6 N HCl, and make up the final volume to 150 mL; store at 4°C after autoclaving).
4. Stacking gel buffer: 0.5 M Tris-HCl (add 6.0 g Tris base in 60 mL water; adjust to pH 6.8 with 6 N HCl; make up the volume to 100 mL; store at 4°C after autoclaving).
5. Sample buffer (SDS reducing buffer): In 3.55 mL water add 1.25 mL 0.5 M Tris-HCl (pH 6.8), 2.5 mL glycerol, 2.0 mL (10% SDS), 0.2 mL (0.5% bromophenol blue). Store at room temperature. Add 50 µL β-mercaptoethanol (βME) to 950 µL sample buffer prior to its use.
6. 10X running buffer (pH 8.3): Dissolve 30.3 g Tris Base, 144.0 g glycine, and 10.0 g SDS in 700 mL water (add more water if not dissolving). Bring up the volume to 1 L and store at 4°C.
7. 10X PBS: Weigh 80 g NaCl, 2 g KCl, 26.8 g $Na_2HPO_4 7H_2O$ (or 14.4 g Na_2HPO_4), 2.4 g KH_2PO_4 in 800 mL water. Adjust pH to 7.4 with HCl and make up the volume to 1 L. Store at room temperature after autoclaving.
8. 20% APS: Dissolve 200 mg ammonium persulfate in 1 mL water (make fresh every 2 wk).
9. Transfer buffer for 1500 mL: Add 300 mL 10X running buffer, 300 mL methanol, 0.15 g SDS in 900 mL water.
10. Plant extraction buffer:

Used Volume	Used Concentration	Final Concentration
60 µL	5 M NaCl	100 mM
60 µL	0.5 M EDTA	10 mM
600 µL	1 M Tris-HCl	200 mM
2 µL	Tween-20	0.05%
30 µL	10% SDS	0.1%
3 µL	βME	14 mM
1.2 mL	1 M sucrose	400 mM
1 mL	Water	
60 µL	100 mM PMSF	2 mM

Add PMSF just before use (vortex to dissolve PMSF crystals).

11. PMSF (phenylmethyl sulfonyl fluoride): Dissolve 17.4 mg of powdered PMSF in 1 mL of methanol by vortexing and store at −20°C for up to a month.

2.7. CTB-GM1-Gangliosides Binding ELISA Assay

1. Bicarbonate buffer: 15 mM Na_2CO_3, 35 mM $NaHCO_3$, pH 9.6.
2. PBS containing 0.05% Tween-20.

2.8. Macrophage Lysis Assay

1. Extraction buffer with CHAPS detergent: 4% CHAPS, 10 mM EDTA, 100 mM NaCl, 200 mM Tris-HCl, pH 8.0, 400 mM sucrose, 14 mM β-mercaptoethanol, 2 mM PMSF.
2. MTT 3-[4,5-dimethylthiazol-2-yl]-2,5-diphenyltetrazolium bromide (Sigma).
3. DMSO.

2.9. Purification of HSA

1. Extraction buffer: 0.2 M NaCl, 25 mM Tris-HCl (pH 7.4), 2 mM PMSF, and 0.1% Triton X-100.
2. Suspension buffer: 6 M Gu-HCl, 0.1 M βME, and 0.25 mM Tris-HCl (pH 7.4).
3. Dilution buffer: 100 mM NaCl, 50 mM Tris-HCl (pH 8.5), and 1 mM EDTA.
4. Polyethyleneglycol.

2.10. Electron Microscopy and Immunogold Labeling of HSA

1. Cacodylate buffer: 0.1 M, pH 7.4 (2.5% glutaraldehyde, 2% paraformaldehyde, and 5 mM CaCl$_2$).
2. 95% ethanol.
3. Glycine: 0.05 M prepared in PBS buffer.
4. Blocking solution: PBS containing 2% nonfat dry milk.
5. Glutaraldehyde: 2% diluted in PBS.

3. Methods

3.1. Isolation of Genomic DNA from Plants

Extract the genomic DNA from fresh green leaves using DNeasy Plant kit (Qiagen) following vendor's instructions.

3.2. Amplification of Chloroplast Flanking Sequence

Species-specific flanking sequences from the chloroplast DNA or genomic DNA of a particular plant species are amplified with the help of PCR using a set of primers that are designed using known and highly conserved sequences of the tobacco chloroplast genome.

Conditions for running PCR reaction are as follows: There are three major steps in a PCR, which are repeated for 30 to 40 cycles. (1) *Denaturation at 94°C*: to separate double-stranded chloroplast DNA. (2) *Annealing at 54 to 64°C*: primers bind to single-stranded DNA with formation of hydrogen bonds and the DNA polymerase starts copying the template. (3) *Extension at 72°C*: DNA Polymerase at 72°C extends to the template that forms strong hydrogen bond with primers. Mismatched primers will not form strong hydrogen bonds and therefore all these temperatures may vary based on DNA sequence homology. The bases complementary to the template are coupled to the primer on the 3' side. The polymerase adds dNTPs from 5' to 3', reading the template in 3' to 5' direction, and bases are added complementary to the template.

3.3. Chloroplast Transformation Vector

The left and right flanks are the regions in the chloroplast genome that serve as homologous recombination sites for stable integration of transgenes. A strong promoter and the 5' UTR and 3' UTR are necessary for efficient transcription and translation of the transgenes within chloroplasts. For multiple gene expression, a single promoter may regulate the transcription of the operon, and individual ribosome binding sites must be engineered upstream of each coding sequence (*2*) (**Fig. 1**). The following steps are used in vector construction:

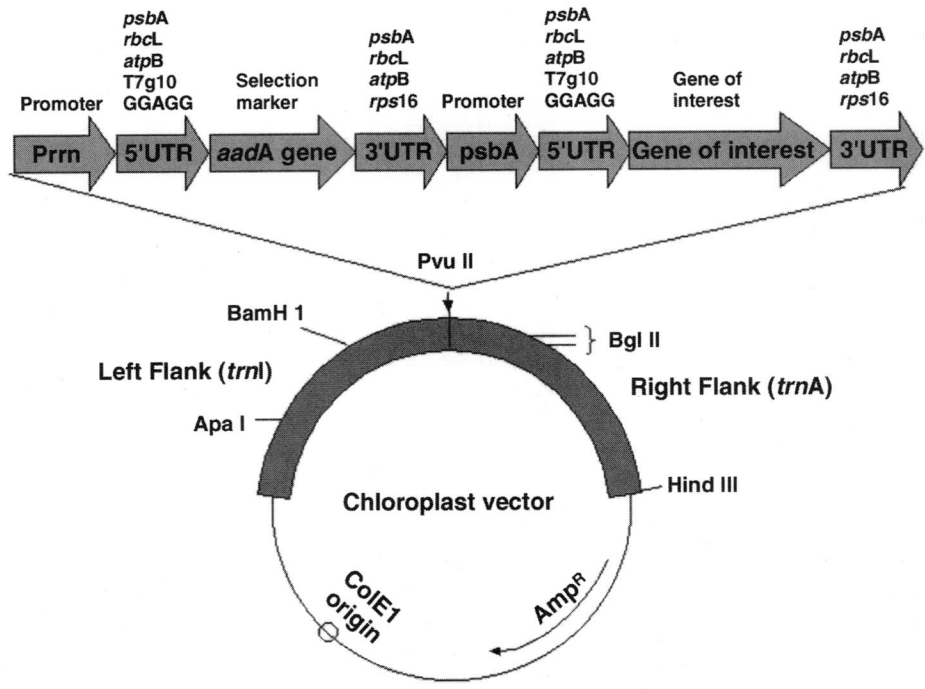

Fig. 1. Generalized schematic view of plastid transformation vector (*see* **Notes 3 and 4**).

1. Amplify flanking sequences of plastid with primers that are designed on the basis of known sequence of the tobacco chloroplast genome (between 16S–23S regions of chloroplast).
2. Insert the PCR product containing the flanking sequence of the chloroplast genome into pUC19 plasmid digested with *Pvu*II restriction enzyme (to eliminate the multiple cloning site), dephosphorylated with the help of alkaline phosphatase (CIP) for 5 min at 50°C (to prevent recircularization of the cloning vector). Inactivate CIP enzyme at 68°C for 10 min.
3. Clone chloroplast transformation cassette (which is made blunt with the help of T4 DNA polymerase or Klenow filling) into a cloning vector digested at the unique PvuII site in the spacer region, which is conserved in all higher plants examined so far.

3.4. Delivery of Foreign Genes Into Chloroplasts Via Particle Gun

This is the most successful and a simple technique to deliver transgenes into plastids and is referred to as the Biolistic PDS-1000/He Particle Delivery System (*18,19*). This technique has proven to be successful for delivery of foreign DNA to target tissues in a wide variety of plant species and integration of transgenes has been achieved in chloroplast genomes of tobacco (*2*), *Arabidopsis* (*20*), potato (*21*), tomato (*22*), and transient expression in wheat (*23*), carrot, marigold, and red pepper (*24*) (*see* **Note 5**).

3.4.1. Preparation of Gold Particle Suspension

1. Suspend 50–60 mg gold particles in 1 mL 100% ethanol and vortex for 2 min.
2. Spin at maximum speed approx 10,000g (using tabletop microcentrifuge) for 3 min.
3. Discard the supernatant.
4. Add 1 mL fresh 70% ethanol and vortex for 1 min.
5. Incubate at room temperature for 15 min and shake intermittently.
6. Spin at 10,000g for 2 min.
7. Discard supernatant, add 1 mL sterile distilled H_2O, vortex for 1 min, leave at room temperature for 1 min, and spin at 10,000 g for 2 min.
8. Repeat above washing process (**step 7**) three times with H_2O.
9. Resuspend the gold pellet in 1 mL 50% glycerol; store stock in – 20°C freezer.

3.4.2. Precipitation of the Chloroplast Vector on Gold Particles for Five Samples

1. Take 50 µL of the gold particles in 1.5-mL tube after vortexing for 1 min.
2. Add 10 µL DNA (about 1 µg/µL plasmid DNA), and vortex the mixture for 30 s.
3. Add 50 µL of 2.5 M $CaCl_2$ and vortex the mixture for 30 s.
4. Add 20 µL of 0.1 M spermidine and vortex the mixture for 20 min at 4°C.

3.4.3. Washing of Chloroplast Vector Coated on Gold Particles

1. Add 200 µL 100 % ethanol and vortex for 30 s.
2. Spin at 3000g for 40 s.
3. Pour off ethanol supernatant.
4. Repeat ethanol washings five times.
5. In the last step, pour off ethanol carefully and add 35–40 µL ethanol (100 %).

3.4.4. Preparation of Macrocarriers

1. Sterilize macrocarriers by dipping in 100% ethanol for 15 min and insert them into sterile steel ring holder with the help of a plastic cap when air-dried.
2. Vortex the gold-plasmid DNA suspension and pipet 8–10 µL in the center of macrocarrier and let it air-dry.

3.4.5. Gene Gun Setup for Bombardment of Samples

1. Wipe the gun chamber and holders with 100% ethanol using fine tissue paper (do not wipe the door with alcohol).
2. Turn on the vacuum pump.
3. Turn on the valve (helium pressure regulator) of helium gas tank (counterclockwise).
4. Adjust the gauge valve (adjustable valve) approx 200 to 250 psi above the desired rupture disk pressure (clockwise) using adjustment handle.
5. Turn on the gene gun.
6. Place the rupture disk (sterilized by dipping in 100% ethanol for 5 min) in the rupture disk retaining steel cap and tightly screw to the gas acceleration tube.
7. Place a stopping screen in the macrocarrier launch assembly and above that, place the macrocarrier with gold particles, with the chloroplast vector facing down toward screen. Screw assembly with a macrocarrier cover lid and insert in the gun chamber.

8. Place an intact leaf or explants to be bombarded on a filter paper (Whatman No. 1) soaked in medium containing no antibiotics. Place sample plate over the target plate shelf, insert in the gun chamber, and close the bombardment chamber door.
9. Press the **Vac** switch to build pressure (up to 28 in. Hg) in the vacuum gauge display. Turn same switch down at hold point and press the **Fire** switch until you hear a burst sound of the ruptured disk.
10. Press the **Vent** switch to release the vacuum and open the chamber to remove the sample.
11. Shut down the system by closing the main valve (helium pressure regulator) on the helium gas cylinder. Create some vacuum in the gene gun chamber and keep using the **Fire** switch on and off until both pressure gauges show zero reading. Release the vacuum pressure and turn off the gene gun and vacuum pump.
12. Incubate the bombarded sample plates in the culture room for two days in the dark (i.e., covered with aluminum foil); on the third day cut explants in appropriate pieces and place on the selection medium.

3.5. Plant Tissue Culture and Chloroplast Transformation

3.5.1. Tobacco Chloroplast Transformation

A highly efficient and reproducible protocol has been established for *Nicotiana tabacum* cv. Petit Havana (*24*) (*see* **Notes 6–9**).

1. Bombard 4-wk-old dark green tobacco leaves on the abaxial (bottom) side with the chloroplast vector and incubate leaves in the dark for 2 d on selection-free medium.
2. On the third day cut bombarded leaf explants into small square pieces (5 mm) and place explants facing abaxial surface toward selection medium containing MS salts, 1mg/L thiamine HCl, 100 mg/L myo-inositol, 3% sucrose, 1 mg/L BAP, and 0.1 mg/IAA, along with 500 mg/L spectinomycin as a selective agent.
3. Transgenic shoots should appear after 3 to 5 wk of transformation. Cut the shoot leaves again into small square explants (2 mm) and subject to a second round of selection for achieving homoplasmy on fresh medium.
4. Regenerate transgenic shoots (confirmed by PCR for transgene integration) on rooting medium containing MS salts, 1 mg/L thiamine HCl, 100 mg/L myo-inositol, 2% sucrose, and 1 mg/L IBA with 500 mg/L spectinomycin.
5. Transfer transgenic plants into pots under high humidity and move them to a greenhouse or growth chamber for further growth and characterization.

3.5.2. Plastid Transformation of Edible Crops

The concept of a universal vector for using the chloroplast DNA from one plant species to transform another species (of unknown sequence) was developed by the Daniell group (*8*). Using this concept, both tomato and potato chloroplast genomes were transformed as described below (*see* **Note 7**).

3.5.2.1. POTATO CHLOROPLAST TRANSFORMATION

Using the tobacco chloroplast vector, leaf tissues of potato cultivar FL1607 were transformed via biolistics, and stable transgenic plants were recovered using the selective *aad*A gene marker and the visual green fluorescent protein (GFP) reporter gene (*21*).

1. Bombard potato leaves (3–4 wk old) and incubate in the dark for 2 d on selection-free medium.
2. On the third day excise leaves into small square pieces (5 mm) and place on MS medium containing B5 vitamins, 0.5 mg/L ZR, 0.1 mg/L NAA, and 3% sucrose. Gradually increase spectinomycin selection pressure (40 to 400 mg/L) and after every 2 wk subculture under diffuse light.
3. Regenerate shoots from transgenic potato calli on MS medium containing B_5 vitamins, 0.01 mg/L NAA, 0.1 mg/L GA3, 2% sucrose, and 40–400 mg/L spectinomycin.
4. Transfer transgenic shoots onto basal MS medium containing B_5 vitamins, 2% sucrose, and 40–400 mg/L spectinomycin for root induction. Transfer transgenic plantlets to growth chamber.

3.5.2.2. TOMATO CHLOROPLAST TRANSFORMATION

Using the tobacco chloroplast vector, tomato (*Lycopersicon esculentum* cv. IAC Santa Clara) plants with transgenic plastids were generated using a very low intensity of light (*25*).

1. Bombard 4-wk-old tomato leaves and incubate in the dark for 2 d on selection-free medium.
2. Excise bombarded leaves into small pieces and place on shoot induction medium containing 0.2 mg/L IAA, 3 mg/L BAP, 3% sucrose, and 300 mg/L spectinomycin.
3. Select spectinomycin-resistant primary calli after a 3- to 4-mo duration without any shoot induction.
4. Regenerate shoots in about 4 wk after transfer of transgenic calli to shoot induction medium containing 0.2 mg/L IAA, 2 mg/L ZR, 2% sucrose, and 300 mg/L spectinomycin, then root on hormone-free medium. Transfer regenerated transgenic plants into the greenhouse.

3.6. Molecular Analysis of Transgenic Plants

3.6.1. PCR Screening of Transgenic Shoots

This method has been used to distinguish between mutants and nuclear and chloroplast transgenic plants. By landing one primer on the native chloroplast genome adjacent to the point of integration and a second primer on the *aadA* gene (*26*), PCR product of an appropriate size should be generated in chloroplast transformants. Since this PCR product cannot be obtained in nuclear transgenic plants or mutants, the possibility of nuclear integration or mutants should be eliminated (*8*).

1. Extract the genomic DNA from transgenic leaf tissue using DNeasy Plant kit (Qiagen) by following vendor's instructions. For lower amounts of transgenic tissues, the volume of buffers may be reduced appropriately.
2. Run PCR reaction with Taq DNA Polymerase (Qiagen) using appropriate primers, following the same conditions as described above for amplification of flanking sequences.

3.6.2. Analysis of Homoplasmy by Southern Blot

In Southern blot analysis, tobacco plastid genome digested with suitable restriction enzymes should produce a smaller fragment (flanking region only) in wild-type plants compared to transgenic chloroplasts that include the transgene cassette as well as the flanking region. In addition, homoplasmy in transgenic plants is achieved when only the transgenic fragment is observed.

3.6.2.1. TRANSFER OF DNA TO MEMBRANE

1. Digest the genomic DNA (approx 2 to 10 µg) with suitable restriction enzymes from transgenic samples (including wild-type as a control) and run digested DNA on 0.8% agarose gel containing 5 µL EtBr (from 10 mg/mL stock) in 100 mL for 4 h at 40 V.
2. Soak gel in 0.25 N HCI (depurination solution) for 15 min and rinse gel twice in distilled H_2O for 5 min.
3. Soak gel for 20 min in transfer buffer to denature DNA.
4. Transfer overnight DNA from gel to nylon membrane (presoak first in water, then in transfer buffer for 5 min) using the transfer buffer.
5. Next day, rinse membrane twice with 2X SSC buffer for 5 min each and air-dry for 5 min on filter papers. Cross-link transferred DNA to membrane using GS GeneLinker UV Chamber (Bio-Rad) at appropriate (C3) setting.

3.6.2.2. PREPARATION OF PROBE

1. Digest any plasmid (containing flanking sequences of the chloroplast genome) with appropriate restriction enzymes.
2. Denature 45 µL flanking DNA fragment (50–250 ng) at 95°C for 5 min, then place on ice for 2–3 min.
3. Add denatured probe to Ready-To-Go DNA Labeling Beads (– dCTP) tube (Amersham Biosciences, USA) and mix gently by flicking the tube.
4. Add 5 µL radioactive $\alpha^{32}P$ (dCTP; Amersham Biosciences, USA) to probe mixture, incubate at 37°C for 1 h, and filter the probe using ProbeQuant G-50 Micro Columns (Amersham Pharmacia Biotech Inc., USA).

3.6.2.3. PREHYBRIDIZATION AND HYBRIDIZATION

1. Place the blot (DNA transfer side facing toward the solution) in a hybridization bottle and add 10 mL Quik-Hyb (Stratagene, USA).
2. Incubate for 1 h at 68°C. Add 100 µL sonicated salmon sperm (10 mg/mL stock; Stratagene, USA) to the labeled probe, heat at 94°C for 5 min, and add to bottle containing membrane and Quik-Hyb solution. Incubate for 1 h at 68°C.

3.6.2.4. WASHING AND AUTORADIOGRAPHY

1. Discard Quik-Hyb solution with probe and wash membrane twice in 50 mL (2X SSC buffer and 0.1% SDS) for 15 min at room temperature.
2. Wash membrane twice in 50 mL (0.1X SSC buffer and 0.1% SDS) for 15 min at 60°C.
3. Wrap the wash membrane in plastic wrap and expose blot to X-ray film in the dark and leave at −70°C until ready for development.

3.6.3. Determination of Transgene Expression by Western Blot

3.6.3.1. EXTRACTION OF PLANT PROTEIN

1. Grind 100 mg of leaf in liquid nitrogen and add 200 µL of extraction buffer to samples on ice.

2. Add appropriate volume of freshly prepared 2X sample loading buffer to an aliquot plant extract (from a stock containing 50 μL β-mercaptoethanol and 950 μL sample loading buffer).
3. Boil samples for 4 min with loading dye and centrifuge for 2 min at 10,000g, then immediately load 20-μL samples into gel.

3.6.3.2. RUNNING GEL

Load samples on gel and run for 30 min at 100 V, then 1 h at 150 V until the marker bands corresponding to your protein are in the middle.

3.6.3.3. TRANSFER OF PROTEIN TO MEMBRANE

Transfer protein from gel to membrane using Mini Transfer Blot Module at 30 V overnight, 65 V for 2 h, or 100 V for 1 h. Membrane wrapped in plastic wrap can be stored at −20°C for a few days if necessary.

3.6.3.4. MEMBRANE BLOCKING

1. After transfer, rinse the membrane with water and incubate it in PTM (100 mL 1X PBS, 50 μL 0.05% Tween-20, and 3 g dry milk (3%) for 1 h at room temperature.
2. Add primary antibody in suitable dilution for 15 mL and incubate for 2 h at room temperature. Wash membrane twice with 1X PBS for 5 min each.
3. Add secondary antibody in proper dilution for 20 mL. Incubate for 1.5 h at room temperature on a shaker.
4. Wash twice with PT (100 mL 1X PBS + 50μL Tween-20) for 15 min and finally with 1X PBS for 10 min.

3.6.3.5 EXPOSURE OF THE BLOT TO X-RAY FILM

1. Mix 750 μL of each chemiluminescent solution (Luminol Enhancer and stable peroxide) in 1.5-mL tube, add to membrane, cover thoroughly.
2. Wipe out extra solution, expose blot to X-ray film for appropriate duration, and develop film.

3.6.4. Seed Sterilization

1. Vortex small amount of seeds into microcentrifuge tube with 1 mL 70% ethanol for 1 min. Discard ethanol after brief spin.
2. Add 1 mL disinfecting solution (15% bleach and 0.1% Tween-20) in tube and vortex intermittently for 15 min. Discard solution after brief spin.
3. Wash the seed three times with sterile distilled water.
4. Spray seeds with sterile water on plate containing RMOP basal medium supplemented with 500 μg/mL spectinomycin to determine maternal inheritance in transgenic chloroplast plants.

3.7. Evaluation of Results

3.7.1. Maternal Inheritance in Chloroplast Transgenic Plants

Transgenes integrated into chloroplast genomes are inherited maternally. This is evident when transgenic seeds of tobacco (as shown in **Fig. 2**) are germinated on RMOP

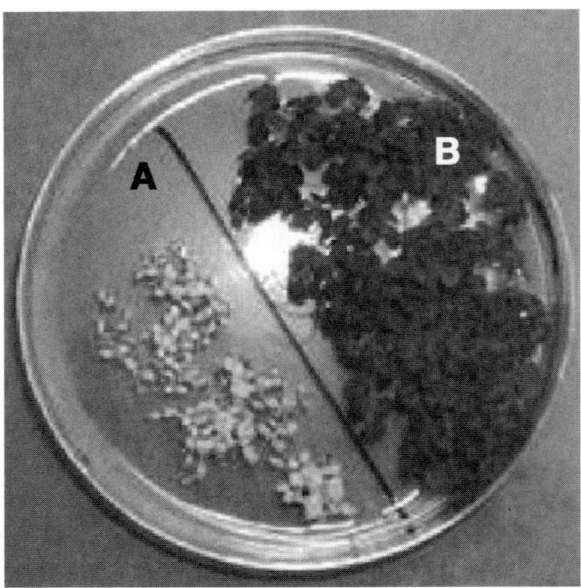

Fig. 2. Germination of control untransformed (**A**) and chloroplast transgenic (**B**) seeds on RMOP basal medium containing 500 µg/mL spectinomycin.

basal medium containing 500 µg/mL spectinomycin. There should be no detrimental effect of the selection agent in transgenic seedlings, whereas untransformed seedlings will be affected. In **Fig. 2**, all transgenic seedlings carry the spectinomycin resistance trait and show maternal inheritance without any Mendelian segregation of introduced transgenes.

3.7.2. Cholera Toxin (CTB) Antigen as an Edible Vaccine

Chloroplast transgenic plants are ideal for production of vaccines. The heat-labile toxin B subunits of *E. coli* enterotoxin (LTB), or cholera toxin of *Vibrio cholerae* (CTB) have been considered as potential candidates for vaccine antigens. Integration of the unmodified native *CTB* gene into the chloroplast genome has demonstrated high levels of CTB accumulation in transgenic chloroplasts (*7*). This new approach not only allowed the high level expression of native *CTB* gene but also enabled the multimeric proteins to be assembled properly in the chloroplast, which is essential because of the critical role of quaternary structure for the function of many vaccine antigens. The expression level of CTB in transgenic plants was between 3.5% and 4.1% total soluble protein (tsp) and the functionality of the protein was demonstrated by binding aggregates of assembled pentamers in plant extracts similar to purified bacterial antigen. Binding assays confirmed that both chloroplast-synthesized and bacterial CTB bind to the intestinal membrane GM1-ganglioside receptor, confirming correct folding and disulfide bond formation of CTB pentamers within transgenic chloroplasts (**Fig. 3**).

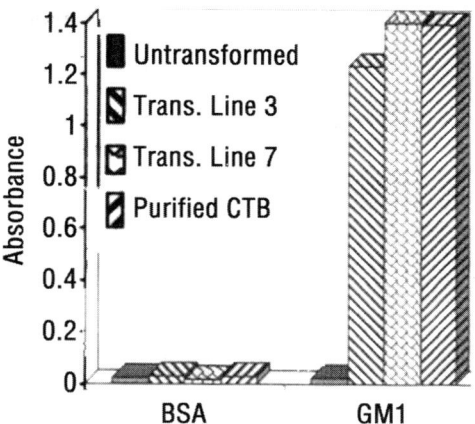

Fig. 3. CTB-GM1-ganglioside binding ELISA assay. Plates, coated first with GM1-ganglioside and BSA, respectively, were plated with total soluble plant protein from transgenic lines (3 and 7) and untransformed plant total soluble protein and 300 ng of purified bacterial CTB. The absorbance of the GM1-ganglioside-CTB-antibody complex in each case was measured at 405 nm.

3.7.2.1. CTB-GM1-Gangliosides Binding ELISA Assay

1. Coat microtiter plate (96-well ELISA plate) with monosialoganglioside-GM1 (3.0 µg/mL in bicarbonate buffer) and as a control, coat BSA (3.0 µg/mL in bicarbonate buffer) in a few wells.
2. Incubate plate overnight at 4°C.
3. Block wells with 1% (w/v) bovine serum albumin (BSA) in 0.01 M phosphate-buffered saline (PBS) for 2 h at 37°C.
4. Wash wells three times with PBST buffer.
5. Incubate plate by adding soluble protein from transformed and untransformed plants and bacterial CTB in PBS.
6. Add primary antibodies (rabbit anticholera serum diluted 1:8000 in 0.01 M PBST containing 0.5% BSA) and incubate plate for 2 h at 37°C.
7. Wash well three times with PBST buffer.
8. Add secondary antibodies diluted 1: 50,000 (mouse anti-rabbit IgG-alkaline phosphatase conjugate in 0.01 M PBST containing 0.5% BSA) and incubate plate for 2 h at 37°C.
9. Develop plate with Sigma Fast pNPP substrate. Stop reaction by adding 3 M NaOH and read plate absorbance at 405 nm.

3.7.2.2. Anthrax-Protective Antigen

Bacillus anthracis is the causative agent for anthrax disease. Using chloroplast transformation technology, the protective antigen (PA) was expressed in higher amounts in plant leaves (**27**). Crude transgenic plant extracts contained up to 78 µg/mL (156 µg PA/g fresh leaf tissue). PA functionality was confirmed by macrophage lysis assays.

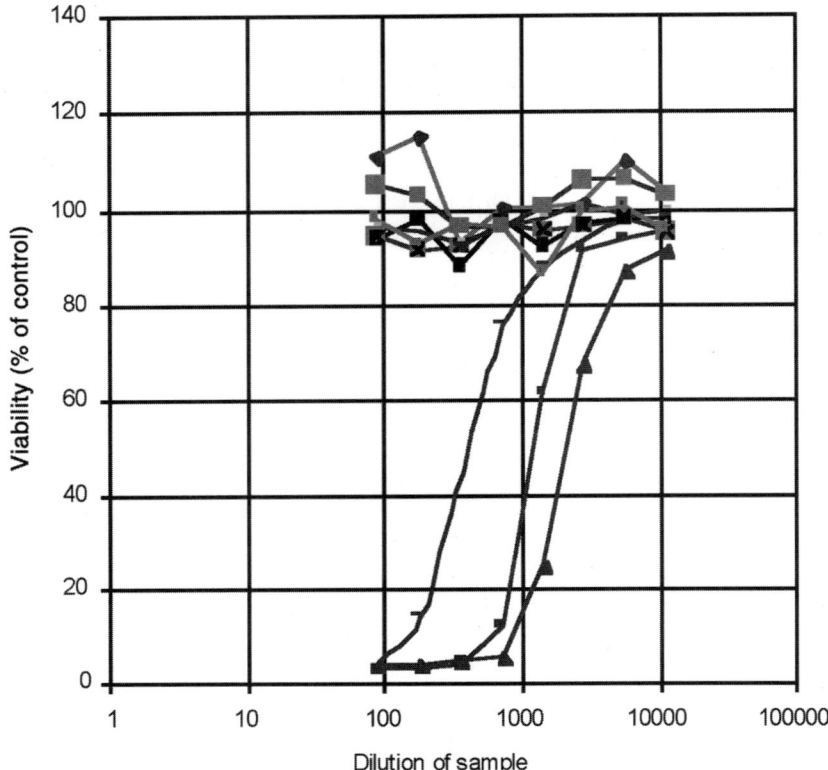

Fig. 4. Macrophage cytotoxic assays for extracts from transgenic plants. Supernatant samples from T₁ pLD-5'-PA tested (proteins extracted in buffer containing no detergent and MTT added after 5 h).

—■— pLD-5'-PA (extract stored 2 d);
—▲— pLD-5'-PA (extract stored 7 d);
———— PA 5 μg/ml;
—✕— Control wild-type (extract stored 2 d);
—■— Control wild-type (extract stored 7 d);
—●— Control wild-type no LF (extract stored 2 d);
—■— Control wild-type no LF (extract stored 7 d);
—◆— Control pLD-5'-PA no LF (extract stored 2 d);
—▲— Control pLD-5'-PA no LF (extract stored 7 d).

Chloroplast-derived PA efficiently bound to anthrax toxin receptor, underwent proper cleavage, heptamerized, and bound the lethal factor, resulting in macrophage lysis (**Fig. 4**). With observed expression levels, 600 million doses of vaccine (free of contaminants) could be produced per acre of transgenic tobacco to combat bioterrorism or outbreaks.

The macrophage lysis assay is as follows:

1. Isolate crude extract protein from 100 mg transgenic leaf using 200 μL of extraction buffer containing CHAPS, and one without CHAPS detergent.
2. Spin samples for five minutes at 10,000g and use both the supernatant and the homogenate for assay.
3. Plate macrophage cells RAW 264.7 (grown to 50% confluence) into a 96-well plate; incubate in 120 μL Dulbecco's Modified Eagle's Medium (DMEM; from Invitrogen Life Technologies).
4. Aspirate medium from wells and add 100 μL medium containing 250 ng/mL proteins in crude leaf extract.
5. In control plate, add only DMEM with no leaf fraction to test toxicity of plant material and buffers.
6. In another plate, add 40μL dilutions onto RAW 264.7 cells from plant samples, which are serially diluted two fold, so that the top row has plant extract at 1:14 dilution.
7. Add 20 μL of MTT 3-[4,5-dimethylthiazol-2-yl]-2,5-diphenyltetrazolium bromide; Sigma) to each well containing cells (from a stock 5mg/mL MTT dissolved in 1X PBS and filter-sterilized) after 5 h to assess the cell death.
8. Incubate the plate at 37°C for 5 h. Remove media with needle and syringe. Add 200 μL of DMSO to each well and pipe up and down to dissolve crystals. Transfer to plate reader and measure absorbance at 550 nm.
9. Active PA was found in both the supernatant and homogenate fractions. However, maximum macrophage lysis activity was noticed in supernatant when extraction buffer was used with CHAPS detergent.

3.7.2.3 ORAL DELIVERY OF VACCINES AND SELECTION OF TRANSGENIC PLANTS WITHOUT THE USE OF ANTIBIOTIC SELECTABLE MARKERS

Betaine aldehyde dehydrogenase (*BADH*) gene from spinach has been used as a selectable marker to transform the chloroplast genome of tobacco (*26*). Transgenic plants were selected on media containing betaine aldehyde (BA). Transgenic chloroplasts carrying BADH activity convert toxic BA to the beneficial glycine betaine (GB). Tobacco leaves bombarded with a construct containing both *aad*A and *BADH* genes showed very dramatic differences in the efficiency of shoot regeneration (**Fig. 5A-C**). Transformation and regeneration was 25% more efficient with BA selection, and plant propagation was more rapid on BA in comparison to spectinomycin. Chloroplast transgenic plants showed 15- to 18-fold higher BADH activity at different developmental stages than untransformed controls. Expression of high BADH level and resultant accumulation of glycine betaine did not result in any pleiotropic effects, and transgenic plants were morphologically normal and set seeds as untransformed control plants.

3.7.3. Production of Human Therapeutic Proteins in Transgenic Chloroplasts

3.7.3.1. HUMAN SERUM ALBUMIN (HSA) PROTEIN

Human serum albumin (HSA) accounts for 60% of the total protein in blood and widely used in a number of human therapies. Chloroplast transgenic plants were generated expressing HSA (*13*). Levels of HSA expression in chloroplast transgenic plants

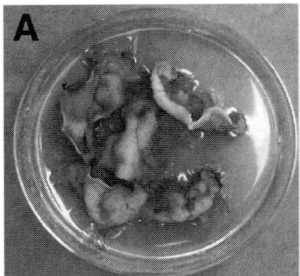
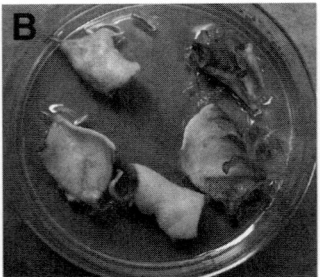

Fig. 5. Nontransgenic control (**A**) and transgenic explants (**B–C**) from *Nicotiana tabacum* cv. Petit Havana transformed with chloroplast vector containing *aadA* and *badh* genes were selected on RMOP medium supplemented with 500 mg/L spectinomycin (**B**) and 15 m*M* betaine aldehyde as a nonantibiotic selection (**C**).

was achieved up to 11.1% tsp. Formation of HSA inclusion bodies (**Fig. 6 A–C**) within transgenic chloroplasts was advantageous for purification of protein. Inclusion bodies were precipitated by centrifugation and separated easily from the majority of cellular proteins present in the soluble fraction with a single centrifugation step. Purification of inclusion bodies by centrifugation may eliminate the need for expensive affinity columns or chromatographic techniques.

3.7.3.2. Purification of HSA

1. Solubilize the HSA inclusion bodies from transformed tissues using extraction buffer.
2. Spin at 10,000*g*. Suspend the pellet in suspension buffer.
3. Dilute plant extract 100-fold in dilution buffer.
4. Concentrate HSA protein by precipitation using a polyethyleneglycol treatment at 37%.
5. Separate protein fractions by running a SDS-PAGE gel and stain gel with silver reagent following vendor's instruction (Bio-Rad).

3.7.3.3. Electron Microscopy and Immunogold Labeling

1. Cut the transformed and untransformed leaf into 1–3-mm squares.
2. Fix them in 0.1 *M* cacodylate buffer, pH 7.4, for 15 min under vacuum and 12 h at 4°C.
3. Rinse samples twice in 0.1 *M* cacodylate buffer (pH 7.4) after fixation.
4. Dehydrate fixed samples through a graded ethanol series to 95%, then implant in LRW resin at 60°C for 24 h.
5. Cut ultra-thin sections using a Leica Ultracut T ultramicrotome and collect sections onto nickel grids.
6. Incubate sections in 0.05 *M* glycine prepared in PBS buffer for 15 min to inactivate residual aldehyde groups.
7. Place grids onto drops of blocking solution (PBS containing 2% nonfat dry milk) and incubate for 30 min.
8. Incubate sections for 1 h in a goat anti-human albumin polyclonal antibody (dilution range from 1:1000 to 1:10,000 in blocking solution).

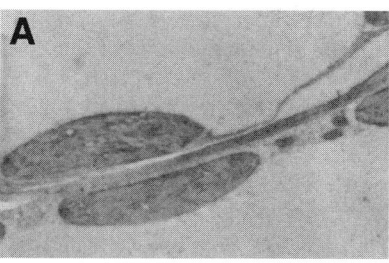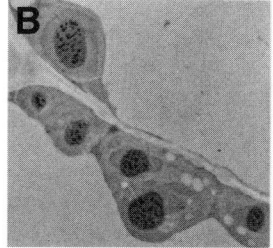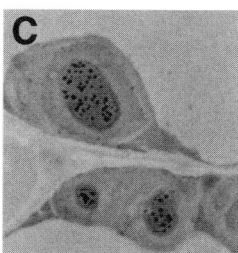

Fig. 6. HSA accumulation as inclusion bodies in transgenic chloroplasts. (**A–C**) Electron micrographs of immunogold labelled tissues from untransformed (**A**) and transformed mature leaves with the chloroplast vector pLDApsbAHSA (**B–C**). Magnifications are A ×10,000; B ×5000; C ×6300.

9. Wash sections with blocking solution 6 × 5 min each.
10. Incubate sections for 2 h with a rabbit anti-goat IgG secondary antibody conjugate to 10 nm gold diluted 1:40 in blocking solution.
11. Wash sections 6 × 5 min in blocking solution and 3 × 5 min with PBS, and fix sections in 2% glutaraldehyde diluted in PBS for 5 min.
12. Wash fixed sections in PBS 3 × 5 min, then in distilled water 5 × 2 min each.
13. Stain sections using uranyl acetate and lead citrate and examine samples under transmission electron microscope at 60 kV.

4. Notes

1. Gold particles suspended in 50% glycerol may be stored for several months at −20°C. Avoid refreezing and thawing spermidine stock; use once after thawing and discard the remaining solution. Use freshly prepared $CaCl_2$ solution after filter sterilization. Do not autoclave.
2. Precipitation efficiency of DNA on gold and spreading of DNA-gold particle mixture on macrocarriers is very important. For high transformation efficiency via biolistics, a thick film of gold particles should appear on the macrocarrier disks after alcohol evaporation. Scattered or poor gold precipitation reduces the transformation efficiency.
3. Generally, a 1000-bp flanking sequence region on each side of the expression cassette is adequate to facilitate stable integration of transgenes.
4. Use of the 5' untranslated region (5' UTR) and the 3' untranslated region (3' UTR) regulatory signals are necessary for higher levels of transgene expression in plastids (*13*). The expression of transgene in the plant chloroplast depends on a functional promoter, stable mRNA, efficient ribosomal binding sites; efficient translation is determined by the 5' and 3' untranslated regions (UTR). Chloroplast transformation elements *Prrn, 5'UT*R, *3'UTR* can be amplified from tobacco chloroplast genome.
5. Bombarded leaves after two days' dark incubation should be excised in small square pieces (5–7 mm) for first round of selection and regenerated transgenic shoots should be excised into small square pieces (2–4 mm) for a second round of selection.
6. Temperature for plant growth chamber should be around 26–28°C for appropriate growth of tobacco, potato, and tomato tissue cultures. Initial transgenic shoot induction in potato and tomato require diffuse light. However, higher intensity is not harmful for tobacco.
7. Transformation efficiency is very poor for both potato and tomato cultivars compared to tobacco.

8. Tobacco chloroplast vector gives low frequency of transformation if used for other plant species. For example, when petunia chloroplast flanking sequences were used to transform the tobacco chloroplast genome (*11*), it resulted in very low transformation efficiency.

9. Under diffuse light conditions, highly regenerating tomato cultivar (Microtom) shoots produce premature flowering that inhibit further growth of transgenic plants. Therefore, after the first shoot induction phase, shoots should be moved to 23 hr light conditions.

Acknowledgments

Investigations reported in this article was supported in part by funding from NIH R 01 GM63879, USDA 3611-21000-017-00D and Chlorogen Inc. to H.D.

References

1. DeCosa, B., Moar, W., Lee, S. B., Miller, M., and Daniell, H. (2001) Overexpression of the *Bt* Cry2Aa2 operon in chloroplasts leads to formation of insecticidal crystals. *Nat. Biotechnol.* **19**, 71–74.

2. Daniell, H., Khan, M. S., and Alison, L. (2002) Milestones in chloroplast genetic engineering: an environmentally friendly era in biotechnology. *Trends in Plant Sci.* **7**, 84–91.

3. Daniell, H. (2002) Molecular strategies for gene containment in transgenic crops. *Nature Biotechnol.* **20**, 581–586.

4. Daniell, H. and Parkinson, C. L. (2003) Jumping genes and containment. *Nature Biotechnol.* **21**, 374–375.

5. Daniell, H. and Dhingra, A. (2002) Multigene engineering: Dawn of an exciting new era in biotechnology. *Curr. Opin. Biotechnol.* **13**, 136–141.

6. Lee, S. B., Kwon, H. B., Kwon S. J., Park, S. C., Jeong, M. J., Han, S. E., et al., (2003) Accumulation of trehalose within transgenic chloroplasts confers drought tolerance. *Mol. Breeding* **11**, 1–13.

7. Daniell, H., Lee, S. B., Panchal, T., and Wiebe, P. O. (2001) Expression of cholera toxin B subunit gene and assembly as functional oligomers in transgenic tobacco chloroplasts. *J. Mol. Biol.* **311**, 1001–1009.

8. Daniell, H., Datta, R., Varma, S., Gray, S., and Lee, S. B. (1998) Containment of herbicide resistance through genetic engineering of the chloroplast genome. *Nat. Biotechnol.* **16**, 345–348.

9. Kota, M., Daniell, H., Varma, S., Garczynski, S. F., Gould, F., and William, M. J. (1999) Overexpression of the *Bacillus thuringiensis* (Bt) Cry2Aa2 protein in chloroplasts confers resistance to plants against susceptible and *Bt*-resistant insects. *Proc. Natl. Acad. Sci. USA* **96**, 1840–1845.

10. McBride, K. E., Svab, Z., Schaaf, D. J., Hogan, P. S., Stalker, D. M., and Maliga, P. (1995) Amplification of a chimeric Bacillus gene in chloroplasts leads to an extraordinary level of an insecticidal protein in tobacco. *Bio/Technology* **13**, 362–365.

11. DeGray, G., Kanniah, R., Franzine, S., John, S., and Daniell, H. (2001) Expression of an antimicrobial peptide via the chloroplast genome to control phytopathogenic bacteria and fungi. *Plant Physiol.* **127**, 852–862.

12. Ruiz, O. N., Hussein, H., Terry, N., and Daniell, H. (2003) Phytoremediation of organomercurial compounds via chloroplast genetic engineering. *Plant Physiol.* **132**, 1344–1352.

13. Fernandez-San Millan, A., Mingo-Castel, A., and Daniell, H. (2003) A chloroplast transgenic approach to hyper-express and purify human serum albumin, a protein highly susceptible to proteolytic degradation. *Plant Biotechnol. J.* **1**, 71–79.

14. Leelavathi, S., Gupta, N., Maiti, S., Ghosh, A. and Reddy, V. S. (2003) Overproduction of an alkali-and thermo-stable xylanase in tobacco chloroplasts and efficient recovery of the enzyme. *Mol. Breeding* **11**, 59–67.
15. Guda, C., Lee, S. B., and Daniell, H. (2000) Stable expression of biodegradable protein based polymer in tobacco chloroplasts. *Plant Cell Rep.* **19**, 257–262.
16. Staub, J. M., Garcia, B., Graves, J., Hajdukiewicz, P. T. J., Hunter, P., Nehra, N., et al. (2000) High-yield production of a human therapeutic protein in tobacco chloroplasts. *Nat. Biotechnol.* **18**, 333–338.
17. Daniell, H., Streatfield, S. J., and Wycoff, K. (2001) Medical molecular farming: production of antibodies, biopharmaceuticals and edible vaccines in plants. *Trends Plant Sci.* **6**, 219–226.
18. Sanford, J. C., Smith, F. D., and Russel, J. A. (1993) Optimizing the biolistic process for different biological applications. *Methods Enzymol.* **217**, 483–509.
19. Daniell, H. (1993) Foreign gene expression in chloroplasts of higher plants mediated by tungsten particle bombardment. *Methods Enzymol.* **217**, 536–556.
20. Sikadar, S. R., Serino, G., Chaudhuri, S., and Maliga, P. (1998) Plastid transformation in *Arabidopsis thaliana*. *Plant Cell. Rep.* **18**, 20–24.
21. Sidorov, V. A., Kasten, D., Pang, S. Z., Hajdukiewicz, P. T., Staub, J. M., and Nehra, N. S. (1999) Technical advance: stable chloroplast transformation in potato: use of green fluorescent protein as a plastid marker. *Plant J.* **19**, 209–216.
22. Ruf, S., Hermann, M., Berger, I., Carrer, H., and Bock, R. (2001) Stable genetic transformation of tomato plastids and expression of a foreign protein in fruit. *Nat. Biotechnol.* **19**, 870–875.
23. Daniell, H., Krishnan, M., and McFadden, B. A. (1991) Expression of β-glucuronidase gene in different cellular compartments following biolistic delivery of foreign DNA into wheat leaves and calli. *Plant Cell Rep.* **9**, 615–619.
24. Hibberd, J. M., Philip, J. L., Khan, M. S., and Gray, J. C. (1998) Transient expression of green fluorescent protein in various plastid types following microprojectile bombardment. *Plant J.* **16**, 627–632.
25. Daniell, H. (1997) Transformation and foreign gene expression in plants mediated by microprojectile bombardment. *Methods in Mol. Biol. Recombinant Gene Expression Protocols* **62**, 463–489.
26. Daniell, H., Muthukumar, B., and Lee, S. B. (2001) Marker free transgenic plants: engineering the chloroplast genome without the use of antibiotic selection. *Curr. Genet.* **39**, 109–116.
27. Daniell, H., Watson, J., Koya, V., and Leppla, S. H. (2004) Expression of *Barillus anthracis* protective antigen in trangenic chloroplasts of tobacco, a non-food/feed crop. *Vaccine,* in press.

26

New Selection Marker for Plant Transformation

Barbara Leyman, Nelson Avonce, Matthew Ramon, Patrick Van Dijck, Johan M. Thevelein, and Gabriel Iturriaga

Summary

A number of systems to insert foreign DNA into a plant genome have been developed so far. However, only a small percentage of transgenic plants are obtained using any of these methods. Stable transgenic plants are selected by co-introduction of a selectable marker gene, which in most cases are genes that confer resistance against antibiotics or herbicides. In this chapter we describe a new method for selection of transgenic plants after transformation. The selection agent used is the nontoxic and common sugar glucose. Wild-type *Arabidopsis thaliana* plantlets that have been germinated on glucose have small white cotyledons and remain petite because the external sugar switches off the photosynthetic mechanism. The selectable marker gene encodes the essential trehalose-6-phophate synthase, AtTPS1, that catalyzes the first reaction of the two-step trehalose synthesis. Upon ectopic expression of *AtTPS1* driven by the 35S promoter, transformed *Arabidopsis thaliana* plants became insensitive to glucose in comparison to wild-type plants. After transformation using *AtTPS1* as a selection marker and 6% glucose as selection agent it is possible to single out the green and normal sized transgenic plants amid the nontransformed plantlets.

Key Words: Transformation; glucose; selection marker; selection agent; *Arabidopsis thaliana*; flower dip; trehalose; trehalose-6-phosphate; AtTPS1.

1. Introduction

A number of techniques for inserting exogenous DNA into a plant genome have been developed so far. However, only a small percentage of transgenic cells are obtained using any of these methods. Therefore, an effective selection system is needed to pick out the rare transgenic plants from a pool of nontransformed plants.

From: *Methods in Molecular Biology, vol. 267:*
Recombinant Gene Expression: Reviews and Protocols, Second Edition
Edited by: P. Balbás and A. Lorence © Humana Press Inc., Totowa, NJ

1.1. Antibiotic and Herbicide Resistance

Originally, selection systems were based exclusively on antibiotic or herbicide resistance acquired by the transgenic plants. The most common selectable marker gene used until now is the *NptII* gene, conferring resistance against kanamycin. Other antibiotics used as selection agents are hygromycin, gentamycin, and the less common bleomycin, methotrexate, spectinomycin, and sulfonamides. Phosphinothricin is a herbicide often used as a selection agent and it can be neutralized by the BASTA gene product (*1*). The negative aspects of these selectable marker genes are clear-cut. There is a risk of creating resistant bugs and weeds against the toxic agents by cross-pollination or horizontal gene transfer in the rhizosphere. Even though these selection systems are very rigorous, the use of toxic compounds and their corresponding resistance genes should be avoided when making new genetically modified crops (*2*).

1.2. Selectable Marker Genes From Microorganisms

As an alternative for the selection systems based on toxic agents, various positive selection protocols were developed using a characteristic from another organism and introducing it into plants. One such trait exploited the metabolism of the glucose derivative 2-deoxyglucose (2-DOG). 2-DOG is phosphorylated in the cytosol of plant cells by endogenous hexokinase yielding 2-deoxyglucose-6-phosphate (2-DOG-6-P), which is toxic for the plant. Constitutive expression of the DOG^R1 gene from *Saccharomyces cerevisiae* encoding a 2-DOG-6-P phosphatase resulted in resistance against 2-DOG in transgenic tobacco plants (*3*).

Sugars such as mannose and xylose that cannot be metabolized by wild-type plants were also used as selection agents. The complementary selectable marker genes encoded the missing metabolic enzymes, giving the transgenic cells an advantage over the starved nontransformed cells. Mannose, like 2-DOG, can be phosphorylated by hexokinase, resulting in the accumulation of the nonmetabolizable mannose-6-phosphate (*4*). The phosphomannose isomerase (PMI) from *E. coli* converts mannose-6-phosphate to the metabolizable fructose-6-phosphate. Constitutive expression of *pmi* resulted in transgenic plants that have a selective advantage on medium with mannose as the sole carbon source. *pmi* was used as a selectable marker gene in sugar beet, maize, and rice (*5–7*). Based on the same principle, the xylose isomerase gene (*xylA*) from *Thermoanaerobacterium thermosulfurogenes* was used as a marker gene. In contrast to wild-type cells, *xylA* transgenic cells of tobacco, tomato, and potato were able to proliferate on D-xylose as a carbohydrate source (*8*). Nevertheless, the marker genes described above are of nonplant origin, which still has been regarded as a potential biosafety concern.

1.3. Removal of Selectable Marker Genes

Another way to resolve the issue of the selectable marker gene is to remove it after selection. After all, this gene becomes redundant once a transgenic plant has been generated and isolated. Several methods have been described in the literature so far to achieve this concept.

1.3.1. Temporal Control of the Inducible Marker Genes

Various chemically inducible promoters could be used to control the expression of the selectable marker gene (*9*). An interesting example is the temporary overexpression of the isopentenyltransferase (*ipt*) gene from the Ti plasmid of *Agrobacterium tumefaciens*. The cytokinin biosynthesis *ipt* gene controlled by the dexamethasone (Dex)-inducible promoter was transformed to tobacco. Under inductive conditions, cells transformed with the *ipt* gene produced elevated cytokinin levels, stimulating the regeneration of transgenic shoots from plant calli or explants (*10*).

1.3.2. Segregation of the Gene of Interest and Selectable Marker Gene

Co-transformation of a single plant genome with two independent T-DNA regions provides the opportunity for genetic separation in subsequent generations (*11,12*). Under certain conditions, integration of different T-DNAs occurred to unlinked sites at reasonably high frequencies. In the progeny of the transgenic plant, the selectable marker gene was eliminated by segregation of the two T-DNAs, resulting in the marker-free insertion of a particular transgene. A similar segregation can be obtained after relocation of the marker gene away from the gene of interest with the help of the Ac transposon system (*13,14*).

1.3.3. Excision of the Selectable Marker Gene

Different approaches have been described to excise the selectable marker gene (*15*). One of these transformation systems consists of the site-specific recombination system Cre*lox*. The marker gene is aimed between the two directly repeated recognition sites (*lox*) of the recombinase (Cre). Subsequent expression of the Cre recombinase results in the recombination between the flanking *lox* sites, thus deleting the marker gene (*16,17*).

The Cre*lox* excision system has also been used in combination with a chemically regulated promoter to control the expression of the Cre recombinase (*18*). In this way, application of the inducer β-estradiol initiated excision. From a different perspective, excision of a selectable marker from the plastid genome also works effectively using the Cre*lox* system (*19,20*).

A similar but improved method based on intrachromosomal recombination between bacteriophage λ attachment (attP) sequences has been developed to remove the selectable marker genes from tobacco transgenes (*21*). As the attP system does not require the expression of helper proteins, it is a useful tool to remove undesired transgene regions. Excision of selectable marker genes by recombination using the above-mentioned methods can also be obtained after transformation of plastids (*22*).

1.4. Why a Different Selection System?

Concern about the use of herbicide and antibiotic selectable marker genes has been raised. Wild weeds and bugs might become resistant through contamination by cross-pollination or horizontal gene transfer, thereby upsetting the balance of the ecosystem. Using microorganism-derived selection may raise the ethical question about inserting genes from a different species into plants. If possible, it is preferable to avoid insertion of a microorganism-derived selection marker to limit this problem to some extent. In an attempt to make a marker-free transgenic plant, alternative selection procedures have been developed to dispose of the selectable marker gene. However, in many of these procedures, the selectable marker gene is either still present, as is the case for temporal control of the gene, or could leave a mark. In many cases excision or translocation leaves a scar in the form of a deletion, especially when recombination takes place in the plastid genome (*19*). Co-transformation is a cleaner method, but segregation is very labor-intensive and time-consuming, and becomes impractical for some plants, such as fruit trees, where regeneration takes years.

Therefore we developed a novel selection system based on a plant-derived selectable marker gene, which is widespread in the plant kingdom and uses a nontoxic, inexpensive selection agent. This way there is no risk to human health and environment and hence no excision or recombination is needed.

One other system in literature is described that meets these criteria. The chloroplast *BADH* from spinach encodes a betaine aldehyde dehydrogenase that converts the toxic betaine aldehyde to the nontoxic glycine betaine. Tobacco chloroplast transformation with *BADH* as a selectable marker gene results in transgenic plants with an increased BADH activity, providing them with an advanced shoot generation on betaine aldehyde in comparison with nontransgenic cells (*23*).

1.5. High Expression of AtTPS1 Decreases Glucose Sensitivity

Trehalose is a nonreducing disaccharide that is very common in bacteria and fungi, but rarely found in plants. Only small amounts could be detected in *Arabidopsis thaliana* (*24*). In addition to its role as storage carbohydrate, trehalose levels have been shown to correlate very well with cellular stress resistance in many cell types and conditions. In the yeast *Saccharomyces cerevisiae*, trehalose is synthesized in two reactions from UDPglucose and glucose-6-phosphate by trehalose-6-phosphate (Tre6P) synthase (Tps1) and Tre6P phosphatase (Tps2). Deletion of *tps1* eliminates growth on glucose because both Tps1 and Tre6P are indispensable for the regulation of glucose influx into glycolysis. The two activities reside in a large complex together with a regulatory subunit redundantly encoded by the *TSL1* and *TPS3* genes. Following the publication of the complete *A. thaliana* genome sequence, we have analyzed the TIGR database with multiple BLAST searches and found 11 *TPS1* homologs (*25*). Interestingly, they can be grouped in two subfamilies, displaying most similarity either to yeast *TPS1* or *TPS2*. No closely related homologs to the yeast *TSL1–TPS3* gene pair have been uncovered. Overexpressing *AtTPS1* complements the yeast *tps1Δ* strain for its growth defect on glucose, which implies an important regulatory role for *AtTPS1* in plant carbon metabolism (*26,27*). In plants, AtTPS1 appears to be essential in the regulation of sugar metabolism during embryo development (*28*). Sugar homeostasis is tightly regulated in plants. Increased levels of sugar in the growth medium causes stunted growth of roots and hypocotyl and inhibition of greening combined with accumulation of anthocyanin in cotyledons of seedlings (*29*). AtTPS1 plays an important regulatory role in this mechanism because ectopic overexpression of *AtTPS1* in *Arabidopsis thaliana* renders those plants significantly less sensitive to glucose than wild-type plants (**Fig. 1**). It is this trait that we want to use in a novel selection method, because trehalose-6-phosphate synthase is widespread in the plant kingdom. We have also indications that the effect of AtTPS1 might be universal as overexpression of *AtTPS1* in *Nicotiana tabacum* has a similar sugar insensitive phenotype as in *Arabidopsis* plants.

2. Materials

1. *Arabidopsis thaliana* ecotype Columbia or Landsberg erecta.
2. Square pots and trays, soil (osmocote, asef including all necessary nutrients).
3. *Agrobacterium* strain C58C1.
4. pTPSM construct with gene of interest cloned into the *Nhe*I site.
5. Sterile LB medium.
6. Transformation solution: 10% sucrose + 0.05% Silwet-77 (Lehle Seeds, Round Rock, TX) freshly made.
7. Transparent aquarium or plastic bag and paper bag or breathable plastic bag.
8. Desiccation jar.
9. Sterilization solution: 100 mL commercial bleach + 3 mL HCl 37% (added in fume hood just before use).
10. Sterile water.
11. MS medium: 1X MS salts and vitamins (Invitrogen, Carlsbad, CA), 0.5 g/L MES pH 5.7 (KOH), 8 g/L phytagar (Invitrogen) and 6% glucose (*see* **Note 1**).
12. Round sterile Petri dishes with 14-cm diameter.

3. Methods

3.1. Construction of the Selection Vector

The new transformation vector pTPSM plasmid with *AtTPS1* as the sole marker gene is a pBin19 derivative (**Fig. 2B**). It is created from the p35S::AtTPS1-NOS construct by deleting the *NptII* gene between the *Nhe*I and *Apa*I restriction sites (**Fig. 2**). Self-ligation of the plasmid without the *NptII* gene resulted in the pTPSM construct containing only the *35S::AtTPS1-NOS* gene between the left and right borders. Outside the T-DNA region, the plasmid contains a prokaryotic origin of replication and the *NptI* gene conferring kanamycin resistance in bacteria (*see* **Note 2**).

3.2. Agrobacterium-*Mediated Transformation of* Arabidopsis

To transform *Arabidopsis* plants, we used the "flower dip" method from Clough and Bent (*30*) with some modifications.

3.2.1. Sowing and Growing of the Parental Plants

1. Resuspend WT *Arabidopsis thaliana* Columbia or Landsberg Erecta seeds in H_2O (*see* **Note 3**).
2. Stratification: Incubate seeds for 2–3 d at 4°C under constant light.
3. Prepare square pots with soil (osmocote, asef includes all necessary nutrients) and soak extensively with water.
4. Dispense the seeds, five per pot (*see* **Note 3**).

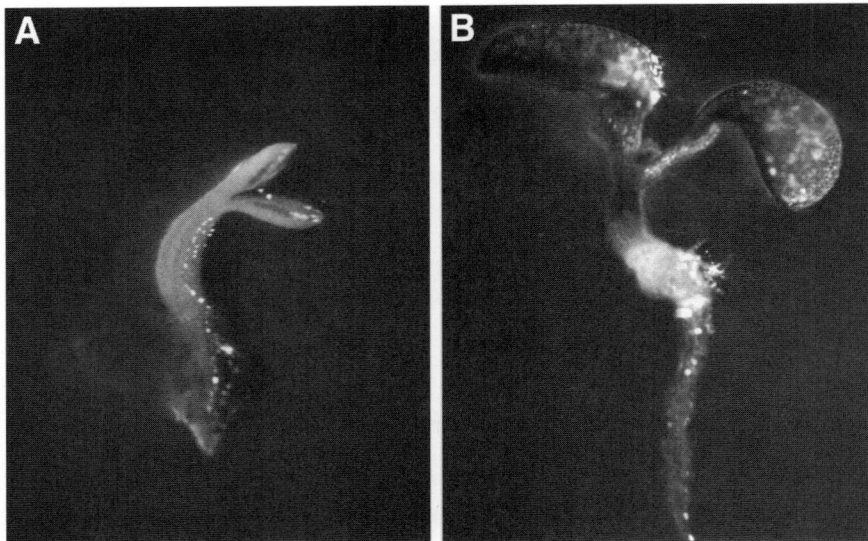

Fig. 1. WT (**A**) and *35S::AtTPS1-NOS* (**B**) seedlings grown on Murashige and Skoog medium containing 6% glucose. Photos were taken 6 d after germination.

5. Growth room conditions: temperatures 22°C/day and 18°C/night; humidity 65%–70%; photoperiod to 12 h/12 h.
6. Germinate plants for 2 wk under high-humidity conditions by covering the trays with plastic wrap.
7. Remove covers and irrigate the pots with water every 2–3 d; check that the soil does not become too flooded to avoid growth of algae.
8. Grow plants until inflorescence appears (*see* **Note 4**).
9. Clip inflorescence close to the base (*see* **Note 5**).
10. After 4–5 d plants have inflorescence of about 10–15 cm long and are ready for transformation (*see* **Note 6**).

3.2.2. Agrobacterium *Growth Conditions and Preparations*

1. Start an *Agrobacterium* culture containing your construct in the morning in a 50-mL tube containing 1 mL LB without antibiotics (*see* **Note 7**).
2. Incubate while shaking for 8–9 h at 28°C.
3. Add 10 mL LB without antibiotic.
4. Incubate overnight while shaking at 28°C.
5. Check OD_{600} in the morning; it should range between 0.5 and 2.0.
6. Add 40 mL of MQ water containing sucrose (10%, 4 g) and Silwet-77 (0.05%; 20 µL).
7. Use immediately.

3.2.3. *Flower Dip*

1. Invert the inflorescences in the *Agrobacterium* culture and agitate gently for 2–3 s (*see* **Note 8**).
2. Use two pots for each transformation.

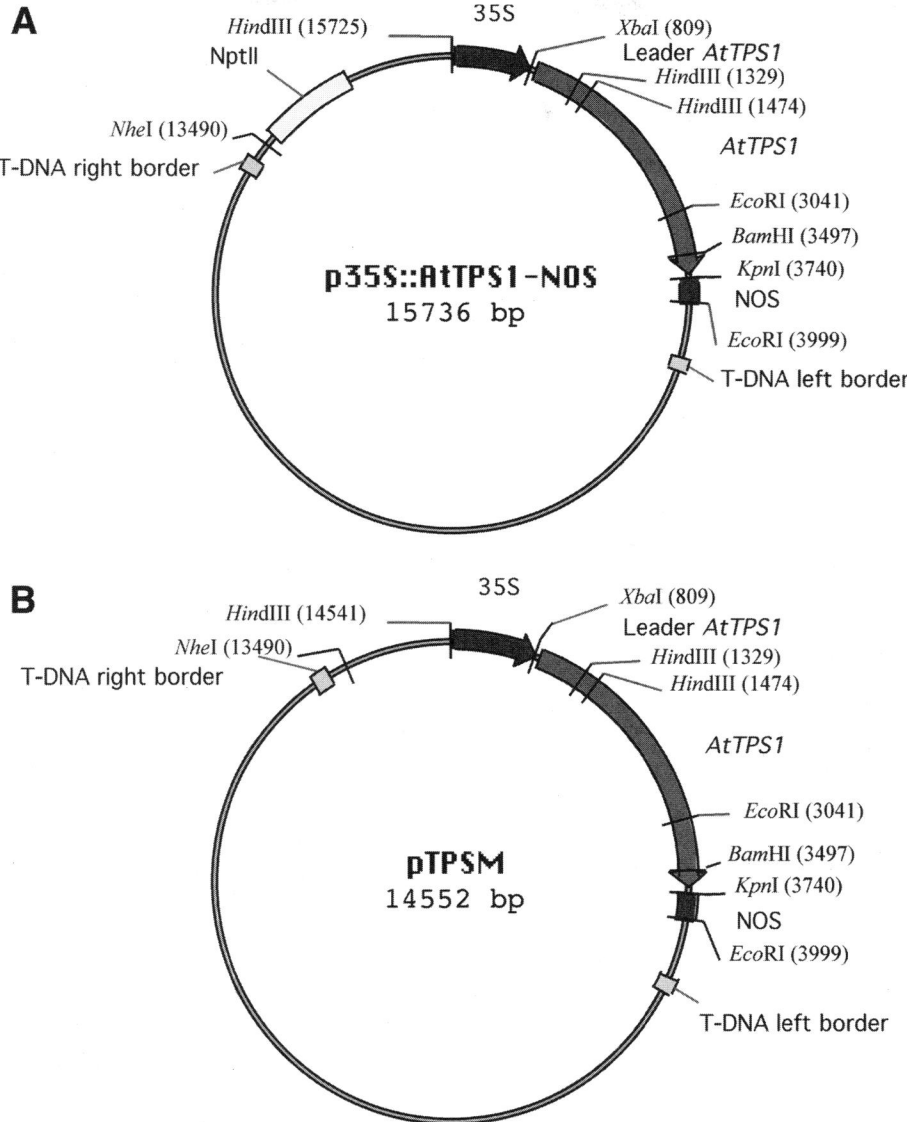

Fig. 2. Map of the p35S::AtTPS1-NOS (**A**) and pTPSM (**B**) vectors.

3. Return the plants to normal growing conditions but cover the plants with a transparent aquarium or plastic bag, so that the humidity is close to 100% (*see* **Note 9**).

4. After 24–48 h remove cover and keep watering plants until siliques are maturing (*see* **Note 10**).

5. Make sure no cross-pollination occurs, by surrounding the plants with a paper bag.

6. Five to six weeks after transformation, stop watering the plants when siliques are getting brown.

7. Harvest seeds when siliques are completely dry and collect the seeds in water-absorbent material, such as paper envelopes, and keep at room temperature.
8. After 2 wk, separate seeds from other plant material and store seeds at 4°C in 1.5-mL microfuge tubes.

3.3. Selection on Glucose

1. Sterilize the seeds by vapor sterilization for 3–5 h (*see* **Note 11**).
2. Obtain a vessel for seed sterilization, typically a dessicator jar. Place in fume hood. Place seeds that are to be sterilized into microcentrifuge tubes. Place open containers of seeds into a rack or stand inside the dessicator jar. Place a 250-mL beaker containing 100 mL bleach into the dessicator jar. Immediately prior to sealing the jar, carefully add 3 mL HCl 37% to the bleach.
3. Seal jar and allow sterilization by chlorine fumes to proceed for a period of 3–16 h. The time needed will vary based on the configuration of the seeds and the extent to which the seeds are contaminated. Three to four hours is often sufficient to obtain reasonably clean seeds. Overnight is usually acceptable, although some seed-killing may occur, especially if the seeds are not fully mature and dry.
4. Open container in fume hood and seal microfuge tubes; surface-sterilized seeds are ready for use.
5. Add sterile water to seeds and incubate at 4°C for 48 h under constant light.
6. Prepare MS medium without sucrose (1X Murashige and Skoog nutrients plus vitamins (Invitrogen), 0.5 g/L MES, pH 5.7 (KOH), 8 g/L phytagar (Invitrogen) with 6% glucose (*see* **Note 1**). Pour round plates with a diameter of 14 cm.
7. Disperse seeds on the plates in a concentration of 1000 seeds for each plate (*see* **Note 12**).
8. Seal the plates with parafilm and incubate them in 24 h light at 22°C.
9. After 6–7 d some transformed seedlings appear green and larger than wild-type seedlings and can be selected (*see* **Notes 13–15**).

3.4. Selection

Figure 3 displays an example of *Arabidopsis* plantlets transformed with the empty pTPSM vector performed using the flower dip method as described above. T0 seeds were plated on 6% glucose containing MS medium for 6 d in 24-h light before selection (**Fig. 3**). Seedlings indicated with an arrow were selected. Selected seedlings were grown further in soil and PCR analysis confirmed the presence of the *35S::AtTPS1-NOS* construct (**Fig. 3C**). Primers were designed against the 35S promoter and the *AtTPS1* gene. The expected amplicon should be 2.1 kb long in the transgenic lines.

4. Notes

1. When preparing glucose-containing medium, it is important to filter sterilize glucose separately from the rest of the MS medium and mix them prior to pouring the plates.
2. So far, there is only one unique restriction site (*Nhe*I) in the T-DNA sequence to clone a gene of interest. Currently we are synthesizing a more user-friendly vector by inserting a Gateway™ recombination cassette.
3. We start with five seeds per pot (9 × 9 cm). After two weeks the two smallest plantlets are discarded. Three plants per pot are ideal for the rosette to have enough space to grow vigorously.
4. Inflorescences usually appear 5–6 wk after sowing. This is longer than usual but in this way plants grow vigorously, which enhances transformation efficiency.

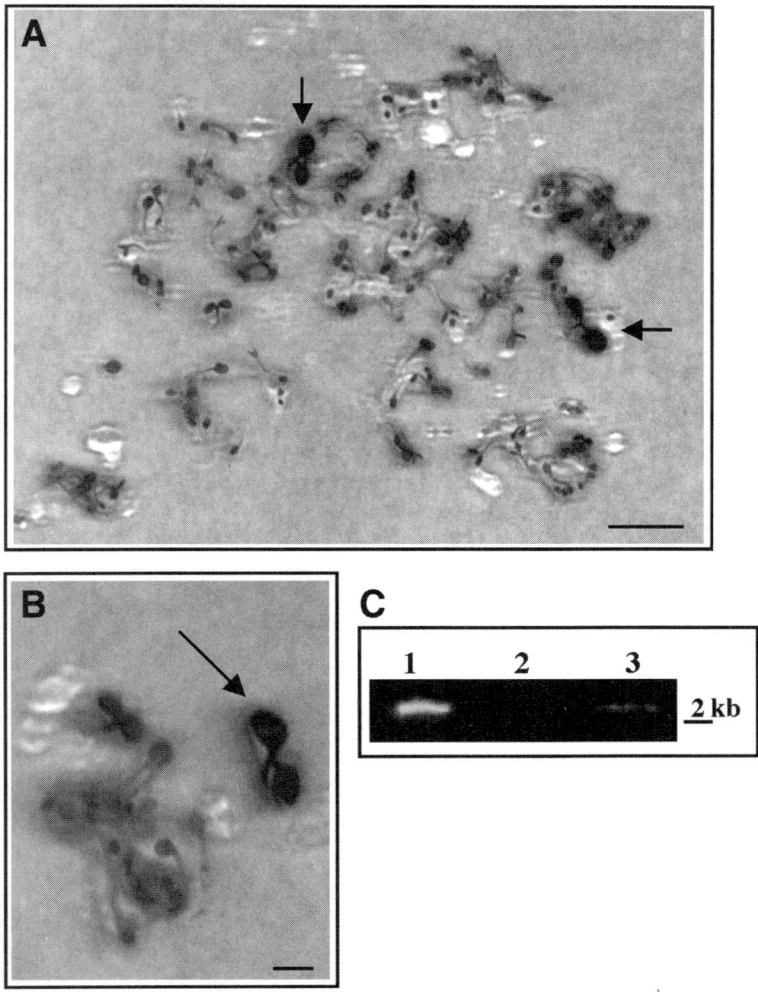

Fig. 3. (**A, B**) Selection of *35S::AtTPS1-NOS* transformed T0 seeds on MS medium containing 6% glucose. The picture was taken 6 d after sowing. Arrow indicates the selected seedlings. Bar = 0.5 cm. (**C**) PCR prducts after amplification of genomic plant DNA with forward primer in the 35S promoter: AAGAAGACGTTCCAACCACG and the reverse primer in the *AtTPS1* gene: CGCTCAGAACAACTATGGTT. Expected PCR product in transgenic lines measures 2.1 kb long. Lanes: 1. homozygous *35S::AtTPS1-NOS* plant; 2. wild type; 3. Selected *35S::AtTPS1-NOS* after transformation.

5. Clipping the inflorescence is not absolutely necessary and one can transform at this stage; however, clipping increases the number of flower buds and it also synchronizes flowering of the different plants.

6. At this stage, there should be many green flower buds and only a few siliques to get the highest efficiency during transformation.

7. It is important NOT to add the antibiotics at this stage because this might be detrimental for the plant during the flower dip process. Overnight growth of *Agrobacterium* without selective pressure does not result in severe vector loss.

8. There are two ways of inverting the flowers into the *Agrobacterium* suspension. The first one is to invert the whole pot; because the rosettes of the plants have grown big, there is no problem with loss of soil. However, we found it more convenient to bend over the inflorescence slightly into the *Agrobacterium* solution. Some very short inflorescences that are hard to reach can be transformed by dripping a small amount of the *Agrobacterium* solution onto the flower buds. But be careful NOT to overdo this, as the plants will die if too much *Agrobacterium* suspension is used. For the same reason, do not leave the flowers longer than a few (2–3) seconds in the suspension. There are high levels of Silwet in the solution that can be lethal when used too abundantly during transformation.

9. Humidity under the aquarium cover can be increased by spraying water into the air.

10. For higher rates of transformation, plants may be dipped two or three times at 7-d intervals. Clough and Bent (*30*) suggest one dip 2 d after clipping, and a second dip 1 wk later. Do not dip less than 6 d apart.

11. This protocol can be found on the web site of Clough and Bent's lab in Illinois: http://plant-path.wisc.edu/afb/vapster.html.

12. It is extremely important here that the seeds are all in contact with the glucose-containing medium. Therefore, the seeds must be well spread over the plate. Ideally water should be used for sowing and not top-agar or other agar, as we noticed that this may conceal the glucose phenotype normally seen in wild-type seeds. However if you are having problems with spreading the seeds properly in water, a sterile Whatman 3 MM filter paper can be used to keep the seeds spread in their position.

13. Some plants become insensitive to glucose, possibly due to a cosupressor effect, and hence the efficiency of the selection is not as rigorous as when, for instance, kanamycin is used. However, in most cases more than 50% of the selected plants turned out to be transgenic plants containing the *35S::AtTPS1-NOS* construct in their genome.

14. Because an endogenous *AtTPS1* gene is present in the *Arabidopsis* genome, it is possible to get co-suppression of the *AtTPS1* gene. In this case, the plants will not be selected, which decreases the transformation efficiency to some extent.

15. The *AtTPS1*-glucose selection procedure has been tested only in *Arabidopsis* so far. Evidently, we will assess these results for selection after transformation of other, economically more important crops. Glucose sensitivity of shoot formation from explants is currently thoroughly tested.

Acknowledgments

The authors thank Martine De Jonge and Miranda Van Meensel for their technical help. This work was supported by the Flemish Interuniversity Institute of Biotechnology (VIB). B.L. is a postdoctoral fellow of the Fund for Scientific Research (Flanders, FWO). M.R. is indebted to the Vlaams Instituut voor de Bevordering van het Wetenschappelijk-Technologisch Onderzoek in de Industrie (IWT) for a predoctoral fellowship. N.A. and G.I. were supported by a studentship and a sabbatical fellowship respectively from CONACYT (Mexico).

References

1. Roger, H., Philip, M., and Harry, K. (2000) A guide to *Agrobacterium* binary Ti vectors. *Trends Plant Sci.* **5**, 446–451.
2. WHO (1993) Health aspects of marker genes in genetically modified plants. Report of a WHO workshop of the food safety unit.
3. Kunze, I., Ebneth, M., Heim, U., Geiger, M., Sonnewald, U., and Herbers, K. (2001) 2-Deoxyglucose resistance: a novel selection marker for plant transformation. *Mol. Breeding* **7**, 221–227.
4. Malca, I., Endo, R. M., and Long, M. R. (1967) Mechanism of glucose counteraction of inhibition of root elongation by galactose, mannose, and glucosamine. *Phytopathology* **57**, 272–278.
5. Lucca, P., Ye, X., and Potrykus, I. (2001) Effective selection and regeneration of transgenic rice plants with mannose as selective agent. *Mol. Breed.* **7**, 43–49.
6. Joersbo, M., Donaldson, I., Kreiberg, J., Petersen, S. G., Brunstedt, J., and Okkels, F. T. (1998) Analysis of mannose selection used for transformation of sugar beet. *Mol. Breed.* **4**, 111–117.
7. Negrotto, D., Jolley, M., Beer, S., Wenck, A. R., and Hansen, G. (2000) The use of phosphomannose-isomerase as a selectable marker to recover transgenic maize plants (zea mays L.) via *Agrobacterium* transformation. *Plant Cell Rep.* **19**, 798–803.
8. Haldrup, A., Petersen, S. G., and Okkels, F. T. (1998) The xylose isomerase gene from *Thermoanaerobacterium thermosulfurogenes* allows effective selection of transgenic plant cells using D-xylose as the selection agent. *Plant Mol. Biol.* **37**, 287–296.
9. Zuo, J. and Chua, N.-H. (2000) Chemical-inducible systems for regulated expression of plant genes. *Curr. Opin. Biotechnol.* **11**, 146–151.
10. Kunkel, T., Niu, Q.-W., Chan, Y.-S., and Chua, N.-H. (1999) Inducible isopentenyl transferase as a high-efficiency marker for plant transformation. *Nat. Biotechnol.* **17**, 916–919.
11. McCormac, A., Fowler, M. R., Chen, D. F., and Elliott, M. C. (2001) Efficient cotransformation of *Nicotiana tabacum* by two independent T-DNAs, the effect of T-DNA size and implications for genetic separation. *Transgenic Res* **10**, 143–155.
12. DeBuck, S., Jacobs, A., Van Montagu, M., and Depicker, A. (1989) Agrobacterium tumefaciens transformation and cotransformation frequencies of Arabidopsis thaliana root explants and tobacco protoplasts. *Mol. Plant Microbe Interact.* **11**, 449–457.
13. Yoder, J. I. and Goldsbrough, A. P. (1994) Transformation systems for generating marker-free transgenic plants. *Bio/Technology* **12**, 263–267.
14. Ebinuma, H., Sugita, K., Matsunaga, E., Endo, S., and Yamada, K. (2001) Systems for the removal of a selection marker and their combination with a positive marker. *Plant Cell Rep.* **20**, 383–392.
15. Hare, P. D. and Chua, N.-H. (2002) Excision of selectable marker genes from transgenic plants. *Nature Biotechnol.* **20**, 575–580.
16. Russell, S. H., Hoopes, J. L., and Odell, J. T. (1992) Directed excision of a transgene from the plant genome. *Mol. Gen. Genet.* **234**, 49–59.
17. Dale, E. C. and Ow, D. W. (1991) Gene transfer with subsequent removal of the selection gene from the host genome. *Proc. Natl. Acad. Sci. USA* **88**, 10558–10562.
18. Zuo, J., Niu, Q. W., Moller, S. G., and Chua, N.-H. (2001) Chemical-regulated, site-specific DNA excision in transgenic plants. *Nat. Biotechnol.* **19**, 157–161.

19. Corneille, S., Lutz, K., Svab, Z., and Maliga, P. (2001) Efficient elimination of selectable marker genes from the plastid genome by the CRE-*lox* site-specific recombination system. *Plant J.* **27**, 171–178.
20. Hajdukiewicz, P. T., Gilbertson, L., and Staub, J. M. (2001) Multiple pathways for Cre/lox-mediated recombination in plastids. *Plant J.* **27**, 161–170.
21. Zubko, E., Scutt, C., and Meyer, P. (2000) Intrachromosomal recombination between attP regions as a tool to remove selectable marker genes from tobacco transgenes *Nat. Biotechnology* **18**, 442–445.
22. Iamtham, S. and Day, A. (2000) Removal of antibiotic resistance genes from transgenic tobacco plastids. *Nat. Biotechnol.* **18**, 1172–1176.
23. Daniell, H., Muthukumar, B., and Lee, S. B. (2001) Marker free transgenic plants: engineering the chloropast genome without the use of antibiotic selection. *Curr. Gen.* **39**, 109–116.
24. Vogel, G., Fiehn, O., L, L. J.-R.-d.-B., Boller, T., Wiemken, A., Aeschbacher, R. A., and Wingler, A. (2001) Trehalose metabolism in *Arabidopsis*: occurrence of trehalose and molecular cloning and characterization of trehalose-6-phosphate synthase homologues. *J. Exp. Bot.* **52**, 1817–1826.
25. Leyman, B., Van Dijck, P., and Thevelein, J. M. (2001) An unexpected plethora of trehalose biosynthesis genes in *Arabidopsis thaliana*. *Trends Plant Sci.* **6**, 510–513.
26. Blasquez, M. A., Santos, E., Flores, C. L., Martinez-Zapater, J. M., Salinas, J., and Gancedo, C. (1998) Isolation and molecular characterization of the *Arabidopsis TPS1* gene, encoding trehalose-6-phosphate synthase. *Plant J.* **13**, 685–689.
27. Van Dijck, P., Mascorro-Gallardo, J. O., Bus, M. D., Royackers, K., Iturriaga, G., and Thevelein, J. M. (2002) Truncation of *Arabidopsis thaliana* and *Selaginella lepidophylla* trehalose-6-phosphate synthase unlocks high catalytic activity and supports high trehalose levels on expression in yeast. *Biochem. J.* **15**, 63–71.
28. Eastmond, P. J., Dijken, A. J. v., Spielman, M., Kerr, A., Tissier, A. F., Dickinson, H. G., et.al., (2002) Trehalose-6-phosphate synthase 1, which catalyses the first step in trehalose synthesis, is essential for *Arabidopsis* embryo maturation. *Plant J.* **29**, 225–235.
29. Sheen, J., Zhou, L., and Jang, J. C. (1999) Sugars as signaling molecules. *Curr. Opin. Plant. Biol.* **2**, 410–418.
30. Clough, S. J. and Bent, A. F. (1998) Floral dip: a simplified method for *Agrobacterium*-mediated transformation of *Arabidopsis thaliana*. *Plant J.* **16**, 735–743.

27

Enhancer Detection and Gene Trapping as Tools for Functional Genomics in Plants

Gerardo Acosta-García,
Daphné Autran, and Jean-Philippe Vielle-Calzada

Summary

Although more than 25,000 genes of *Arabidopsis thaliana* have been sequenced and mapped, adequate expression or functional information is available for less than 15% of them (*1*). In the case of *Oryza sativa* (rice), about half of more than 55,000 predicted genes have been assigned to a vague functional category on the basis of their sequence, but fewer than 100 have been ascribed a precise, verified function after the identification of a mutant phenotype caused by the molecular disruption of the corresponding gene (*2*). Enhancer detection and gene trapping represent insertional mutagenesis strategies that report random expression of many genes and often generate loss-of-function mutations. Several trapping vectors have been designed in a limited number of species, and large-scale enhancer detection and gene trap screens that aim to generate a wide range of spatially and temporally restricted expression patterns have been initiated in both *Arabidopsis* and rice. These strategies are proving to be essential to the functional annotation of completely sequenced genomes, enabling the analysis of gene function in the context of the entire plant life cycle and substantially expanding our understanding of plant growth and development.

Key Words: Enhancer detection; gene traps; functional genomics; *Arabidopsis*; rice.

1. Introduction

Recombinant technologies offer the possibility to explore gene function by manipulating fragments of DNA and then introducing them back into cells. The range of recombinant DNA applications is largely dependent on the isolation of genes and the understanding of their biological role. In flowering plants, a functional analysis of genetic information is under way for several species of biological and agronomical impor-

From: *Methods in Molecular Biology, vol. 267:*
Recombinant Gene Expression: Reviews and Protocols, Second Edition
Edited by: P. Balbás and A. Lorence © Humana Press Inc., Totowa, NJ

tance. Among them, *Arabidopsis thaliana* (a member of the *Brassicaceae*), and *Oryza sativa* (rice) have been chosen as model systems to unravel the molecular mechanisms of development used by the two largest groups of flowering plants (monocots and eudicots). The ultimate goal of genome research on both species is the identification of all their genes and the elucidation of their functions. The sequence of the genome of *Arabidopsis* was virtually completed by December 2000 (*1*) and the completion of a draft sequence of the rice genome was announced in April 2002 (*2*). Even if sequencing projects are close to being completed for both species, functional studies of newly discovered genes have progressed at a much slower pace. The first obvious approach has been to compare the molecular structure of newly predicted genes to sequences of genes and proteins already reported in public-domain databases. In the case of *Arabidopisis*, the function of 69% of the genes was classified in categories according to comparisons to sequences of known function in all other organisms; however, only 9% of those genes have been characterized experimentally, and close to 30% of the predicted genes could not be assigned to a functional category (*1,3*). It is estimated that 80.6% of the predicted *Arabidopsis* genes have a homolog in rice, but that only 49.4% of predicted rice genes have a homolog in *Arabidopsis* (*1–3*). Whereas homology searches can be used to indicate a likely function of many of these genes, it is clear that the role of individual family members will not be determined without systematically assessing gene function by genetic analysis.

Recent progress in large-scale insertional mutagenesis opens new possibilities for functional genomics. The use of insertional mutagenesis in genetic screens has traditionally relied on the elucidation of gene function based on the direct identification of phenotypic abnormalities. By inserting them within the genome, insertional elements can cause genetic lesions and alter the molecular structure of a gene. A successful approach to systematically disrupting genes in *Arabidopsis* and rice has been the use of transferred-DNA (T-DNA) from *Agrobacterium tumefasciens* as an insertional mutagen (*4*). One of the assumptions underlying the use of T-DNA as a mutagen is that the insertion of these DNA elements into the genome occurs at randomly selected locations. Although a small bias toward recovering T-DNA insertions within intergenic regions appears to be the case in *Arabidopsis* (*5*), this bias does not limit the utility of T-DNA as an effective insertional mutagen for use in forward or reverse genetic strategies. Additionally, two transposable element systems from maize have been successfully used for generating loss-of-function mutations in the genome of *Arabidopsis*: the *Activator/Dissociator* (abbreviated *Ac/Ds*), and the *Suppressor-mutator* (abbreviated *Spm/dSpm* or *En/dEn*). The classical studies of Barbara McClintock (reviewed in *6,66,67*) allowed a clear understanding of the genetic properties of these transposons, and therefore the timing, range, and frequency of transposition can be carefully monitored and controlled. In each case, a specific sequence of the autonomous elements *Ac* and *Spm* encodes transposase genes that are required to mobilize the nonautonomous elements *Ds* and *dSpm*, respectively. The nonautonomous elements are deletion derivatives of autonomous elements that retain terminal sequences but lack transposase genes. The subsequent remobilization of these transposons allows for the recovery of a wild-type sequence and the confirmation that its insertion within a particular genomic site is the cause of an associated mutant phenotype. Transposons also allow for the isolation of additional stable or reversible alleles that can be directly associated with specific molecular lesions caused by excision or re-insertions within the genomic sequence of a given locus.

For *Arabidopsis*, the overall number of T-DNA and transposon insertion lines from many different laboratories will soon represent insertions into most predicted coding sequences (*7*). These large populations are being routinely screened for insertions into specific genes, allowing the systematic isolation of knockout lines. Although many genes have been disrupted through this approach, few have been reported to present informative phenotypes that provide a direct clue to gene function. Functional redundancy may explain the absence of obvious phenotypes in some cases, but it is becoming clear that many loss-of-function mutations are conditional or cause only subtle phenotypes that are difficult to identify in large conventional screens (*8*). In many instances, the success of genetic screens based on the identification of mutant phenotypes is largely dependent on the accessibility of the developmental phases of a limited number of tissues or cells. While the use of transposons in heterologous systems has proven to represent an efficient, flexible, and versatile technology to determine the function of genes in a wide variety of organisms, the function of only a limited number of genes has been elucidated based on genetic approaches that rely on the discovery of obvious mutant phenotypes.

A powerful alternative to traditional genetic screens is the elucidation of gene function based on the investigation of gene expression patterns. Modified insertional elements, such as transposons or T-DNA insertions, can result in the expression of a reporter gene under the spatial and temporal control of the regulatory sequences of a potential target gene, either directly by the establishment of a gene fusion (the so-called "gene traps") or indirectly by enhancer action (enhancer traps, better called "enhancer detectors"). Enhancer detection and gene trapping have been implemented successfully in many organisms, not only providing the possibility of identifying the function of unknown genes but also establishing the basis for novel technologies of gene manipulation (*8*). This type of approaches represent direct strategies to monitor and manipulate the activity of specific genes within tissues, a small group of cells, or even a single cell at specific moments of the life cycle of an organism. The wide success of enhancer detection and gene trapping as tools for the identification, isolation, and characterization of many genes has prompted their adoption in many current initiatives of functional genomics. In this chapter, we discuss the development of enhancer detection and gene trapping and place it in the context of current plant functional genomic strategies. We also discuss their current advantages and disadvantages and point to some potential directions that these strategies might take in a near future.

2. Vectors for Trapping

Promoter traps, gene traps, and enhancer detectors are three classes of reporter constructs that are not normally expressed unless they are integrated near or within a chromosomal gene (*20*). In the case of plants, they can be introduced into the genome by conventional transformation approaches (*Agrobacterium*-based transformation or particle gun bombardment). Promoter trap vectors consist of a promoterless reporter gene and a selectable marker that allows the identification of transformed individuals that integrated at least one copy of the construct (**Fig. 1A**). Reporter gene expression occurs when the vector inserts into an exon to generate a fusion transcript that comprises upstream endogenous exonic sequence and the reporter gene. In contrast, a gene trap vector contains splice acceptor sites immediately upstream of the promoterless reporter gene (**Fig. 1A**). After transcriptional activation of an endogenous *cis*-acting promoter and/or enhancer element of the "trapped" gene, a fusion transcript is generated from the upstream coding sequence and the corresponding expression pattern is reported. Enhancer detection vectors rely on an insertional genetic element carrying a reporter gene under the control of a constitutive but minimal promoter (**Fig. 1A**) (*10,11*). If this promoter comes under the control of a genomic *cis*-regulatory element, the reporter gene is expressed in a specific temporal and spatial pattern. Such a reporter construct is not "trapping" genes but rather integrating into genomic sequences to serve as a "detector" of any given regulatory sequence that is acting as an enhancer of promoter activity at the specific location of the insertion. All these strategies are now used in conjunction with methods that allow rapid and efficient isolation and cloning of genomic DNA fragments flanking the enhancer detector or gene trap construct (*12,26*). Therefore, all these approaches represent powerful strategies to directly associate the molecular nature of a specific gene with its functional role in the organism.

The first "trap" vector was developed by Casadaban and Cohen by modifying a *Mu* bacteriophage carrying a promoterless *lac* operon in *Escherichia coli* (*14*). Activation of the *lacZ* reporter gene required the integration of the phage downstream of a bacterial promoter in the correct orientation. The strategy allowed the identification of genes whose expression was induced under specific growth conditions; this same approach was later implemented in *Bacillus subtilis* (*15*). A few years later, enhancer detection was used in *Drosophila melanogaster* to more efficiently identify the complexities of expression patterns in an eukaryotic genome (*16*). The genome of prokaryotes is more densely occupied by genes that the genome of eukaryotes that contains sequences to regulate promoters over large genomic distances. These eukaryotic enhancers do not act in a specific orientation, and therefore the activation of a reporter gene can take place even if the insertion is not located downstream from the site of transcriptional initiation of an endogenous gene.

In *Drosophila*, the first enhancer detectors were constructed from endogenous P transposable elements. The transposase gene promoter of this element is small (87 bp) and appears to be of easy access to nearby enhancers. In flies, large phenotypic screens were substantially facilitated by the implementation of an endogenous source of transposase that allowed the generation of many enhancer detection fly strains from a single parental P-element insertion (*17,18*). Using this approach, 5 to 10% of strains containing single transposon insertions showed a specific pattern of expression during fly development (*8*). In *Caenorhabditis elegans*, where transposable elements cannot be easily manipulated due to the endogenous expression of transposase in the germline, conventional strategies for enhancer detection or gene trapping have not been implemented yet. As an alternative, random genomic DNA fragments have been cloned

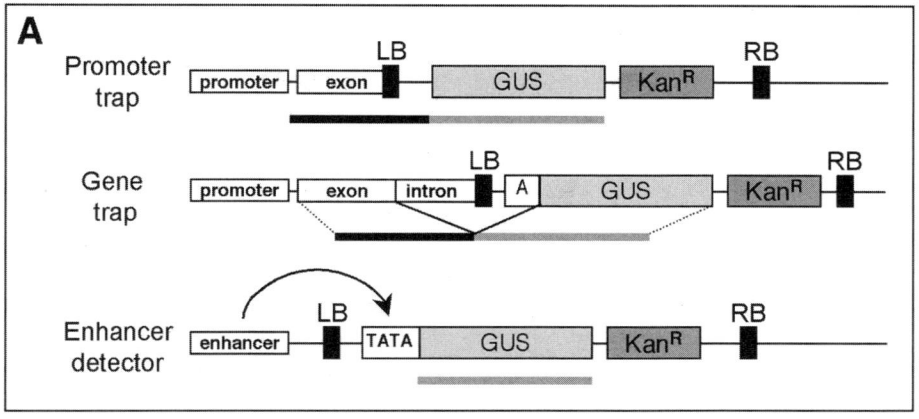

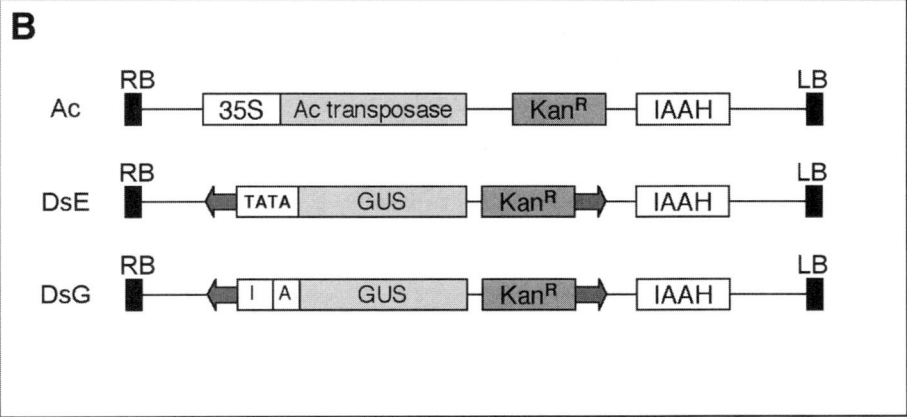

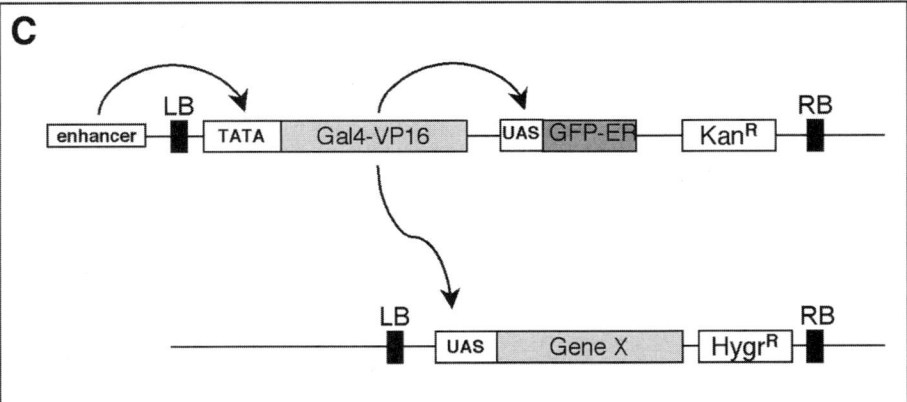

Fig. 1. Enhancer detection and gene trapping strategies in plants. (**A**) Principle of T-DNA based of promoter trap, gene trap, and enhancer detection vectors. In a promoter trap system, the promoterless GUS reporter gene (beta-glucuronidase *uidA)* will be expressed if the insertion occurs in an exon in frame with the target gene sequence. In a gene trap system, splice acceptors (A) are added to the GUS reporter gene to allow expression even if the insertion occurs in an intron. In the enhancer detector system, the reporter gene (GUS) fused to a minimal promoter (TATA)

upstream of the *lacZ* gene to identify specific expression patterns, but active protein fusions conferring reporter gene expression have been observed in only about 1% of all genomic fragments tested (*19*). Finally, several large-scale enhancer detection and gene trap screens are being carried out in *Mus musculus* (mouse) with various vectors introduced after transformation of embryonic stem (ES) cells. Current resources are almost completely public and represent more than 9000 mutagenized ES cell lines freely available to the research community (*20*).

3. Current Strategies for *Arabidopsis* and Rice

The first attempts to create transcriptional fusions that allowed expression of selectable markers in flowering plants were conducted in *Nicotiana tabacum* (tobacco) soon after the first reports demonstrating stable transformation with T-DNA vectors (*21,22*). These experiments allowed the identification of plant constitutive promoters *in situ* by T-DNA-mediated transcriptional fusions to the *npt*-II gene, conferring resistance to the antibiotic kanamycin. Prompted by their spectacular success in animal systems, several groups modified the approach to include reporter genes that would act as promoter traps and allow the identification of expression patterns *in situ* in several species, including *Arabidopsis* (*22–25*).

Fig. 1. *(continued)* is activated when insertion of the T-DNA occurred near cis-acting native enhancer sequences. KanR: neomycin phosphotransferase gene (NPT II) which confers resistance to kanamycin. LB, RB: left and right T-DNA borders. Black and gray bars below each construct represent the resulting transcript. (**B**) Ac/Ds transposon-based enhancer detector and gene trap systems. The maize Ac/Ds transposable element has been adapted to *Arabidopsis* to generate random insertions across the genome. *Ac* and *Ds* elements are introduced into the genome of parental lines by T-DNA mediated transformation. Transposition is induced by genetic crossing of the parental lines. In the F2 generation, independent lines are selected for unlinked transposition events by the use of both positive (resistance to kanamycin) and negative (absence of sensitivity to NAM conferred by the IAAH gene) (*see* **Subheading 3.1.**). The *Ac* construct, conferring sensitivity to NAM, is also counterselected. In the *Ds* enhancer detector transposon (DsE), GUS is fused to a minimal promoter (TATA). In the *Ds* gene trap transposon (DsG), the GUS gene is promoterless, and contains splice acceptor and donor sites (I and A) that will permit GUS expression by transcriptional fusion if the insertion occurs within gene introns or exons. The system was developed by the groups of V. Sundaresan and R. Martienssen at Cold Spring Harbor Laboratory. KanR: *npt*II gene. LB, RB: left and right T-DNA borders. Filled gray arrows: *Ds* inverted repeat sequences. (**C**) T-DNA-based *GAL4-GFP* enhancer detector for targeted misexpression. Insertion of a T-DNA near native enhancer sequences allows expression of the yeast transcription activator Gal4-VP16. The Gal4-VP16 protein activates transcription of a linked reporter gene, the green fluorescent protein (GFP), targeted to endoplasmic reticulum (mGFP5-ER), which is fused to the GAL4 upstream activator sequences (UAS). By genetic crossing, specific expression of any chosen gene(s) (Gene X) fused to UAS sequences can be induced in the same cells/tissues where the GFP reporter is expressed in the Gal4-VP16 enhancer trap line. This strategy was developed by the group of J. Haseloff at University of Cambridge (http://www.plantsci.cam.ac.uk/Haseloff). KanR: *npt*-II gene conferring resistance to kanamycin. HygrR: *hpt* gene conferring resistance to hygromycin. LB, RB: left and right T-DNA borders.

There are now several international, large-scale efforts under way that take advantage of the limited number of vectors available to generate enhancer detector or gene trap lines in *Arabidopsis* and rice (**Table 1**). Overall, and although efforts are not yet coordinated and often require material transfer agreements (MTAs) before making the lines available for research purposes, these initiatives promise to generate thousands of enhancer detector, promoter trap, and gene trap lines in the near future, an overall collection perhaps sufficiently large to ensure at least one insertion in each gene of *Arabidopsis*. Current strategies can be classified into two groups depending on the insertional element that was used to engineer the "trapping" vector: maize transposons or T-DNA.

3.1. Maize Transposons as Enhancer Detectors and Gene Traps

One of the most widely adopted enhancer detection/gene trap strategies was developed by Sundaresan et al. (**26**) at Cold Spring Harbor Laboratory (CSHL). Using *Arabidopsis thaliana* as a model system, their approach relies on the utilization of engineered transposons of maize carrying a reporter gene that is used to identify patterns of expression that mirror the activity of endogenous plant genes. The system uses two independent transposable elements of the *Ac/Ds* family. In practice, *Arabidopsis* parental plants carrying an immobilized *Ac* element are crossed to plants carrying a nonautonomous *Ds* element containing the reporter gene (**Fig. 1B**). The immobilized *Ac* element lacks the inverted repeat termini, and therefore cannot transpose but can supply transposase constitutively and therefore allow the transposition of nonautonomous elements of the *Ds* family. The *Ac* construct carries a small deletion of GC-rich sequences in the untranslated leader of the transposase gene, as well as a strong 35S promoter from the cauliflower mosaic virus to drive its transcription. In contrast, engineered *Ds* elements contain a deletion of the transposase gene but have intact termini, and they can excise when transposase is provided by the *Ac* element. Additionally, these *Ds* elements carry the reporter gene *uidA* and *npt*II that confers kanamycin resistance (KanR). The reporter gene *uidA* (also known as GUS) encodes for an *E. coli* β-glucuronidase (**27**). Unlike β-galactosidase, GUS has low endogenous activity in plants. It is also a stable enzyme capable of hydrolysing a wide range of β-glucuronides, and it can be easily assayed for histochemical analysis using the specific substrate X-Gluc (5-bromo-4-chloro-3-indoyl β-glucuronide). After cleavage, oxidation of the indole derivative causes dimerisation and the production of an insoluble indigo dye. The *npt*II gene is used as a selection marker to identify plants that contain a *Ds* element.

Two different versions of the *Ds* transposon were engineered. For the gene trap vector (called DsG), the reporter gene has no promoter driving its expression; instead, it is preceded by three adjacent plant splice acceptor sequences in different reading frames inserted immediately upstream of the ATG start codon. These splice acceptors ensure that an insertion into an intron will generate an in-frame fusion 50% of the time, leading to a fusion protein expressing reporter activity. For the enhancer detector *Ds* element (called DsE), the reporter gene is preceded by a minimal promoter (a TATA box), and the expression of the reporter gene will depend on the *Ds* element landing near regulatory sequences (enhancers or promoters) from endogenous plant genes. All three of

Table 1
Arabidopsis and Rice Enhancer Detector and Promoter/Gene Trap Collections Initiatives

Project	Trapping vector and reporter gene	Insertion element	Number of Lines (lines available)	Plant material	Reference
University of Cambridge, UK Jim Haseloff and col. *Gal4+UASmGFP5 ER*	Enhancer detector Gal4-GFP *Gal4+UASmGFP5 ER*	T-DNA	8000 (250)	*Arabidopsis* C24	ABRC http://www.plantsci.cam.ac.uk
University of Pennsylvania, USA Scott Poethig and col. *Gal4+UASmGFP5 ER*	Enhancer detector Gal4-GFP	T-DNA	5000 (2000)	*Arabidopsis* Col-0	ABRC http://enhancertraps.bio.upenn.edu
Dartmouth College, USA Tom Jack and col.	Enhancer detector GUS *pD991*	T-DNA	12000 (11300)	*Arabidopsis* Col-6, gl1-*l*	ABRC [1] Campisi et al., 1999
University of Leicester, UK K. Lindsey and col.	Enhancer detector GUS *pΔ35S GUS bin19*	T-DNA	96 (96)	*Arabidopsis* C24	NASC [2] Topping et al., 1991
Universite de Perpignan, France P. Gallois and col.	Enhancer detector *GUS pΔ35S GUS bin19*	T-DNA	283 (283)	*Arabidopsis* C24	ABRC [3] Devic et al., 1995
Cold Spring Harbor Laboratory, USA Rob Martiensen and col.	Ds—Enhancer detector GUS Ds—Gene trap GUS *DsE, DsG*	transposon *Ac/Ds*	33967 (30000) 19612 sequenced	*Arabidopsis* Ler	ABRC http://genetrap.chsl.org
Institute of Molecular Agrobiology, Singapore IMA lines V. Sundaresan and col.	Ds—Enhancer detector GUS Ds—Gene trap GUS *DsE, DsG*	transposon *Ac/Ds* transposon	931 (931) all sequenced	*Arabidopsis* Ler	NASC [5] Parinov et al., 1999 [4] Sundaresan et al., 1995
John Innes Centre, UK J. Clarke and M. Bevan	Ds—Enhancer detector GUS *DsE*	*Ac/Ds*	209 (209)	Ler	NASC [13] Musket et al., 2003 http://www.jic.bbsrc.ac.uk/staff/caroline-dean/mapped.htm

Project	Trapping vector and reporter gene	Insertion element	Number of Lines (lines available)	Plant material	Reference
FGT collection / GARNET (EU consortium) **John Innes Centre, UK** J. Clarke and M. Bevan	Ds—Gene trap GUS DsG	transposon Ac/Ds	24000 (4000) 4000 sequenced	*Arabidopsis* *Ler*	NASC / ABRC http://www.jic.bbsrc.ac.uk/staff/michael-bevan/atis/index.htm [4] Sundaresan et al., 1995
RIKEN, Japan K. Shinosaki and col.	Ds- Enhancer detector GUS $Ds\text{-}GUS_{T\text{-}DNA}$	transposon Ac/Ds	1173 (1125) 1124 sequenced	*Arabidopsis* *No*	Ito et al., 2002 http://www.brc.riken.go.jp/Eng/index.html [6] Fedoroff and Smith, 1993
INRA, Versailles, France URGV—FST project G. Pelletier, N. Bechtold and col.	Promoter trap GUS *pGKB5*	T-DNA	50000 (12300) 28998 sequenced	*Arabidopsis* *Ws*	NASC/ABRC http://flagdb-genoplante-info.infobiogen.fr/projects/fst/fst.html [7] Bechtold et al., 1993; [8] Bouchez et al., 1996, [9] Mollier et al., 1995
Syngenta / MOGEN P.C. Sijmons and col.	Promoter trap GUS *PMOG533*	T-DNA	1250 (50)	*Arabidopsis* *C24*	NASC [10] Goddijn et al., 1993
Pohang University, Korea G. An and col.	Gene trap GUS *pGA1633, pGA2144*	T-DNA	22090 (nd)	*Oryza sativa*	[11] Jeon et al., 2000

(continued)

Table 1 (continued)
Arabidopsis and Rice Enhancer Detector and Promoter/Gene Trap Collections Initiatives

Project	Trapping vector and reporter gene	Insertion element	Number of Lines (lines available)	Plant material	Reference
Gyeongsang National University C.D. Han and col.	Ds- Gene trap GUS *Gene trap Ds*	transposon Ac/Ds	221 (nd)	*Oryza sativa*	[12] Chin et al., 1999

References of table 1:

[1] Campisi, L., Yang, Y., Yi, Y., Heilig, E., Herman, B., Cassista, A. J., et al. (1999) Generation of enhancer trap lines in *Arabidopsis* and characterization of expression patterns in the inflorescence. *Plant J.* **17**, 699–707.

[2] Topping, J.F., Wei, W., and Lindsey, K. (1991) Functional tagging of regulatory elements in the plant genome. *Development* **112**, 1009–1019.

[3] Devic, M., Hecht, V., Berger, C., Delseny, M., and Gallois, P. (1995) An assessment of promoter trapping as a tool to study plant zygotic embryogenesis. *C.R. Acad. Sci. Life Science* **318**, 121–128.

[4] Sundaresan, V., Springer, P., Volpe, T., Haward, S., Jones, J. D., Dean, C., et al. (1995) Patterns of gene action in plant development revealed by enhancer trap and gene trap transposable elements. *Genes Dev.* **15**, 1797–1810.

[5] Parinov, S., Sevugan, M., Ye, D., Yang, W. C., Kumaran, M., and Sundaresan, V. (1999) Analysis of flanking sequences from dissociation insertion lines: a database for reverse genetics in *Arabidopsis. Plant Cell* **11**, 2263–2270.

[6] Fedoroff, N. and Smith, D.L. (1993) A versatile system for detecting transposition in *Arabidopsis. Plant J.* **3**, 273–289.

[7] Bechtold, N., Ellis, J., and Pelletier, G. (1993) In planta *Agrobacterium* mediated gene transfer by infiltration of adult *Arabidopsis thaliana* plants. *C. R. Acad. Sci. Paris, Sciences de la vie/Life Sciences* **316**, 1194–1199.

[8] Bouchez, D. Camilleri, C., and Caboche, M. (1993) A binary vector based on Basta resistance for in planta transformation of *Arabidopsis thaliana C. R. Acad. Sci. Paris, Sciences de la vie/Life sciences* **316**, 1188–1193.

[9] Mollier, P., Montoro, P., Delarue, M., Bechtold, N., Bellini, C., and Pelletier, G. (1995) Promoterless *gusA* expression in a large number of *Arabidopsis thaliana* transformants obtained by the in planta infiltration method. *C.R. Acad. Sci. Paris, Sciences de la vie/Life sciences* **318**, 465–474.

[10] Goddijn, O.J., Lindsey, K., van der Lee, F. M., Klap, J. C. and Sijmons, P. C. (1993) Differential gene expression in nematode-induced feeding structures of transgenic plants harbouring promoter-*gusA* fusion constructs. *Plant J.* **4**, 863–873.

[11] Jeon, J. S., Lee, S., Jung, K. H., Jun, S. H., Jeong, D. H., Lee, J., et al. (2000) T-DNA insertional mutagenesis for functional genomics in rice. *Plant J.* **22**, 561–570.

[12] Chin, H. G., Choe, M. S., Lee, S. H., Park S. H., Koo J. C., Kim N. Y., et al. (1999) Molecular analysis of rice plants harboring an Ac/Ds transposable element-mediated gene trapping system. *Plant J.* **19**, 615–623.

[13] Muskett, P. R., Clissold, L., Marocco, A., Springer, P. S., Martienssen, R., and Dean, C. (2003) A resourse of mapped Dissociation launch pads for targeted insertional mutagenesis in the Arabidopsis genome. *Plant Physiol.* **132**, 506–516.

these constructs were inserted into T-DNA vectors that carry the *IAAH* gene (indole acetic acid hydrolase) that is used as a dominant marker for negative selection.

When seedlings carrying the *IAAH* gene are germinated on medium containing the auxin analog NAM (naphtalene acetamide), their roots wilt and do not grow due to the conversion of NAM to toxic levels of naphthalene acetic acid. The *IAAH* gene remains within the T-DNA border after *Ds* excision and can be segregated out in the F2 generation. When *Ds* elements excise in *Arabidopsis*, they frequently fail to reinsert, and when they do, it is typically close to their original location. Selection schemes based on transposon excision will therefore recover many lines that have transposed elements that are concentrated in a small region of the genome. To avoid this, plants heterozygous for *Ac* and *Ds* are generated by crossing parent *Ac* and *Ds* together. Transposition of *Ds* occurs in these heterozygotes and results in individual flowers that carry new transposed *Ds* elements in their germline. When these flowers set seed by self-pollination, the only F2 progeny that are resistant to both kanamycin and NAM are those that carry a *Ds* element but not the *Ds* donor locus (which confers NAM sensitivity). These progeny carries an unlinked transposed element that is allowed to self-pollinate to produce a full transposant line. Therefore, the selection scheme allows for independent recovery of transpositions that are unlinked to the donor site. After selection of informative transpositions events on the basis of expression patterns, DNA is prepared from each individual plant and the genomic location of the transposon insertion is determined by TAIL (thermal assymetric interlaced) PCR (*12*).

The Plant Group at CSHL has now produced more than 30,000 lines that are made available through a searchable database (http://genetrap.cshl.org) that provides the sequence at the genomic insertion site for more than 15,000 insertions (**Table 1**). Massive screening has revealed that more than 45% of enhancer detector insertions give rise to a GUS expression pattern in seedlings, but up to 80% give rise to reporter expression somewhere in the plant (*28*). In contrast, only 30% of gene trap insertions give rise to reporter expression in seedlings and up to 45% somewhere in the plant (*28*). Whereas enhancer detector insertions are able to act at a close distance from detected genes, only half of all gene trap insertions within genes will be in the appropriate orientation to result in a pattern of expression.

Seeds for any of the CSHL lines can be obtained after agreeing to comply with rules and regulations that restrict their commercial utilization through Material Transfer Agreement (MTA). Other groups have made a portion of their own collections freely available through the Arabidopsis Biological Resource Center (ABRC; www.biosci.ohio-state.edu/plantbio/facilities/abrc) or the Nothingam Arabidopsis Stock Center (NASC; www.nasc.nott.ac.uk) after having sequenced most of the transposon insertion sites (*29*) (**Table 1**). A consortium in Japan has also initiated the generation of a collection based on transposable elements that act as enhancer detectors (*30*). The FTG collection of the Genome Arabidopsis Resource Network initiative GARNet (*31*) has 1000 gene trap lines available through NASC with no MTAs attached to it as of May 2003. This collection is part of the Exon Trapping Insert Consortium (EXOTIC), a network of European laboratories formed to generate at least 3000 gene trap lines with the Ac/Ds vectors generated at CSHL.

3.2. T-DNA for Enhancer Detection and Gene Trapping

The rapid and efficient generation of T-DNA insertions by massive transformation of independent plants has been widely used to systematically disrupt genes (*32–34,55,56*). In *Arabidopsis*, the utilization of vacuum infiltration and later "floral dipping" has dramatically simplified transformation protocols and allowed the possibility of producing large collections of T-DNA that should ensure genome saturation in a very near future. Systematic efforts are now under way to use these collections for "reverse genetic" screens to identify insertions in any cloned gene. Additionally, several groups have also used T-DNA as an insertional element to generate promoter trap, gene trap and enhancer detection vectors (*53,54,57,58*). These lines are not always freely released but some have been progressively made available for noncommercial purposes through ARBC or NASC (see previous section). All of these initiatives follow the technical trends set by the first promoter traps developed in *Arabidopsis* (*22–25*): they rely on GUS as the reporter gene, and on a small region of the 35S cauliflower mosaic virus as the minimal promoter of the enhancer detection vector. In rice, a T-DNA-based binary vector has been engineered to contain the promoterless GUS reporter gene next to the right border, and multimerized transcriptional enhancers from the cauliflower mosaic virus 35S promoter next to the left border. The left border is designed to serve as a transcriptional activator of any sequence located downstream of the insertion site (*35,36*). Thus, the vector can be used to independently screen for identifying expression patterns by promoter trapping and looking for dominant mutant phenotypes caused by activation tagging.

The use of T-DNA as an insertional element for enhancer detection and gene trapping has several disadvantages over transposable elements. Since "trapping" strategies rely on the possibility of directly associating a reporter gene expression pattern with a "tagged gene," single insertions offer the advantage of unambiguously allowing the identification of the genomic location of the insertion site. In contrast, multiple genomic insertions (quite commonly obtained after T-DNA transformation) can drastically complicate the identification and genetic analysis of the corresponding gene. Insertional events can often include T-DNA in direct or inverted tandem repeats, with occasional rearrangements of adjacent chromosomal DNA (*37,38*). These complexities may result in reporter gene expression patterns derived by promoters included in the vector and rearranged by the T-DNA, and not from endogenous promoters. Complex insertions may also complicate subsequent molecular analyses, making it difficult to isolate the chromosomal genes driving reporter gene expression. In contrast to transposable elements, the remobilization of genomic T-DNA insertions is not possible and limits the subsequent genetic analysis of newly identified mutations.

3.3. Enhancer Detection as a Tool to Drive Targeted Misexpression

The implementation of an enhancer detection approach can also serve as the basis for developing more elaborated strategies to manipulate cell fate and cause dominant phenotypes by ectopically expressing selected genes during a specific developmental pathway. In yeast, the transcriptional activator *GAL4* is specifically recognized by independent upstream activations sequences (*UAS*) that remain silent in the absence of *GAL4* (*59*). A method based on the GAL4/*UAS* system has been widely adopted in *Drosophila* to ectopically transactivate genes of interest (*60*). In *Arabidopsis*, a different vector was constructed by including *GAL4* within an enhancer detection construction (**Fig. 1C**; *40,41,61*). With this vector, the identification of a reporter gene expression pattern results also in the expression of *GAL4* in an identical temporal and spatial context. In an independent construction, a chosen target gene can be placed under the control of the *UAS* and transformed into plants. By a simple genetic cross, specific expression of the chosen gene can be induced in the same pattern of expression that was identified by the *GAL4*-containing enhancer detector, causing conditional dominant phenotypes that can be the consequence of misexpression (or overexpression). The system has been designed to include the green fluorescent protein (GFP) from *Aequoria victoria* (jellyfish) as the reporter gene (*41,61*). It was modified to allow efficient expression by removing a cryptic intron that prevented adequate sensitivity in plants (*62*). The fluorescence of GFP is due to a chromophore that is active in all plant tissues examined to date.

4. Additional Applications and Future Perspectives

One of the major advantages of enhancer detection in several multicellular organisms has been the possibility of implementing developmental markers in tissues and cells that were morphologically unidentified or untractable. In *Drosophila*, enhancer detectors have contributed to establish cellular origins and lineages in a multitude of developmental pathways and tissues (reviewed in **ref. 8**), and extensive use of the transposon-based enhancer detection system confirms that reporter gene expression is in most cases confined to the cells or tissues that express the endogenous gene. In *Arabidopsis*, the few cases analyzed to date confirmed these observations (*63*); however, evidence showing that introns can influence the pattern of mRNA localization of endogenous genes suggests that the detection or trapping of regulatory elements could also be conditioned by the location of the trap insertion within a coding sequence (*64*). Compared to the mutagenic efficiency of chemical agents such as ethyl-methanesulfonate (EMS), the mutagenic frequency of transposon-based approaches is poor in all multicellular organisms (*5,8,28,42*).

Molecular epistasis experiments can also be performed to establish possible genetic interactions and hierarchical relationships between a previously characterized mutation and a series of genes identified by a collection of enhancer detector or gene trap lines. In the case of recessive mutations, each of the lines can be crossed to the mutant and F1 progeny allowed to self-pollinate. In the F2 generation, the patterns of GUS expression can be determined in plants heterozygous and homozygous for the selected mutation. A change in the expected expression pattern in a mutant background could indicate either the absence or the modification of a particular element, or a regulatory relationship between the two genes analyzed. This approach allows the simultaneous epistatic analysis of phenotypical effects and changes of gene expression caused by genetic interactions between two or more loci.

All trapping strategies allow the identification of genes that have a redundant function and for which a mutant phenotype is difficult to identify. They are particularly useful for studying gene expression if the juvenile lethal phenotype impedes the determination of a possible function during adult stages of development, In those cases, a pattern of gene expression in heterozygous individuals for the enhancer trap insertion can still be identified even if the homozygous recessive mutation is lethal during embryogenesis (*28,42*). The system also offers several advantages to identify genes that act in the haploid phase of the plant life cycle, during male and female gametophytic development. During female gametogenesis, many genes have the potential to cause a lethal phenotype if mutated, but the characterization of their activity can be achieved if they are tagged by an enhancer detector or gene trap. Whereas housekeeping genes should have a tendency to be constitutively expressed, genes involved in specification or differentiation of haploid cells should be characterized by a specific pattern of expression (*63*). Finally, all trapping strategies are ideally suitable to identify regulatory sequences and subsequently isolate specific promoters acting during a given developmental context. With the advent of RNA interference (RNAi) as the long-awaited tool to target gene disruption in plants (*44,65*), enhancer detection and gene trapping are the ideal complement to isolate a multitude of specific promoters and generate systematic mutant phenotypes during selected developmental mechanisms.

In *Drosophila*, P-element mutagenesis screens cause only 11–13% of lethal phenotypes, and similar frequencies can be expected in *Arabidopsis* in the case of single insertion trapping strategies. However, the versatility of insertional mutagenesis approaches should continue to ensure their popularity among the community of plant biologists. The implementation of enhancer detection and gene trapping is well under way in species such as *Lycopersicon aesculentum* (tomato) (*45*), *Medicago truncatula* (*46*), or the moss *Physcomitrella patens* (*47*). Following the trend of their applications in animal organisms, enhancer detection and gene trapping are proving to be essential for systematically isolating and characterizing the expression of novel genes, but also for determining lineages of specific organs or developmental pathways (*42,48,49*), ablating specific cells by molecular strategies (*50,61*), or ectopically expressing random genes to alter cell fate (*40,41,51,61*). Another interesting application is their use as a teaching resource. A collection of expression patterns can serve for the elementary understanding of patterning and organization of plant tissues and cells in plant developmental biology courses (*52*).

In years to come, a thorough functional annotation of eukaryotic genomes will require the use of a combination of strategies, including high-throughput phenotypic screens of chemically or insertionally mutagenized plants gene targeting, conditional mutagenesis, but also complementary strategies that provide a rapid determination of gene expression patterns *in situ*. Enhancer detection and gene trapping are proving to be the method of choice to systematically determine the expression of a multitude of genes and simultaneously attempt their knockout. The possibility to directly correlate gene expression and function ensures the success of enhancer detection and gene trapping strategies in most if not all new functional genomics initiatives.

Acknowledgments

Research in our laboratory is supported by the Consejo Nacional de Ciencia y Tecnología (CONACyT-México; grants 34324-B and Z-029), and the Howard Hughes Medical Institute through the International Research Scholars Program. D.A. is being supported by a postdoctoral fellowship from the Mexican Ministry of Foreign Affairs (Secretaria de Relaciones Exteriores, México D.F.).

References

1. The *Arabidopsis* Genome Initiative. (AGI) (2000) Analysis of the genome sequence of the flowering plant *Arabidopsis thaliana. Nature* **408**, 796–815.
2. Goff, S. A., Ricke, D., Lan, T. H., Prosting, G., Wang, R., Dunn, M., et al. (2002) A draft sequence of the rice genome (*Oryza sativa* L. ssp. japonica). *Science* **5**, 92–100.
3. Yu, J., Hu, S., Wang, J., Wong, G. K., Li, S., Liu, B., et al. (2002) A draft sequence of the rice genome (*Oryza sativa* L. ssp. indica). *Science* **5**, 79–92.
4. Breyne, P. and Zabeau, M. (2001) Genome-wide expression analysis of plant cell cycle modulated genes. *Curr. Opin. Plant Biol.* **4**, 136–142.
5. Krysan, P. J., Young, J. C., Jester, P. J., Monson, S., Copenhaver, G., Preuss, D., et al. (2002) Characterization of T-DNA insertion sites in *Arabidopsis thaliana* and the implications for saturation mutagenesis. *OMICS* **6**, 163–174.
6. Bancroft, I. and Dean, C. (1993) Transposition pattern of the maize element Ds in *Arabidopsis thaliana. Genetics* **134**, 1221–1229.
7. Pruitt, R. E., Bowman, J. L., and Grossniklaus, U. (2003) Plant genetics: a decade of integration. *Nat Genet.* **33**, 294–304.
8. Bellen, H. J. (1999) Ten years of enhancer detection: lessons from the fly. *Plant Cell* **11**, 2271–2281.
9. O'Kane, C. J. and Gehring, W. J. (1987) Detection *in situ* of genomic regulatory elements in *Drosophila. Proc. Natl. Acad. Sci. USA* **84**, 9123–9127.
10. Wilson, C., Pearson, R. K., Bellen, H. J., O'Kane, C. J., Grossniklaus, U., and Gehring, W. J. (1989) P-element-mediated enhancer detection: an efficient method for isolating and characterizing developmentally regulated genes in *Drosophila. Genes Dev.* **3**, 1301–1313.
11. Grossniklaus, U., Bellen, H. J., Wilson, C., and Gehring, W. J. (1989) P-element-mediated enhancer detection applied to the study of oogenesis in *Drosophila. Development* **107**, 189–200.
12. Liu, Y. G., Mitsukawa, N., Oosumi, T., and Whittier, R. F. (1995) Efficient isolation and mapping of *Arabidopsis thaliana* T-DNA insert junctions by thermal asymmetric interlaced PCR. *Plant J.* **8**, 457–463.
13. Balzergue, S., Dubreucq, B., Chauvin, S., Le-Clainche, I., Le Boulaire, F., de Rose, R., et al. (2001) Improved PCR-walking for large-scale isolation of plant T-DNA borders. *Biotechniques* **30**, 496–498.
14. Casadaban, M. J. and Cohen, S. N. (1979) Lactose genes fused to exogenous promoters in one step using a *Mu*-lac bacteriophage: in vivo probe for transcriptional control sequences. *Proc. Natl. Acad. Sci. USA* **76**, 4530–4533.
15. O'Kane, C., Stephens, M. A., and McConnell, D. (1986) Integrable alpha-amylase plasmid for generating random transcriptional fusions in *Bacillus subtilis. J. Bacteriol.* **168**, 973–981.

16. Bellen, H. J., O'Kane, C. J., Wilson, C., Grossniklaus, U., Pearson, R. K., and Gehring, W. J. (1989) P-element-mediated enhancer detection: a versatile method to study development in *Drosophila. Genes Dev.* **3**, 1288–1300.

17. Cooley, L., Kelley, R., and Spradling, A. (1988) Insertional mutagenesis of the *Drosophila* genome with single P elements. *Science* **4**, 1121–1128.

18. Robertson, H. M., Preston, C. R., Phillis, R. W., Johnson-Schlitz, D. M., Benz, W. K., and Engels, W. R. (1988) A stable genomic source of P element transposase in *Drosophila melanogaster. Genetics* **18**, 461–470.

19. Young, J. M. and Hope, I. A. (1993) Molecular markers of differentiation in *Caenorhabditis elegans* obtained by promoter trapping. *Dev. Dyn.* **196**, 124–132.

20. Stanford, W. L., Cohn, J. B., and Cordes, S. (2001) Gene-trap mutagenesis: past, present and beyond. *Nat. Rev. Genet.* **2**, 756–768.

21. André, D., Colau, D., Schell, J., Van Montagu, M., and Hernalsteens, J. P. (1986) Gene tagging in plants by a T-DNA insertion mutagen that generates APH(3')II-plant gene fusions. *Mol. Gen. Genet.* **204**, 512–518.

22. Teeri, T. H., Herrera-Estrella, L., Depicker, A., Van Montagu, M., and Palva, E. T. (1986) Identification of plant promoters *in situ* by T-DNA–mediated transcriptional fusions to the *npt*-II gene. *EMBO J.* **5**, 1755–1760.

23. Fobert, P. R., Miki, B. L., and Iyer, V. N. (1991) Detection of gene regulatory signals in plants revealed by T-DNA–mediated fusions. *Plant Mol. Biol.* **17**, 837–851.

24. Kertbundit, S., De Greve, H., Deboeck, F., Van Montagu, M., and Hernalsteens, J.-P. (1991) *In-vivo* random β-glucuronidase gene fusions in *Arabidopsis thaliana. Proc. Natl. Acad. Sci. USA* **88**, 5212–5216.

25. Topping, J. F., Wei, W., and Lindsey, K. (1991) Functional tagging of regulatory elements in the plant genome. *Development* **112**, 1009–1017.

26. Sundaresan, V., Springer, P., Volpe, T., Haward, S., Jones, J. D., Dean, C., et al. (1995) Patterns of gene action in plant development revealed by enhancer trap and gene trap transposable elements. *Genes Dev.* **15**, 1797–1810.

27. Jefferson, R. A., Burgess, S. M., and Hirsh, D. (1986) Beta-glucuronidase from *Escherichia coli* as a gene-fusion marker. *Proc. Natl. Acad. Sci. USA* **83**, 8447–8451.

28. Martienssen, R. A. (1998) Functional genomics: probing plant gene function and expression with transposons *Proc. Natl. Acad. Sci. USA* **3**, 2021–2026.

29. Parinov, S., Sevugan, M., Ye, D., Yang, W. C., Kumaran, M., and Sundaresan, V. (1999) Analysis of flanking sequences from dissociation insertion lines: a database for reverse genetics in *Arabidopsis. Plant Cell* **11**, 2263–2270.

30. Fedoroff, N. V. and Smith, D. L. (1993) A versatile system for detecting transposition in *Arabidopsis. Plant J.* **3**, 273–289.

31. Veale, M., Dupree, P., Lilley, K., Beynon, J., Trick, M., Clarke, J., et al. (2002) *GARNet*, the Genomic *Arabidopsis* Resource Network. *Trends Plant Sci.* **7**, 145–147.

32. Pan, X., Liu, H., Clarke, J., Jones, J., Bevan, M., and Stein, L. (2003) ATIDB: *Arabidopsis thaliana* insertion database. *Nucleic Acids Res.* **15**, 1245–1251.

33. Bouche, N. and Bouchez, D. (2001) *Arabidopsis* gene knockout: phenotypes wanted. *Curr. Opin. Plant Biol.* **4**, 111–117.

34. Bouchez, D., Camilleri, C., and Caboche, M. (1993) A binary vector based on Basta resistance for *in planta* transformation of *Arabidopsis thaliana C. R. Acad. Sci. Paris, Sciences de la vie/Life sciences* **316**, 1188–1193.

35. Jeon, J. S., Lee, S., Jung, K. H., Jun, S. H., Jeong, D. H., Lee, J., et al. (2000) T-DNA insertional mutagenesis for functional genomics in rice. *Plant J.* **22**, 561–570.

36. Jeong, D. H., An, S., Kang, H. G., Moon, S., Han, J. J., Park, S., et al. (2002) T-DNA insertional mutagenesis for activation tagging in rice. *Plant Physiol.* **130,** 1636–1644.
37. Laufs, P., Autran, D., and Traas, J. (1999) A chromosomal paracentric inversion associated with T-DNA integration in *Arabidopsis. Plant J.* **18,** 131–139.
38. Nacry, P., Camilleri, C., Courtial, B., Caboche, M., and Bouchez, D. (1998) Major chromosomal rearrangements induced by T-DNA transformation in *Arabidopsis. Genetics* **149,** 641–650.
39. Phelps, C. B. and Brand, A. H. (1998) Ectopic gene expression in *Drosophila* using GAL4 system. *Methods* **14,** 367–379.
40. Kiegle, E., Moore, C. A., Haseloff, J., Tester, M. A. and Knight, M. R. (2000) Cell-type-specific calcium responses to drought, salt and cold in the *Arabidopsis* root. *Plant J.* **23,** 267–278.
41. Tanahashi, H., Ito, T., Inouye, S., Tsuji, F. I. and Sakaki, Y. (1990) Photoprotein aequorin: use as a reporter enzyme in studying gene expression in mammalian cells. *Gene* **15,** 249–255.
42. Springer, P. S. (2000) Gene traps: tools for plant development and genomics. *Plant Cell* **12,** 1007–1020.
43. Lindsey, K., Topping, J. F., Muskett, P. R., Wei, W., and Horne, K. L. (1998) Dissecting embryonic and seedling morphogenesis in *Arabidopsis* by promoter trap insertional mutagenesis. *Symp. Soc. Exp. Biol.* **51,** 1–10.
44. Carthew, R. W. (2001) Gene silencing by double-stranded RNA. *Curr. Opin. Cell Biol.* **13,** 244–248.
45. Meissner, R., Chague, V., Zhu, Q., Emmanuel, E., Elkind, Y., and Levy, A. A. (2000) Technical advance: a high throughput system for transposon tagging and promoter trapping in tomato. *Plant J.* **22,** 265–274.
46. Bell, C. J., Dixon, R. A., Farmer, A. D., Flores, R., Inman, J., Gonzales, R. A., et al. (2001) The Medicago Genome Initiative: a model legume database. *Nucleic Acids Res.* **1,** 114–117.
47. Hiwatashi, Y., Nishiyama, T., Fujita, T., and Hasebe, M. (2001) Establishment of gene-trap and enhancer-trap systems in the moss *Physcomitrella patens. Plant J.* **28,** 105–116.
48. Holding, D. R. and Springer, P. S.(2002) The vascular prepattern enhancer trap marks early vascular development in *Arabidopsis. Genesis* **33**(4), 155–159.
49. Malamy, J. E. and Benfey, P. N. (1997) Analysis of SCARECROW expression using a rapid system for assessing transgene expression in *Arabidopsis* roots. *Plant J.* **12,** 957–963.
50. Thorsness, M. K., Kandasamy, M. K., Nasrallah, M. E., and Nasrallah, J. B. (1993) Genetic ablation of floral cells in *Arabidopsis. Plant Cell* **5,** 253–261.
51. Nakajima, K., Sena, G., Nawy, T., and Benfey, P. N. (2001) Intercellular movement of the putative transcription factor SHR in root patterning. *Nature* **20,** 307–311.
52. Geisler, M., Jablonska, B., and Springer, P. S. (2002) Enhancer trap expression patterns provide a novel teaching resource. *Plant Physiol.* **130,** 1747–1753.
53. Campisi, L., Yang, Y., Yi, Y., Heilig, E., Herman, B., Cassista, A. J., et al. (1999) Generation of enhancer trap lines in *Arabidopsis* and characterization of expression patterns in the inflorescence. *Plant J.* **17,** 699–707.
54. Devic, M., Hecht, V., Berger, C., Delseny, M., and Gallois, P. (1995) An assessment of promoter trapping as a tool to study plant zygotic embryogenesis. *C.R. Acad. Sci. Life Science* **318,** 121–128.
55. Bechtold, N., Ellis, J., and Pelletier, G. (1993) In planta *Agrobacterium* mediated gene transfer by infiltration of adult *Arabidopsis thaliana* plants. *C. R. Acad. Sci. Paris, Sciences de la vie/Life Sciences* **316,** 1194–1199.

56. Mollier, P., Montoro, P., Delarue, M., Bechtold, N., Bellini, C., and Pelletier, G. (1995) Promoterless *gusA* expression in a large number of *Arabidopsis thaliana* transformants obtained by the *in planta* infiltration method. *C.R. Acad.Sci. Paris, Sciences de la vie/Life sciences* **318**, 465–474.

57. Goddijn, O. J., Lindsey, K., van der Lee, F. M., Klap, J. C., and Sijmons, P. C. (1993) Differential gene expression in nematode-induced feeding structures of transgenic plants harbouring promoter-gusA fusion constructs. *Plant J.* **4**, 863–873.

58. Kim, S. C., Choi, H. C., Cho, M. J. and Han, C. D. (1999) Molecular analysis of rice plants harboring an Ac/Ds transposable element-mediated gene trapping system. *Plant J.* **19**, 615–623.

59. Fischer, J. A., Giniger, E., Maniatis, T., and Ptashne, M. (1988) GAL4 activates transcription in *Drosophila*. *Nature* **332**, 853–865.

60. Brand, A. and Perrimon, N. (1993) Targeted gene expression as means of altering cell fates and generating dominant phenotypes. *Development* **118**, 401–415.

61. Boisnard-Lorig, C., Colon-Carmona, A., Bauch, M., Hodge, S., Doerner, P., Bancharel, E., et al. (2001) Dynamic analyses of the expression of the HISTONE::YFP fusion protein in arabidopsis show that syncytial endosperm is divided in mitotic domains. *Plant Cell* **13**, 495–509.

62. Haseloff, J., Siemering, K. R., Prasher, D. C., and Hodge, S. (1997) Removal of a cryptic intron and subcellular localization of green fluorescent protein are required to mark transgenic *Arabidopsis* plants brightly. *Proc. Natl. Acad. Sci. USA* **94**, 2122–2127.

63. Vielle-Calzada, J-Ph., Baskar, R., and Grossniklaus, U. (2000) Delayed activation of the paternal genome during seed development. *Nature* **404**, 91–94.

64. Sieburth, L. E. and Meyerowitz, E. M. (1997) Molecular dissection of the AGAMOUS control region shows that cis elements for spatial regulation are located intragenically. *Plant Cell* **9**, 355–365.

65. Chuang, C. F. and Meyerowitz, E. M. (2000) Specific and heritable genetic interference by double-stranded RNA in *Arabidopsis thaliana*. *Proc. Natl. Acad. Sci. USA* **25**, 4485–4490.

66. Banks, J. A., Masson, P., and Fedoroff, N. (1988) Molecular mechanisms in the developmental regulation of the maize suppressor-mutator transposable element. *Genes Dev.* **2**, 1364–1380.

67. Fedoroff, N. V. (1995) Barbara McClintock—June 16, 1902–September 2, 1992. *Biogr. Mem. Natl. Acad. Sci.* **68**, 211–235.

V

ANIMALS AND ANIMAL CELLS

28

Gene Transfer and Expression
in Mammalian Cell Lines and Transgenic Animals

Félix Recillas-Targa

Summary

Manipulation of the eukaryotic genome not only has contributed to the progress in our knowledge of multicellular organisms but has also ameliorated our experimental strategies. Biological questions can now be addressed with more efficiency and reproducibility. There are new and varied strategies for gene transfer with improved methodologies that facilitate the acquisition of results. Cellular systems and transgenic animals have demonstrated their invaluable benefits. This chapter presents an overview of the methods of gene transfer, with particular attention to cultured cell lines. Alternative strategies of gene transfer are also shown and the applications of such methods are discussed. Finally, several comments are made about the influence of chromatin structure on gene expression. Recent experimental data have shown that for convenient stable transgene expression, the influence of chromatin structure should be seriously taken into account. Novel chromatin regulatory and structural elements are proposed as an alternative for proper and sustained gene expression. These chromatin elements are facing a new era in transgenesis and we are probably beginning a new generation of gene and cancer therapy vectors.

Key Words: Gene transfer; gene expression; antisense; transient transfection; stable expression; retroviral infection; gene therapy; microcell fusion; position effects; chromatin; locus control region; insulator.

1. Introduction

Starting with recombinant DNA technology, a large spectrum of applications has emerged. In particular, intense efforts have been made to transfer genetic information to cell lines, primary cell cultures, diverse organisms, and tissues, or to generate genetically modified organisms. Gene transfer technology requirements vary depending on

From: *Methods in Molecular Biology, vol. 267:*
Recombinant Gene Expression: Reviews and Protocols, Second Edition
Edited by: P. Balbás and A. Lorence © Humana Press Inc., Totowa, NJ

the transfer method and cell type. Without a doubt, gene transfer has been critical in the advance of our knowledge related to general phenomena such as gene regulation, post-translational events, recombinant protein production, and gene therapy, among others.

One particular fact to be taken into account is that there is no general methodology for gene transfer. Each cell type and organism needs prior careful characterization to ensure optimal transfer conditions to reach the highest efficiencies and reproducibility in terms of gene expression. But this is not the only obstacle that we face when we decide to transfer genetic information. In the great majority of experimental biological systems, we are constantly confronted by two variables. The first one has to do with the great variability in the expression levels of transgenes, and the second with a less studied phenomenon, the progressive extinction of transgene expression, which is present in the great majority of cases *(1)*. In addition, a less clear general phenomenon has to do with multiple copies of the same transgene which, once integrated into the host genome, induces a phenomenon called co-suppression in plants, causing gene expression silencing when multicopies of transgenes are integrated in tandem *(2)*.

In this chapter I summarize the experimental strategies and applications of recombinant DNA transfer. I analyze the most general methods for gene transfer, taking the two major experimental strategies to study transgene expression into account; that is, transient (also called episomal) strategies and stable integration into the host genome. Chromosomal transfer will also be included in this overview as a strategy to study gene expression in a natural chromosomal environment. I discuss transgene expression in detail and its association with regulatory elements, and conclude with some prospects and a recently discovered way to possibly overcome some transgene expression difficulties.

1.1. Goal of Recombinant DNA Transfer

The lack of accessible in vivo systems necessitates the search for alternatives to gene expression in eukaryotic cells. Over the years a long list of gene transfer applications has arisen and the constant appearance of new methodologies has made gene transfer more accessible and reproducible for research scientists.

1.1.1. Regulation of Gene Expression

One of the most common applications for gene transfer is the study of gene expression patterns. Developmental studies can be carried out in different types of cultured cell lines or in transgenic animals. Such kinds of studies can be supported by the use of primary cell cultures from various organisms and tissues, but we can also take advantage of transformed cell lines derived from viral infections or even different types of tumors. Overexpression of gene products can be an alternative to define gene function, to interfere with and search for a particular signal transduction pathway, or even to titrate posttranslational modifications of an endogenous peptide. At the present time, gene transfer methodologies allow controlled levels of gene expression, even though some problems remain unsolved.

Gene regulation studies are probably the most frequent way to investigate the activity of eukaryotic gene regulatory elements. Actually, the list of such elements is still growing, with classical examples being promoters, enhancers, locus control regions (LCRs), and, more recently, insulators *(3,4)*. All these studies are based on two main components—the use of measurable reporter genes and subsequent cell transfer in a transient or stable way. Plasmids carrying different reporter genes, such as chloramphenicol acetyl-transferase *(CAT)*, and more recently luciferase genes, β-galactosidase, and the green fluorescence proteins, are commercially available, and they facilitate and are less time-consuming when we are interested in defining the activity of a control element *(see* **Table 1**). Such reporters are frequently used in transient transfection experiments when the activity can be assayed enzymatically or colorimetrically between 24 and 48 h post-transfection. These kinds of experiments are not restricted to the use of those reporters. Specific genes can be employed with particular developmental and differential expression patterns. One of the classical examples emerged from the use of the human adult β-globin gene in the presence of its own promoter and with distinct versions of the LCR *(5,6)*. In this case, no enzymatic activity can be followed but molecular biology protocols can be applied like RT-PCR, RNase protection, S1-mapping, or even Northern blotting to estimate transcription levels *(7)*.

At this point it is very important to mention that transient transfection experiments, though very instructive, can give inconsistent results. In particular, this happens when compared to the same reporter vectors integrated into the genome of the host cell. In other words, the chromatin environment of an integrated reporter plasmid can give totally different results compared to episomal vectors. This fact is further supported by the fact that chromatin structure needs to be remodeled to allow regulatory factors to recognize their target sequences. Nonetheless, there is both direct and indirect experimental evidence suggesting that circular test plasmids acquire a chromatin-like organization with nucleosomes positioned along the plasmidic DNA every 200 base pairs (bp) or more, instead of the usual 147 bp *(8)*. Indirect evidence has shown that the use of histone deacetylase inhibitors in transiently transfected cells can cause the reactivation of silenced reporter genes *(9)*. In any case, the use of transiently transfected vectors represents a useful and direct way to analyze the activity of regulatory elements over reporter genes. Yet in the integrated context, where chromatin plays a central role, permanent and more realistic results are obtained.

1.1.2. Peptide Production

The generation of peptides from an isolated cDNA has great possibilities for diverse applications. One of the principal applications of peptide production is the use of tagged sequences to assist purification of a particular fusion protein for subsequent immunizations in order to generate antibodies for co-immunoprecipitation experiments. The use of eukaryotic cells can be particularly attractive when we take post-translational peptide modifications into account. Technically, transfection conditions for protein production should be performed in medium supplemented with serum or in medium without serum. However, it is preferable to express proteins in a serum-free medium since serum proteins interfere with the subsequent purification protocols of the expressed protein. A large list of research topics can take advantage of such experimental strategies, ranging from cell membrane receptors to signal transduction and

Table 1
Commonly Used Reporter Genes

Gene	Detection
β-Galactosidase (β-*gal*)	Enzymatic and colorimetric
Chloramphenicol acetyl transferase (*CAT*)	Enzymatic and radioactive
Luciferase	Luminiscence
Green fluorescent protein (*GFP*)	Fluorescence

chromatin structure. Recombinant peptide expression and maturation can also be useful for transport studies coupled with immunolocalization. More recently, a number of improved methods for particular peptide production have emerged. Among them, insect (bacculovirus), *Drosophila* cell lines, and bacteria are alternative options to mammalian cell lines, with a direct impact on the cost and on the amounts of recombinant peptides that are produced.

1.1.3. Antisense Expression

A more recent and everyday more effective way to manipulate gene expression is by the transfer of expression vectors that transcribe and target specific antisense RNAs to interfere with gene translation *(10,11)*. This approach allows the study of gene function in a more natural context, but several examples have shown a significant cellular cytotoxic effect *(12)*. A recent publication argues in favor of the use of short interfering RNAs (siRNAs) between 21 and 24 nucleotides *(13)*, but one of the encountered difficulties is the transient effect of such interfering RNA sequences. More recently, Agami and colleagues designed a novel mammalian interference expression vector which requires a specific 19-nucleotide sequence in the context of a well-defined transcription start site and a highly specific stem-loop structure under the control of the H1-RNA gene promoter *(11)*. The advantage of this pSUPER-siRNA vector is that it can be integrated into mammalian cells, thereby ensuring a constant rate of interference. These gene transfer strategies based on RNA interference represent, at least at the level of cultured cell lines, a real alternative to knockdown gene products as an experimental strategy to address reverse genetics questions *(14)*.

1.1.4. Tissue-Specific Expression

One of the principal goals in gene therapy design is the definition of the activity of highly specific regulatory elements in terms of their particular patterns of gene expression. Transient and stable transfections have demonstrated their utility, but with some limitations. At the organism level, transgenic animals and targeting strategies are also very instructive. The definition of tissue-specific patterns of gene expression remains crucial from the viewpoint of not only understanding the development of an organism, but also for future gene and cancer therapy protocols. At this point, it is critical to understand gene expression profiles but it is urgent to mention that it is not a simple task due to the different networks and redundancies used in nature to reach a highly specific pattern of gene expression. If we add the complexity and the participation of cell nucleus compartments and their associated chromatin structures, it remains clear that a lot of work and new tools are needed to start manipulating the genome *(15)*.

2. Methods for Gene Transfer

2.1. Transient Transfection

Successful transient transgene transfer is generally subject to several parameters. The gene transfer method, the cell physiology and cell type (suspension or adherent cultures), the enzymatic or colorimetric method of reporter activity quantification, the DNA amount, and the data normalization all require particular attention. The time period of gene expression in transient transfections ranges from 24 to 72 h. The cell conditions are critical for success in terms of healthy and actively dividing cells being best as efficient recipients of recombinant plasmids. Optimal conditions can be reached using a high-quality plasmid purification procedure. In our hands the best method is the one offered by Qiagen. Transient transfection efficiencies and reproducibility can be determined by taking advantage of a second reporter that expresses an easily detectable and measurable gene product. Such a reporter gene should not be present in the genome of the cell type to be tested, or in the worst case, present and expressed at low levels. The most commonly used reporter genes are shown in **Table 1.**

It is important to outline that plasmid integrity and its ability to reach the nucleus drastically diminishes the efficiency of reporter gene expression *(16)*. Southern blot analysis of 24 and 48 h posttransfection plasmids demonstrated that almost 50% of transiently transfected circular vectors are degraded (Recillas-Targa, unpublished observations). Over the years, a long discussion has been recurrently addressed concerning the location of transiently transfected plasmids inside the cell and/or the cell nucleus. Transient transfection and reporter gene expression inside the cell remain a controversial issue. Gene transfer methods usually direct plasmid DNA to the cytoplasm and it is not clear what percentage of these circular or linear molecules reach the cell nucleus. In addition, careful parameter definition is needed to reach a reasonable degree of reproducibility. It has been suggested that, among different mechanisms, passive diffusion of plasmids through the cytoplasm and nuclear pores represents the way in which reporter genes can migrate to the cell nucleus and be transcriptionally activated. New evidence argues in favor of a novel insight to explain the transfer of a DNA plasmid to the cell nucleus based on import DNA signals *(16,17)*. This new proposition is established through the idea of protein-plasmid DNA interactions in the cytoplasm, and that such transcription factor interactions mediate nuclear localization signals, allowing the importation of exogenous DNA into the nucleus.

If this vision is true, we now face an additional obstacle since the transfected plasmid will be conditioned to the tissue- and stage-specific factors and their associated nuclear import signals. One clear conclusion is that the procedure to analyze the expression of a transiently transfected gene can be oversimplified and we can face misinterpretations.

2.1.1. Lipofection

Cationic liposome-mediated transfection (lipofection) has been a more recent and efficient method for gene transfer both in vivo and in vitro *(18,19)*. Positively charged cationic liposomes form complexes with the negatively charged DNA molecule, allowing their fusion to cellular membranes. This complex is able to enhance gene delivery in diverse cell types, tissues, and even in combination with viral particles *(20,21)*. Furthermore, in vivo, cationic liposomes have been used successfully to deliver genes into different tissues, including lung, endothelium, and muscle, as well as for drug and vaccine delivery *(22,23)*. Lipofection is routinely performed in the absence of serum directly in the cell culture media. At the present time a growing list of commercial and noncommercial liposome formulations are now available *(24,25)*, outlining those that can be used in the presence of serum that facilitates the overall procedure (such as Effectine from Qiagen, Valencia, CA). Particular attention is needed to establish the liposome:DNA ratio for each cell type and DNA molecule. An excess of cationic liposomes could be toxic for a particular cell line, an observation that suggests the need to fix a low amount of liposomes and gradually increase the plasmid DNA concentration. Additional care is needed during the coupling procedure, since it needs to be done very slowly, with addition of progressive amounts of DNA, in order to wrap each liposome particle with the largest amount of plasmid DNA. In this way, optimal transient or stable transfections can the reached. Finally, in our hands lipofection has yielded the best results when adherent cells are used.

2.1.2. Particle Bombardment

Biolistic technology has been born as an alternative to gene transfer, particularly based on the need for plant transformation. In this procedure, target cells and tissues are held under vacuum for bombardment at variable distances. The DNA to be transferred is usually coupled to gold particles and bombardment occurs in a chamber that absorbs the shock waves and gases of the gunpowder explosion. The most recent modification to biolistic technology is the use of a helium-powdered acceleration system. Gene transfer has been widely used in plants, for example, maize suspension cells and embryos *(26,27)*. In any case, this method of gene transfer is not restricted to plants; there are examples of *in situ* transfections in diverse animal tissues *(28)*. It is worthwhile to mention that Bio-Rad offers the best technology and variety of instruments for this procedure.

2.1.3. Electroporation

An alternative and widely used method for gene transfer is based on the delivery of an electrical pulse that decays exponentially through time. Such a pulse permeabilizes cell membranes, allowing DNA molecules to be transferred inside the cell. Electroporation, as a strategy for gene transfer, involves several parameters that require particular attention, forcing the researchers to calibrate each experiment as a function of the DNA molecules to be transferred, and in particular, for the cell type to be transfected. The electrical pulse for an optimal transfection needs to take into account, the cell diameter or size and the composition of the cell membrane. In addition, it is well known that the electrical pulse induces a high degree of cell death due to the discharge. Of course, electroporation conditions should be established carefully and for each cell type to find an equilibrium between the highest efficiencies of gene transfer and the minimal amount of cell death. More recently, a new design of eletroporation apparatus allows in vivo transfection in whole-embryo culture. This improvement represents an attractive alternative for *in situ* gene expression, particularly during embryonic stages, that certainly complements genetic approaches like the generation of transgenic and mutant mice *(29)*.

2.1.4. Direct DNA Injection

Direct naked DNA transfer to cells has been widely used. It remains clear that the delivery to specific cell types remains limited with this strategy. Nonetheless, the direct delivery of plasmid DNA has been successfully applied to gene transfer into muscle cells, and particularly into epithelial cells *(30,31)*. One clear limitation of this procedure is the lack of control of critical parameters like plasmid DNA concentration, degradation, and efficiencies of gene transfer. In any case, and for specific applications like ectopic gene transfer, direct naked DNA transfer demonstrates some advantages.

2.2. Stable Expression

I have previously discussed the advantages of the generation of stable cell lines. Briefly, for the generation of stable lines, the transgene of interest should be co-transfected with the same circular plasmids or separate plasmids carrying a selectable marker gene that gives drug resistance in order to select positive integrants. At the present time a growing list of selectable marker genes is available. In any case, for every stable transfection assay, transgene integration into the target cell should be carefully monitored. For example, when the transgene of interest and the selectable marker are transfected on separate plasmids we can obtain resistant cells but without integration of the test plasmid. Another scenario is the integrity of the transgene. We have observed that in a range of 1% to 10%, integrated vectors are not intact, showing deletions that hamper their further analysis. In addition to all these careful characterizations of the integrant genes, Southern blot analysis can at the same time be very helpful to determine the copy number of the integrated transgenes. This could be valuable information depending on the type of experiment planned. In addition, antibiotics or drugs required for selection need to perform in a dose-response curve to determine the optimal concentration for each cell type and for each particular cell culture condition. One of the most frequently used marker genes is the amninoglycoside phosphotransferase gene (APH or neo^r), which confers resistance to the geneticin-selective antibiotic (G418 sulfate) by directing the synthesis of the APH enzyme that renders the drug inactive through phosphorylation.

In a typical stable transfection experiment, we first need to allow the cells to recover from transfection for 24 to 72 h in the absence of selection. This permits the cells to express adequate amounts of the resistance enzyme to protect the stably transfected cells against the drug. Forty-eight to 72 h post-transfection, the culture media should be removed and replaced with fresh media complemented with the drug. Selection can pursue for 1 to 3 wk posttransfection with frequent changes of the selectable culture media. Finally, the integrity and copy number should be determined by Southern blotting.

An easy alternative is the generation of stable pools that consist of groups of stable lines derived from different integration events of a particular transgene. With this strategy, we do not analyze one integration event, but instead we measure a group of events that reflect the overall activity of the transgene. This is a less time-consuming strategy compared to the selection of individual clones, which for some cell lines is difficult, particularly when they are not able to grow in semisolid media where clonal selection is carried out.

2.3. Retroviral Infection and Gene Therapy

Gene or genetic therapy faces a most exciting time. Novel strategies for gene transfer take into account the use of diverse types of viral vectors. Viral vectors demonstrated their ability to stably transform cultured cells and even to produce tumors in animal models that represent a useful strategy for probing the mechanisms of oncogenesis. Viral vectors have proved to be more efficient than other gene transfer strategies. Unfortunately, viral vectors are also subject to position effects and to their own viral structural and regulatory elements that cause progressive gene silencing, thereby preventing sustained transgene expression (32). Another consideration is that viral gene expression faces an additional problem in the ability of the viral particle to reach the desired target cell type (33).

2.4. Pro-Nucleus Micro-Injection and Knockout Technology (ES Cells)

Transgenic animals have been created with the aim of studying gene function and generating models of gene expression patterns during development, and to correlate dose observation with diseases. Transgenic animals, mice in particular, are unique for their production of peptides in an in vivo context, particularly in milk. Transgenic animals still have several technical limitations, starting with the low frequencies in obtaining founder animals. More recently, site-specific homologous recombination systems have allowed the elimination of a given sequence from the mouse genome and/or the site- and time-specific gene targeting of foreign sequences, and this has become a more frequently used experimental strategy (34,35). On the other hand, inducible systems such as tetracycline are more commonly used to control the activation of gene expression in transgenic animals at a particular moment, eliminating the risk of lethal phenotypes.

A more commonly used strategy for the manipulation of the mouse genome is by the introduction of mutations into embryonic stem (ES) cells by homologous recombination and the subsequent generation of chimeric mice. In this way, the mouse genome can be manipulated in favor of the loss of developmental gene expression patterns (36). One of the main limitations has to do with early lethal phenotypes, making the targeting approach very difficult to evaluate. Detailed analysis in a particular cell lineage can be very difficult, since some phenotypes affect multiple tissues in the mouse. At the present time, particular attention has been focused on the generation of conditional manipulations of the mouse genome in order to avoid the previously mentioned problems, and this allows phenotype characterizations in the cell or tissue type of interest at a specific developmental stage (37,38).

3. Chromosome Transfer

3.1. Microcell Fusion

Due to the central role of chromatin structure in the regulation of gene expression, undoubtedly genetic studies need to be done in the context of the intact chromatin environments *(15)*. Homologous recombination and transgenesis have been demonstrated as powerful tools for modifying and manipulating mammalian genetic loci. The great majority of such studies are performed by micro-injection of transgenes into the pro-nucleus of a fertilized mouse oocyte, or by the use of mouse embryonic stem cells where specific mutations or deletions can be generated by homologous recombination. In both cases we are confronted by two main problems. First, both approaches are limited by their low frequencies and large screening procedures are needed, and second, we are restricted to some mammalian organisms, particularly mice, limiting the manipulation of the human genome in a more natural chromatin environment, for example. To overcome such problems, intact chromosomes can be transferred from one cell type to another, allowing their genetic manipulation *(39–41)*.

Complementary to such chromosomal transfer, we can manipulate the transferred chromosome in the chicken B-cell line, DT40, which is an avian leukosis virus (ALV) transformed cell line where high targeting efficiencies by homologous recombination can be reached (ranging from 10% to 90%) *(41,42)*. Much less is known about the molecular mechanisms for the unusual high efficiencies of homologous recombination seen in this chicken cell line. The critical point with the use of the DT40 cell line is that chromosomes from other species can be transferred into these cells, where targeting efficiencies of the transferred chromosomes are of an unusually higher magnitude than in mammalian cells. Thus, genetic manipulation can be done in DT40 cells in the presence of the appropriate selectable markers, and then the modified chromosome can again be transferred to another cell line with an appropriate or convenient genetic background where gene expression studies can be performed (*see* **Fig. 1**).

Thus, a typical experiment takes into account two steps that involve microcell fusion of particular human chromosomes that contain the genetic domain of interest and a previously integrated selectable marker gene. At this point it is important to outline that for each microcell fusion event it is extremely important to verify the integrity of the transferred chromosomes, since during these procedures the loss of entire chromosomal regions have been observed with some frequency *(41,43)*. To this end, systematic PCR analysis all along the chromosome with previously defined genetic markers, complemented with fluorescent *in situ* hybridization (FISH), are necessary to confirm chromosome integrity. Microcell fusion procedures coupled with homologous recombination techniques have been successfully applied to chromatin studies, particularly to the human and mouse β-globin loci *(41,43)*.

In summary, microcell fusion coupled to the DT40 cell capacity to perform homologous recombination represents a clear alternative to knockout mice, and more broadly, the spectrum of cell types that can be analyzed.

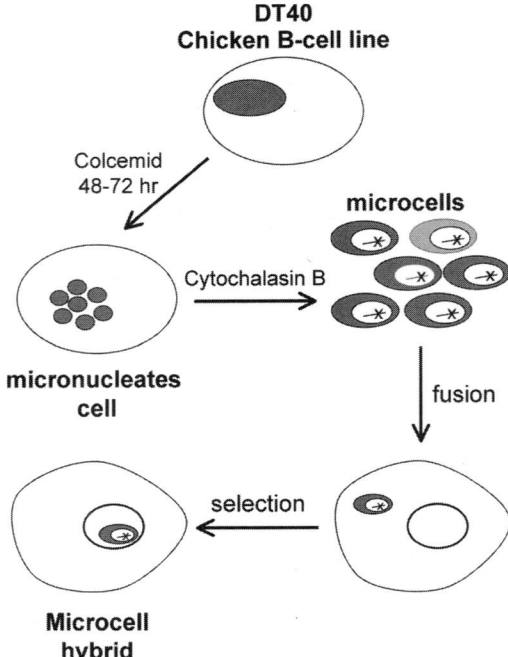

Fig. 1. Experimental procedure for the generation of microcell hybrids and modification of chromosomal DNA using homologous recombination in chicken B-cell tumor line DT40. The chromosome carrying the homologous recombination event is represented by an *X*. Homologous recombination events are selected and transferred into an appropriate mammalian cell line in which functional and chromatin structure can be done. Interestingly the cell donor can come from mammalian cells or directly from the chicken DT40 cells. For example, the first donor cell could be from human origin, modified in DT40 and the modified human chromosome analyzed in a mouse cell line *(41,43)*. At the present time DT40 technology is well established and documented at: http://swallow.gsf.de/dt40.html.

4. Gene Transfer and Expression Problems

Gene transfer and expression are two experimental procedures that are confronted by a couple of problems. The gene transfer method and its associated efficiency is one of the key factors for a gene to reach the cell nucleus in the case of cell lines, or in the most difficult scenario, to access a particular cell type in a tissue. On the other hand, transgene expression represents a key component for experimental or therapeutic purposes. As mentioned previously, gene expression faces two obstacles: the first one consists of a frequent variability in transgene expression levels, and the second, which is a less-studied phenomenon, is the progressive extinction of expression or silencing of the transgene. In both cases, the main component responsible for such effects, also known as chromatin position effects, is chromatin structure *(1)*.

4.1. Position Effects and Chromatin

There are two types of position effects: chromatin position effects caused by different integration sites and position effect variegation induced by a rearrangement and subsequent silencing of an active gene, frequently due to its inactivation because of its proximity to heterochromatin *(44–46)*. Briefly, position effect variagation has been defined as a stochastic and heritable silencing of gene expression. Originally described in *Drosophila*, it has been observed that the probability of gene expression is dependent on the strength of the regulatory elements of the transgene on the one hand, and on the other hand, on the chromatin environment surrounding the integration site *(45–47)*. Such an effect is particularly accentuated when the integration of the transgene occurs close to heterochromatin, where the silencing pressure is even stronger.

On the other hand, chromatin position effects are basically understood to be the variability in gene expression due to the random insertion of each transgene in varied chromatin environments in the genome. Yet, how does chromatin structure affect transgene expression? In recent years, basic research and therapeutic strategies have inspired the study of the causes of transgene silencing over time in vivo. The great majority of transgenes are subject to progressive silencing due to modifications in their chromatin organization, generating a highly repressive structure mediated by histone deacetylation and DNA methylation *(1,48,49)*. At the present time it remains unclear which of these chromatin modifications occurs first. In any case, initial observations show that in long-term cell culture, the histones of stably transfected genes are highly deacetylated and the CpGs located in their control elements are hypermethylated *(1,49)*. Interestingly, hypermethylation of the promoter regions of tumor suppressor genes appears to be an alternative mechanism of gene silencing with direct consequences in the origins of tumorigenesis *(50)*. As an indirect way to confirm transgene silencing by repressive chromatin conformation, histone deacetylase inhibitors like trichostatin A have been widely used, demonstrating a varied range of transgene expression reactivation *(50,51)*. 5-azacytidine or 5-aza-2'-deoxycytidine are DNA methylation inhibitors that are also capable of reactivating silenced transgenes *(1,52)*. Interestingly, a connection between DNA methylation and histone deacetylation has been demonstrated, first by the interaction of methyl-CpG-binding proteins with methylated DNA, followed by the attraction of histone deacetylases by co-repressors *(53)*. Based on such observations, when histone deacetylases and DNA methylation inhibitors are used simultaneously, the highest levels of gene expression reactivation are observed in the majority of cases *(50,54)*. All of this evidence confirms the influence of chromatin structure on transgene expression. As a consequence, studies of chromatin-remodeling mechanisms have been very useful in beginning to counteract the silencing effect of chromatin on integrated genes in order to reach sustained and homogeneous transgene expression.

4.2. Tissue-Specific Regulatory Elements

For any successful gene transfer and expression, the tissue or cell specificity should be carefully determined. In addition, the strength of the associated regulatory elements represents a complementary factor. Chromatin position effects had been overcome by the use of strong and dominant regulatory elements. Certain viral enhancer-promoter combinations are largely sufficient to overcome position effects, such as the CMV promoter-enhancer elements (Rincón-Arano and Recillas-Targa, unpublished observations). Unfortunately, these elements commonly possess a large spectrum of cell-type activity, making transgene expression too general. With the discovery of the locus control region (LCR) in the human β-globin locus, particular excitement was created with the possibility of overcoming chromatin position effects when those sequences were included in transgene vectors *(55)*. LCRs are defined as a group of DNase I hypersensitive sites located far upstream in the 5' non-coding region of the human β-globin domain *(5)*. Two main functions for LCRs have been suggested: first, they have strong and specific enhancer activity, and second, they contribute to open chromatin structure at the domain level *(5,6)*. The β-globin LCR provides strong erythroid-specific gene expression, and in its absence, the genes are now subject to strong chromatin position effects. When linked to a reporter gene, there is a copy-number-dependent gene expression in erythroid cells independently of the integration site *(55)*. In other words, the presence of the LCR has a positive and dominant effect on transgene expression independently of the chromatin integration context *(5,6,56)*. Actually, a significant number of LCRs have been discovered, all of them being tissue-specific *(57)*. The principal restriction for the use of LCRs in recombinant gene expression is their tissue specificity. In some way this is not a problem, particularly when individual cell types need to be reached.

Based on its properties, the LCR has been incorporated into transgene design in retroviral vectors and in the generation of transgenic mice *(32,48,58)*.

4.3. Sustained Expression and Chromatin Insulators

A more recent and attractive alternative for transgene expression, once again based on basic research on chromatin, is the use of insulator or boundary elements *(4)*. Initially discovered in *Drosophila,* vertebrate insulators have emerged in different chromatin domains. At the present time the best-characterized insulator is the chicken β-globin insulator *(59)*. Insulators are functionally defined based on two experimental properties: (1) they are able to interfere with enhancer-promoter communication exclusively when located between them, and (2) they have the capacity to protect a transgene, when located on each side of the vector, against chromatin position effects independent of the genomic integration site *(4,59)*. Usually, insulators are found to be constitutively hypersensitive to the action of nucleases, and in general they are neutral elements, i.e., they are not activators or repressors of transcriptional activity. All these features, and particularly the ability to shield transgenes against chromatin position effects and progressive extinction of expression, show the real potential of insulators in transgenesis and gene therapy. A convincing example of the capacity of insulators to improve transgene expression comes from the chicken β-globin cHS4 insulator, originally defined as the 5' boundary of the locus *(59)*. Placing the cHS4 insulator on each side of a reporter vector contributes to a more constant level of gene expression, much more dependent on the control elements of the introduced transgene. In addition, the expression is maintained over long periods of time in stably transfected cell lines *(1,60)*. More recently, it has been clearly confirmed that the chicken cHS4 insulator recruits high levels of histone acetylation through the action of histone acetyltransferases over the protected transgene; at the same time, it impedes DNA methylation of transgene control elements *(49,61)*. For the present discussion, we can say that insulators are becoming a realistic option to offset chromatin position effects in a great majority of cell types and organisms without tissue-specific restrictions, as seen with LCRs. The cHS4 insulator has been successfully used to improve the number of transgenic mice founders and to obtain homogeneous gene expression in transgenic mice *(62,63)*. The same insulator has also been used in *Drosophila*, culture cell lines, and retrovirus vectors *(62–65)*. It is also worthwhile to mention that the cHS4 and other insulators do not always protect a transgene against position effects. Therefore, much work remains to be done to better understand insulator mechanisms and their spectrum of action. Thus, the use of chromatin insulators in combination with tissue-specific regulatory elements is getting closer to reality as a means of protecting against position effects with direct consequences in the expression patterns of transgenes and gene therapy vectors.

4.4. Alternatives to Chromatin Insulators

Not only insulators have been shown to be successful in shielding transgenes against position effects, but matrix attachment regions (MARs) have also demonstrated their utility. MARs are A+T-rich sequences biochemically defined based on their capacity to interact with the nuclear matrix *(66–68)*. MAR sequences have been subject to certain skepticism in terms of their properties and contribution to gene regulation. They have frequently been proposed to drive chromatin loop formation, facilitating a topological configuration of a particular domain that indirectly favors gene activation. Consistent with this idea, the great majority of MAR sequences have been found in noncoding regions flanking certain domains. The best example is the chicken lysozyme locus, where two MAR sequences delimit the domain and show insulator properties, particularly by their capacity to improve gene expression in transgenic mice *(69,70)*. More recently, MAR sequences have been used in transgenic plants, dramatically improving gene expression *(71,72)*. At this point it is important to note that not all MAR sequences are able to improve transgene expression, and MARs from distinct loci do not give consistent results.

5. Conclusions and Prospects

Gene transfer and transgene expression represents one of the most frequently used approaches to addressing different scientific and practical questions. As mentioned above, the particular method used to transfer DNA molecules, chromatin environment, and specific regulatory elements associated with a transgene turn out to be critical factors in obtaining convenient patterns of expression. Alternative strategies are recommended to confirm the obtained results. In any case, a growing number of laboratories are concentrating their efforts on developing strategies for a more uniform and sustained level of transgene expression. MAR and insulator sequences require particular attention in the near future. At this point, there is no question that nuclear organization and dynamics, combined with chromatin structure and remodeling activities, need to be taken into account to establish better gene transfer protocols in the future.

Acknowledgments

I would like to thank Catherine M. Farrell for suggestions and critical reading of the manuscript. This work was partially supported the Dirección General de Asuntos del Personal Académico-UNAM (IN203200), the Consejo Nacional de Ciencia y Tecnología, CONACyT (33863-N), the Third World Academy of Sciences (TWAS, Grant 01-055 RG/BIO/LA), and the Fundación Miguel Alemán, A.C.

References

1. Pikaart M. J., Recillas-Targa, F., and Felsenfeld, G. (1998) Loss of transcriptional activity of a transgene is accompanied by DNA methylation and histone deacetylation and is prevented by insulators. *Genes Dev.* **12**, 2852–2862.
2. Garrick, D., Fiering, S., Martin, D. I., and Whitelaw, E. (1998) Repeat-induced gene silencing in mammals. *Nat. Genet.* **18**, 56–59.
3. Blackwood, E. M. and Kadonaga, J. T. (1998) Going the distance: a current view of enhancer action. *Science* **281**, 61–63.
4. West, A. G., Gaszner, M. and Felsenfeld, G. (2002) Insulators: many functions many mechanisms. *Genes Dev.* **16**, 271–288.
5. Bulger, M. and Groudine, M. (1999) Looping versus linking: toward a model for long-distance gene activation. *Genes Dev.* **13**, 2465–2477.
6. Li, Q., Peterson, K. R., Fang, X., and Stamatoyannopoulos, G. (2002) Locus control regions. *Blood* **100**, 3077–3086.
7. Carey, M. and Smale, S. T. (2000) Transcriptional regulation in eukaryotes. Cold Spring Harbor Laboratory Press, Cold Spring Harbor, NY p. 640.
8. Reeves, R., Gorman, C. M. and Howard, B. (1985) Minichromosomes assembly of nonintegrated plasmid DNA transfected into mammalian cells. *Nucl. Acids Res.* **13**, 3599–3615.
9. Krumm, A., Madisen, L., Yang, X. J., Goodman, R., Nakatani, Y., and Groudine, M. (1998) Long-distance transcriptional enhancement by the histone acetyltransferase PCAF. *Proc. Natl. Acad. Sci. USA* **95**, 13501–13506.
10. Sharp, P. A. (1999) RNAi and double-strand RNA. *Genes Dev.* **13**, 139–141.
11. Brummelkamp, T. R., Bernards, R., and Agami, R. (2002) A system for stable expression of short interfering RNAs in mammalian cells. *Science* **296**, 550–553.
12. Hunter, T., Hunt, T., Jackson, R. J., and Robertson, H. D. (1975) The characteristics of inhibition of protein synthesis by double-stranded ribonucleic acid in reticulocyte lysates. *J. Biol. Chem.* **250**, 409– 417.
13. Elbashir, S. M., Harborth, J., Lendeckel, W., Yalcin, A., Weber, K. and Tuschl, T. (2001) Duplexes of 21-nucleotide RNAs mediate RNA interference in cultured mammalian cells. *Nature* **441**, 494–498.
14. Varambally, S., Dhanasekaran, S. M., Zhou, M., Barrette, T. R., Kumar-Sinha, C., Sanda, M. G., et al. (2002) The polycomb group protein EZH2 is involved in progression of prostate cancer. *Nature* **419**, 624– 629.
15. Recillas-Targa, F. and Razin, S. V. (2001) Chromatin domains and regulation of gene expression: familiar and enigmatic clusters of chicken globin genes. *Crit. Rev. Eukaryot. Gene Exp.* **11**, 227–242.
16. Zohar, M., Mesika, A. and Reich, Z. (2001) Analysis of genetic control elements in eukaryotes: transcriptional activity or nuclear hitchhiking? *BioEssays* **23**, 1176–1179.
17. Vacik, J., Dean, B. S., Zimmer, W. E., and Dean, D. A. (1999) Cell-specific nuclear import of plasmid DNA. *Gene Ther.* **6**, 1006–1014.

18. Felgner, P.L. and Ringold, G.M. (1998) Cationic liposome-mediated transfection. *Nature* **337**, 387–388.

19. Pautz, G.E., Yang, Z.Y., Wu, B.Y., Gao, X., Huang, L., and Nabel, G.J. (1993) Immunotherapy of malignancy by *in vivo* gene transfer into tumors. *Proc. Natl. Acad. Sci. USA* **90**, 4645–4649.

20. Buttgereit, P., Weineck, S., Ropke, G., Marten, A., Brand, K., Heinicke, T., et al. (2000) Efficient gene transfer into lymphoma cells using adenoviral vectors combined with lipofection. *Cancer Gene Ther.* **7**, 1145–1155.

21. Pampinella, F., Lechardeur, D., Zanetti, E., MacLachlan, I., Benharouga, M., Lukacs, G.L., et al. (2002) Analysis of differential lipofection efficiency in primary and established myoblasts. *Mol. Ther.* **5**, 161–169.

22. Ledley, F.D. (1994) Non-viral gene therapy. *Curr. Opin. Biotech.* **5**, 626–636.

23. Gregoriadis, G. (1995) Engineering liposomes for drug delivery: progress and problems. *Trends Biotechnol.* **13**, 527–537.

24. Regelin, A.E., Fankhaenel, S., Gurtesch, L., Prinz, C., von Kiedrowski, G., and Massing, U. (2000) Biophysical and lipofection studies of DOTAP analogs. *Biochim. Biophys. Acta* **1464**, 151–164.

25. Dass, C.R., Walker, T.L., and Burton, M.A. (2002) Liposomes containing cationic dimethyl dioctadecyl ammonium bromide: formulation, quality control and lipofection efficiency. *Drug Deliv.* **9**, 11–18.

26. Klein, T., Fromm, M., Wiessinger, A., Tornes, D., Schaef, S., Slerren, M., et al. (1988) Transfer of foreign gene into intact maize cells with high-velocity microprojectiles. *Proc. Natl. Acad. Sci. USA* **85**, 4305–4309.

27. Klein, T., Harper, E.C., Svab, Z., Sanford, J., Fromm, M., and Maliga, P. (1988) Stable genetic transformation of intact *Nicotiana* cells by the particle bombardment process. *Proc. Natl. Acad. Sci. USA* **85**, 8502–8505.

28. Muangmoonchai, R., Wong, S.C., Smirlis, D., Phillips, I.R., and Shephard, E.A. (2002) Transfection of liver *in vivo* by biolistic particle delivery: its use in the investigation of cytochrome P450 gene regulation. *Mol. Biotechnol.* **20**, 145–151.

29. Osumi, N. and Inoue, T. (2001) Gene transfer into cultured mammalian embryos by electroporation. *Methods* **24**, 35–42.

30. Davis, H.L., Demeneix, B.A., Quantin, B., and Coulomb, J. (1993) Plasmid DNA is superior to viral vectors for direct gene transfer into adult mouse skeletal muscle. *Hum. Gen. Ther.* **4**, 733–740.

31. Sawamura, D., Akiyama, M., and Shimizu, H. (2002) Direct injection of naked DNA and cytokine transgene expression: implications for keratinocyte gene therapy. *Clin. Exp. Dermatol.* **27**, 480–484.

32. Pannell, D. and Ellis, J. (2001) Silencing of gene expression: implications for design of retrovirus vectors. *Rev. Med. Virol.* **11**, 205–217.

33. Nettelbeck, D.M., Jerone, V., and Muller, R. (2000) Gene therapy: designer promoters for tumour targeting. *Trends Genet.* **16**, 174–181.

34. Metzger, D. and Chambon, P. (2001) Site- and time-specific gene targeting in the mouse. *Methods* **24**, 71–80.

35. Kwan, K.-M. (2002) Conditional alleles in mice: practical considerations for tissue-specific knockouts. *Genesis* **32**, 49–62.

36. Yu, T. and Bradley, A. (2001) Mouse genomic technologies engineering chromosomal rearrangements in mice. *Nat. Rev. Genet.* **2**, 780–790.

37. Lobe, C.G. and Nagy, A. (1998) Conditional genome alteration in mice. *BioEssays* **20**, 200–208.

38. Kaartinen, V. and Nagy, A. (2001) Removal of the floxed neo gene from a conditional knockout allele by the adenoviral Cre recombinase *in vivo*. *Genesis* **31**, 126–129.

39. Killary, A. McN. and Lott, S. T. (1996) Production of microcell hybrids. *Methods* **9**, 3–11.

40. Dieken, E. S. and Fournier, R. E. K. (1996) Homologous modification of human chromosomal genes in chicken B-cell X human microcell hybrids. *Methods* **9**, 56–63.

41. Dieken, E. S., Epner, E. M., Fiering, S., Fournier, R. E. K., and Groudine, M. (1996) Efficient modification of human chromosomal alleles using recombination-proficient chicken/human microcell hybrids. *Nat. Genet.* **12**, 174–182.

42. Buerstedde, J. M. and Takeda, S. (1991) Increased ratio of targeted to random integration after transfection of chicken B cell lines. *Cell* **67**, 179–188.

43. Epner, E., Reik, A., Cimbora, D., Telling, A., Bender, M. A., Fiering, S., et al. (1998) The β-globin LCR is not necessary for an open chromatin structure or developmentally regulated transcription of the native mouse β-globin locus. *Mol. Cell* **2**, 447–455.

44. Robertson, G., Garrick, D., Wu, W., Kearns, M., Martin, D., and Whitelaw, E. (1995) Position-dependent variegation of globin transgene expression in mice. *Proc. Natl. Acad. Sci. USA* **92**, 5371–5375.

45. Henikoff, S. (1996) Dosage-dependent modification of position-effect variegation in *Drosophila*. *BioEssays* **18**, 401–409.

46. Wakimotot, B. T. (1998) Beyond the nucleosome: epigenetic aspects of position-effect variegation in *Drosophila*. *Cell* **93**, 321–324.

47. Karpen, G. H. (1994) Position-effect variegation and the new biology of heterochromatin. *Curr. Opin. Genet. Develop.* **4**, 281–291.

48. Rivella, S. and Sadelain, M. (1998) Genetic treatment of severe hemoglobinopathies: the combat against transgene variegation and transgene silencing. *Sem. Hematol.* **35**, 112–125.

49. Mutskov, V. J., Farrell, C. M., Wade, P. A., Wolffe, A. P. and Felsenfeld, G. (2002) The barrier function of an insulator couples high histone acetylation levels with specific protection of promoter DNA from methylation. *Genes Dev.* **16**, 1540–1554.

50. Jones, P. A. and Baylin, S. B. (2002) The fundamental role of epigenetic events in cancer. *Nat. Rev. Genet.* **3**, 415–428.

51. Marks, P. A., Rifkind, R. A., Richon, V. M., Breslow, R., Miller, T., and Kelly, W. (2001) Histone deacetylation and cancer: causes and therapies. *Nat. Rev. Cancer* **1**, 194–202.

52. Christman, J. K. (2002) 5-Azacytidine and 5-aza-2'-deoxycytidine as inhibitors of DNA methylation: mechanistic studies and their implications for cancer therapy. *Oncogene* **21**, 5483–5495.

53. Bird, A., (2002) DNA methylation patterns and epigenetic memory. *Genes Dev.* **16**, 6–21.

54. Magdinier, F. and Wolffe, A. P. (2001) Selective association of the methyl-CpG binding protein MBD2 with the silent *p14/p16* locus in human neoplasia. *Proc. Natl. Acad. Sci. USA* **98**, 4990–4995.

55. Grosveld, F., van Assendelf, G. B., Greaves, D. R. and Kollias, B. (1987) Position-independent high-level expression of the human β-globin gene in transgenic mice. *Cell* **51**, 975–985.

56. Festenstein, R. and Kioussis, D. (2000) Locus control regions and epigenetic chromatin modifiers. *Curr. Opin. Genet. Dev.* **10**, 199–203.

57. Bonifer, C. (2000) Developmental regulation of eukaryotic gene loci: which *cis*-regulatory information is required? *Trends Genet.* **16**, 310–315.

58. Neff, T., Shotkoski, F., and Stamatoyannopoulos, G. (1997) Stem cell gene therapy, position effects and chromatin insulators. *Stem Cells* **15**, 265–271.

59. Burgess-Beusse, B., Farrell, C., Gaszner, M., Litt, M., Mutskov, V., Recillas-Targa, F., et al. (2002) The insulation of genes from external enhancers and silencing chromatin. *Proc. Natl. Acad. Sci. USA* **9,** 16,433–16,437.

60. Recillas-Targa, F., Pikaart, M. J., Burgess-Beusse, B., Bell, A. C., Litt, M. D, West, A. G., et al. (2002) Position-effect protection and enhancer blocking by the chicken β-globin insulator are separable activities. *Proc. Natl. Acad. Sci. USA* **99**, 6883–6888.
61. Litt, M. D., Simpson, M., Recillas-Targa, F., Prioleau, M. N., and Felsenfeld, G. (2001) Transitions in histone acetylation reveal boundaries of three separately regulated neighboring loci. *EMBO J.* **20**, 2224–2235.
62. Wang, Y., DeMayo, F. J., Tsai, S. Y., and O'Malley, B. W. (1997) Ligand-inducible and liver-specific target gene expression in transgenic mice. *Nat. Biotechnol.* **15**, 239–243.
63. Potts, W., Tucker, D., Wood, H., and Martin, C. (2000) Chicken β-globin 5'HS4 insulator function to reduce variability in transgenic founder mice. *Biochem. Biophys. Res. Commun.* **273**, 1015–1018.
64. Ciana, P., DiLuccio, G., Belcredito, S., Pollio, G., Vegeto, E., Tatangelo, L., et al. (2001) Engineering of a mouse for the *in vivo* profiling of estrogen receptor activity. *Mol. Endocrinol.* **15**, 1104–1113.
65. Emery, D. W., Yannak, E., Tubb, J., Nishino, T., Li, Q., and Stamatoyannopoulos, G. (2002) Development of virus vectors for gene therapy of beta chain hemoglobinopathies: flanking with a chromatin insulator reduces gamma-globin gene silencing *in vivo*. *Blood* **100**, 2012–2019.
66. Hart, C. M. and Laemmli, U. K. (1998) Facilitation of chromatin dynamics by SARs. *Curr. Opin. Genet. Dev.* **8**, 519–525.
67. Cremer, T., Kreth, G., Koester, H., Fink, R. H., Heintzmann, R., Cremer, M., et al. (2000) Chromosome territories, interchromatin domain compartment, and nuclear matrix: an integrated view of the functional nuclear architecture. *Crit. Rev. Eukaryot. Gene Expr.* **10**, 179–212.
68. Razin, S. V. (2001) The nuclear matrix and chromosomal DNA loops: Is their any correlation between partitioning of the genome into loops and functional domains? *Cell. Mol. Biol. Lett.* **6**, 59–69.
69. Stief, A., Winter, D. M., Strätling, W. H., and Sippel, A. E. (1989) A nuclear DNA attachment element mediates elevated and position-independent gene activity. *Nature* **341**, 343–345.
70. Phi-Van, L. and Strätling, W. H. (1996) Dissection of the ability of the chicken lysozyme gene 5' matrix attachment region to stimulate transgene expression and to dampen position effects. *Biochemistry* **35**, 10735–10742.
71. Mlynarova, L., Jansene, R. C., Conner, A. J., Stiekema. W. J., and Nap, J. P. (1995) The MAR-mediated reduction in position effect can be uncoupled from copy number-dependent expression in transgenic plants. *Plant Cell* **7**, 599–609.
72. Brouwer, C., Bruce, W., Maddock, S., Avramova, Z., and Bowen, B. (2002) Suppression of transgene silencing by matrix attachment regions in maize: dual role for the maize 5'ADH1 matrix attachment region. *Plant Cell* **14**, 2251–2264.

29

Sustained Heterologous Transgene Expression in Mammalian and Avian Cell Lines

Héctor Rincón-Arano and Félix Recillas-Targa

Summary

Chromosomal position effects are the cause of variegation of transgene expression relating to the eukaryotic genome environment. These effects modify heterologous transgene expression, presumably due to dominant epigenetic marks that cause variegated or gradual silencing of transgene expression. As a solution, chromosomal insulators have arisen as a way to protect transgene expression against repressive marks (position effects), allowing homogeneous and consistent expression. Sustainable expression enables us to gain insight into in vivo protein function and protein stability, or to study position effects. Here, we show that the chicken cHS4 β-globin insulator is able to protect transgene expression in cell lines, demonstrating its potential as a tool to improve the heterologous expression of any desired transgene. In addition, we show that green fluorescent protein (GFP) can be used as a reporter protein in the study of chromosomal position effects.

Key Words: Stable transfection; insulator; locus control region (LCR); position effects; chromatin; gene resistance; silencing; GFP.

1. Introduction

Over the last decade, transient and stable heterologous gene expression in cultured cell lines has become one of the best methods to define the estimated function of genes in vivo. However, the stable expression of transgenes in mammalian cells, as in transgenic animals, varies between independent clones, where transgene expression traits are diminished over time, and in extreme cases are completely silenced (*1,2*). The silencing of transgene expression is a frequent problem in sustained heterologous protein expression, and mainly in gene therapy (*3*). For example, transcriptional silencing of retrovirus

From: *Methods in Molecular Biology, vol. 267:*
Recombinant Gene Expression: Reviews and Protocols, Second Edition
Edited by: P. Balbás and A. Lorence © Humana Press Inc., Totowa, NJ

vectors is one of the major obstacles in gene therapy *(3)*. These problems are, in part, due to position effects (positive or negative) of the chromatin environment at the integration site of the transgene. To overcome these problems in mammalian cell lines, we should consider two main issues: (1) stable transfection experiments must be optimized to identify and select clones with a robust and sustained expression, or (2) transgenic constructs should be protected from position effects.

Regulatory elements such as enhancers were considered as the first tool to maintain sustained expression of a transgene. These elements are able to stimulate transcription and are tissue-specific. However, the great majority of enhancers are sensitive to position effects, giving a variegated expression pattern *(4)*. Recently, locus control regions (LCRs) and insulators have become a realistic option to obtain constant and sustained expression. LCRs and insulators fall into the category of higher-order chromatin organization elements *(5,6)*. They favor a relaxed chromatin environment and their effects seem to be dominant. Transgenes that are protected by LCRs or insulators maintain sustained expression for long time periods in a copy number-dependent manner and independent of the integration site *(7,8)*. Several of these elements have been described and are found in different transcriptionally active chromatin domains *(6,9)*. LCRs and insulators have therefore arisen as an alternative option to maintain sustained transgene expression. They can be used as a potential tools in transgene expression and gene therapy.

1.1. Chromatin Position Effects

Position effects (PE) are a set of processes that positively or negatively influence the transcriptional expression of transgenes. Sooner or later after transgene integration, the progressive extinction of expression starts and sustained expression is compromised *(2,4)*. On occasions, the effects of the integration site can be positive because transgenes can be integrated into transcriptionally active sites or into an open chromatin environment, favoring their proper and sustained expression. However, transgenes can also be integrated in close proximity to heterochromatic regions, like telomeres, affecting transgene transcription and silencing their expression. Even though the precise mechanism of transcriptional silencing is still unknown, molecular mechanisms of transcriptional silencing are suspected to be due mainly to the invasion of heterochromatin as an epigenetic mark *(10)*. This invasion, also known as "heterochromatin spreading," has been suggested to happen through progressive DNA methylation and histone deacetylation of the transgene *(2)*. For instance, in pre-erythroblast cells, silenced transgene expression can be rescued by treatment with trichostatin A (TSA), a histone deacetylase inhibitor, and with 5-azacytidine (5-azaC), a nonmethylatable homolog of cytosine *(2)*. On the other hand, transcriptional silencing is suspected to be the result of relocalization of transgenes near heterochromatic domains or to inactive nuclear territories *(9)*. Repressive chromatin modification and gene relocalization may be prevented by several chromatin regulatory and structural elements. Expression of stably integrated β-globin transgenes in the absence of the locus control region, which is necessary to keep the chromatin in an open state and enhance transgene expression, occurs at low levels and varies with the site of integration, and thus, it is subject to position effects.

In the past few years, regulatory sequences, such as enhancers or LCRs, were used to overcome negative position effects. Unfortunately, their action is restricted to specific tissues. Recently, boundary elements, such as matrix attachment regions (MARs) and chromatin insulators, have emerged as new tools to protect transgenes from chromatin position effects *(7,11)*. These elements are able to maintain histone hyperacetylation in a directional manner and to protect against DNA methylation of the shielded domain in vivo, allowing homogeneous and sustained transgene expression *(12)*.

1.2. Locus Control Regions

Physiological expression of cellular genes is regulated through the action of several elements, such as enhancers, silencers, insulators, and locus control regions *(13)*. LCRs are long-range regulatory elements that confer tissue-specific, copy-number-dependent and position-independent transcriptional expression on linked genes *(7,8,14)*. Transgenes protected with an LCR are shielded against negative or positive position effects. These elements are able to maintain a transcriptionally active locus in a tissue-specific manner, and together with their enhancer function, allow sustained and tissue-specific expression levels. Because of their characteristics, LCRs have arisen as an alternative to protect transgene expression against position effects. However, the main obstacle in considering LCRs as a general tool to protect transgenes from PE is their tissue specificity *(8)*. LCRs could be used as cell type-specific tools in the sustained expression of heterologous genes in particular tissues. Recently, LCRs have been considered in the development and design of tissue-specific gene therapy vectors *(15)*.

1.3. Insulators

It has been proposed that genetic information within the eukaryotic genome is organized into functionally discrete and independent domains *(16)*. Each individual domain is separated from the others by chromatin components that serve to organize and distribute the genome into active or inactive transcriptional clusters *(17)*. The main idea is that insulators, also known as boundary elements, are chromatin elements that contribute to the formation and maintenance of a particular chromatin domain. Functionally, they are defined as elements that are able to modulate enhancer-promoter interactions and are capable of overcoming chromatin position effects *(18)*. These elements can protect stably integrated transgenes against strong adjacent regulatory elements and heterochromatic regions in a wide range of cell types *(6)*. Experimental evidence supports the fact that chromatin insulators function in a tissue-independent manner *(7)*. Nonetheless, their exact mechanism of overcoming position effects is not yet clear. The best-characterized insulator is chicken hypersensitive site 4 (cHS4), which defines the 5' end of the β-globin domain. The cHS4 β-globin insulator protects transgenes from position effects. Recently, the Felsenfeld group has suggested that the cHS4 β-globin insulator protects a transgene against promoter methylation, which acts as an epigenetic mark that induces a repressive and permanent chromatin structure over the transgene *(12)*. At the present time there is clear evidence showing that chromatin insulators possess different activities and mechanisms of action *(6)*. Some insulators have clearly arisen as a potential tool to be used for the sustained expression of transgenes

(7,19). Their effects over regulatory elements and chromatin have generated great interest in various research fields.

1.4. Stable Transfection and Sustained Expression

A well-designed strategy to get efficient amounts of transgene-expressing cells is the requirement to obtain homogeneous and sustained expression. This strategy will depend on two main issues that we must consider: endogenous and exogenous problems. Endogenous conditions have to do with the efficiency of expressing a heterologous protein and with overcoming position effects from the chromatin environment in each cell line. Transgene expression can be improved by introducing all the necessary cis-regulatory elements that maintain expression in vivo (i.e., to maintain stable and sustained transgene expression). On the other hand, exogenous or experimental conditions, such as the gene-transfer procedure and the cell type used, as well as the plasmid features or the promoter used, must be optimized and standardized for every cell line.

For sustained expression with stable transfections of heterologous genes, it has been necessary to co-express a resistance marker (antibiotic selection). However, it has been suggested that some antibiotics may alter specific experimental results. On the other hand, the absence of negative selection shows that transcription decays over time (**Fig. 1**). Insulators can be used to protect transgene expression, avoiding regulatory interference from the resistance gene and allowing sustained expression over long time periods in cell culture.

This chapter describes the sustained expression maintained by the chicken β-globin insulator and a method to follow transgene expression over long time periods. We do not attempt to characterize the β-globin insulator or to endorse this insulator as the only insulator to be used. Here, we used the enhanced green fluorescent protein (EGFP) as a reporter transgene in clonal and stable cell lines to be monitored by fluorescent cytometry over more than 80 d in culture. EGFP fluorescence could be analyzed as an alternative to the use of fluorescein-labeled antibodies, avoiding the use of expensive and specific antibodies (a time-consuming procedure). Flow cytometry analysis of GFP has been used for monitoring expression of inducible reporters (20) and for monitoring promoter activity (21).

2. Materials
2.1. Plasmid DNA and Linearization

1. 1.5-mL microtubes.
2. Escherichia coli strain DH5α.
3. Plasmids: Chicken α-globin enhancer and α^D promoter were cloned upstream of the EGFP coding sequence in the plasmid pEGFP-1 (Clontech) to make the construct named $pGE\alpha^D$. The human retinoblastoma promoter was amplified from human leukocyte genomic DNA and cloned upstream of the EGFP gene (pEGFP-1) to construct pERb. Standard reagents and methods were used for subcloning DNA fragments into plasmids.
4. Maxipreps or midipreps were carried out using a commercially available DNA extraction kit (e.g., the Qiagen Midiprep Plasmid DNA extraction kit).

5. 10X reaction buffer: 500 mM potassium acetate, 200 mM Tris-acetate, 100 mM magnesium acetate, 10 mM dithiothreitol (pH 7.9).
6. Single-cut restriction endonuclease for the plasmid (Clontech), e.g., *Apa*LI or *Eco*0190I.
7. 10X bovine serum albumin (BSA): dissolve 100 mg of albumin in 100 mL of sterile water to get a final concentration of 1 mg/mL.
8. Pre-cooled absolute ethanol.
9. Pre-cooled 70% ethanol.
10. Water or TE (10 mM Tris-HCl, 1 mM EDTA, pH 7.8).
11. Phenol/chloroform/isoamyl alcohol (25:24:1).
12. 3 M sodium acetate, pH 5.3.

2.2. Tissue Culture for Cell Lines

2.2.1. Pre-Erythroid HD3 Cell Line (Suspension Cells)

1. Dulbecco's Modified Eagle Medium (DMEM, Invitrogen).
2. Inactivated fetal bovine serum (Invitrogen).
3. Inactivated chicken serum (Invitrogen).
4. Cell culture dishes.

2.2.2. Cervical HeLa Cell Line

1. Dulbecco's Modified Eagle Medium (DMEM, Invitrogen).
2. Inactivated fetal bovine serum.
3. 25% Trypsin-1 mM EDTA (Invitrogen or Sigma).
4. Cell culture dishes.

2.3. DNA Transfection

2.3.1. Electroporation Procedure

1. Dulbecco's Modified Eagle Medium (DMEM, Invitrogen) with 10% fetal bovine serum.
2. DMEM (Invitrogen), serum-free.
3. Cold phosphate-buffered saline (PBS): 137 mM NaCl, 2.7 mM KCl, 10 mM Na$_2$HPO$_4$ and 2 mM KHPO$_4$, pH 7.2, supplemented with 6 mM glucose.
4. Electroporation apparatus (e.g., Bio-Rad Gene Pulser II apparatus).
5. Electroporation cuvets (e.g., 0.4 cm Bio-Rad gene pulser cuvetts).
6. Plasmid pGEα^D.
7. 10-cm and 25-cm cell culture dishes.

2.3.2. Lipofection Procedure

1. DMEM (Invitrogen) with 10% fetal bovine serum.
2. DMEM, serum-free.
3. Fetal bovine serum.
4. 0.25% Trypsin/ EDTA.
5. Liposome reagent (e.g., Lipofectamine Plus reagent, Invitrogen).
6. 25-cm cell culture dishes.
7. 6-well plates.
8. 15-mL tubes (e.g., Falcon tubes).

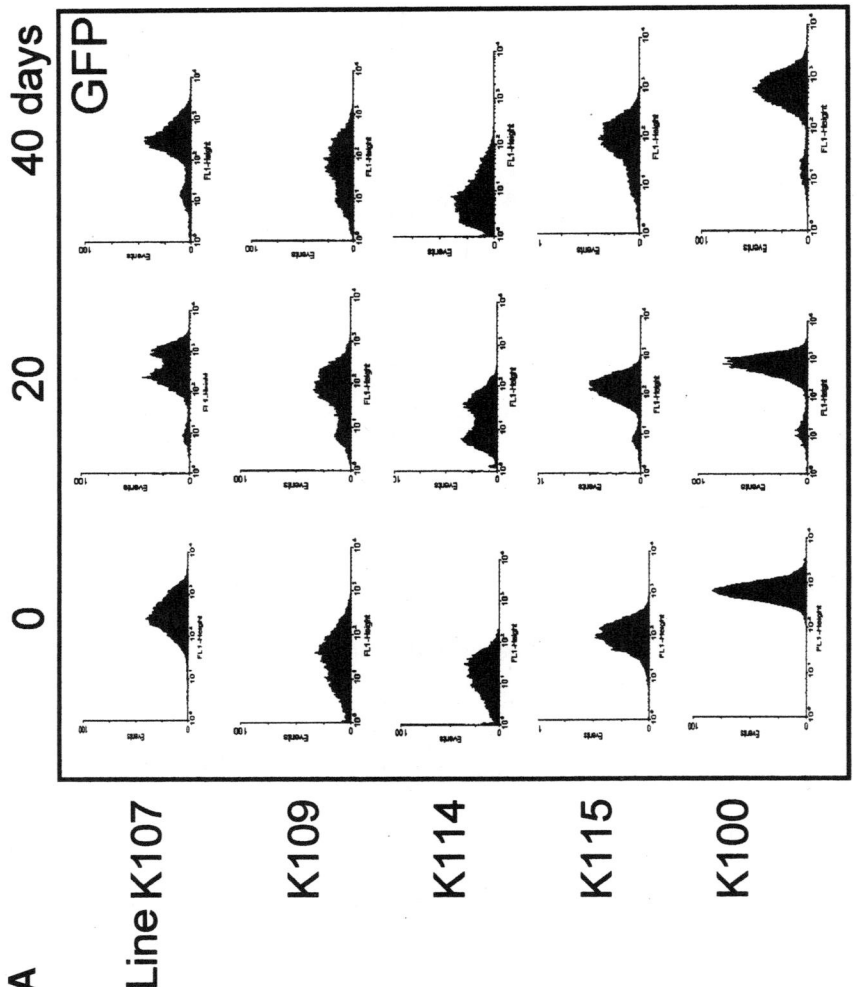

440

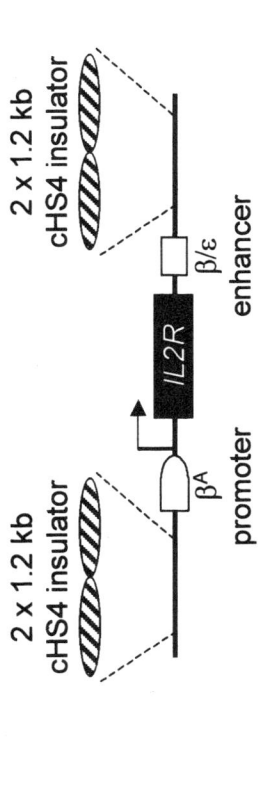

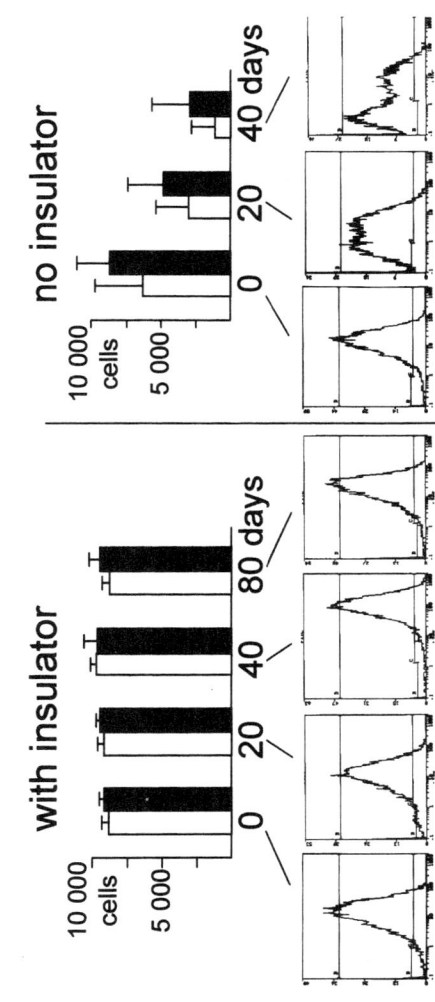

Fig. 1. Progressive extinction of transgene expression and insulator action. (**A**) GFP extinction of expression in the chicken HD3 cell line. HD3 clones were transfected with pGEα^D (*see* **Subheading 2.3.**) and were incubated in the absence of drug selection in continuous cell culture. (**B**) Several clones of the chicken pre-erythroblast 6C2 cell line transfected with an IL2R-based transgene show silencing over time. The bar graph represents the average of 10 independent stable lines and shows the IL2R-expressing cells (white bars: single-copy integrants; black bars: multi-copy integrants). Notice that the size of the error bars reflects the large variability in expression of individual lines, in particular on the uninsulated vector, caused by position effect derived from the influence of different integration sites. Cells were transfected with the plasmid in the absence or presence of two copies of 1.2-kb cHS4 β-globin insulator (data kindly provided by Gary Felsenfeld). The presence of the cHS4 β-globin insulator protects transgene expression over time. On the bottom a representative graph profile is shown.

2.4. Clone Selection

2.4.1. Suspension-Growing Cells

1. Semisolid medium: We used HD3 cells growing in medium with 10% methylcellulose (Methocel MC, Fluka). Dissolve the medium (DMEM) in 400 mL of dH_2O and adjust the pH to 7.2. Add serum and antibiotics, 8% fetal bovine serum, 2% chicken serum, and 0.9 mg/mL geneticin (active fraction). Adjust the volume to 500 mL to give a 2X medium. In another water-containing sterile bottle, the methylcellulose should be dissolved in 500 mL. The bottle must be heated on a thermoblock to completely dissolve the methylcellulose. Once dissolved, the methylcellulose-containing bottle should be cooled down to reach room temperature, and after that, the complemented 2X medium should be added. Shake the medium overnight at 4°C, and finally, make 40-mL aliquots in 50-mL tubes. Store at 4°C until ready for use.
2. Chicken serum.
3. Fetal bovine serum.
4. 50 mg/mL geneticin (Invitrogen).
5. 25-cm cell culture dishes.
6. 96-well plates.
7. Selection growth medium: DMEM supplemented with 8% fetal bovine serum, 2% chicken serum, and 0.9 mg/mL geneticin.

2.4.2. Cloning Antibiotic-Resistant Adherent Cells

1. Growth medium for HeLa cells: DMEM supplemented with 10% fetal bovine serum.
2. Selection growth medium: DMEM supplemented with 10% fetal bovine serum and 0.5 mg/mL geneticin.
3. 10- and 25-cm cell culture dishes.
4. Cloning discs (PGC Scientifics).
5. 0.25% Trypsin/1 mM EDTA.
6. 24-well or 6-well plates.

2.5. Transgene Integrity Determination by Southern Blot Analysis

2.5.1. Rapid DNA Extraction

1. Sterile 1.5-mL tube.
2. T1 buffer: 200 mM Tris-HCl, pH 8.5, 400 mM NaCl.
3. T2 lysis buffer: 0.4% sodium dodecyl sulfate (SDS), 10 mM EDTA, and just before use add 50 μg RNase A and 50 mg proteinase K.
4. Isopropanol.
5. Pre-cooled 70% ethanol.
6. Water or TE (1 mM EDTA, 10 mM Tris-HCl, pH 7.8).

2.5.2. Southern Transfer

1. Genomic DNA digestion: genomic DNA must be incubated with a restriction endonuclease that cuts asymmetrically in the reporter plasmid to allow copy number determination and integrity.
2. 1% agarose gel in 1X TBE.
3. 1X TBE.

4. Ethidium bromide.
5. Radiolabeled DNA size markers.
6. Probe.
7. Electrophoresis apparatus.
8. Denaturation buffer (1 L): 3 M NaCl (175.5 g NaCl) and 0.4 N NaOH (16 g NaOH).
9. Filtered hybridization buffer (1 L): 7.5 mL 37% formamide, 1.5 mL 10% SDS, 3 mL 5 M NaCl, 3 mL 50% dextran sulfate. Add the denatured probe and 200 μL denatured salmon sperm DNA (dssDNA). The DNA can be denatured by incubating for 5 min at 95°C and then cooling down immediately on ice.
10. Pre-hybridization buffer: 6 mL 20X SSC, 2 mL 50X Denhardt's solution, 1 mL 10% SDS, 200 μL 10 mg/mL dssDNA, 10 μL 37% formamide and 1 mL H_2O (filtered).
11. Transfer buffer (1 L): 3 M NaCl (175.5 g NaCl), 8 mM NaOH (0.32 g NaOH), 2 mM sarkosyl (0.63 g sarkosyl, not necessary).
12. Neutralization buffer (5X, 1 L): 1 M phosphate buffer, pH 6.8, 71 g Na_2HPO_4 and 60 g $NaH_2PO_4 \cdot H_2O$.

2.6. Flow Cytometer Analysis

1. Phosphate buffered solution (PBS).
2. Cytometer (FACScalibur microflow cytometer, Becton Dickinson, Franklin Lakes, NJ). The argon laser was tuned at 488 nm, and the fluorescent cells were evaluated with a 525 nm band-pass filter.
3. Calibration beads (Becton Dickinson).

3. Methods

3.1. DNA Linearization

Obtaining high-purity DNA is a necessary step to improve transfection efficiency. DNA extraction with a CsCl-gradient protocol is one of the best ways to get DNA with the highest purity, but it is more labor-intensive and time-consuming. Several kits are commercially available, and DNA from these gives the required purity. DNA extraction kits are easy to use and reduce the labor time. Several companies offer different kinds of DNA extraction kits; we suggest that the DNA quality from different kits should be compared for the transfection protocol chosen. Excellent-quality DNA will ensure easy manipulation and will improve transfection efficiency.

Frequently, stable transfection requires the linearization of the DNA with a single-cut restriction endonuclease. The cut site should be located outside the coding sequence of the reporter or the selection gene. DNA linearization protects and ensures better transgene integrity. The transgene will be integrated using the digested ends of the plasmid to maintain the integrity of the transferred sequences. Previous evidence shows that circular plasmids can integrate into the genome, but the integration sequence of the plasmid is random, with the risk of deleterious events occurring with the transgene sequences.

1. CaCl$_2$-competent DH5α cells were transformed by heat shock as reported (**22**).
2. Plasmids were purified from a 25 mL culture using the Qiagen Midiprep Kit, according to the manufacturer's protocol.

3. In a sterile 1.5-mL Eppendorf tube, mix 50–100 μg of supercoiled DNA plasmid (pEα^D or pERb, see materials), 10 μL of a 10X buffer (specific for the selected restriction endonuclease), 20 U of *Apa*L1 or *Eco*0190I, 10 μL of 10X albumin (if it is necessary) and adjust the volume to 100 μL (*see* **Note 1**).
4. Incubate 4–8 h at 37°C.
5. Add 1 volume of phenol/chloroform/isoamyl alcohol and mix.
6. Centrifuge for 5 min at room temperature in a microcentrifuge at 16,000*g* and transfer the upper phase to a new tube.
7. Add 1/10 vol of 3 *M* sodium acetate, pH 5.3, and 3 vol of pre-cooled absolute ethanol.
8. Incubate 10 min at –20°C.
9. Centrifuge at >11,000*g* and discard the supernatant.
10. Wash the pellet with pre-cooled 70% ethanol.
11. Centrifuge at >11,000*g* and discard the supernatant.
12. Air-dry the pellet and resuspend in 50 μL of sterile water or TE.
13. Check DNA concentration (*see* **Note 2**).

3.2. Tissue Culture

HeLa and HD3 cells were kept at a semiconfluent level in their respective media (*see* **Subheading 2**). HeLa and HD3 cell lines were selected with 0.5 and 0.9 mg/mL geneticin (G418), respectively. The first clonal subconfluent pass was divided in three vials, one to be stored in liquid nitrogen, another for genomic analysis, and the third vial for the first fluorescent cytometer analysis; after that, clones were incubated in their respective media without geneticin. Clonal cells were passed every three days and maintained for more than 80.

3.3. Transfection

3.3.1. Electroporation

This method is frequently used to obtain a high range of expressing cells. However, the cell death rate is very high and specific electroporation conditions are necessary for each cell. Therefore, transfection conditions must be standardized in terms of the amount of cells, DNA concentration, and electroporation conditions before using with the aim of improving transgene expression for individual cell lines.

3.3.2. HD3 Transfection

1. 5×10^6 cells of a 24 h previously subdivided culture are used for transfection, considering that a 60–80% semiconfluent cell concentration is needed for the transfection day.
2. Pellet cells by centrifugation at 300*g* for 5 min.
3. Resuspend the pellet in DMEM without serum to wash the cells.
4. Centrifuge the cells at 300*g* for 5 min.
5. Resuspend the pellet in 500 μL of PBS + 6 m*M* glucose for conditioning.
6. Incubate on ice for 10 min in an 0.4-cm electroporation cuvette.
7. Apply a pulse of 0.3 kV with 960 μFa.
8. Immediately transfer the cell suspension to DMEM with serum.
9. Incubate the transfected cells at 37°C for 48 h.

10. Assay the cells with the adequate procedure or transfer the cells to semisolid medium for selecting individual clones.

3.3.3. Lipofection Procedure

At the present time, a growing list of synthetic liposome transfection reagents is commercially available. Each one has its own success on each cell line. It should always be taken into account that transfection conditions must be standardized to yield high-efficiency transfections. Compounds such as Lipofectamine Plus reagent (Invitrogen), Fungene (Invitrogen), or Effectene (Qiagen) have been shown to yield high-efficiency transfections in most adherent cell lines. The following procedure uses Lipofectamine Plus, as recommended in the manufacturer's procedure (*see* **Note 3**).

1. One day before transfection, trypsinize and count the cells and plate them on a 6-well plate, considering that 50% confluency should be present on the day of transfection. This will allow higher transfection efficiencies.
2. On the day of transfection, pre-complex the DNA with the Plus reagent: 1 µg of DNA (pERb) is incubated with 100 µL of serum-free DMEM in a sterile 1.5-mL Eppendorf tube. Add 6 µL of pre-warmed Plus reagent (room temperature); mix and incubate for 20 min at room temperature.
3. In another tube, dilute 3 µL of lipofectamine in 100 µL of serum-free DMEM.
4. Add the DNA mix to the lipofectamine-containing tube slowly, pipetting up and down (this takes 5 min).
5. Incubate for 30 min at room temperature.
6. When the DNA-lipofectamine complex is formed, wash the cells twice with 2 mL of serum-free medium in each well (*see* **Note 4**).
7. Add 800 µL of serum-free DMEM to each well.
8. Add 200 µL of DNA-lipofectamine complex to each well and incubate for 4 h.
9. After incubation, add 1 mL of completed medium and serum to bring it to the normal culture concentration. The DNA-lipofectamine complex is considered not to be hazardous to the cells during long incubations. However, we recommend changing the culture medium with fresh complete culture medium 24 h after transfection.
10. Incubate 24–48 h to assay transgene expression or to replate the cells in cloning medium.

3.4. Clone Selection

3.4.1. Cell Suspension (HD3 Cell Line)

1. Once transfected, cells are incubated for 48 h at 37°C.
2. 1/5 or 1/10 of the cell suspension is mixed with 40 mL of complemented methylcellulose media.
3. Mix in a 50-mL tube by inversion for 2 min.
4. Add the methylcellulose with the transfected cells to a 25-cm tissue culture dish, trying to avoid air bubbles.
5. Incubate for 1–2 wk at 37°C, or until visible colonies are formed. (Colonies will be seen as clear and round spots over the methylcellulose.)
6. Recover the colony carefully with a micropipet tip.
7. Transfer the colony to a 96-well plate with 200 mL of selective growth medium.
8. Incubate for 1 wk and expand the clone to a desired volume.

3.4.2. Cloning Antibiotic-Resistant Adherent Cells

1. After 48 h of incubation, 1/5 and 1/10 of the transfected cells are incubated on a 25-cm tissue culture dish overnight (*see* **Note 5**).
2. The next day, replace the medium with fresh medium containing geneticin (G418).
3. After 1 wk, spots of clonal cells can be identified visually on the bottom of the plate.
4. On the bottom of the plate, mark the circular colonies.
5. Remove tissue culture medium from the plate.
6. Wash the plate with medium without serum.
7. Dip cloning disks in trypsin solution to wet them and remove the excess trypsin.
8. Place cloning disks on the marked colonies.
9. Incubate for 10 min to complete trypsinization.
10. While plate is incubating, label a prepared 24- or 6-well plate and add culture medium.
11. Remove the cloning disks from the plate with sterile forceps and place them in the 24- or 6-well plate with culture medium.

3.5. Genotype Determination by Southern Blot Analysis

The first step in the analysis of a stably expressed transgene is the characterization of its integrity to ensure that the recombinant protein is biologically active. Every clone should be analyzed to determine the presence, copy number, and state of the transgene. Frequently, PCR has been used to verify transgene presence in the genome. This technique is rapid and allows easy analysis of a large number of samples. However, the generation of false positives is an inherent hazard in this technique. Southern blot analysis is the best way to know the transgene state, but it is more labor-intensive and time-consuming. A well-designed Southern blot strategy will give valuable and additional information about transgene integrity and transgene copy number.

To determine transgene integrity, genomic DNA must be digested with a restriction endonuclease that cuts a single site in the transfected plasmid. This cut will give a minimal fragment according to its distance from the site of linearization. This strategy will allow a clear determination of single-copy integrants (with the presence of a single band) or multiple integrants (with several bands). The probe can be made with the transgene coding sequence, and particular caution should be taken to avoid endogenous sequences.

3.5.1. Rapid DNA Extraction

1. Extract genomic DNA from clone, usually from 1 mL of suspension cells or a confluent well of a 6-well plate for adherent cells.
2. Harvest cells in a 1.5-mL Eppendorf tube, usually using 1 mL of semiconfluent suspension cells or a semiconfluent well for adherent cells.
3. Resuspend the cells in 100 μL of T1 buffer.
4. Add 100 μL of T2 lysis buffer and incubate for 2 h at 50°C.
5. Add 1 mL of isopropanol, and invert the tube until a white precipitate is seen.
6. Centrifuge at 16,000g for 5 min at room temperature.
7. Wash the pellet with 500 μL of 70% ethanol and re-centrifuge.
8. Dry the pellet and resuspend in 100 μL of water or TE (*see* **Note 6**). DNA concentration could be assayed at 260 nm. An OD_{260} of 1 equals 50 μg/mL.

3.5.2. Southern Transfer

1. Genomic DNA extraction from individual clones are usually done with 1 mL of suspension cells or a confluent well of a 6-well plate for adherent cells.
2. Digest genomic DNA (10 µg) with a single-cut restriction enzyme in the transgene construct. The total volume of the digest should not exceed the maximum volume held by a single well of the agarose gel (*see* **Note 7**).
3. Add 1/10 vol of sample buffer to the digestion.
4. Load digested genomic DNA and radiolabeled DNA size marker on a large 0.7-1% TBE agarose gel containing 1 µg/µL of ethidium bromide and run at 4 V/cm (*see* **Note 8**). Stop the run when the bromophenol blue dye reaches the end of the gel (*see* **Note 9**).
5. Take a picture of the DNA on a UV light transilluminator, together with a fluorescent ruler placed alongside the edge of the gel next to the DNA size marker.
6. Wash four times (15 min each time) with denaturing buffer.
7. Wash once with transfer buffer.
8. While the washes are being performed, prepare the membrane and absorbent paper.
9. Mount the transfer apparatus in the following order: 3 MM paper, gel, membrane, two sheets of 3 MM paper, one sheet of 3-MM paper, and 10 cm of compressed absorbent paper. Weigh down the sandwich with a bottle or a book (approx 500 g, *see* **Note 10**).
10. Transfer by capillary action overnight (8–12 h).
11. Dismantle the blot; mark the wells with a pencil and mark the orientation of the gel.
12. Transfer the membrane to neutralization buffer (1X).
13. Remove excess neutralization buffer from the membrane on a sheet of 3 MM paper.
14. Cross-link the DNA to the membrane (auto cross-linking program, Stratagene apparatus at 1200 µJ/30 s).
15. Pre-hybridize 4–6 h at 42°C.
16. Hybridize with the specific probe overnight at 65°C (*see* **Note 11**).
17. Next day, wash once at room temperature with 2X SSC, mixing strongly.
18. Wash once at 65°C with 2X SSC for 10 min.
19. Wash twice with 2X SSC, 1 % SDS (pre-warmed) at 65°C (30–45 min each wash).
20. Wash twice with 0.2X SSC, 0.1 % SDS at 65°C (30–45 min each wash).
21. Wash once with 0.2X SSC at room temperature.
22. Expose the blot with an autoradiograph film (Kodak).

3.6. Analyzing Sustained Expression by Flow Cytometry

Flow cytometry analysis is a powerful tool to detect transgene expression through both surface molecules (**2**) as well as intracellular fluorophores. EGFP can be analyzed with cytometry as an intracellular reporter protein (**21**). This methodology allows the analysis of a cell population or of individual cells. However, to analyze a surface molecule it is necessary to use specific antibodies, and whether or not the detection is indirect, it requires a secondary antibody. As a consequence, the procedure becomes expensive and utilizes a time-consuming protocol. The use of GFP as a reporter molecule facilitates data analysis by fluorescence cytometry. GFP emits fluorescence at 530 nm, as fluorescein molecules do, and its emission through a plasmatic membrane can be detected by the cytometer. Actually, GFP has been modified through mutations to make it less toxic for cell lines, and in fact, enhanced green fluorescence protein can be maintained for months without any cellular damage. Thus, through the use of GFP, the expression analysis becomes simplified. Cells are detached and washed, and they are then ready to be analyzed using GFP emission. Once in the cytometer, we define the forward scatter and side scatter parameters, and using these parameters, death cell is eliminated. Therefore, we were cautious to discard death cell, thus restricting our study to live cells.

1. Harvest 1×10^6 cells in a 1.5-mL Eppendorf tube, usually 0.5 mL of suspension cells or a semiconfluent well for adherent cells.
2. Centrifuge for 5 min at 400g in a microfuge.
3. Carefully resuspend the pellet in 1 mL of PBS.
4. Wash twice with PBS.
5. Resuspend the pellet in PBS.
6. Use calibration beads to adjust the cytometer
7. Analyze 10^4 cells in a calibrated cytometer for conditions to detect green fluorescent light (*see* **Note 12**).

4. Notes

1. The cut site must be located in sequences that do not include the transgene or antibiotic-resistance coding sequences. High-quality DNA plasmid is necessary to improve enzyme-mediated DNA digestion and high-efficiency transfection. The digestion of the plasmid can be checked on a 1% agarose gel.
2. The amount of DNA recovered must be above 50% of the input. The DNA concentration can be checked at 260 nm in a spectrophotometer, considering that 1 absorbance unit = 50 µg/ml, and checking that the absorbance ratio at 260 nm/280 nm is at least 1.6–1.8.
3. This is a starting protocol that must work in the majority of adherent cells. However, it is recommended that several conditions (DNA/lipofectamine ratio) should be assayed to determine the highest transfection efficiency. In our hands, DNA/lipofectamine ratios of 1:1, 1:2, and especially 1:3 give excellent results. In some cell lines, high liposome concentrations can be toxic.
4. Adherent cells can be transfected through a modified procedure: the cells of each well are detached with 0.5 mL of trypsin. To inactivate the trypsin, add 3 volumes of DMEM with 10% of fetal bovine serum. The cells are then centrifuged at 300 g at room temperature. Wash the cells once with serum-free medium and centrifuge again. Resuspend the pellet in 800 µL of serum-free DMEM and replate the cells into their respective wells. Mix the 200 µL of DNA/lipofectamine-coupled particles with the 800 µL of cell suspension. Allow the cells to attach to the plate for 5 h, and after that, follow the procedure as described.
5. We recommend using large plates (25 cm) to get perfectly isolated colonies. This can be improved through extensive cell mixing and by avoiding the presence of cell clots. If a high density of colonies is observed, thus making the cloning difficult, proceed by making higher dilutions.
6. This procedure gives a standard DNA quality. This is sufficient to perform digestions with the desired restriction endonuclease. Do not overdry the pellet, and remove excess 70% ethanol by pipetting. Once resuspended, let the pellet dissolve in water overnight at 37°C.
7. The presence of smeared bands, which are derived from a salt excess in the digest, can be avoided by 100% ethanol precipitation and by one wash with 70% ethanol. We recommend resuspending the pellet in a reduced volume (30–50 µL).
8. Radiolabeled DNA size markers were prepared first by dephosphorylating the DNA for 1 h at 37°C as follows: in a 1.5-mL tube, add 1 µg of DNA size markers, 2 µL (2 U) of Calf

intestine alkaline phosphatase (CIAP, Invitrogen), and adjust the volume to 20 μL. The CIAP is then inactivated at 65°C for 20 min. After that, add to the tube: 10 μL of forward kinase reaction buffer, 5 U of T4 kinase, 1 μL of [^{32}P]ATP at 10 mCi/mL and adjust the volume to 100 μL. Inactivate the kinase by adding EDTA to a final concentration of 1 mM or by incubating for 10 min at 65°C. Load 1–3 μL of the mix into a well of the agarose gel.

9. Gel should have at least a 10- to 15-cm running length. Generally, 0.7 to 1% agarose gels are convenient, giving good resolution between 500 bp and 15 kb.

10. Avoid the presence of bubbles in the sandwich and during the assembly. Use gloves because any grease contamination could give a signal on the film.

11. Specific probes can be made by nick-translation, PCR, or random priming. Here, we used random priming to produce a probe specific for the EGFP gene. The coding sequence was obtained from a *Bam*HI/*Not*I digestion from pEGFP-1, isolated by gel purification, and 50 ng was mixed with 20 μCi of [^{32}P] dCTP, and adjusted to a final volume of 50 μL with water. Labeling beads were used (DNA Labelling Beads, Pharmacia Biotech). The mix was incubated for 30 min at 37°C. The probe was cleaned up with a probe purification spin column (NAP-25 column, Pharmacia Biotech). Half of the probe was added to the membrane for hybridization.

12. The cytometer must be calibrated in advance to obtain reproducible results. The cytometer can be calibrated by using FACScomp software and calibration beads (Becton Dickinson). Avoid the formation of cell clots, as these clots can obstruct the cytometer conducts and produce damage or a stop in the flow. Clots can be avoided by the use of a fixative buffer, enabling the formation of cell groups (avoid use of paraformaldehyde).

Acknowledgments

We are grateful to Catherine M. Farrell for suggestions and critical reading of the manuscript. We thank Georgina Guerrero for excellent technical assistance. The present work was supported by the Dirección General de Asuntos del personal Académico-UNAM (IN203200), the Consejo Nacional de Ciencia y Tecnología, CONACyT (33863-N), the Third World Academy of Sciences (TWAS, grant 01-055RG/BIO/LA), and the Fundación Miguel Alemán, A. C.

References

1. Opsahl, M. L., McClenaghan, M., Springbett, A., Reid, S., Lathe, R., Colman, A., et al. (2002) Multiple effects of genetic background on variegated transgene expression in mice. *Genetics* **160**, 1107–1112.

2. Pikaart, M. J., Recillas-Targa, F., and Felsenfeld, G. (1998) Loss of transcriptional activity of a transgene is accompanied by DNA methylation and histone deacetylation and is prevented by insulators. *Genes Dev.* **12**, 2852–2862.

3. Pannell, D. and Ellis, J. (2001) Silencing of gene expression: implications for design of retrovirus vectors. *Rev. Med. Virol.* **11**, 205–217.

4. Wilson, C., Bellen, H. J., and Gehring, W. J. (1990) Position effects on eukaryotic gene expression. *Annu. Rev. Cell. Biol.* **6**, 679–714.

5. Bulger, M. and Groudine, M. (1999) Looping versus linking: toward a model for long-distance gene activation. *Genes Dev.* **13**, 2465–2477.

6. West, A. G., Gaszner, M., and Felsenfeld, G. (2002) Insulators: many functions, many mechanisms. *Genes Dev.* **16**, 271–288.

7. Neff, T., Shotkoski, F., and Stamatoyannopoulos, G. (1997) Stem cell gene therapy, position effects and chromatin insulators. *Stem Cells* **15 Suppl 1**, 265–271.
8. Chow, C. M., Athanassiadou, A., Raguz, S., Psiouri, L., Harland, L., Malik, M., et al. (2002) LCR-mediated, long-term tissue-specific gene expression within replicating episomal plasmid and cosmid vectors. *Gene Ther.* **9**, 327–336.
9. Gerasimova, T. I., and Corces, V. G. (1996) Boundary and insulator elements in chromosomes. *Curr. Opin. Genet. Dev.* **6**, 185–192.
10. Whitelaw, E., Sutherland, H., Kearns, M., Morgan, H., Weaving, L., and Garrick, D. (2001) Epigenetic effects on transgene expression. *Methods Mol. Biol.* **158**, 351–368.
11. Recillas-Targa, F., and Razin, S. V. (2001) Chromatin domains and regulation of gene expression: familiar and enigmatic clusters of chicken globin genes. *Crit. Rev. Eukaryot. Gene Expr.* **11**, 227–242.
12. Mutskov, V. J., Farrell, C. M., Wade, P. A., Wolffe, A. P., and Felsenfeld, G. (2002) The barrier function of an insulator couples high histone acetylation levels with specific protection of promoter DNA from methylation. *Genes Dev.* **16**, 1540–1554.
13. Bonifer, C. (2000) Developmental regulation of eukaryotic gene loci. *Trends Genet.* **16**, 310–315.
14. Kioussis, D. and Festenstein, R. (1997) Locus control regions: overcoming heterochromatin-induced gene inactivation in mammals. *Curr. Opin. Genet. Dev.* **7**, 614–619.
15. Indraccolo, S., Minuzzo, S., Roccaforte, F., Zamarchi, R., Habeler, W., Stievano, L., et al. (2001) Effects of CD2 locus control region sequences on gene expression by retroviral and lentiviral vectors. *Blood* **98**, 3607–3617.
16. Wolffe, A. P. (1994) Gene regulation. Insulating chromatin. *Curr. Biol.* **4**, 85–87.
17. Prioleau, M. N., Nony, P., Simpson, M., and Felsenfeld, G. (1999) An insulator element and condensed chromatin region separate the chicken β-globin locus from an independently regulated erythroid-specific folate receptor gene. *EMBO J.* **18**, 4035–4048.
18. Recillas-Targa, F., Pikaart, M. J., Burgess-Beusse, B., Bell, A. C., Litt, M. D., West, A. G., et al. (2002) Position-effect protection and enhancer blocking by the chicken β-globin insulator are separable activities. *Proc. Natl. Acad. Sci. USA* **99**, 6883–6888.
19. Emery, D. W., Yannaki, E., Tubb, J., Nishino, T., Li, Q., and Stamatoyannopoulos, G. (2002) Development of virus vectors for gene therapy of beta chain hemoglobinopathies: flanking with a chromatin insulator reduces gamma-globin gene silencing in vivo. *Blood* **100**, 2012–2019.
20. Anderson, M. T., Tjioe, I. M., Lorincz, M. C., Parks, D. R., Herzenberg, L. A., and Nolan, G. P. (1996) Simultaneous fluorescence-activated cell sorter analysis of two distinct transcriptional elements within a single cell using engineered green fluorescent protein. *Proc. Natl. Acad. Sci. USA* **93**, 8508–8511.
21. Ducrest, A. L., Amacker, M., Lingner, J., and Nabholz, M. (2002) Detection of promoter activity by flow cytometry analysis of GFP reporter expression. *Nucleic Acids Res.* **30**, e65.
22. Sambrook, J., Fritsch, E. F., and Maniatis, T. (1989) *Molecular Cloning: a Laboratory Manual*. Cold Spring Harbor Laboratory Press, Cold Spring Harbor, NY.

30

Inducible Gene Expression in Mammalian Cells and Mice

Wilfried Weber and Martin Fussenegger

Summary

Inducible expression of desired transgenes in mammalian cells and animals is a current priority in basic and applied research, biopharmaceutical manufacturing, gene therapy, and tissue engineering, as well as in drug discovery. Among the most prominent human-compatible transgene control technologies are engineered promoter/transactivator configurations that adjust heterologous target gene transcription in response to clinically licensed antibiotics (tetracyclines, streptogramins, macrolides). In this chapter we provide a detailed case study on macrolide-inducible expression of the human model glycoprotein SEAP (human placental secreted alkaline phosphatase) in transgenic Chinese hamster ovary (CHO) cell cultures or following implantation of microencapsulated CHO cells into mice.

Key Words: Inducible gene expression; inducible promoter; TET system; PIP system; E.REX system; SEAP; encapsulation; CHO; tetracycline; pristinamycin; erythromycin.

1. Introduction

In recent years, technologies for adjustable expression of desired transgenes in mammalian cells and animals have generated tremendous impact in gene-function analyses (*1*) and biopharmaceutical manufacturing (*2*), and by pioneering prototype strategies for conditional molecular interventions in gene therapy and tissue engineering (*3,4*). A variety of different heterologous gene control concepts have been developed and extensively studied in the past decades (*1*). However, many of these established transcription-adjusting technologies systems are incompatible with gene therapy scenarios, as their regulating compounds elicit severe side effects in humans (*1*).

To date, five heterologous gene regulation systems have been designed with a focus to maximize human compatibility, which includes responsiveness to clinically licensed

From: *Methods in Molecular Biology, vol. 267:*
Recombinant Gene Expression: Reviews and Protocols, Second Edition
Edited by: P. Balbás and A. Lorence © Humana Press Inc., Totowa, NJ

small molecule drugs: (1) rapamycin-induced dimerizing technology (**5**); (2) steroid-hormone receptor-based systems (**6**); (3) tetracycline- (TET system) (**7**); (4) streptogramin-(PIP system) (**8**); and (5) macrolide- (E.REX system) (**9**) adjustable transcription control configurations. All antibiotic-responsive gene regulation systems (TET, PIP, E.REX) are derived from bacterial antibiotic (tetracycline, streptogramin, macrolide) resistance regulons. Generic adaptation of key resitance regulon components for use as mammalian gene regulation systems included: (1) Fusion of the antibiotic biosensor (TetR, Pip, E) to a eukaryotic transactivation domain (e.g., VP16 derived from *Herpes simplex*) to produce an artificial transactivator (tTA, TetR-VP16; PIT, Pip-VP16; ET, E-VP16). (2) These transactivators bind and activate artificial promoters ($P_{hCMV*-1}$; P_{PIR}; P_{ETR}), assembled by cloning tTA-, PIT-, ET-specific operators (*tetO*, *pir*, *etr*) adjacent to a minimal eukaryotic promoter (e.g., $P_{hCMVmin}$ derived from the human cytomegalovirus) in the absence of regulating antibiotics. However, in the presence of regulating antibiotics, the specific transactivator is released from its cognate promoter, which results in full repression of transgene expression. Since the aforementioned gene regulation systems are fully expressed in the absence of regulating antibiotics, they are often referred to as "OFF" systems (TET_{OFF}; PIP_{OFF}; E_{OFF}). However, for therapeutic applications, gene regulation scenarios that enable co-induction of transgene expression concomitantly with antibiotic administration, so-called "ON" systems (TET_{ON}; PIP_{ON}; E_{ON}) would be highly desirable. ON-type gene regulation configurations consist of antibiotic biosensors (TetR; Pip; E) that function as transrepressors by binding to their conate operators (*tetO*; *pir*; *etr*) and repressing strong constitutive promoters placed immediately upstream. Addition of regulatory antibiotics results in release of the repressor from its operator module and induction of gene expression (TET_{ON}) (**10**); PIP_{ON}, (**8**); E_{ON}, (**9**) (**Fig. 1**). Here we exemplify the use of heterologous antibiotic-inducible transcription control systems by a detailed coverage of the E_{ON} system from expression vector design, construction of stable CHO cell lines to its in vivo validation by implantation of microencapsulated cells into mice and macrolide-inducible as well as adjustable expression of the human model glycoprotein SEAP in these animals.

2. Materials

2.1. Reagents/Solutions

1. FMX-8 medium (Cell Culture Technologies GmbH, Zürich, Switzerland; http://www.cell-culture.com).
2. FMX-8 complete medium: FMX-8 medium containing 10% fetal calf serum (FCS; PAA Laboratories GmbH, Linz, Austria; cat. no. A15-022; lot no. A01129-242) and 1% penicillin-streptomycin solution (Sigma, St. Louis, MO, cat. no. P0906).
3. Trypsin solution: Trypsin-EDTA (1X) in HBSS (Hank balanced salt solutuion) without calcium and magnesium (Gibco, now Invitrogen, Carlsbad, CA; cat. no. 25300-054).
4. Sterile $CaCl_2$ stock solution, 1 *M*.
5. Phosphate solution: 50 m*M* HEPES, 280 m*M* NaCl, 1.5 m*M* Na_2HPO_4, pH 7.05, filter-sterilized and stored at 4°C for up to 6 mo.
6. Sterile glycerol, sterile DMSO.

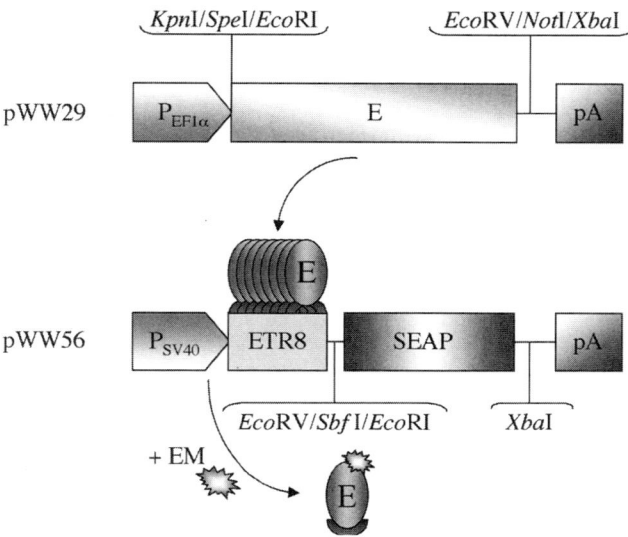

Fig. 1. Scheme of the macrolide-inducible gene regulation system (E_{ON}). Plasmid pWW29 contains the expression unit for the macrolide-dependent repressor E, which binds to its cognate ETR operator engineered downstream of the P_{SV40} promoter encoded on pWW56. Binding of E, in the absence of macrolide antibiotics, represses P_{SV40}-driven SEAP (human placental secreted alkaline phosphatase) expression. Addition of erythromycin (EM) abolishes the E-ETR interaction and results in derepression of P_{SV40} and full induction of SEAP expression. Selected restriction sites are indicated. Plasmids pWW29 and pWW56 contain ampicillin resistance determinants for maintenance and selection in *E. coli*. $P_{EF1}\alpha$, human elongation factor 1α promoter; P_{SV40}, simian virus 40 promoter; pA, polyadenylation site.

7. Zeocin (Invitrogen, Carlsbad, CA) stock solution of 100 mg/mL in water; store at $-20°C$ for up to 6 mo.

8. Erythromycin (Fluka, Buchs, Switzerland, cat. no. 4573) stock solution of 10 mg/mL in 96% ethanol; store at $-20°C$ for up to 2 mo.

9. Neomycin (G 418 sulfate, Calbiochem, La Jolla, CA, cat. no. 345810) stock solution of 100 mg/mL in water; store in aliquots at $-20°C$ for up to 6 mo.

10. Lipid-based transfection reagents (Fugene 6, Roche Diagnostics, Mannheim, Germany, cat. no. 1 814 443).

11. Bioencapsulation kit for animal cells (Intotech Encapsulation AG, Dottikon, Switzerland, cat. no. IE1300).

12. Great EscAPe™ SEAP reporter system 2, chemiluminescence-based SEAP quantification kit (Roche Diagnostics AG, Mannheim, Switzerland, cat no. 1779 842).

13. 2X SEAP buffer: 20 m*M* homoarginine, 1 m*M* MgCl$_2$, 21% (w/v) diethanolamine/HCl, pH 9.8; store in the dark at 4°C for up to 6 mo.

14. pNPP solution: 120 m*M* para-nitrophenylphosphate (Sigma 104® Phosphatase Substrate, Sigma, St. Louis, MO, cat. no. 104-0) in 2X SEAP buffer. Store in single-use aliquots at $-20°C$ (thaw only once).

15. Phosphate-buffered saline solution without magnesium and calcium (PBS, Sigma, cat. no. D-5652), adjust to pH 7.2.
16. Sterile physiological salt solution (0.9% NaCl).
17. MOPS washing buffer (Inotech Encapsulation AG; http://www.inotech.ch).
18. Nembutal (Abbott, Abbott Park, IL).

2.2. Other Materials/Hardware

1. CHO-K1 cells (ATCC CCL 61).
2. Plasmids pWW29 and pWW56 encoding the macrolide-inducible gene regulation system (**Fig. 1**) (*9,11*).
3. Plasmid pZeoSV2 (Invitrogen).
4. Plasmid pSV2*neo* (Invitrogen).
5. Silica-based anion-exchange DNA purification kits (Genomed Jetstar 2.0 Midiprep, Genomed AG, Bad Oeynhausen, Germany).
6. Microtainer serum separation tubes SST™ (Becton Dickinson, Franklin Lakes, NJ, cat. no. 365968).
7. Cell counting device (Casy1® counter, Schärfe System, Reutlingen, Germany; alternatively, a hemacytometer can be used).
8. Cell encapsulation device (Inotech Encapsulator Research, Inotech Encapsulation AG; http://www.inotech.ch).
9. Luminometer (TD-20/20, Turner Designs, Sunnyvale, CA).
10. Fluorescence-activated cell sorter (FACStar^Plus and Cell QuestTM software, Becton Dickinson, San Jose, CA) for single-cell sorting and cloning of stable transgenic cell lines.
11. 50-μm filters (DAKO Diagnostics AG, Zug, Switzerland; cat. no. 150475).
12. 1-mL Luer Lock syringe and needle (Becton Dickinson, Franklin Lakes, NJ, Microlance 3 needle, 23G A 1/4 0.6 × 30.
13. 20-mL Luer Lock syringe for encapsulation (Norm–Jcet 20 ml, Sigma-Aldrich, St. Louis, MO, cat. no. Z248037-1PAK).
14. Incubator (e.g., Nuaire Autoflow, NU-4500 E, Nuaire, Plymouth, MN) for cultivation of mammalian cells at 37°C in a humidified atmosphere containing 5% CO_2.
15. StrataCooler cryopreservation module (Stratagene, La Jolla, CA; cat. no. 400005).
16. Cell culture-certified disposable plasticware: T-75 flasks, T-150 flasks, 6-well plates, 96-well plates, pipets, 15-ml and 50-ml Falcon tubes, cryotubes.

3. Methods
3.1. Selection Criteria for a Suitable Gene Regulation System

First criterion: "ON" vs "OFF"-type regulation system. The prime decision when implementing gene regulation technology in mammalian cells is whether to use an inducible "ON"-type (target gene expression is induced upon addition of the regulating agent; TET_{ON} (*10*); PIP_{ON} (*8*); E_{ON} (*9*); rapamycin-induced dimerizing system (*5*); steroid-hormone receptor-based systems (*6*)) or a repressible "OFF"-type system (target gene expression is repressed upon addition of the regulating agent; TET_{OFF} (*7*); PIP_{OFF} (*8*); and E_{OFF} (*9*)). Whereas "ON"-type regulation systems are ideal for short-term expression scenarios predominant in the clinics, "OFF"-type configurations are preferred for long-term induction profiles with occasional repression by addition of the regulating molecule. Owing to their genetic configuration combining a transactivator-specific operator and a TATA-box, promoters of "OFF"-type gene regulation systems are enhancer traps and may experience undesired expression interferences by adjacent transcription-modulating chromosomal loci. By contrast, inducible promoters assembled by cloning a constitutive promoter in front of the transactivator-specific operator modules are considered to be largely inert with respect to undesired *in cis* transcription-initiation by flanking chromatin. Due to their potential to mediate interference-reduced or free titration of transgene expression, "ON"-type expression systems are the preferred configuration for implementation in gene therapy and tissue engineering scenarios.

Second criterion: The right gene control system for the right cell line. The regulation performance of individual gene regulation systems varies dramatically between different cell lines. For most types of gene regulation systems (e.g., TET, PIP, E.REX) an entire family of transactivator/transrepressor and promoter derivatives have been cloned and need to be used in a cell-specific combination to enable optimized transgene control in a desired cell line (*12–15*).

Third criterion: one-gene vs multigene expression adjustment. While regulated one-gene expression is expected to be sufficient for most applications, expression of oligomeric protein complexes requires simultaneous as well as coordinated expression of different transgenes. For adjustable expression of two different transgenes bidirectional expression cassettes have been designed (*16*). However, latest-generation operonlike multicistronic expression configurations enable coordinated fine-tuning of three and more transgenes (rational reprogramming of mammalian cells to achieve desired therapeutic or cell phenotypes (*11,17,18*) (*see* **Note 1**) are the method of choice), where up to three genes are expressed in an operonlike configuration under control of one inducible promoter.

Fourth criterion: independent control of different transgenes. Complex molecular interventions in mammalian cells require independent control of different transgenes. So far, the TET, PIP and E.REX systems have been shown to be functionally compatible and could be used for independent control of up to three different sets of transgenes (*11, 19–21*) (*see* **Note 1**).

The following generic protocol exemplifies engineering of the biotechnologically important Chinese hamster ovary (CHO) cell line for macrolide-inducible (E_{ON}-based) gene expression. However, with minor modifications, these procedures can be adapted to other gene regulation systems as well as cell lines.

3.2. Expression Plasmid Construction

The basic E_{ON} system is encoded on two different plasmids pWW29 and pWW56 (**Fig. 1**) (*9*). pWW29 enables constitutive expression of the macrolide-responsive transrepressor E, which binds to its cognate pWW56-encoded artificial operator module (ETR8) placed between the strong constitutive promoter P_{SV40} and the model product gene SEAP (human placental secreted alkaline phosphatase) and represses P_{SV40}-driven SEAP expression in the absence of macrolide antibiotics (erythromycin, clarithromycin, roxithromycin). Upon addition of macrolides to pWW29/pWW56-

containing cells, E is released from ETR8, which results in derepression of P_{SV40} and induction of SEAP expression in a dose-dependent manner. In principle, pWW56 can be modified to express any desired transgene by exchanging SEAP using standard DNA techniques (*22*).

3.3. Construction of a CHO-K1 Cell Derivative Transgenic for pWW56

Installation of the E_{ON} system in CHO cells requires a two-step procedure (*see* **Note 2**): (1) stable chromosomal integration of the macrolide-inducible SEAP expression vector pWW56 in CHO-K1 cells. (2) Chromosomal integration of the pWW29-encoded transrepressor E into the pWW56-transgenic CHO-K1 cell derivative (*see* **Subheading 3.4.**) Basic cell culture techniques including sterile cultivation and splitting of cells are described elsewhere (*23*).

3.3.1. Stable Transfection of pWW56 into CHO-K1

CHO-K1 cells were stably cotransfected with the $P_{ETR}ON8$-driven SEAP expression vector pWW56 and the zeocin resistance conferring selection plasmid pZeoSV2 (*see* **Note 3**) using the following protocol:

1. Isolate pWW56 and pZeoSV2 plasmid DNA using silica-based anion-exchange DNA purification kits according to the manufacturers' protocols (Genomed Jetstar 2.0 midiprep kit).
2. Seed 200,000 CHO-K1 cells resuspended in 2 mL FMX-8 complete medium into each of two wells of a 6-well plate 15 h prior to transfection.
3. Optimized $CaPO_4$-based transfection protocol for CHO-K1 cells:

    a. Prepare a 60-µL solution containing 5.75 µg pWW56, 0.25 µg pZeoSV2, and 500 m*M* $CaCl_2$ in a 50-mL Falcon tube.
    b. Add 60 µL phosphate solution and vortex immediately for 5 s to allow $CaPO_4$-DNA complex formation.
    c. Add 2 mL FMX-8 medium supplemented with 2% FCS exactly after 25 s to stop $CaPO_4$-DNA complex formation.
    d. Aspirate the medium from one CHO-K1-containing well.
    e. Add the $CaPO_4$-DNA complex from step c. to the cells while rocking gently.
    f. Put the cells in an incubator for 5 h.
    g. Aspirate the medium and add 2 mL FMX-8 supplemented with 2% FCS and 15% glycerol.
    h. After exactly thirty replace the glycerol-containing medium with 5 mL FMX-8 complete medium. Rock gently for optimal washing.
    i. Replace the washing medium with 2 mL FMX-8 complete medium.
    j. Continue cultivation in an incubator.

4. Perform the same transfection procedure (a–j) in parallel for the second well using 6 µg pWW56 to provide a negative selection control in the absence of pZeoSV2.
5. Forty-eight hours posttransfection, aspirate the medium, add 0.5 mL trypsin solution, and place in incubator for 10 min. Tap the 6-well plate to completely detach the cells.
6. Transfer detached cells into a T-75 flask containing 15 mL selection medium (FMX-8 complete medium supplemented with 100 µg/ml zeocin).

7. Exchange selection medium every other day and transfer transfected cell populations into a new T-75 flask every 4 d until the control population transfected in the absence of pZeoSV2 is completely eliminated, typically after 8–10 d:

   a. Aspirate the (selection) medium.
   b. Add 3 mL trypsin solution and incubate for 10 min until cells detach (place in incubator for 10 min: tap the T-75 flask to completely detach the cells).
   c. Transfer cells into a 15-mL Falcon tube containing 7 mL FMX-8 complete medium.
   d. Centrifuge 3 min at 300g and aspirate supernatant.
   e. Resuspend the cell pellets in 10 mL FMX-8 complete medium and transfer the cells into T-75 flasks.
   f. Add zeocin to a final concentration of 100 µg/ml.

8. Cultivate the transfected cell population in selection medium until zeocin-resistant colonies of at least 2 mm in diameter emerge.

9. Quantify SEAP activity in the supernatant of the transfected population to validate stable integration of pWW56 and proceed with single-cell cloning.

3.3.2. FACS-Mediated Single-Cell Sorting

Transgenic monoclonal cell lines are generated from stable mixed populations by FACS-mediated single-cell sorting (*see* **Note 4**).

Preparation of FACS sorting:

1. Prepare five 96-well tissue culture plates containing 200 µL FMX-8 complete medium supplemented with 100 µg/ml zeocin per well. Prewarm 96-well plates in an incubator for 5 h.
2. Perform **steps 7 a–d** of **Subheading 3.3.1**.
3. Resuspend the cell pellet in 10 mL sterile PBS without calcium and magnesium.
4. Centrifuge the cell suspension, aspirate the supernatant, and resuspend the pellet in 1 mL PBS. Filter the cell suspension using a 50-µm filter to remove cell aggregates.
5. FACS-mediated single-cell sorting into the nonperipheral wells of the 96-well plates (1 cell/well; 60 cells/plate).
6. Put the 96-well plates in an incubator for up to 1 wk to allow for clonal expansion.
7. Microscopic inspection of individual wells for single one-cell-derived colony. Discard wells with more than one colony.

3.3.3. Selection of Transgenic Cell Clones

Upon clonal expansion to 50% confluence in a 96-well plate, the monoclonal populations are analyzed for their desired transgenic status (stable integration of pWW56).

1. Select 60 cell clones for quantitative SEAP expression analysis

   a. Pipet 50 µl cell culture supernatant into an Eppendorf cup.
   b. Chemiluminescence-based quantification of SEAP expression using the Great EscAPe™ SEAP reporter system 2 (*see* **Note 5** for an alternative SEAP quantification protocol).

2. Select five clones showing different SEAP expression levels.
3. Remove the remaining cell culture medium from the 96 wells containing the desired clones.
4. Add 50 µL trypsin solution per well.

5. Following detachment of the cells, transfer them into individual wells of a 6-well plate containing 2 mL FMX-8 complete medium per well supplemented with 100 µg/mL zeocin.

6. Cultivate the cell clones in an incubator to near 90% confluence. Transfer the expanded monoclonal cell populations into T-75 flasks by following steps 5 and 6 of **Subheading 3.3.1**.

7. Cryopreservation of the desired cell clones.

   a. Follow **steps 7 a–d** of **Subheading 3.3.1**.
   b. Resuspend the cell pellet in 5 mL FCS containing 10% DMSO.
   c. Transfer the cell suspension into cryotubes (1 mL/tube), put into a StrataCooler cryo preservation module and freeze at −80°C overnight.
   d. Transfer the cryotubes into a liquid nitrogen tank for long-term preservation.

   SEAP expression levels of 40 randomly selected clones are shown in **Fig. 2**. Cell clones number 9, 10, 17, 24, and 28 were cryopreserved for further use. Cell clone number 24, designated CHO-SEAP$_1$, was considered for the second round of transfection.

3.4. Stable Transfection of the Macrolide-Responsive Repressor E (pWW29)

3.4.1. Transfection of pWW29

Stable transfection of pWW29 into the pWW56-containing CHO-SEAP$_1$ is performed in analogy to the protocol described in **Subheadings 3.3.1.–3.3.3**. Therefore, pWW29 and pSV2*neo*, which encodes the G418 resistance gene, were stably cotransfected into CHO-SEAP$_1$. Throughout the entire second transfection process the cells were cultivated in FMX-8 complete medium containing 100 µg/mL zeocin to co-select for pWW56. Mock-transfected CHO-SEAP$_1$ were used as control cells.

1. Perform transfection as described in **steps 1–4** of **Subheading 3.3.1**.
2. Forty-eight hours posttransfection, the cells are transferred into T-75 flasks (*see* **step 5** of **Subheading 3.3.1**) containing FMX-8 medium supplemented with 100 µg/mL zeocin (co-selection for pWW56), 400 µg/mL neomycin (coselection for pWW29), and 10 µg/mL erythromycin (*see* **Note 6**). Always use these antibiotics throughout the entire selection process unless otherwise stated.
3. Perform the selection procedure as outlined in **steps 7–9** of **Subheading 3.3.1**. followed by single-cell cloning (**steps 1–7** of **Subheading 3.3.2.**) until monoclonal colonies have grown to 50% confluence in 96-well plates.
4. Replace the culture medium of clonal populations by 200 µL FMX-8 complete medium supplemented with 100 µg/mL zeocin and 400 µg/mL neomycin (the medium is devoid of erythromycin). Continue to cultivate the cell populations for another 2 d.

3.4.2. Analysis of Dual-Stable E$_{ON}$-Containing CHO Cell Clones

E$_{ON}$-containing dual-stable cell clones were validated by cultivating each clone in the presence and absence of erythromycin, followed by analysis of their SEAP expression profiles:

1. Remove the culture medium of 60 96-well plates containing clonal cell populations. Detach the cells by addition of 120 µL trypsin solution and transfer 50 µL of the trypsin solution/cell suspension in each of two wells of a 12-well plate containing 1 mL FMX-8 complete medium,

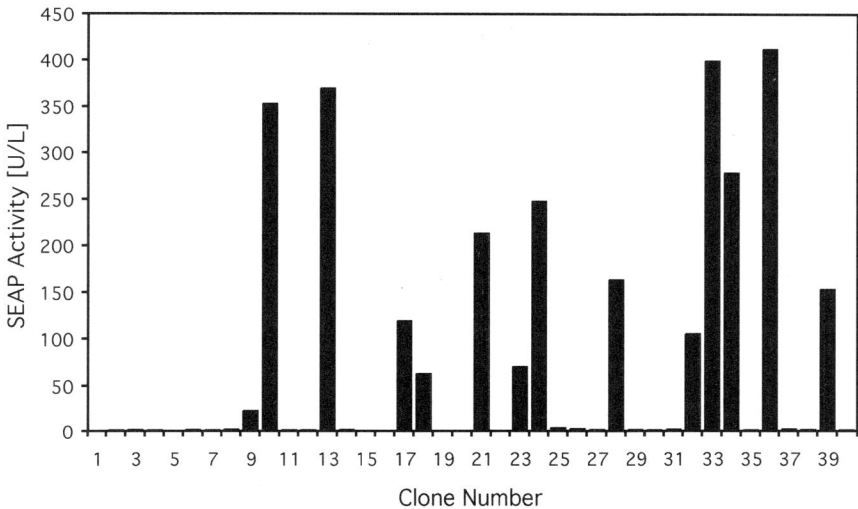

Fig. 2. SEAP expression levels of 40 clones analyzed after transfection of CHO-K1 cells with pWW56 and pZeoSV2 and FACS-mediated single-cell sorting.

supplemented with 100 μg/mL zeocin and 400 μg/mL neomycin (detach cells by carefully pipetting up and down in order to transfer a maximum of cells).

2. Add 10 μg/mL erythromycin to one well of each clone. The second well remains erythromycin-free.

3. Following 72 h cultivation under aforementioned conditions, the SEAP expression profiles of all cell clones grown in the presence and absence of erythromycin is quantified using a chemiluminescence-based Great EscAPe™ SEAP reporter system 2. Cell clones showing maximum induction ratios (high SEAP expression in the presence of erythromycin and near-undetectable SEAP expression in the absence of macrolides) were chosen for further analysis.

4. Expand two to three positive cell clones in T-75 flasks and cryopreserve at least 10 aliquots of each clone as outlined in step 7 of **Subheading 3.3.3.**

5. E_{ON}-containing dual-stable cell clones must be cultivated at all times in the presence of 100 μg/mL zeocin and 400 μg/mL neomycin to co-select for both transgenes. These cell lines may be cultivated continuously for up to 40 passages (corresponding to a continuous three-month cultivation time) before they have to be replaced by a new batch.

The SEAP expression profiles of 2 representative pWW29- and pWW56-containing cell clones, designated CHO-E_1-SEAP$_1$ and CHO-E_2-SEAP$_2$, are shown in **Fig. 3A**. CHO-E_1-SEAP$_1$ and CHO-E_2-SEAP$_2$ were used for further analysis.

3.5. Dose-Response Characteristics of CHO-E_1-SEAP$_1$ and CHO-E_2-SEAP$_2$

Cell clones CHO-E_1-SEAP$_1$ and CHO-E_2-SEAP$_2$ expressing SEAP in an erythromycin-inducible manner were subjected to a dose-response study to assess the adjustability of SEAP expression in response to a wide range of macrolide concentrations.

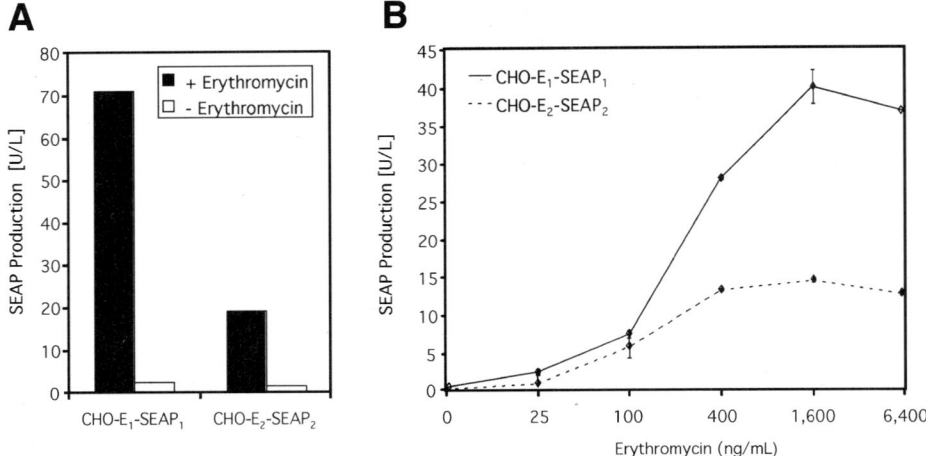

Fig. 3. (**A**) SEAP expression levels of two clones stably transfected with pWW29 and pWW56 (CHO-E$_1$-SEAP$_1$ and CHO-E$_2$-SEAP$_2$) in the presence (+) and absence (−) of regulating erythromycin (10 µg/mL). (**B**) Erythromycin-dependent dose-response response characteristics of SEAP expression in CHO-E$_1$-SEAP$_1$ and CHO-E$_2$-SEAP$_2$ cell cultures grown for 48 h in culture medium supplemented with indicated erythromycin concentrations (reprinted from *9*).

Therefore, cell clones CHO-E$_1$-SEAP$_1$ and CHO-E$_2$-SEAP$_2$ were cultivated in 6-well plates (100,000 cells/well) containing 2 mL FMX-8 complete medium supplemented with 100 µg/mL zeocin and 400 µg/mL neomycin, as well as different concentrations of erythromycin ranging from 25 ng/mL to 6400 ng/mL). The SEAP expression readings taken after 48 h are shown in **Fig. 3B**.

3.6. Microencapsulation of Stable CHO Cells

The CHO cell clone CHO-E$_1$-SEAP$_1$, which stably expresses SEAP under control of the macrolide-inducible expression system E$_{ON}$, was encapsulated in alginate-PLL-alginate (PLL: poly-L-lysine) microcapsules for injection into mice. The alginate-PLL-alginate encapsulation shields the transgenic CHO cell derivatives from the mouse immune systems and prevents elicitation of a targeted immune response following transplantation of heterologous microencapsulated cells into immunocompetent animals. Monodisperse multilayer capsules were produced using an Inotech Research Encapsulator (Inotech Encapsulation AG) in combination with Inotech's ready-to-use bioencapsulation kit for animal cells. Microencapsulation was performed following Inotech's protocol and the following specific settings:

1. The desired cell clones were expanded for encapsulation:

    a. CHO-E$_1$-SEAP$_1$ was expanded in a T-75 flask containing 10 mL FMX-8 complete supplemented with 100 µg/mL zeocin and 400 µg/mL neomycin.
    b. Upon reaching 90% confluence, the medium was removed and cells were detached using 3 mL trypsin solution (*see* **step 5, Subheading 3.3.1.**).

  c. Cells were then split into 6 T-150 flasks containing 20 mL FMX-8 complete medium supplemented with 100 µg/mL zeocin and 400 µg/mL neomycin and grown to near 90% confluence for encapsulation.

2. Equip encapsulator with a 200-µm nozzle, prepare and prewarm to 37°C all solutions required for encapsulation (polymerization solution, MOPS wash buffer, alginate 1.5%, PLL solution, depolymerization solution); pour 200 mL polymerization solution into the encapsulation reservoir.

3. Remove the culture medium in the 6 T-150 flasks. Detach cells using 5 mL trypsin per flask (*see* point 5, **Subheading 3.3.1.**).

4. Transfer the cells into 50-mL Falcon tubes and add 2 volumes of FMX-8 complete medium (15 mL cells and 30 mL medium per Falcon tube).

5. Mix by pipetting the cell suspension gently up and down. Take a sample to determine the cell concentration. Perform cell counting using a Casy1® counter (alternatively, a hemacytometer or similar device can be used) (*23*). The total cell number achieved is typically around 90 × 10^6 cells.

6. Centrifuge the cells at 300g for 3 min and remove the supernatant.

7. Resuspend the cell pellet in MOPS-washing buffer while adjusting the cell concentration to 36 × 10^6 cells per mL (perform **steps 8–10** without interruption at maximum speed).

8. Transfer 2 mL cell suspension into a new 50-mL Falcon tube.

9. Add 10 mL of 1.5% prewarmed alginate solution and mix gently with the cells. Avoid formation of air bubbles. Fill the cell-alginate suspension into a Luer lock-compatible 20-mL syringe.

10. Connect the syringe to the encapsulation device and start capsule formation using the following specific settings: flow rate 23.5 units, vibration frequency 1088 Hz, voltage for bead dispersion 1300 V.

11. Follow the protocol as recommended by Inotech (http://www.inotech.ch). In brief:

   a. Allow for capsule polymerization for 5 min.
   b. Remove the polymerization solution.
   c. Add 75 mL of 0.05% PLL solution and incubate for 10 min to allow formation of the outer membrane.
   d. Remove the remaining PLL solution.
   e. Wash twice with 100 mL and 150 mL MOPS washing buffer for 1 min and 5 min, respectively.
   f. After formation of the outer alginate layer following addition of 100 mL 0.03 % alginate for 10 min, wash the capsules once with 150 mL MOPS buffer for 1 min.
   g. Dissolve the alginate core by addition of 150 mL depolymerization solution for 10 min.
   h. After two washing steps (100 mL and 150 mL MOPS buffer for 1 and 5 min, respectively), the capsules are washed once with 50 mL FMX-8 complete medium (when adding any solution into the encapsulation container via the sterile filter, make sure to press all the liquid out of the filter in each step).

12. Resuspend the capsules in 50 mL FMX-8 medium and pipet 10 mL capsule solution into each of 5 T-75 flasks containing 15 mL FMX-8 complete medium. Cultivate in an incubator until use (typically overnight).

   Capsules can be visualized by a light transmission microscope. A representative picture of microencapsulated CHO cells in shown in **Fig. 4A**.

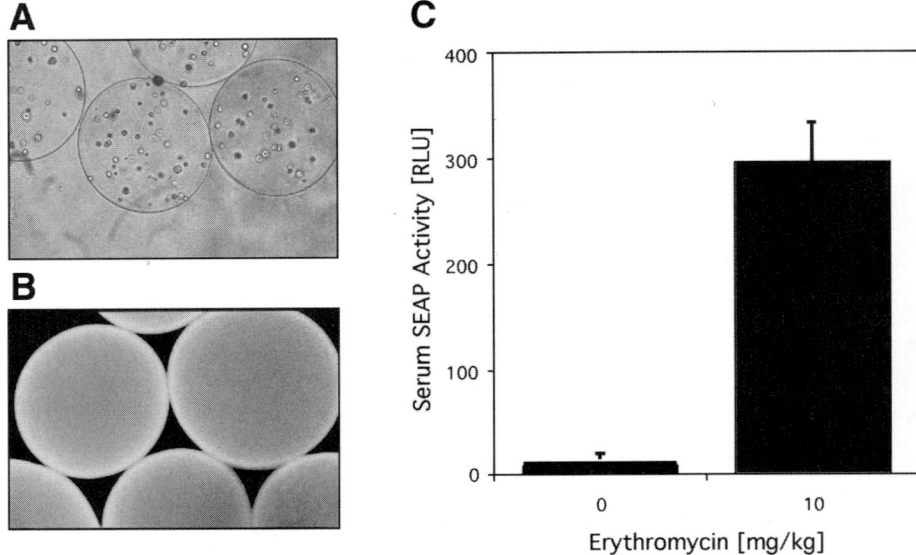

Fig. 4. (**A**) Monodispersed microencapsulated CHO-K1 cells stably transfected with plasmids pWW29 and pWW56 (clone CHO-E_1-$SEAP_1$). The cells were encapsulated in a three-layer alginate-PLL-alginate membrane (PLL: poly-L-lysine). (**B**) Fluorescence micrograph of FITC-labeled capsules (*see* **Note 8**). (**C**) Serum SEAP levels in mice 72 h after implantation of microencapsulated CHO-K1 cells stably transfected with pWW29 and pWW56. Equal mouse populations were injected with either 10 mg/kg body weight erythromycin or 0.9% NaCl as control (0 mg/kg body weight erythromycin).

3.7. Capsule Injection Into Mice

The microencapsulated cells are now ready for injection into mice. The mice have free access to food and water throughout the experiment. For positive and negative control experiments, use capsules containing wild-type CHO-K1 cells or cells that constitutively express SEAP (e.g., CHO-$SEAP_1$ cells stably containing pWW56 (*see* **Subheading 3.3.3.**). At least 10 mice should be used for each experimental group.

1. Collect all capsules in one T-75 flask; allow them to settle.
2. Remove the cell culture medium.
3. Add an equal volume of MOPS washing buffer (capsules: liquid = 1:1) and resuspend capsules by shaking gently.
4. Inject 700 µL capsule suspension (corresponding to 10,000 capsules or 2×10^6 cells) intraperitoneally per mouse using a 1-mL Luer lock syringe and a 0.6 × 30 mm needle. Let the mice recover for 1 h.
5. Inject regulating erythromycin adjusted to desired concentrations in physiological salt solution (maximum injection volume per mouse is 200 µL when injected intraperitoneally). Use a 100 mg/ml erythromycin stock solution in ethanol and dilute appropriately in physiological salt solution. Mix well prior to injection, as erythromycin may precipitate.

3.8. Analysis of Serum SEAP Levels in Mice

SEAP serum levels of mice were quantified 72 h post-implantation of microencapsulated E_{ON}-transgenic CHO cell derivatives. Mice were sacrificed and blood was collected by heart puncture. SEAP levels were assessed using a chemiluminescence-based Great EscAPe SEAP reporter system 2.

1. Put the mice to sleep by intraperitoneal injection of 30 mg/kg body weight nembutal (sodium pentobarbital).
2. Open the thorax to expose the beating heart. Collect 200 μL blood using a 1-ml syringe and a 0.6 × 30 mm needle.
3. Pour the blood sample into a microtainer serum separation tube; invert six times and incubate 30 min at room temperature for coagulation.
4. Open the mouse peritoneum and check for the presence of intact capsules and the absence of inflammation.
5. Centrifuge the microtainer serum separation tubes at 10,000g for 2 min.
6. Collect the serum and use 12.5 μL for SEAP quantification following the protocol provided with Great EscAPe SEAP reporter system 2, except for a minor modification: Centrifuge the sample at 20,000g for 2, min following addition of the substrate solution to remove possible serum precipitates. Resume the manufacturer's protocol using the sample supernatant.

The SEAP serum levels of mice following administration of different erythromycin concentrations is shown in **Fig. 4B**.

4. Notes

1. A huge expression vector portfolio encoding macrolide-, streptogramin-, and tetracycline-responsive promoters is available for straightforward adaptation of antibiotic-responsive transgene expression control technology to almost any desired experimental setting (*11,17,20*).
2. In principle, the order of transfection of pWW56 and pWW29 does not matter. However, transfecting the target gene-encoding vector pWW56 first facilitates subsequent screening for optimal cell clones. To avoid potential interferences between the expression units encoded on pWW56 and pWW29, we recommend sequential rather than simultaneous installation of these plasmids on the host cell chromosome.
3. The protocols provided here can easily be adapted to other cell types and other selection markers (*see* **Note 7**). Prior to construction of stable cell lines, the effective concentration of the selective agents has to be determined for each cell type. Thereby, 50,000 target cells/well of a 6-well plate are cultivated in 2 mL medium supplemented with increasing concentrations of the selective agent (e.g., for neomycin: 10, 20, 50, 100, 200, 500, 1000 μg/ml). The medium is replaced every 3 d during a total cultivation time of 7–10 d. The selective antibiotic concentration corresponds to the minimal concentration that kills all adherent cells.
4. Besides FACS-mediated single-cell sorting, "limiting dilution" or manual cloning of resistant cell colonies using cloning rings could be used as alternative methods to generate monoclonal transgenic cell populations (*23*).
5. Alternative SEAP quantification protocol: This alternative SEAP assay is based on SEAP-catalyzed conversion of para-nitrophenylphosphate to para-nitrophenolate, which has a strong absorbance at 405 nm. Take 100 μL supernatant, incubate for 10 min at 65°C to inactivate endogenous phosphatases, and spin down for 2 min to remove cell debris. In the meantime,

prepare a flat-bottom 96-well plate (e.g., Microtest™ Fat Bottom, Becton Dickinson, Franklin Lakes, NJ, cat. no. 353070) containing 100 μL 2X SEAP buffer. Add 80 μL of SEAP-containing cell culture supernatant as well as 20 μL pNPP solution and immediately determine the light absorbance time course at 405 nm. Determine the increase in optical density per minute within the linear part of the absorbance timecourse and calculate SEAP according to Lambert-Beer's law, $E = a \times c \times d$, where $a = 18600$ M^{-1} cm^{-1}; c = increase of pNP concentration per minute [M/min] and d = length of the light path in the liquid [cm] (typically 0.5 cm in this 96-well setup). E is the increase in optical density per minute. The activity in units per liter [U/L; μmol/min/L] is calculated as follows: Activity [U/L] = $c \times 10^6 \times 200/80$ (dilution factor). Besides SEAP, a variety of other mammalian reporter genes are available (*24*).

6. During the selection procedure for stable cell lines containing antibiotic-inducible gene regulation components, the regulating antibiotic should always be present in the culture medium. Empirically, the presence of regulation antibiotics during the selection process increased the frequency of finding transgenic cell clones that show desired gene regulation characteristics.

7. In direct comparison, the calcium-phosphate-based transfection protocol yields cell clones with superior regulation characteristics compared to lipid-based transfection (e.g., Fugene 6). Also, even when considering lipid-based transfection protocols (Fugene 6 or equivalent) for transient transfection of antibiotic-responsive expression units, add regulating antibiotics (e.g., erythromycin and pristinamycin) 5 h posttransfection to prevent interferences between antibiotics and the transfection reagents.

8. Fluorescent staining of microcapsules facilitates their tracing in animals post-implantation. FITC-labeled microcapsules are produced using the standard protocol (*see* **Subheading 3.6.**) and a PLL solution supplemented with 0.1% (w/w) FITC-labeled PLL (Sigma, St. Louis, MO, cat. no. P3543). Fluorescently labeled cells can be visualized using a fluorescence microscope (e.g., Leica DM-RB; Leica, Heerbrugg, Switzerland) equipped with an appropriate filter (e.g., XF104, Omega Optical, Brattleboro, VT).

Acknowledgments

We thank Marie Daoud-El Baba for animal studies, Bettina Keller for construction of stable cell lines, Christoph Heinzen (Inotech Encapsulation AG) for providing an encapsulator and for advice in this study, Eva Niederer (Institute of Biomedical Engineering, ETH Zurich) for FACS sorting, and Cornelia Fux and Cornelia C. Weber for critical comments on the manuscript. All experiments involving animals were approved by the French Ministry of Agriculture and Fishery (Paris, France) and performed by Marie Daoud-El Baba at the Institut Universitaire de Technologie, IUTA, F-69200 Villeurbanne Cedex, France. This work was supported by the Swiss National Science Foundation (grant 631-065946) and Cistronics Cell Technology GmbH, Einsteinstrasse 1-5, CH-8093 Zürich, Switzerland.

References

1. Fussenegger, M. (2001) The impact of mammalian gene regulation concepts on functional genomic research, metabolic engineering, and advanced gene therapies. *Biotechnol. Prog.* **17**, 1–51.

2. Fussenegger, M., Schlatter, S., Datwyler, D., Mazur, X., and Bailey, J. E. (1998) controlled proliferation by multigene metabolic engineering enhances the productivity of Chinese hamster ovary cells. *Nat. Biotechnol.* **16**, 468–472.

3. Clackson, T. (1997) Controlling mammalian gene expression with small molecules. *Curr. Opin. Chem. Biol.* **1**, 210–218.
4. Weber, W. and Fussenegger, M. (2002) Artificial mammalian gene regulation networks—novel approaches for gene therapy and bioengineering. *J. Biotechnol.* **98**, 161–187.
5. Rivera, V. M., Clackson, T., Natesan, S., Pollock, R., Amara, J. F., Keenan, T., et al. (1996) A humanized system for pharmacologic control of gene expression. *Nat. Med.* **2**, 1028–1032.
6. Beerli, R. R., Schopfer, U., Dreier, B., and Barbas, C. F. (2000) Chemically regulated zinc finger transcription factors. *J. Biol. Chem.* **275**, 32617–32627.
7. Gossen, M. and Bujard, H. (1992) Tight control of gene expression in mammalian cells by tetracycline-responsive promoters. *Proc. Natl. Acad. Sci. USA* **89**, 5547–5551.
8. Fussenegger, M., Morris, R. P., Fux, C., Rimann, M., von Stockar, B., Thompson, C. J., et al. (2000) Streptogramin-based gene regulation systems for mammalian cells. *Nat. Biotechnol.* **18**, 1203–1208.
9. Weber W., Fux, C., Daoud-El Baba, M., Keller, B., Weber, C. C., Kramer, B. P., et al. (2002) Macrolide-adjustable transgene control technology in mammalian cells and mice. *Nat. Biotechnol.* **20**, 901–907.
10. Yao, F., Svensjo, T., Winkler, T., Lu, M., Eriksson C., and Eriksson, E. (1998) Tetracycline repressor, TetR, rather than the TetR-mammalian cell transcription derivatives, regulates inducible gene expression in mammalian cells. *Hum. Gene Ther.* **9**, 1939–1950.
11. Weber, W., Marty, R. R., Keller, B., Rimann, M., Kramer, B. P., and Fussenegger, M. (2002) Versatile macrolide-responsive mammalian expression vectors for multiregulated multigene metabolic engineering. *Biotechnol. Bioeng.* **80**, 691–705.
12. Freundlieb, S., Schirra-Muller, C., and Bujard, H. (1999) A tetracycline controlled activation/repression system with increased potential for gene transfer into mammalian cells. *J. Gene Med.* **1**, 4–12.
13. Weber, W., Kramer, B. P., Fux, C., Keller, B., and Fussenegger, M. (2002) Novel promoter/transactivator configurations for macrolide- and streptogramin-responsive transgene expression in mammalian cells. *J. Gene Med.* **4**, 676–686.
14. Urlinger, S., Helbl, V., Guthmann, J., Pook, E., Grimm, S., and Hillen, W. (2000). The p65 domain from NF-kappaB is an efficient activator in the tetracycline-regulatable gene expression system. *Gene* **247**, 103–110.
15. Pollock, R., Issner, R., Zoller, K., Natesan, S., Rivera, V. M., and Clackson, T. (2000) Delivery of a stringent dimerizer-regulated gene expression system in a single retroviral vector. *Proc. Natl. Acad. Sci. USA* **97**, 13221–13226.
16. Baron, U., Freundlieb, S., Gossen, M., and Bujard, H. (1995) Co-regulation of two gene activities by tetracycline via a bidirectional promoter. *Nucleic Acids Res.* **23**, 3605–3606.
17. Fussenegger, M., Mazur, X., and Bailey, J. E. (1998) pTRIDENT, a novel vector family for tricistronics gene expression in mammalian cells. *Biotechnol. Bioeng.* **57**, 1–10.
18. Moser, S., Schlatter, S., Fuc, C., Rimann, M., Bailey, J. E., and Fussenegger, M. (2000) An update of pTRIDENT multicistronics expression vectors: pTRIDENTs containing novel streptogramin-responsive promoters. *Biotechnol. Prog.* **16**, 724–35.
19. Fux, C., Moser, S., Schlatter, S., Rimann, M., Bailey, J. E., and Fussenegger, M. (2001) Streptogramin- and tetracycline-responsive dual regulated expression of p27Kip1 sense and antisense enables positive and negative growth control of Chinese hamster ovary cells. *Nucl. Acids. Res.* **29**, e19.
20. Moser, S., Rimann, M., Fux, C., Schlatter, S., Bailey, J. E., and Fussenegger, M. (2001) Dual-regulated expression technology, a new era in adjusting heterologous gene expression in mammalian cells. *J. Gene Med.* **3**, 1–23.

21. Fussenegger, M. (2001) Dual-regulated gene expression in mammalian cells, a novel approach to gene therapy. *Gene Therapy and Regulation*, **1**, 233–264.
22. Sambrook, J., Fritsch, E. F., and Maniatis, T. (2000) *Molecular Cloning, A Laboratory Manual*, 3rd ed. Cold Spring Harbor Laboratory Press, Cold Spring Harbor, NY.
23. Doyle, A., Griffiths, J. B., and Newell, D. G. (1995) *Cell & Tissue Culture: Laboratory Procedures*. Update 9. John Wiley & Sons Ltd., Hoboken, NJ.
24. Schlatter, S., Rimann, M., Kelm, J., and Fussenegger, M. (2002) SAMY, a novel mammalian reporter gene derived from *Bacillus stearothermophilus* α-amylase. *Gene* **282,** 19–31.

31

Flp-Mediated Integration of Expression Cassettes into FRT-Tagged Chromosomal Loci in Mammalian Cells

Dagmar Wirth and Hansjörg Hauser

Summary

The creation of recombinant mammalian cells with defined expression levels requires extensive screening. This is mainly due to the unpredictable site and copy number of the recombinant DNA in the hosts' chromosomal DNA. The method presented here is based on the exchange of expression cassettes in a previously tagged site. Since the site of chromosomal integration remains the same in all targeted cells and since the copy number is reduced to one, the level of expression is highly predictable. The use of this method includes two steps. The first one, the tagging step, makes use of retroviral reporter constructs, which ensure single-copy integration of the tagging vector. Upon screening for cell clones with appropriate expression strength these cells will be targeted. The targeting construct harboring the gene of interest will precisely replace the tagging reporter cassette making use of the Flp recombinase.

Key Words: Mammalian cells, predictable expression, Flp, FRT, site-specific recombination, targeted exchange, recombinase-mediated cassette exchange (RMCE).

1. Introduction

Most transfection protocols for transduction of recombinant genes are accompanied by random integration of the expression cassettes into the hosts' genome. The nature of the integration site has a major impact on transgene expression since it provides genetic elements that modulate transgene expression (*1–5*). As a consequence, significant clone-to-clone variability with regard to the level and stability of expression is observed. Thus, much effort usually has to be made in order to screen cell clones for appropriate expression levels.

From: *Methods in Molecular Biology, vol. 267:*
Recombinant Gene Expression: Reviews and Protocols, Second Edition
Edited by: P. Balbás and A. Lorence © Humana Press Inc., Totowa, NJ

Homologous recombination is a commonly applied method for site-specific integration into precloned integration sites in mouse embryonal stem cells. Unfortunately, this method has gained no significance for targeted integration in permanent cell lines because these cells have elevated levels of nonhomologous, i.e., illegitimate, recombination. As a consequence homologous recombination events are superposed by the much more frequent random integration events (**6**). In recent years site-specific recombination systems like Cre/loxP and Flp/FRT have been adapted for specific integration into mammalian cells (**7,8**). However, the efficiency of the classical approach is largely limited due to the fact that the backward reaction, i.e., site-specific excision, is not suppressed. Here, we present an advanced method for efficient targeting of precharacterized integration sites (**Fig. 1**). The method is based on the use of FRT spacer sequences that can recombine only with identical counterparts but are refractory to recombination between heterologous sites. In addition, a stringent selection procedure is applied that both simplifies and accelerates the screening for correct targeting events. With this technique, a routine exchange of expression cassettes is feasible (**9,10**).

2. Materials

2.1. Cell Culture Conditions

Standard mammalian cell culture equipment is required. Cells are handled according to standard protocols. The cells described in this method are cultivated in DME medium supplemented with 10% of fetal calf serum (FCS), 20 mM glutamine, 60 µg/mL penicillin, and 100 µg/mL streptomycin. This medium is called "standard medium" in the following.

2.2. Tagging

2.2.1. Establishment of Producer Cells for Production of Tagging Viruses

1. PT67 packaging cells (Clontech).
2. pTag-1 (**Fig. 2**).
3. 2X HEBS solution: 280 mM NaCl, 50 mM HEPES, 1.5 mM Na$_2$HPO$_4$, pH 7.1; aliquot and conserve at −20°C.
4. 2.5M CaCl$_2$, sterile-filtered; conserve at −20°C.
5. Hygromycin B: 431660 U/mL (batch-dependent), sterile-filtered; conserve at 4°C.
6. 25-cm^2 T flasks.
7. 75-cm^2 T flasks.

2.2.2. Determination of the Virus Titer

1. Polybrene 4 mg/mL, sterile-filtered; conserve at −20°C.
2. Hygromycin B: 431660 U/mL (batch-dependent), sterile-filtered; conserve at 4°C.
3. Crystal violet solution: 5 g crystal violet, 8.5 g NaCl, 143 mL formaldehyde, 500 mL ethanol adjusted to 1000 mL with H$_2$O.
4. 25-cm^2 T flasks.
5. 24-well plates.
6. 0.45-µm syringe filter.

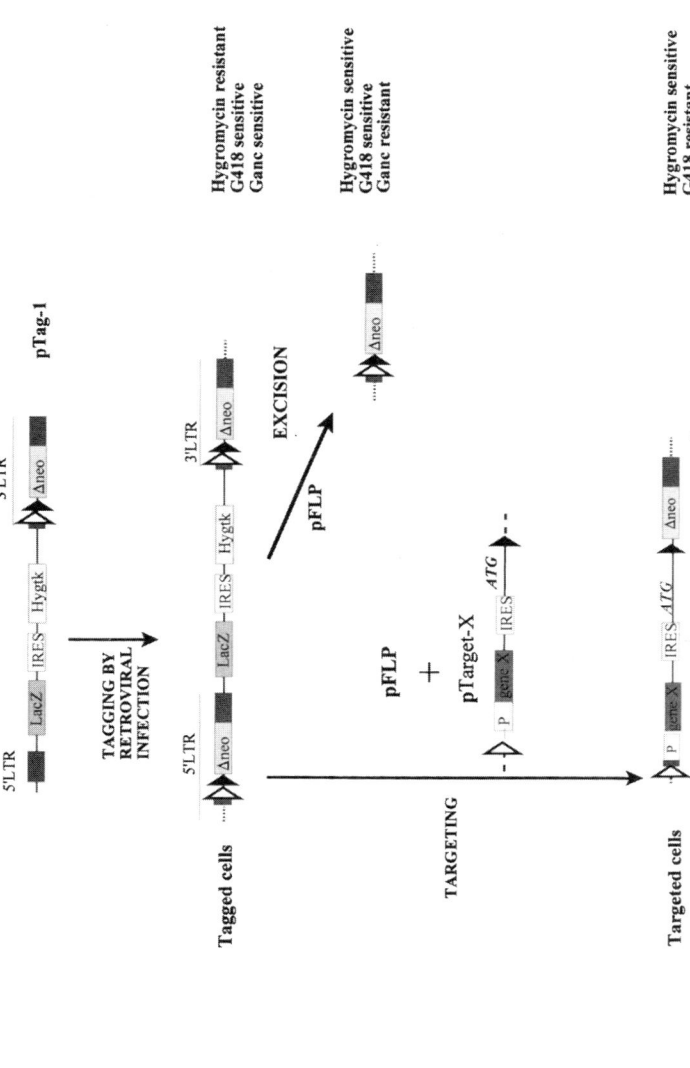

Fig. 1. Strategy for efficient reuse of tagged and characterized chromosomal loci by means of the site-specific Flp/FRT recombination system. The retroviral vector pTag-1 encodes the reporter gene (β-galactosidase/*lacZ*) and a selection marker (*hygtk*, a fusion gene of hygromycin B phosphotrans-ferase and thymidine kinase). pTag-1 contains two FRT sites in tandem in the 3′ LTR, a wild-type (white triangle) and a noninteracting spacer mutant FRT form (black triangle), followed by an ATG defective neomycin phosphotransferase (*neo*) gene. Infection with this vector results in duplication of the U3 region of the 3′ LTR to the 5′ LTR. The tagged clones are resistant hygromycin B but are G418-sensitive. These cells are screened for appropriate levels of the reporter gene. Cotransfection of the Flp recombinase-expressing plasmid (pFlp) and the targeting vector pTarget-X leads to sequence-specific recombination via the outer FRT sites. Since the defective neomycin phophotransferase gene in the 3′-LTR is complemented by the missing IRES element and the ATG positioned in frame, recombination events are detected by G418 selection.

2.2.3. Tagging of Cells by Retroviral Infection

1. Polybrene 4 mg/mL, sterile-filtered; conserve at −20°C.
2. Hygromycin B: 431660 U/mL (batch-dependent), sterile-filtered; conserve at 4°C.
3. 25-cm² T flasks.
4. 55-cm² plates.
5. 24-well plates.
6. 0.45-µm syringe filter.

2.2.4. Screening of Cells for β-Galactosidase Expression

1. Fluorescence reader, excitation wave length 365 nm, emission 450 nm.
2. 96-well optical plate.
3. PBS: 140 mM NaCl, 27 mM KCl, 7.2 mM Na$_2$HPO$_4$, 14.7 mM KH$_2$PO$_4$ pH 6.8–7.0.
4. TEN: 10 mM Tris-HCl, 1 mM EDTA, 100 mM NaCl, pH 7.4.
5. Substrate buffer: 60 mM NaH$_2$PO$_4$, 40 mM Na$_2$HPO$_4$, 10 mM HCl, 1 mM MgSO$_4$, 50 mM β-mercaptoethanol.
6. MUG: 4-methylbelliferryl-β-D-galactoside, 10 mM in stock solution in dimethylformamide, store at −20°C.
7. Liquid nitrogen.
8. Eppendorf tubes.
9. Centrifuge for Eppendorf tubes.

2.3. Targeting of Preselected Clones

1. pTarget-1/X (**Fig. 2**).
2. pFLP [pFLP corresponds to pOG44M in ref. *11*]
3. 2X HEBS solution: 280 mM NaCl, 50 mM HEPES, 1.5 mM Na$_2$HPO$_4$, pH 7.1; aliquot and conserve at −20°C.
4. 2.5 M CaCl$_2$, sterile-filtered; conserve at −20°C.
5. G418: 100 mg/mL stock solution in H$_2$O, sterile-filtered; conserve at −20°C.
6. Gancyclovir: 10 mM in H$_2$O, sterile-filtered; conserve at 4°C.
7. 25-cm² T flasks.
8. 6-well plates.
9. 24-well plates.

3. Methods

3.1. Overall Strategy

The overall outline of the tag-and-target strategy is described in **Fig. 1**. In a first step, cells are tagged by random integration of a reporter cassette (β-galactosidase), which allows for evaluation of the intrinsic expression characteristics of the genomic locus (level and stability). Retrovirus-mediated gene transfer is used to precisely adjust the number of tags that are introduced into the genome to one (see below). The tagging vector carries a set of two noninteracting FRT sites in the 3' LTR. Upon retroviral infection, the tagged cells will harbor the β-galactosidase reporter cassette flanked by sets of noninteracting FRT sites. At the same time, a nonfunctional—i.e., a promoter and start codon-deficient neomycin-resistance—gene is integrated. This gene will also be duplicated.

pTag-1

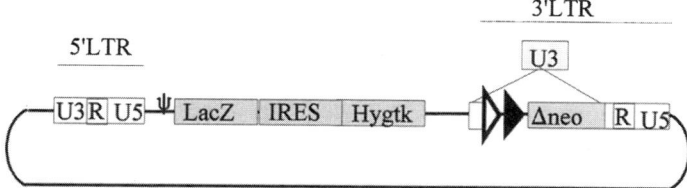

pTarget-1

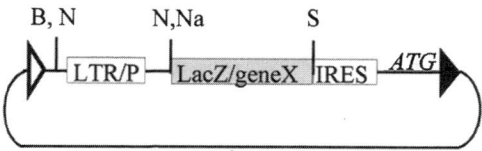

Fig. 2. The retroviral vector pTag-1 is used for tagging individual chromosomal sites and to measure the intrinsic expression properties. The reporter gene β-galactosidase/*lacZ* and the fusion gene *hygtk* comprising the hygromycin B phosphotransferase and thymidine genes are translated from a bicistronic message via the EMCV IRES element. pTag-1 carries all retroviral elements required for retroviral transduction including two LTRs (comprising U3, R, and U5) and the packaging signal χ. In the 3' U3 region the Tag- and Target-cassette is integrated. It consists of two heterologous—i.e., noninteracting—FRT sites (symbolized by the white and black triangles) and a promoter and ATG deficient neomycin resistance gene (Δ*neo*).

The targeting plasmid pTarget-1 is used for targeted integration of expression cassettes of choice into tagged chromosomal loci. The expression cassette is flanked by the heterologous, noninteracting FRT sites (white and black triangles). By cloning via the restriction sites *Bst*XI (B), *Not*I(N), *Nar*I(Na) and *Sal*I(S), genes of interest and—if appropriate—promoter elements can be integrated. An EMCV IRES element together with an ATG start codon positioned in frame renders targeted cells resistant to G418.

Cells tagged with the reporter cassette are screened for resistance hygromycin B. Isolated cell clones are analyzed for appropriate expression of the reporter gene (β-galactosidase).

In a second step selected cell clones are targeted. For this purpose a specific targeting vector has to be prepared in which the expression cassette is flanked by corresponding outer FRT sequences (*see* **Note 1**). This vector can be obtained by substituting the reporter cassette of pTarget-1 with the expression cassette of choice. Flp-mediated targeting of this vector into the tagged cells will result in exchange of the expression cassettes within the defined chromosomal site. Since recombination results in the activation of the nonfunctional neomycin-resistance gene, targeted cell clones are screened by cultivation of cells in G418-containing medium.

A protocol is given below which has been established for tagging and targeting NIH3T3 cells. Principally, the protocol should be transferable to any cell that is susceptible to the gene transfer and selection methods used here (*see* **Note 2**).

3.2. Tagging of Cells by Retroviral Infection

3.2.1. Establishment of Producer Cells by Transfection With Calcium-Phosphate-DNA Coprecipitation

1. The retroviral packaging cell line PT67 is transfected with the tagging vector pTag-1.
2. On the day before transfection, the cells are seeded in culture flasks ($1-3 \times 10^5$ cells/25 cm²).
3. 4 h before transfection, a medium exchange with standard medium is performed.
4. 5 µg of pTag-1 is suspended in 250 µL 250 mM CaCl$_2$.
5. The solution is added drop by drop to 250 µL 2X HEBS buffer under continuous vortexing. In this solution the DNA coprecipitates with calcium phosphate.
6. After 5–10 min the precipitates are pipetted to the cells and 15 h later the medium on the cells is replaced by fresh medium.
7. Two days after transfection, selection of stably transfected cells is started.
8. The cells are transferred to a 75-cm² flask and are selected for resistance to hygromycin B (200 U/mL).
9. Medium exchange is done every 3–5 d.
10. In general, the selection is finished after 7–12 d.
11. Pools of cells (called producer cells below) are collected. An aliquot is frozen and cells are further passaged to collect the virus supernatant.

3.2.2. Determination of the Virus Titer

1. Producer cells are seeded to reach confluence on the day of infection.
2. 24 h before infection the standard DME medium (virus production medium) is renewed.
3. The supernatant of the producer cells containing the tagging virus is harvested and filtered (0.45-µm filter) to eliminate producer cells.
4. The number of producer cells is determined.
5. Polybrene is added to the supernatant (8 µg/mL final concentration) to increase the efficiency of the infection.
6. The supernatant is diluted in steps of ten down to 10^{-6}. Duplicates of each dilution in 1 mL standard medium provided with 8 µg/mL polybrene, as well as a negative control (virus-free supernatant), are prepared.
7. The samples are added to NIH3T3 test cells that have been seeded on 24-well plates (5000 cells/well) the day before infection.
8. On the day after infection the medium is exchanged for standard medium supplemented with 200 µg/mL hygromycin B.
9. After 7–10 d noninfected cells should be eliminated. Appearing cell clones can be stained with crystal violet and the number of virus particles per mL can be calculated.
10. The virus titer is calculated as infectious particles per 10^6 producer cells.

The calculation of the virus titer is important to limit the number of transgene copies per cell to one. Using a low multiplicity of infection (MOI) of, e.g., 0.01 (corresponding to 1 infectious virus particle per 100 cells to be infected), more than 99% of infected cells integrate a single copy of the transduced cassette.

Thus, to reach an infection of one virus per cell, a low multiplicity of infection should be chosen. With the Poisson distribution one can calculate how many cells are transfected with 0, 1, 2, and more viruses at a given MOI.

$$\text{Poisson distribution: } P(k) = x^k/k! \times e^{-x},$$

where $P(k)$ is the probability of k events, k the number of events (here, the number of viruses that infect a cell) and x determines the average number of events (here, MOI). For this case, the Poisson distribution is $P(k) = MOI^k/k! \times e^{-MOI}$. MOI = 0.01 ($e^{-0.01}$ = 0.99) gives the following distribution:

$P(0) = 1/1 \times 0.99 = 0.99$	99% of the cells are not infected
$P(1) = 0.01/1 \times 0.99 = 9.9 \times 10^3$	0.99% of the cells are infected with one virus.
$P(k > 1) = 1 - P(0) - P(1) = 1 \times 10^{-4}$	0.01% of the cells are infected with two or more viruses

If 10^5 cells are infected under these conditions (MOI = 0.01), 990 cells are infected with one virus and 10 cells with two or more viruses. one percent of the infectants will harbor more than one proviral integrate.

3.2.3. Tagging of Chromosomal Sites
Using Retroviral Transfer of the Tagging Vector

1. Cells to be tagged are seeded in a concentration of $1–2 \times 10^5$ cells/25 cm² flask the day before infection.
2. For infection, the 24-h production medium is filtered through a 0.45-μm syringe filter. The volume of virus supernatant corresponding to an MOI of 0.01 according to the previously calculated titer is replenished with standard medium to a final volume of 4 mL, supplemented with 8 μg/mL polybrene is added to the cells to be infected.
3. Twenty-four hours later the cells are transferred to a 55-cm² plate and selected in hygromycin-containing medium (200 U/mL). Seven to 10 days later, cell clones will appear that are picked and transferred to 24-well plates for further expansion. An aliquot should be frozen.

3.2.4. Characterization and Screening
of Tagged Cells for β-Galactosidase Expression

Isolated cell clones are analysed for β-galactosidase expression (modification of Kalos and Fournier [12]).

1. Subconfluent cells in a 6-well plate are washed with PBS, harvested in 1 mL TEN, and transferred to Eppendorf tubes.
2. After centrifugation (5 min, 1000 rpm, Hettich Rotanta) the pellet is resuspended in 150 μl Tris-HCl, pH 7.8.
3. To lyse the cells they are frozen four times in liquid nitrogen, followed each time by thawing the cells in a 37°C water bath.
4. The lysed material is centrifuged for 10 min at 4°C (13,000 rpm) and the supernatant is measured.

5. In a 96-well optical plate, 20 μl for each sample in the desired dilutions is pipetted.
6. By adding 100 μl substrate buffer supplemented with 30 μg/mL MUG, the reaction is started.
7. Measurement is done in a fluorescence-ELISA reader with an excitation wavelength of 365 nm. The emission is measured at 450 nm in different time intervals (5 min). To compare the different independent measurements, a standard plot for purified enzyme (β-galactosidase, Boehringer Mannheim) has to be made in parallel. The β-galactosidase activity is normalized to the protein content.
8. In addition, the integrity of the expression cassette and its single-copy status should be confirmed by Southern blot analysis.
9. Cell clones that reach the criteria concerning the level and stability of expression will be used for targeted integration of the gene of interest.

3.3. Targeting of Selected Clones With the Expression Cassette of Choice

3.3.1. Cloning of the Specific Targeting Vector pTarget-X

For targeting the gene of interest into the preselected genomic loci a specific targeting vector (pTarget-X) must be prepared in which the expression cassette of choice is flanked by the outer FRT sites. This vector can be obtained by cloning the expression cassette into pTarget-1 cleaved with appropriate restriction enzymes (e.g., for exchange of the reporter gene, pTarget-1 will be restricted with *Nar*I and *Sal*I). Cloning is performed according to standard protocols.

3.3.2. Flp-Mediated Targeting Integration of the Expression Cassette

For Flp-mediated recombination, the targeting vector pTarget-X and the Flp expression plasmid pFLP are cotransfected using DNA-calcium phosphate coprecipitation according to the method described in **Subheading 3.2.1.** (*see* **Note 3**). The day before targeting, the cells tagged with a single copy of the expression cassette are seeded in 6-well plates ($0.5–1 \times 10^5$ cells/well, 3 wells per cell clone) in the presence of hygromycin B (200 U/mL). The next day transfection is performed. Three transfections are made in parallel:

(A) 12 μg of pFLP + 4 μg of pTarget-X (for targeted integration).
(B) 4 μg of pTarget-X (will result in random integration only and serves as a negative control).
(C) 12 μg of pFLP (excision of the tagging cassette).

On the day after transfection, the medium is renewed and the cells are left for 4 d in the incubator. Within this time the cells should undergo the cassette exchange, thereby losing their hygromycin resistance and thymidine kinase activity. On day 5 any remaining activity of the resistance proteins should be gone and selection can be started. Successful targeting will render the cells resistant to G418. Samples (A) and (B) are selected with G418 (750–1000 μg/mL). While (A) will give the successfully targeted cells, transfection (B) serves as a negative control. To determine the efficiency of excision the cells (C) are selected with gancyclovir (10 μM/mL). The cells are transferred to a culture plate for the selection process to pick the appearing clones and expand them.

3.4. Final Characterization

The cell clones obtained by the procedure described in **Subheading 3.3.** are characterized with respect to the integrity of the expression cassette and the specific targeting. Previous results showed that the efficiency and accuracy of this method are high: Usually, far more than 90% of G418-resistant clones have exchanged the cassettes correctly. Thus, there is a high probability that single clones show the exchanged phenotype. Nevertheless, this should be confirmed by Southern blot analysis, by a biological assay proving the activity of the integrated cassette, and/or by Northern blot analysis to prove the activity of the integrated gene of interest.

4. Notes

1. Conserve the structure of the exchange cassettes; otherwise, unexpected results may be obtained (*10*).
2. Other cell lines to be tagged; infection conditions have to be adapted and optimized. Other packaging cell lines might be more appropriate; test concentration of polybrene and drug selection conditions.
3. Alternatively, Flp gene expression can be linked to a marker gene (e.g., puromycin resistance gene, GFP) in a bicistronic message (*13*). Cassette exchange is performed as described above; however, 2 d after transfection, cells are sorted for their GFP expression. Alternatively, they are seeded in puromycin-containing medium for 48 h. This procedure allows enrichment of cells that have taken up the Flp gene. However, due to the early moment in which the enrichment is performed, a preferential enrichment of cells that have stably integrated the Flp gene is circumvented.

References

1. Bell, A. C. and Felsenfeld, G. (2001) Gene regulation: insulators and boundaries: versatile regulatory elements in the eukaryotic genome. *Science* **291**, 447–450.
2. Bode, J., Benham, C., Knopp, A., and Mielke, C. (2000) Transcriptional augmentation: modulation of gene expression by Scaffold/matrix attached regions (S/MAR elements). *Crit. Rev. Eukaryot. Gene Expr.* **10**, 73–90.
3. Bode, J. Schlake, M., Ríos-Ramírez, M., Mielke, C., Stengert, M., Kay, V., et al. (1995) Scaffold/matrix-attached regions: structural properties creating transcriptioally active loci, in *Structural and Functional Organization of the Nuclear Matrix (International Review of Cytology.)* (Berezney, R. and Jeon, K. W., eds). Academic Press, San Diego, CA, pp. 389–453.
4. Li, Q. L., Harju, S., and Peterson, K. R. (1999) Locus control regions—coming of age at a decade plus. *Trends Genet.* **15**, 403–408.
5. Bell, A. C. and Felsenfeld, G. (1999) Stopped at the border: boundaries and insulators. *Curr. Opin. Genet. Develop.* **9**, 191–198.
6. Cappechi, M. R. (1990) Gene targeting: how efficient can you get? *Nature* **348**, 109.
7. Kilby, N. J., Snaith, M. R., and Murray, J. A. H (1993) Site-specific recombinases: tools for genome engineering. *Trends Genet.* **9**, 413–421.
8. Baer, A. and Bode, J. (2001) Coping with kinetic and thermodynamic barriers: RMCE, an efficient strategy for the targeted integration of transgenes. *Curr. Opin. Biotechnol.* **12**, 473–480.
9. Verhoeyen, E., Hauser, H., and Wirth, D. (1998) Efficient targeting of retrovirally FRT-tagged chromosomal loci. *Techn. Tips online* T01515 URL: http://tto.trends.com.

10. Verhoeyen, E., Hauser, H., and Wirth, D. (2001) Evaluation of retroviral vector design in defined chromosomal loci by Flp-mediated cassette replacement. *Hum. Gene Ther.* **12**, 933–944.
11. O'Gorman, S., Fox, D. T., and Wahl, G. M. (1991) Recombinase mediated gene activation and site-specific integration in mammalian cells. *Science* **251**, 1351–1355.
12. Kalos, M. and Fournier, R. E. K (1995). Position-independent transgene expression mediated by boundery elements from the Apolipoprotein B chromatin domain. *Mol. Cell. Biol.* **15**, 198–207.
13. Taniguchi, M., Sanbo, M., Watanabe, S., Naruse, I., Mishina, M., and Yagi, T. (1998) Efficient production of Cre-mediated site-directed recombinants through the utilization of the puromycin resistance gene, pac: a transient gene-integration marker for ES cells. *Nucleic Acids. Res.* **26**, 679–680.

32

Generation of High-Recombinant-Protein-Producing Chinese Hamster Ovary (CHO) Cells

Jean-Louis Goergen and Lucía Monaco

Summary

The purpose of this chapter is to propose a practical procedure for the generation and selection of Chinese hamster ovary cells producing high levels of recombinant protein by combining in vitro and in vivo amplification of the foreign gene. A detailed description of the expression and amplification plasmids utilized for the in vitro generation of long DNA concatenamers, as well as the cell transfection and selection protocols are given. The procedure required for in vivo gene amplification using the dihydrofolate reductase/methotrexate system is also described.

Key Words: Recombinant protein; Chinese hamster ovary cells; in vitro amplification; in vivo amplification; dihydrofolate reductase; methotrexate; DNA concatenamers.

1. Introduction

Heterologous gene expression in mammalian cells is the system of choice for the production of therapeutic or diagnostic recombinant proteins when posttranslational modifications affect the biological activity or the physicochemical parameters (such as stability or clearance) of the protein of interest. However, specific rates of recombinant protein production by mammalian cells remain relatively low compared to other expression systems such as bacteria (e.g., *Escherichia coli)* or yeast (e.g., *Pichia pastoris* or *Saccharomyces cerevisiae*). Therefore, sophisticated genetic manipulations are necessary to reach high-level expression of foreign proteins.

Since 1980, when Urlaub and Chasin (*1*) developed a Chinese hamster ovary (CHO) cell line mutated in the dihydrofolate reductase gene (*dhfr*), and therefore devoid of DHFR activity, a growing number of recombinant proteins have been expressed with this expression system (*2–5*). DHFR converts folate to tetrahydrofolate, a precursor of

From: *Methods in Molecular Biology, vol. 267:*
Recombinant Gene Expression: Reviews and Protocols, Second Edition
Edited by: P. Balbás and A. Lorence © Humana Press Inc., Totowa, NJ

glycine, thymidine, and purines. After transfection of DHFR-deficient cells with the *dhfr* gene, it is possible to isolate recombinant DHFR-positive cell lines after about 2 wk of culture in selective media (lacking thymidine and hypoxanthine, or ribonucleosides and deoxyribonucleosides).

Methotrexate (MTX) is a folate analog, an inhibitor of the DHFR activity; addition of MTX to the culture medium will generally result in the amplification at the chromosome level of the DNA region containing the *dhfr* gene and the surrounding genes, including the cDNA coding for the recombinant protein (up to 1000 copies per cell) (**6**).

As an additional strategy to improve expression levels, it is possible to directly transfect cells with long DNA concatenamers constituted by a large number of copies of expression cassettes for the gene of interest, linked to one or a few copies of a cassette for the expression of a selectable marker gene. This method has been named in vitro amplification, to indicate the fact that the large, multicopy DNA molecules are obtained in the test tube prior to transfection, as opposed to the amplified DNA resulting from the MTX-mediated in vivo amplification inside the cell. The in vitro amplification protocol involves the use of two plasmids: one containing a selectable marker (M) and the other bearing the gene of interest (G); the in vitro ligation of these two linearized plasmids, mixed at an M/G molar ratio of at least 1/10, will yield long DNA concatenamers, which can further be transfected into the cells.

The in vitro amplification method results in the integration of the whole concatenamers in the host cell genome; consequently, a large proportion of selected clones displays high expression levels of the transgene. To further increase expression levels, this method can be combined to the MTX-mediated in vivo amplification system, provided the *dhfr* gene is used as a selectable marker.

The present contribution proposes to perform first the in vitro amplification of the recombinant gene (generation of long DNA concatenamers of marker and recombinant genes) and then, after transfection of those polymers into the cells, to apply the DHFR/MTX amplification procedure.

2. Materials

1. Ligation buffer : 50 mM Tris-HCl, pH 8.0, 10 mM MgCl$_2$, 1 mM DTT, 0.5 mM ATP.
2. Restriction enzymes, T4 DNA ligase.
3. Pulsed-field electrophoresis apparatus (Bio-Rad, model Chef DRII).
4. LB medium, ampicillin.
5. Chloroform, ethanol, phenol/chloroform, and 3 M sodium acetate, pH 5.2.
6. Agarose and DNA electrophoresis equipment.
7. Nonselective αMEM medium containing nucleosides (αMEMw).
8. Selective αMEM medium lacking nucleosides (αMEMw/o).
9. 25 kDa polyethyleneimine (PEI).
10. Methotrexate (MTX).
11. Phosphate-buffered saline (PBS) w/o Ca^{2}+ and Mg^{2}+, pH 7.4, glutamine (200 mM) and glucose (450 g/L) stock solutions.
12. Fetal calf serum and dialyzed fetal calf serum versus buffer (e.g., PBS) (dialyzation aims at avoiding bovine ribonucleosides and deoxyribonucleosides).
13. Cloning cylinders.

3. Methods

The methods presented in this chapter describe (1) the construction of the expression DNA concatenamers and (2) the selection of high producing clones.

3.1. Expression and Amplification Plasmids

3.1.1. pCISfiT and pSfiSVdhfr

Both pCISfiT and pSfiSVdhfr (**Fig. 1**) are necessary for the in vitro amplification of the gene of interest. These two plasmids bear nonpalindromic *Sfi*I sites flanking the expression unit. The sticky ends generated by *Sfi*I digestion force a head-to-tail orientation of the different nucleotidic sequences during the ligation process (**Fig. 2**). pSfiSVdhfr is the plasmid carrying the selection marker *dhfr* under the control of the SV40 early promoter. pCISfiT is the plasmid bearing the cDNA coding for the recombinant protein. This plasmid has a human cytomegalovirus (CMV) promoter upstream and a transcription terminator sequence from the human gastrin gene downstream of the polycloning site.

3.1.2. Cloning

Insert the cDNA coding for the recombinant protein into the polylinker of the pCIS-fiT vector by standard subcloning techniques (**7**).

3.2. Polymers Preparation

1. 10 µg pCISfiT and 50 µg pSfiSVdhfr are digested with respectively 20 and 100 units of *Sfi*I overnight at 50°C.
2. Digested DNAs are phenol/chloroform-extracted and ethanol-precipitated in the presence of 1/10 (v/v) NaOAc 3 *M*.
3. pCISfiT and pSfiSVdhfr are mixed at a molar ratio of 1:20 in ligation buffer and incubated with 2 units of T4 DNA ligase/µg DNA. Use a DNA concentration of 200 ng/µL, to favor intermolecular ligation events producing concatenated instead of recircularized DNA molecules. Incubate the ligation overnight at 16°C. Concatenamers of the two expression cassettes will on the average contain 1 copy of the selection marker for 20 copies of the cDNA cassette.
3. The length of ligation products is checked by pulsed-field electrophoresis on 1.2% agarose gel, at 4.8 V/cm, applying 2.5–7.5-s pulses over 16 h at 14°C. A smear or ladder of ligation products should appear (**Fig. 3**). The size of the ligation products will generally be between 50 and 200 kDa, depending on the length of the expression cassette and the success of the ligation.
4. After ligation, concatenamers are phenol/chloroform, chloroform, and diethyl ether extracted. Ethanol precipitation is avoided, as the DNA pellet might be difficult to dissolve. Concatenamers are then dissolved in sterile water for CHO transfection (*see* **Note 1**).

3.3. Cell Transfection and Selection

1. Exponentially growing *dhfr⁻* CHO cells in nonselective medium are harvested from a culture flask. Viability of those cells is checked to be higher than 90% (typically cells are passaged 24 h earlier). This cell line can grow only when the culture medium is supplemented with the necessary salvage route components (e.g., adenosine, deoxyadenosine, and thymidine).

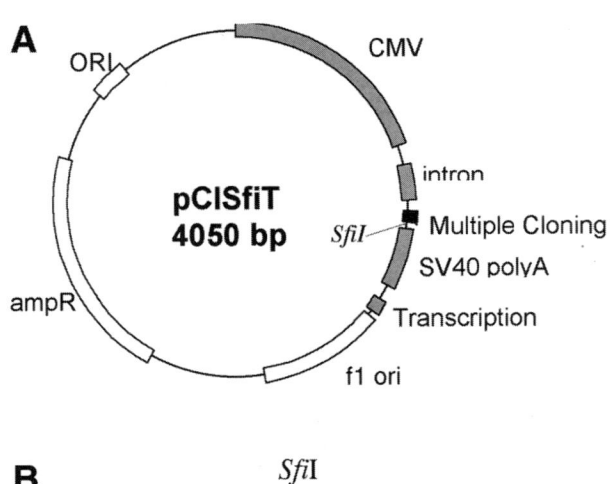

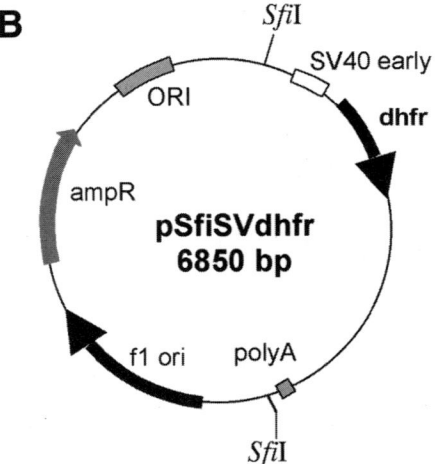

Fig. 1. Plasmids used for the cloning and in vitro amplification of the gene of interest. pCIS-fiT possesses a multiple cloning site terminated by a *Sfi* I cloning site; the gene of interest is under the control of the strong CMV promoter; a SV40 late polyadenylation sequence and a transcription terminator from the human gastrin gene are located downstream to the cDNA coding for the r-protein. pSfiSVdhfr bears the *dhfr* gene under the control of the SV40 early promoter; a SV40 early polyadenylation sequence is used.

2. CHO cells are effectively transfected by a number of transfection protocols, including calcium phosphate coprecipitation, lipofection, and the like (*see* **Notes 2** and **3**). Here, transfection using the cationic polymer polyethyleneimine (PEI) is described.

3. The day before transfection, detach CHO cells from a cell culture flask and plate them in 35-mm dishes (or in a well of a 6-well culture plate) in the nonselective medium. The next day, cells should be 50–70% confluent. This is usually achieved by plating approximately two to three times 10^5 cells per dish; the number of cells, however, needs to be empirically determined.

4. Transfect each well with 5 µg DNA complexed with PEI (*see* **Note 4**). Dilute DNA (5 µg) in 150 mM NaCl to a final volume of 250 µL. Dilute 15 µL of 10 mM PEI in 150 mM NaCl to

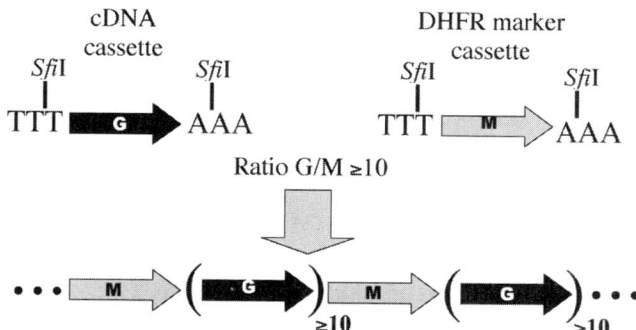

Fig. 2. Schematic representation of the in vitro amplification system, using pCISfiT and pSfiSVdhfr. The cDNA cassette (G) is represented as a black arrow; the DHFR marker cassette (M) is represented as a gray arrow. The in vitro ligation is performed with a G/M ratio of at least 10/1.

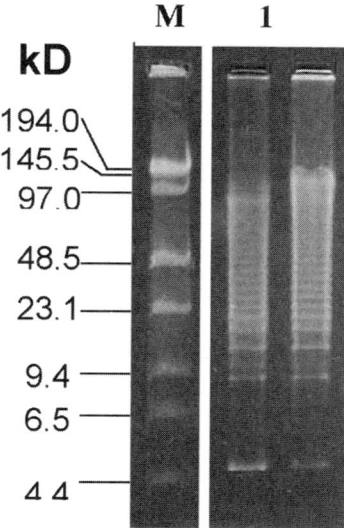

Fig. 3. Lane M: molecular weight marker; lanes 1 and 2: typical pulsed-field electrophoresis of in vitro ligated mixtures of cassettes containing the cDNA of interest and the selective/amplifiable gene *dhfr*.

a final volume of 250 μL. Add the PEI solution into the DNA solution, mix by briefly vortexing, and leave at room temperature for 15 min. Then distribute the DNA/PEI mix onto the cells. It is not necessary to remove the medium before transfection. To increase transfection efficiency, centrifuge cells at 1500*g* for 5 min. Return cells to the CO_2 incubator at 37°C.

5. After a 48-h-period in the nonselective medium, cells are passaged into two 100-mm dishes and grown in selective medium (e.g.,αMEM without ribonucleosides and deoxyribonucleosides) supplemented with dialyzed fetal calf serum.

6. After 3–4 wk in this medium, isolated foci should appear on the dishes.
7. Remove the medium from the dishes (one by one to avoid drying out the cells) and place cloning cylinders on about 10 foci or more per dish.
8. Remove carefully the remaining medium contained in each cloning cylinder and detach the cells from each cloning cylinder by adding an appropriate volume of trypsin (30–100µL on average = 1 volume).
9. Leave at 37°C for about 1–3 min.
10. Add 1 volume of fresh medium and pipet up and down the cell suspension.
11. Transfer the cells to 96-well plates and amplify the different clones until 24-well plates can be seeded.
12. Add to the selective medium 10 nM MTX as a first step for dhfr amplification.
13. Change the medium every 2–3 d with the same selective/amplification culture medium.
14. When cells recover and can grow again, they will be submitted to a higher concentration of MTX.
15. MTX titer in the culture medium can be increased by a factor of 2–4 at each step until a few colonies remain in the different culture plates. Depending on the cell line, rounds of amplification can be performed up to several hundred micromolar MTX in the culture medium.
16. The amplification procedure is time consuming; each amplification step requires 2–4 wk in order to have growing cells again.
17. Check for the recombinant protein expression all over the amplification process (*see* **Note 5**).

4. Notes

1. DNA concatenamers can be stored at –20°C.
2. Electroporation of DNA concatenamers did not prove advantageous when compared to electroporation of the two G and M plasmids at the same molar ratio (*8*). It is possible that the electroporation process does not favor the entry of large DNA molecules, such as the concatenated DNA used in the in vitro amplification method.
3. Transfection of small cDNAs (200–400 bp) might be performed by electroporation. We did not measure any difference, in terms of RNA level, between co-lipofection and co-electroporation of vectors pSfiSVdhfr and pCISfiT containing antisense fragments of 200 or 400 bp.
4. 100 m*M* PEI stock solution is prepared in H_2O, brought to pH 7.4 with HCl, filter-sterilized, and stored at 4°C. PEI MW is assumed to be 43 g, corresponding to the MW of the protonated monomer. The 10 mM solution is prepared by dilution in H_2O.
5. The combined in vitro/in vivo amplification procedures were used for the production of granulocyte colony-stimulating factor (G-CSF) (*8*) and the production of antisense RNAs to inhibit the translation of CMP-*N*-acetylneuraminic acid hydroxylase (*9*) in CHO cells. G-CSF production could be improved 30-fold compared to the in vitro amplification protocol alone. In case of antisense RNAs, an 80% stable inhibition of the CMP-*N*-acetylneuraminic acid hydroxylase was obtained with 200-nucleotide antisense fragments.

Acknowledgments

The authors would like to thank Stefano Rocco from Keryos S.p.A for expert advice regarding the preparation of the polymers and the operation of the pulse-field electrophoresis.

References

1. Urlaub, G. and Chasin, L. A. (1980) Isolation of Chinese hamster cell mutants deficient in dihydrofolate reductase activity. *Proc. Natl. Acad. Sci. USA* **77**, 4216–4220.
2. Benkovic, S. J., Fierke, C. A., and Naylor, A. M. (1988) Insights into enzyme function from studies on mutants of dihydrofolate reductase. *Science* **239**, 1105–1110.
3. Hamlin, J. L. (1992) Amplification of the *dhfr* gene in methotrexate-resistance CHO cells *Mutat. Res.* **276**, 179–187.
4. Peroni, C. N., Soares, C. R. J., Gimbo, E., Morganti, L., Ribela, M. T. C. P., and Bartolini, P. (2002) High-level expression of human thyroid-stimulating hormone in Chinese hamster ovary cells by co-transfection of dicistronic expression vectors followed by a dual-marker amplification strategy. *Biotechnol. Appl. Biochem.* **35**, 19–26.
5. Kito, M., Itami, S., Fukano, Y., Yamana, K., and Shibui, T. (2002) Construction of engineered CHO strains for high-level production of recombinant proteins. *Appl. Microbiol. Biotechnol.* **60**, 442–448.
6. Schimke, R. T. (1988) Gene amplification in cultured cells. *J. Biol. Chem.* **263**, 5989–5992.
7. Sambrook, J., Fritsch, E. F., and Maniatis, T. (1989) *Molecular Cloning, a Laboratory Manual*, 2 ed. Cold Spring Harbor Laboratory Press, Cold Spring Harbor, NY.
8. Monaco, L., Tagliabue, R., Giovanazzi, S., Bragonzi, A., and Soria, M. R. (1996) Expression of recombinant human granulocyte colony-stimulating factor in CHO dhfr- cells—new insights into the in vitro amplification expression system. *Gene* **180**, 145–150.
9. Chenu, S., Grégoire, A., Malykh, Y., Visvikis, A., Monaco, L., Shaw, L., et al. (2003) Reduction of CMP-N-acetylneuraminic acid hydroxylase activity in engineered chinese hamster ovary cells using an antisense-RNA strategy. *Biochim. Biophys. Acta* **1622,** 133–144.

33

Preparation of Recombinant Proteins in Milk

Louis-Marie Houdebine

Summary

Using transgenic animals as the source of recombinant proteins has several specific advantages. Large amounts of proteins can be obtained, essentially from milk. These proteins are often properly processed. They are in a number of cases correctly folded, assembled, cleaved, glycosylated, γ-carboxylated, and so on. Purification of recombinant proteins from milk is not a particularly difficult task. The level of expression of foreign genes in milk cannot be predicted in all cases and appropriate vectors must be used. Generation of transgenic mice is popular but their production is quite limited. Transgenesis in larger animals, rabbits and farm animals, is achieved essentially by a few companies. Some recombinant proteins may be found in blood circulation and alter animal health. Milk from transgenic animals has become a quite attractive alternative to other sources of recombinant proteins.

Key Words: Recombinant proteins; milk; expression vectors; transgenic animals; post-translational modifications; purification.

1. Introduction

The preparation of recombinant proteins is currently achieved for different purposes. Researchers need proteins to study their biological activity and their structure using native molecules and mutants obtained by genetic engineering. In a number of cases, this approach is essential to designing new drugs interacting with proteins and potentially utilizable as pharmaceuticals.

For centuries, our ancestors have used plant extracts as pharmaceuticals. Proteins remained out of the game until their major roles in living organisms were identified. For decades a small number of proteins were used as pharmaceuticals. Only a few naturally abundant proteins that can be extruded from human or animal organs or blood were

From: *Methods in Molecular Biology, vol. 267:*
Recombinant Gene Expression: Reviews and Protocols, Second Edition
Edited by: P. Balbás and A. Lorence © Humana Press Inc., Totowa, NJ

used. Insulin, human growth hormone, and blood-clotting factors are good examples of this approach. Genetic engineering rapidly offered quite new possibilities. Human insulin and growth hormone are now prepared from recombinant bacteria. A number of proteins, however, are too complex to be prepared from recombinant bacteria or yeast. Animal cells are needed to proceed to the numerous and subtle posttranslational modifications that are required to obtain active stable and nonimmunogenic proteins. Cultured animal cells are currently used to prepare some recombinant proteins. Recombinant Chinese hamster ovary (CHO) cells are presently the only source of several recombinant proteins. The idea of using transgenic animals as bioreactors was first suggested in 1982. Several biological fluids from animals can be used as the source of recombinant proteins (*1*). Milk is probably the most attractive. It is the system which is presently the closest to industrial application.

Using transgenic animals as bioreactors raises specific problems. The first is to construct vectors leading to high expression of the recombinant proteins. A second problem is to generate, breed, and milk animals. The purification of the recombinant protein from milk is the next step, which is not particularly difficult but requires strict protocols if recombinant proteins are to be used as pharmaceuticals.

This chapter describes the main steps of recombinant protein preparation from milk of different animals.

2. Materials and Methods

2.1. Gene Constructions

Targeting recombinant proteins to milk implies that a milk protein promoter is fused to the coding sequence of the protein of interest. The design of the expression vectors is discussed below. Vectors may be conventional and contain a few regulatory elements.

Alternatively, long genomic DNA fragments may be used to optimize transgene expression. The conventional methods of genetic engineering are to be used to reach this goal.

2.2. Design of Expression Vectors

Expressing a transgene in an appropriate manner is not an easy task. The first experiments carried out in the early 1980s revealed that a gene construct that is quite active in transfected cells may be only weakly expressed in transgenic animals. Moreover, each line of animals shows a specific pattern of transgene expression. In addition, a number of transgenes remain silent. This is a clear limitation for the use of transgenic animals. In a majority of cases, however, researchers prepare an exceeding number of transgenic founders until they find the lines expressing the foreign gene as they expected. This approach implies a wastage of time and animals but it appears sufficient to obtain 1000–10,000 copies of recombinant protein per cell exhibiting the expected biological action. This relative but real success of the expression vectors dissuaded researchers for years from improving them.

The situation is different when recombinant proteins are to be prepared in milk. Highly efficient expression vectors are then needed to obtain the expected large amounts of proteins. Costly transgenic farm animals must often be used and reliable expression vectors are highly preferable. On the other hand, the recombinant proteins must be restricted to milk as much as possible to avoid any potential deleterious effects on animals. The groups involved in recombinant proteins in milk and a few others have done systematic studies to improve expression vectors. Designing the ideal vector for transgene expression remains an impossible task. Yet basic studies on the mechanisms of gene expression and empirical observations have led to defined rules for vector construction and to identification of pitfalls that can be avoided.

A functional transcription unit is composed of several elements that cooperate with each other. These elements and their mode of action are not all known. A gene construction consists of associating more or less known elements. This may generate nonfunctional vectors for unknown reasons. The major recommendations for designing reliable vectors have been described elsewhere (*2*). They may be summarized as follows.

2.2.1. Promoters

The promoter region contains the DNA sequences that allow the formation of the transcription initiation complex and, in most cases, those that give the specificity of expression. Promoters are restricted to the 100–200 bp preceding the transcription start site. Enhancers have a global amplification effect. It is now admitted that they do not increase promoter potency by enhancing the chance of having active transcription complex. Enhancers may reinforce tissue specificity of expression. In most cases, they are found in the 0.5–10 kb region preceding the promoter. Enhancers may be found in the transcribed region and after the transcription terminator, and they can be added to a construct in the region upstream of the promoter but also in the transcribed region, even within an intron (*3*).

An increasing number of promoters are becoming available. They may be chosen using published data to drive a cell specific expression in vivo.

2.2.2. Insulators

Numerous experiments have shown that a transgene is generally poorly expressed (1) when it contains a cDNA rather than the corresponding genomic DNA; (2) when it is integrated as multiple copies; (3) when it contains nonmammalian DNA sequences, namely CpG-rich sequences; and (4) when it is integrated in a silent region of chromatin. The last phenomenon is known as "position effect." Vector constructions may be performed taking these observations into account.

Studies carried out during the 1990s contributed to establishing the concept of insulators (*see* Chapters 28 and 29). It is now acknowledged that groups of genes, acting or not acting in a coordinated manner, are bordered by DNA sequences that insulate them from the rest of the chromosome. These sequences have been named insulators or LCRs (locus control regions) (*4*). Insulators are composed of multiple elements. Some are silencers, which block the action of the enhancers present in the vicinity of the integrated transgene. Other elements are considered as chromatin openers (*5*), which bind protein complexes inducing histone acetylation. This is accompanied by a local histone and DNA demethylation (**6**). Insulators also often contain AT-rich matrix-attached regions (MAR).

The addition of insulators to gene constructs greatly enhances transgene expression. A higher number of lines express the transgene and the mean level of expression is also enhanced. The addition of the 5'HS4 element from the LCR of the chicken β-globin locus—preferably on both sides, but also only upstream of the promoter—allowed cDNAs driven by the human EF1α gene (**7**) or the rabbit WAP (whey acidic protein) gene promoter (**8**) to be expressed in all rabbit and mouse lines, respectively.

Several long genomic DNA fragments allow their genes to be expressed at a high levels (corresponding to the known potency of the promoters) in a tissue-specific manner and as a function of the integrated copy number (**5**). This was also observed for two milk protein genes, human and goat α-lactalbumin gene (**9,10**) and pig WAP gene (**11**). This long genomic cDNA fragment can be used to express foreign genes with high efficiency (**12**). In the future, identified insulator elements might be used instead of the whole BAC to construct shorter vectors, which are easier to manipulate.

2.2.3. Transcribed Regions

The transcribed regions of the genes contain a number of regulatory elements that control exon splicing, mRNA transfer to cytoplasm translation, mRNA targeting, and mRNA half-life.

A vector for transgene expression must contain a least one intron, which is required for the transfer of the mRNA to cytoplasm, although several introns may be present. In this respect, minigenes are generally more efficient than vectors containing a single exogenous intron. This may be due to multiple factors, including the presence of transcription enhancers within introns.

The intron may be added before or after the cDNA. However, the presence of the intron before the cDNA, rather than after, prevents the nonsense-mediated decay (NMD) mechanism to inactivate the mRNA. This destruction mechanism is considered as a quality control system, which eliminates all the mRNA that has a premature terminator codon located more than 50 nucleotides from the last acceptor site of an intron (**13**). This situation is never found in natural mRNA. When it happens it is recognized by cells at the result of a mutation generating a terminator codon. An appropriate construction may avoid the NMD mechanism to inactivate a mRNA when an intron is added after the cDNA.

Alternative splicing is a quite frequent event in higher-vertebrate mRNA. It is a way to control gene expression and to enhance the diversity of the cellular mechanisms using a limited number of genes. The association of introns and cDNAs often leads to alternative splicing, which prevents the transgene from driving the synthesis of the expected protein. *In silico*, studies may predict the combination of donor and acceptor splicing sites. However, the splicing of exon is a complex phenomenon implying a number of cellular proteins that are present in some cell types only and induced by multiple cellular events. Hence, the use of splicing sites cannot be predicted only *in silico*. An evaluation of the construct may be done using transfection into cultured cells. The mouse mammary cell line HC11 may be helpful for this purpose. One way to avoid inappropriate splicing is to mutate the splicing sites of the construct.

2.2.4. Nontranslated Regions (UTR)

The 5'UTR may contain sequences that control translation. The elements known as internal ribosome entry sites (IRES) seem to play this role (*14*). Selected 5'UTR may be added before the cDNAs to favor their translation.

The 3'UTR often contain elements that control mRNA stability and mRNA targeting (*3*). AU-rich regions are known to be potent destabilizers of mRNAs. On the contrary, the motif C/UCCANXCCCU/A PyX.UCC/UCC stabilizes various mRNAs.

2.2.5. Codon Optimization

Each group of living organism uses isocodons preferentially. Using mammalian codons, and preferably those found in the mRNA of milk protein genes, may greatly increase transgene expression by favoring translation and stability of its mRNA.

2.2.6. Protein Secretion

To be found in milk, a protein must contain a signal peptide. Naturally secreted proteins have such a signal. Cellular proteins may be secreted with good efficiency in many cases when a sequence coding for signal peptide is added to the cDNA. However, the protein may contain sequences targeting it to a cellular compartment other than the endoplasmic reticulum. This may create a conflict and reduce the secretion rate of the protein.

2.2.7. Vectors for the Simultaneous Expression of Two Coding Sequences

A number of proteins are composed of several subunits that must be coexpressed and assembled in the Golgi apparatus.

The coinjection of two or three independent gene constructs leads to a cointegration and coexpression of the different coding sequences. This has been achieved successfully with different constructs and in several tissues of mammals. This approach is expected to be less efficient with long vectors such as BAC, which are often integrated as a single copy.

Another possibility is to associate the two constructs by ligating them in vitro before microinjection. This may be difficult when each construct is long. On the other hand, the vicinity of the two constructs may induce a phenomenon known as transcription interference, which may reduce expression of both genes.

A third approach may consist of using IRES. These sequences are known to allow the expression of the second cDNAs of bicistronic constructs. Numerous IRESs have been identified and are available. Their mechanism of action is still not well understood (*14*). A fact that has been regularly observed is that the IRESs commonly used allow the simultaneous expression of two cDNAs of bicistronic mRNAs but at a level that is lower or much lower than this observed with each cDNA used alone. Hence, IRESs are appropriate to express cDNAs at a low level but perhaps not to produce recombinant proteins in milk at a sufficient rate.

A systematic study revealed that the size of the region between the terminator codon and IRES should be about 80 nucleotides to optimize expression of both cDNAs of bicistronic mRNAs (*14*).

2.3. DNA Preparation for Microinjection

The quality of DNA is essential to obtain transgene animals with good chance of success. Plasmid sequences must be eliminated because they may considerably reduce transgene expression. Inserts containing only the elements of expression vectors must be separated from the plasmid in agarose gel and purified. The conventional inserts (not exceeding 30 kb) may be purified from agarose by classical methods; Gene Clean (Bio101) is one of them. Centrifugation at 10,000*g* for at least 30 min is then required to eliminate particles that may block DNA flow in micropipets. Alternatively, filtration using 0.22-μ filters can be achieved.

For long fragments (50–500 kb) contained in bacterial artificial chromosomes (BAC) or yeast artificial chromosomes (YAC), the inserts must be released using agarose and purified by dialysis (*11*). The presence of polyamines is required to protect the longer fragments from shearing (*15*).

2.4. Generation of Transgenic Animals

Microinjection of DNA into embryo pronuclei remains the most frequently used method to generate transgenic mammals. The protocol to generate transgenic mice has been described in several books. The readers may refer to two of them (*16,17*). The preparation of larger transgenic mammals by DNA microinjection requires more specific methods, which have been described earlier (*18*).

Alternative methods to generate transgenic animals can be implemented. Transposons and retroviral vectors may be used in mice. Gene transfer into somatic cells further used as nucleus donors for cloning is presently the most powerful method to generate transgenic ruminants (*19*).

2.5. Purification of Recombinant Proteins From Milk

Milk collection from ruminants is achieved using conventional milking machines. In mice, rabbits, and pigs, the mammary gland has no cisternae. Milk stored in the mammary gland is ejected by an active process. Oxytocin (up to 5 IU) must be injected intraperitoneally in mice. Up to 1 mL of milk may be obtained by applying mild vacuum to the teats. Up to 100 mL milk can be obtained per day from lactating rabbits using essentially the same protocol. Alternatively, the mammary gland of mice (or rabbits) separated from their pups for one day may be collected and kept for a few hours in a Petri dish on ice. Quantitative milk collection is obtained in this way (up to 1 mL from mice and 200 mL from rabbits). Although this protocol implies the sacrifice of the animals (*20*), it is appropriate to obtain easily limited amounts of milk.

Recombinant proteins are found in the lactoserum, which is the milk fraction remaining after the elimination of lipids and caseins. Lipids are removed after centrifugation. Casein micelles can be eliminated by different methods, including high-speed centrifugation, specific precipitation at pH 4.6, calcium excess, or polyethyleneglycol (*21*). Available chromatographic methods may then be used to purify the proteins of interest from the lactoserum.

3. Conclusions

The major reason that justifies the use of milk as a source of recombinant proteins is the production of pharmaceuticals. Indeed, for limited production, CHO cells in fermentor may be sufficient. However, this technique is not so easy to manage and it may not be implemented by an academic laboratory. The use of transgenic mice may appear easier.

Only a few proteins—mainly monoclonal antibodies—are candidates to become pharmaceuticals, although their number is rapidly growing. Yet, many proteins are needed for experimenters to study their properties, including their potential clinical use. For very small amounts of proteins (not exceeding 10–100 mg), transgenic mice may be used. Their milk starts being available 4 mo after DNA microinjection into embryos. For larger amounts, rabbits are quite appropriate. The lactating animals start being available 6–7 mo after DNA microinjection, and one female can provide up to 2 L of milk over a period of 1 mo (corresponding to one lactation).

One month later, the female is again lactating. Hence, rabbits can provide researchers with recombinant proteins in sufficient amounts to study their properties in depth. Rabbits may reproduce at a high rate and herds are rapidly available. Hundreds of rabbits may be bred to produce 1–10 kg of recombinant proteins per year.

For higher amounts of proteins, goats, sheep, cows, and potentially pigs must be used. This is justified only for the production of pharmaceuticals or proteins, such as spider silk, having specific properties.

The methods to be implemented tend to be standardized. The expression vectors have been and are still being improved. One of them, known as pBC1 (Invitrogen), is sufficient in most cases to produce small quantity of proteins in mouse milk. Other vectors are available in laboratories studying milk protein gene promoters. The available efficient promoters are from the following genes: ruminant $\alpha S1$- and β-caseins, rabbit and mouse WAP, and ovine β-lactoglobulin. All these promoters are patented by the companies that use them.

The techniques to generate transgenic mammals are standardized, even if they are still being improved.

One major problem of recombinant proteins is their posttranslational modifications (*22*). The available information indicates that the recombinant proteins found in milk may be not fully sialylated (*22*). Yet the proteins produced in milk may be slightly different from those secreted by CHO cells, but of similar quality. This was seen with human EC-superoxide dismutase prepared in CHO cells and in rabbit milk (*23*).

The incomplete glycosylation, cleavage γ-carboxylation, is clearly due to the lack of the appropriate enzymes in the mammary cells (*22*). A higher production of a recombinant protein may lead to a lower achievement of posttranslational modifications. This is clearly due to a saturation of the enzymatic systems.

This shortage may be theoretically compensated by the overexpression of enzymes after transfer of the corresponding genes to embryos. It is interesting to note that the overexpression of the furin gene allowed human protein C in mouse milk to be almost quantitatively cleaved and transformed into an active protein (22).

The major posttranslational modifications to optimize are the following: sialylation, fucosylation, N-acetylglycosylation of the constant part of antibodies to stimulate killer lymphocytes, the addition of sialic acid as N-acetylneuraminic acid rather than N-glycosylneuraminic acid, which is not present in human proteins, and cleavage by furin (24).

The mammary cell proved to be able to assemble quite different protein subunits, such as fibrinogen, collagen, and antibodies (25). EC-superoxide dismutase is also an interesting case. This protein is a glycosylated homotetramer containing one copper ion per monomer. This protein was found at the concentration of several grams per liter in rabbit milk in a quite functional state (23); thus the mammary gland was capable of collecting large amount of copper from blood and adding it to EC-superoxide dismutase. Hence, the capacity of the mammary gland to associate proteins is probably not limiting. This may reflect its high capacity to assemble caseins to form micelles.

The implementation of transgenic animals to prepare recombinant proteins currently seems inevitable. This is clearly due to the specific advantages of this production system. It should be recalled that two of its major advantages are certainly its robustness and flexibility. Indeed, transgenic animals are not fragile in comparison to cells in a fermentor. They can be stored as living animals, as frozen embryos, and as sperm as well. The scaling up of a large fermentor is a long and costly process in comparison to the multiplication of animals.

To allow safe production of pharmaceuticals, the transgenic animals must be bred in narrowly controlled conditions. The points to consider defined by the US FDA and EU EMEA are not dissuading for the companies that develop this process (26).

Some recombinant proteins that are present in blood at a low concentration and not only in milk may alter the health of the animals. This was the case for human erythropoietin (27). In other cases, such as the production of EC-superoxide dismutase or of human protein C, the development of the mammary gland may be impaired, for unknown reasons.

Despite these minor drawbacks, the implementation of milk for the production of recombinant proteins appear inevitable. Indeed, all the available production systems, even if they are fully utilized, are expected to be unable to prevent a shortage of the overall production of recombinant proteins in this decade (28,29).

References

1. Houdebine, L. M. (2000) Transgenic animal bioreactors. *Transgenic Res.* **9**, 305–320.
2. Houdebine, L. M., Attal, J., and Vilotte, J. L. (2002) Vector design for transgene expression, in *Transgenic Animal Technology, Second Edition* (Pinkert, C. A., ed.). Academic Press, pp. 419–458.

3. Petitclerc, D., Attal, J., Theron, M. C., Bearzotti, M., Bolifraud, P., Kann, G., et al. (1995) The effect of various introns and transcription terminators on the efficiency of expression vectors in various cultured cell lines and in the mammary gland of transgenic mice. *J. Biotechnol.* **40**, 169–178.

4. Li, Q., Harju, S., and Peterson, K. R. (1999) Locus control regions: coming of age at a decade plus. *Trends Genet.* **15**, 403–408.

5. Recillas-Targa, F., Bell, A. C., and Felsenfeld, G. (1999) Positional enhancer-blocking activity of the chicken beta-globin insulator in transiently transfected cells. *Proc. Natl. Acad. Sci. USA* **96**, 14354–14359.

6. Mutskov, V. J., Farrell, C. M., Wade, P. A., Wolffe, A. P., and Felsenfeld, G. (2002) The barrier function of an insulator couples high histone acetylation levels with specific protection of promoter DNA from methylation. *Genes Dev.* **16**, 1540–1554.

7. Taboit-Dameron, F., Malassagne, B., Viglietta, C., Puissant, C., Leroux-Coyau, M., Chereau, C., et al. (1999) Association of the 5'HS4 sequence of the chicken beta-globin locus control region with human EF1 alpha gene promoter induces ubiquitous and high expression of human CD55 and CD59 cDNAs in transgenic rabbits. *Transgenic Res.* **8**, 223–235.

8. Rival-Gervier, S., Pantano, T., Viglietta, C., et al. (2003) The insulator effect of 5HS4 region from the β-globin chicken locus on the rabbit *WAP* gene promoter activity in transgenic mice. *Transgenic Res.* **12**, 723–730.

9. Fujiwara, Y., Miwa, M., Takahashi, R., Kodaira, K., Hirabayashi, M., Suzuki, T., et al. (1999a) High-level expressing YAC vector for transgenic animal bioreactors. *Mol. Reprod. Dev.* **52**, 414–420.

10. Stinnakre, M. G., Soulier, S., Schibler, L., Lepourry, L., Mercier, J. C., and Vilotte, J. L. (1999) Position-independent and copy-number-related expression of a goat bacterial artificial chromosome alpha-lactalbumin gene in transgenic mice. *Biochem. J.* **339**, 33–36.

11. Rival-Gervier, S., Viglietta, C., Maeder, C., Attal, J., and Houdebine, L. M. (2002) Position-independent and tissue-specific expression of porcine whey acidic protein gene from a bacterial artificial chromosome in transgenic mice. *Mol. Reprod. Dev.* **63**, 161–167.

12. Fujiwara, Y., Takahashi, R. I., Miwa, M., Kameda, M., Kodaira, K., Hirabayashi, M., et al., (1999b) Analysis of control elements for position-independent expression of human alpha-lactalbumin YAC. *Mol. Reprod. Dev.* **54**, 17–23.

13. Maquat, L. E. (2001) The power of point mutations. *Nat. Genet.* **27**, 5–6.

14. Houdebine, L. M. and Attal, J. (1999) Internal ribosome entry sites (IRESs): reality and use. *Transgenic Res.* **8**, 157–177.

15. Umland, T., Montoliu, L., and Schütz, G. (1997.) The use of yeast artificial chromosomes for transgenesis, in *Transgenic Animals. Generation and Use.* (Houdebine, L. M., ed.), Harwood Academic Publishers, Amsterdam, pp. 289–298.

16. Clarke, A. R. (ed.) (2002) Transgenesis techniques. Principles and protocols, in *Methods Mol. Biol.*, 2nd ed. Vol **180**, Humana Press, Totowa, NJ.

17. Pinkert, C. A. (2002). *Transgenic Animals Technology. Second Edition.* Academic Press.

18. Houdebine, L. M. (ed.) (1997) *Transgenic Animals. Generation and Use.* Harwood Academic Publishers, Amsterdam.

19. Houdebine, L. M. (2002) The methods to generate transgenic animals and to control transgene expression. *J. Biotechnol.* **98**, 145–160.

20. Stinnake, M. G., Devinoy, E., Chene, N., Bayat-Samardi, M., Grabowiski, H., and Houdebine, L. M. (1992) Quantitative collection of milk and active recombinant proteins from mammary glands of transgenic mice. *Animal Biotechnol.* **3**, 245–255.

21. Morcöl, T., He, Q., and Bell, J. D. (2001) Model process for removal of caseins from milk of transgenic animals. *Biotechnol. Prog.* **17**, 577–582.
22. Lubon, H. (1998) Transgenic animal bioreactors in biotechnology and production of blood proteins. *Biotechnol. Annu. Rev.* **4**, 1–54.
23. Stromqvist, M., Stromqvist, M., Houdebine, M., Andersson, J. O., Edlund, A., Johansson, T., et al. (1997) Recombinant human extracellular superoxide dismutase produced in milk of transgenic rabbits. *Transgenic Res.* **6,** 271–278.
24. Houdebine, L. M. (2002) Antibody manufacturing: transgenic animals. *Curr. Opin. Biotechnol.* 13, 625–629.
25. Pollock, D. P., Kutzko, J. P., Birck-Wilson, E., Williams, J. L., Echelard, Y., and Meade, H. M. (1999) Transgenic milk as a method for the production of recombinant antibodies. *J. Immunol. Methods* **231**, 147–157.
26. Gavin, W. G., (2001) The future of transgenics. *Regulatory Affairs Focus*, 13–18.
27. Massoud, M., Attal, J., Thépot, D., Pointu, H., Stinnakre, M. G., Theron, M. C., et al. (1996) The deleterious effects of human erythropoietin gene driven by the rabbit whey acidic protein gene promoter in transgenic rabbits. *Reprod. Nutr. Dev.* **36**, 555–563.
28. Gura, T. (2002) Therapeutic antibodies: magic bullets hit the target. *Nature* **417**, 584–586.
29. Andersen, D. C. and Krummen, L. (2002) Recombinant protein expression for therapeutic applications. *Curr. Opin. Biotechnol.* **13**, 117–123.

Index